Lecture Notes in Computer Science 7107

Commenced Publication in 1973
Founding and Former Series Editors:
Gerhard Goos, Juris Hartmanis, and Jan van Leeuwen

Daniel J. Bernstein   Sanjit Chatterjee (Eds.)

Progress in Cryptology – INDOCRYPT 2011

12th International Conference on Cryptology in India
Chennai, India, December 11-14, 2011
Proceedings

 Springer

Volume Editors

Daniel J. Bernstein
University of Illinois at Chicago
Department of Computer Science
Chicago, IL 60607–7053, USA
E-mail: djb@math.uic.edu

Sanjit Chatterjee
Indian Institute of Science
Department of Computer Science and Automation
Bangalore 560 012, India
E-mail: sanjit@csa.iisc.ernet.in

ISSN 0302-9743   e-ISSN 1611-3349
ISBN 978-3-642-25577-9   ISBN 978-3-642-25578-6 (eBook)
DOI 10.1007/978-3-642-25578-6
Springer Heidelberg Dordrecht London New York

Library of Congress Control Number: 2011941501

CR Subject Classification (1998): E.3, K.6.5, D.4.6, C.2, J.1, G.2.1

LNCS Sublibrary: SL 4 – Security and Cryptology

Typesetting: Camera-ready by author, data conversion by Scientific Publishing Services, Chennai, India

Printed on acid-free paper

Springer is part of Springer Science+Business Media (www.springer.com)

Preface

Indocrypt 2011, the 12th International Conference on Cryptology in India, took place December 11–14, 2011. It was co-hosted by the Institute of Mathematical Sciences in Chennai and the Chennai Mathematical Institute. Indocrypt has been held every year since 2000, and has been held under the aegis of the Cryptology Research Society of India since 2003.

We followed the Indocrypt 2006 idea of splitting the submission deadline into two. Authors submitting papers were required to register titles and abstracts by the first deadline, 31 July 2011. A total of 127 submissions had been received by this deadline, although some were withdrawn before review. Authors were allowed to continue working on their papers until the second deadline, August 7.

Submissions were evaluated in three phases over a period of nearly two months. The selection phase started on August 1: Program Committee members began evaluating abstracts and volunteering to handle various papers. We assigned a team of people to each paper. The review phase started on August 9: Program Committee members were given access to the full papers and began in-depth reviews of 98 submissions. Most of the reviews were completed by August 29, the beginning of the discussion phase. Program Committee members were given access to other reviews once they had completed all of their own reviews, and built consensus in their evaluations of the submissions. In the end the discussions included 353 full reports and 219 additional comments. The submissions, reviews, and subsequent discussions were handled smoothly by iChair.

On September 18 we sent out comments from the reviewers, 2 notifications of conditional acceptance, and 20 notifications of unconditional acceptance. The conditionally accepted papers eventually met their acceptance conditions; the final program contains 22 contributed papers, 3 invited talks, and 3 tutorials. The authors prepared final versions of the 22 contributed papers by September 30.

It is our pleasure to thank the other 56 Program Committee members for lending their expertise to Indocrypt 2011 and for putting tremendous effort into detailed reviews and discussions. We would also like to thank Thomas Baignères and Matthieu Finiasz for writing the iChair software; Tanja Lange for modifying the software to handle split submissions and managing the software on her Web server in Europe; the General Chairs, R. Balasubramaniam from the Institute of Mathematical Sciences in Chennai and Rajeeva Laxman Karandikar from the Chennai Mathematical Institute, for smoothly handling all of the local arrangements; 80 external referees who reviewed individual papers upon request from the Program Committee; and, most importantly, all authors for submitting interesting new research papers to Indocrypt 2011.

December 2011   Daniel J. Bernstein and Sanjit Chatterjee
                                                Program Chairs, Indocrypt 2011

Organization

General Chairs

R. Balasubramaniam   Institute of Mathematical Sciences, Chennai, India

Rajeeva Laxman Karandikar   Chennai Mathematical Institute, India

Program Chairs

Daniel J. Bernstein   University of Illinois at Chicago, USA

Sanjit Chatterjee   Indian Institute of Science, Bangalore, India

Program Committee

Roberto Avanzi	Ruhr University Bochum, Germany
Rana Barua	Indian Statistical Institute, India
Lejla Batina	Radboud University Nijmegen, The Netherlands, and KU Leuven, Belgium
Daniel J. Bernstein	University of Illinois at Chicago, USA
Sanjay Burman	Centre for Artificial Intelligence and Robotics, India
Debrup Chakraborty	CINVESTAV-IPN, Mexico
Sanjit Chatterjee	Indian Institute of Science, India
Chen-Mou Cheng	National Taiwan University, Taiwan
Ashish Choudhury	Indian Statistical Institute, India
Sherman S.M. Chow	University of Waterloo, Canada
Christophe De Cannière	Google, Switzerland
Yvo Desmedt	University College London, UK
Christophe Doche	Macquarie University, Australia
Matthieu Finiasz	ENSTA, France
Praveen Gauravaram	Technical University of Denmark, Denmark
Vipul Goyal	Microsoft Research, India
Tim Güneysu	Ruhr University Bochum, Germany
Shay Gueron	University of Haifa, Israel, and Intel Corporation, Israel
Kishan Chand Gupta	Indian Statistical Institute, India
Helena Handschuh	Intrinsic-ID, USA, and KU Leuven, Belgium
Thomas Johansson	Lund University, Sweden
Antoine Joux	DGA and Université de Versailles Saint-Quentin-en-Yvelines, France
Koray Karabina	University of Waterloo, Canada

Tanja Lange	Technische Universiteit Eindhoven, The Netherlands
Vadim Lyubashevsky	ENS, France
Subhamoy Maitra	Indian Statistical Institute, India
Keith Martin	Royal Holloway, University of London, UK
David McGrew	Cisco, USA
Payman Mohassel	University of Calgary, Canada
Michele Mosca	University of Waterloo, Canada
Debdeep Mukhopadhyay	Indian Institute of Technology Kharagpur, India
Michael Naehrig	Technische Universiteit Eindhoven, The Netherlands
Mridul Nandi	Indian Statistical Institute, India
Roger Oyono	Université de la Polynesie Française, French Polynesia
Daniel Page	University of Bristol, UK
Kenny Paterson	Royal Holloway, University of London, UK
Josef Pieprzyk	Macquarie University, Australia
Manoj Prabhakaran	University of Illinois at Urbana-Champaign, USA
Bart Preneel	Katholieke Universiteit Leuven, Belgium
Christian Rechberger	ENS Paris, France
Vincent Rijmen	KU Leuven, Belgium, and TU Graz, Austria
Kouichi Sakurai	Kyushu University, Japan
Palash Sarkar	Indian Statistical Institute, India
P.K. Saxena	DRDO, India
Peter Schwabe	National Taiwan University, Taiwan
Mike Scott	Dublin City University, Ireland
Nicolas Sendrier	INRIA, France
Francesco Sica	
Martijn Stam	University of Bristol, UK
François-Xavier Standaert	Université Catholique de Louvain, Belgium
Damien Stehlé	CNRS and ENS de Lyon, France
Christine Swart	University of Cape Town, South Africa
Michael Szydlo	Akamai, USA
Berkant Ustaoglu	Sabanci University, Turkey
C.E. Veni Madhavan	Indian Institute of Science, India
Huaxiong Wang	Nanyang Technological University, Singapore
Michael J. Wiener	Irdeto, Canada
Bo-Yin Yang	Academia Sinica, Taiwan

Referees

Mohamed Ahmed Abdelraheem	Sk. Subidh Ali
David Adam	Elena Andreeva
Shweta Agrawal	Josep Balasch

Table of Contents

Secret-Key Cryptography, Part 1

Invited Talk 2

Secret-Key Cryptography, Part 2

Hash Functions

Pairings

Invited Talk 3

Protocols

Tutorial 3

Tor and the Censorship Arms Race: Lessons Learned

Roger Dingledine

The Tor Project, USA
arma@freehaven.net

Abstract. Tor is a free-software anonymizing network that helps people around the world use the Internet in safety. Tor's 2500 volunteer relays carry traffic for several hundred thousand users, including ordinary citizens who want protection from identity theft and prying corporations, corporations who want to look at a competitor's website in private, and soldiers and aid workers in the Middle East who need to contact their home servers without fear of physical harm.

Tor was originally designed as a civil liberties tool for people in the West. But if governments can block connections to the Tor network, who cares that it provides great anonymity? A few years ago we started adapting Tor to be more robust in countries like China. We streamlined its network communications to look more like ordinary SSL, and we introduced "bridge relays" that are harder for an attacker to find and block than Tor's public relays.

Through the Iranian elections in June 2009, the periodic blockings in China, the demonstrations in Tunisia and Egypt, and whatever's coming next, we're learning a lot about how circumvention tools work in reality for activists in tough situations. This talk will start with a brief overview of the Tor design and its diverse users, and then jump into the technical and social problems we're encountering, what technical approaches we've tried so far (and how they went), and what approaches I think we're going to need to try next.

D.J. Bernstein and S. Chatterjee (Eds.): INDOCRYPT 2011, LNCS 7107, p. 1, 2011.
© Springer-Verlag Berlin Heidelberg 2011

Elliptic Curves for Applications

Tanja Lange

Technische Universiteit Eindhoven, The Netherlands
tanja@hyperelliptic.org

Abstract. More than 25 years ago, elliptic curves over finite fields were suggested as a group in which the Discrete Logarithm Problem (DLP) can be hard. Since then many researchers have scrutinized the security of the DLP on elliptic curves with the result that for suitably chosen curves only exponential attacks are known. For comparison, the RSA cryptosystem is broken if large numbers can be factored; factoring is possible in subexponential time. As a consequence the parameters for elliptic-curve cryptography (ECC) can be chosen significantly smaller than for RSA at the same level of security and arithmetic becomes faster, too.

The NaCl library (Networking and Cryptography library) uses ECC as the public-key component for authenticated encryption (using symmetric-key cryptography for the authenticator and for generating the bulk of the ciphertext) and for signatures. On all levels the algorithms are chosen to simplify implementation without leaking information through software side channels. All implementations in NaCl are timing-invariant and do not have data-dependent branches.

This tutorial explains how to compute on elliptic curves over fields of odd characteristic; how to make the arithmetic efficient; how to avoid data-dependent branches in single-scalar multiplication in the variable-base-point and in the fixed-base-point scenario; how the algorithms in NaCl are designed; and how to use NaCl.

NaCl is joint work with Daniel J. Bernstein and Peter Schwabe. Software and documentation are available at http://nacl.cr.yp.to.

D.J. Bernstein and S. Chatterjee (Eds.): INDOCRYPT 2011, LNCS 7107, p. 2, 2011.

PKDPA: An Enhanced Probabilistic Differential Power Attack Methodology

Dhiman Saha*, Debdeep Mukhopadhyay**, and Dipanwita RoyChowdhury***

Abstract. The paper presents an enhancement of univariate Differential Power Analysis (DPA), referred to as *Probable Key Differential Power Analysis (PKDPA)*. The proposed analysis uses the standard Difference of Means (DoM) test as the distinguisher and employs its enhancement strategy to reduce the number of power traces required to mount the attack. Theoretical analysis for the developed attack has been furnished to justify the efficiency of the proposed attack in retrieving the key using significantly less number of traces compared to conventional DPA attacks. The theoretical claims have been supported by extensive experiments on real life attacks mounted on Field Programmable Gate Array (FPGA) implementations of the Data Encryption Standard (DES), Triple-DES (3-DES) and the Advanced Encryption Standard (AES). The efficacy of the proposed method is further proved by attacking a masked implementation of AES using only 13,000 power traces.

Keywords: Side channel cryptanalysis, differential power attacks, block ciphers.

1 Introduction

Differential Power Analysis (DPA) is a statistical test to reveal secret keys of a cipher based on its power consumption. It was introduced by P. Kocher et al. in his pioneering paper [1]. Power analysis exploits two facts. First, power consumption at any instant of time depends on the operation being carried out. Secondly, power consumption of the same operation at different instants of time depends on the data being processed. DPA exploits the second fact. Researchers have proposed different schemes to measure the efficiency of DPA, one of the popular[2,3] of which, has been to reduce number of traces/measurements required to mount a successful attack. Authors in [4] have emphasized that considering number of leakage traces as an evaluation criterion is better suited while considering practical adversaries. A DPA enhancement is proposed in [2] which exploits the bias of a linear combination of output bits of the Substitution-Box (Sbox) of the

---

* Dhiman Saha is a Senior Software Engineer at Interra Systems India Pvt. Ltd., India. crypto@dhimans.in

** Debdeep Mukhopadhyay is an Assistant Professor in the Department of Computer Sc. and Engg, IIT Kharagpur. debdeep@cse.iitkgp.ernet.in

*** Dipanwita RoyChowdhury is a Professor in the Department of Computer Sc. and Engg, IIT Kharagpur. drc@cse.iitkgp.ernet.in

D.J. Bernstein and S. Chatterjee (Eds.): INDOCRYPT 2011, LNCS 7107, pp. 3–21, 2011.
© Springer-Verlag Berlin Heidelberg 2011

block ciphers. It has been demonstrated that a micro-controller based design of DES [5] may be attacked using 2048 traces [2]. Simultaneous testing of various hypotheses to classify the power traces has been suggested in [3]. However, the attack has been mounted using simulated power traces and not on hardware. To enhance DPA some authors [6] have introduced multi-bit DPA attacks which use multiple bits instead of a single bit to devise a distinguisher. In [7], the authors present a DPA enhancement whereby they find the keys registering maximum DPA bias for all target bits and then perform a majority voting to arrive at the correct key.

One of the popular attack strategies is Correlation Power Analysis (CPA) which exploits Pearson's correlation coefficient. These attacks [8,9,10,11,12] have been extensively studied in literature. In [8,9], authors have given good illustrations of CPA and have showcased CPA to be a better method than DPA. However, a recent work [4] by *Mangard et. el.* has shown that for standard univariate DPA, distinguishers like DoM test [1] or a Pearsons correlation coefficient(CPA) or Gaussian templates [13] are in fact asymptotically equivalent, given that they are provided with the same a priori information about the leakages and they differ in terms, which if properly estimated become key-independent. The results of [4] are further complimented by [14] where *Doget. et al* study if the statement in [4] also hold in non-asymptotic contexts (i.e. when the number of measurements is reasonably small). Their work focuses on standard non-profiled side channel distinguishers. They show that all these distinguishers exploit essentially the same statistics and that any difference can be expressed as a change of model. Additionally, the authors compare various power attacks with what they call *robust side channel attacks* based on [15] and having few general assumptions about the leakage. They confirm that while in the presence of a perfect leakage model CPA performs very well, in the absence of a good model, the efficiency decreases considerably. In [9], some interesting results are reported regarding CPA and a CPA enhancement strategy. An electro-magnetic radiation attack on an ASIC implementation of DES is presented. It would be worthwhile to scrutinize the performance of these strategies on architectures operating on 128-bit or higher data at a time. Though [10] mounts an attack on an AES FPGA core, the attack does not recover the entire key but reduces the key search space for each Sbox to 10 at the cost of 5,000 measurements. In [16], the authors have tried to address problem of evaluating side channel key recovery attacks in the presence of heterogeneous implementations platforms.

In this work, the FPGA platform has been chosen to validate the theoretical claims which will also be applicable to other implementation paradigms. With the increasing popularity of FPGA devices, security analysis of encryption algorithms implemented on such devices becomes imperative. There are debates about the challenge and difficulty in attacking crypto implementations on FPGA platforms. While previous results [17] highlighted FPGAs to be noisy targets for side channel attacks, the works of [18,19,20] confirmed the possibility to apply power analysis to FPGAs. It is shown in [20] that with improved measurement the power traces of an FPGA can be almost perfectly correlated to the predicted values. However, the authors of [20] agree that the power consumption of the FPGA depends on resources like LUTs etc. and if the target points of the adversary are connected to low effective capacitances within an FPGA, the attacks become more challenging. Thus, in order to develop more realistic DPA attacks it is important to use power models like the bit model which make the least possible assumptions about the platform architecture [21,14]. Also the attacks should be investigated with real-life traces, as simulations often cannot account for the unpredictability of the FPGA platform.

In this work, we propose Probable Key Differential Power Analysis (PKDPA), a new attack strategy for mounting differential power attacks and show that it is more efficient than contemporary enhancement schemes. The key concept of the work is to maintain a window of probable keys as opposed to conventional DPA which targets to find the correct key. Experiments show that at the cost of some computational overhead, a significant reduction in the number of power traces occurs under the proposed attack. It can be noted that the proposed strategy has no limitation like [2], where all bits need to influence the power consumption at the same time. Our scheme need not be confused with multi-bit DPA attacks where multiple output bits of the Sbox are targeted at the same time. We show that our scheme performs better than the majority voting [7] scheme and can arrive at the correct key even if the correct key does not produce the highest DPA bias for *any* of the target bits, a scenario which would have led to the failure of [7]. It is worth mentioning that the majority voting scheme can be considered as a special case of our strategy where the window-size is selected to be one. A comparative study with contemporary schemes is also furnished.

The paper is organized as follows. The proposed analysis strategy is introduced and illustrated in Section 2. In Section 3 we give an analytical reasoning for ideal value of one of the parameters of PKDPA. Section 4 explains the properties of the probable key matrix. The PKDPA attack on AES is demonstrated in 5.1. The attack on masked AES is reported in Section 5.2. We provide a comparative study of the schemes proposed here with other contemporary strategies in Section 6. Section 7 concludes the paper.

2   Probable Key Differential Power Analysis

The central idea behind the proposed attack is to obtain a set of keys, which have a high probability of being the correct key. Like a class of conventional DPA attacks the difference of means (DoM) value is used as a parameter to decide the probability of a key being the actual one. DPA attacks use a Divide-and-Conquer philosophy, attacking one Sbox at a time. The number of key bits revealed is equal to the number of inputs to the Sbox under attack. The attack is then extended to the remaining Sboxes thereby revealing the entire round key. One of the key differences between the classical mono-bit DPA attack and the proposed Probable Key Differential Power Analysis (PKDPA) strategy is that while mono-bit DPA always looks for the key with the highest DoM value, PKDPA looks for a set of keys with high DoM values. The cardinality of the set is defined as the *window-size.* So PKDPA analyzes a window of keys instead of a single key. The specifications of PKDPA for finding the correct key corresponding to a single Sbox are given below:

Algorithm. *PKDPA*
Input: Plaintext/Ciphertext,power traces,target Sbox,window-size,key guesses
Output: Correct Key
1.   Choose target Sbox, specify window-size and obtain power traces
2.   **for** All target bits (* Each output bit of the Sbox *) and all possible key guesses
3.   Perform DoM test.
4.   **end**
5.   Extract probable keys (* the keys whose DoM value lies inside the chosen window-size *) based on the corresponding DoM values.
6.   **for** Each probable key
7.   Find frequency of occurrence
8.   **end**
9.   **if** Frequency of occurrence of a key $\geq \left\lceil \frac{1}{2}(\# \text{ of output bits of the Sbox}) \right\rceil$,
10.   **then**
11.   **return** correct key.
12.   **else**   (* If no such key is found *)
13.   Obtain more power traces and goto step 2.

Our analysis shows that as the number of traces goes on increasing the DoM value for the correct key also increases. Thus, the correct key *bubbles* up among all incorrect ones and makes its way into the set of keys PKDPA observes. Looking at a larger window of keys helps us capture the correct key with quite a lesser number of traces than mono-bit DPA attack. Before the onset of PKDPA, all guessed keys are equally probable. From these equally probable keys, the attack filters out a set of keys which are claimed to have high probability of being the correct key. So, these keys are referred to as 'probable keys'. This filtering is done for all target bits giving us a set of sets of probable keys. PKDPA then analyzes this master set to reveal the correct key. It is worth mentioning that PKDPA can reveal the correct key in a scenario when the classical mono-bit DPA fails. We now study PKDPA in details.

2.1   Initialization of the Attack

In this step, the input parameters of PKDPA are set. The Sbox to be attacked i.e., the *target Sbox* is chosen. The window size is fixed. PKDPA is a known-plaintext or ciphertext-only attack. So either the plaintexts or the ciphertexts are required. The corresponding power traces must also be acquired. The attack is then initiated by performing the DoM test for all possible keys guessed. The key space K is limited to all possible combinations of the inputs bits to the Sbox.

$$|K| = 2^m, m = \text{number of inputs to an Sbox.}$$

2.2   Iterative DoM Test

Before the onset of this step, one of the output bits of the Sbox is chosen as the *target bit*. For all $|K|$ possible key guesses the DoM test is carried out based on the output of the target bit as explained in the preliminaries section. We call this step iterative because the DoM test has to be repeated for all the possible targets bits of the chosen Sbox. The number of target bits is equal to the number of output bits of the Sbox. So, in the following paragraphs the terms 'target bit' and 'output bit' has been used synonymously. The DoM values (also referred to as the DPA bias) obtained corresponding to all key guesses for all target bits are stored for the next step of *Probable Key Extraction*.

2.3   Probable Key Extraction

Before proceeding with this step we define the notion of *Probable Keys*.

Let i denote one of the output bits of the Sbox i.e., the target bit in the partition function and $d_i^t(k)$ denote the difference of means corresponding to the key guess k after the tth trace. Then $d_i^t(k)$ is used as a measure for the probability $p_i^t(k)$ of key k to be the correct one after t traces. Let S_i^t denote the set of $p_i^t(k)$ for all $k \in K$. Hence,

$$S_i^t = \{p_i^t(k) : k \in K\} \qquad (1)$$

Also let V be a set of real values and $max_w(V) \subset V$ be the set of the first w maximum values from the set V. Then $max_w(S_i^t)$ is the set of first w maximum $p_i^t(k)$ values. A *probable key* is a key corresponding to which the DPA bias or DoM value is quite large but not necessarily the maximum. Probable keys form a set which is defined as

$$P_i^t = \{k : p_i^t(k) \in max_w(S_i^t)\} \qquad (2)$$

Here $w \leq |S_i^t|$ is referred to as the *window size*. It decides the number of probable keys to be considered i.e., the number of probable keys we iterate our experiments with. We term them as 'probable keys' because one of these keys is expected to be the actual or correct key. In classical DPA we go on increasing the traces, until we find a sharp peak in the differential traces of the keys and the key corresponding to which the differential trace has the highest peak is concluded to be the correct key. However, in the probable key analysis we can find the correct key even if the differential traces are not distinguishable. Instead of finding the key with the highest DPA bias, i.e. the key which has the highest peak in the differential trace among all the keys, we take into consideration w keys with DPA biases among the first w maximum DPA biases. The aim of the analysis is to find out the correct key when we are not in a position to get the key by mere visual inspection of the differential trace. Thus we are able to retrieve the key with lesser number of traces as compared to classical DPA. Also with inadequate number of traces, classical DPA fails to recover the correct key. Our attack may succeed in such a scenario thereby enhancing the strength of power based attacks.

Now that the concept of probable keys is understood we can proceed with the analysis. For probable key extraction we first arrange the DoM values obtained for each target bit in the previous step in decreasing order of magnitude. Then the keys corresponding to the first w (window-size) maximum DoM values for a particular target bit are used to form *probable key set* for that bit. This has to be repeated for all target bits. If the number of target bits be m then at the end of this step we will have m probable key sets each having w probable keys. All these sets together can be visualized as an $(w \times m)$ matrix called the *probable key matrix*. In the next step we analyze this matrix.

2.4 Key Frequency Analysis

Here we analyze the probable key matrix in terms of the frequency of occurrence of a key in the matrix. The number of probable key sets for an Sbox depends on the number of its output bits. Let $f_i(k)$ represent the occurrence of the key k in the probable key set P_i^t of the ith output bit. Then $f_i(k)$ is given as:

$$f_i(k) = \begin{cases} 1 & \text{if } k \in P_i^t \\ 0 & \text{otherwise} \end{cases} \tag{3}$$

We now give the expression for the total frequency of occurrence for a key in all the m probable key sets i.e., in the probable key matrix.

$$F(k) = \sum_{i=1}^{m} f_i(k), \quad \forall k \in K, \quad F(k) \le m \tag{4}$$

Let A represent the set of all such frequencies. Then according to *Probable Key Analysis* the key k is the correct one if it satisfies the following criteria:

1. It has the maximum frequency of occurrence.

$$k \in K_c, \quad \text{where} \begin{cases} K_c = max_1(A) \\ |K_c| = 1 \end{cases} \tag{5}$$

   Here, $max_1(A)$ denotes the subset of elements of A with maximum frequency.
2. The frequency of occurrence is:

$$F(k) \ge \left\lceil \frac{m}{2} \right\rceil, \quad k \in K_c \tag{6}$$

If $|K_c| > 1$ i.e., if there are multiple keys which satisfy the above criteria or if for $k \in K_c$, $F_i(k) < \frac{m}{2}$, then the experiment is inconclusive and must be repeated with higher number of traces. It can be argued that attaching a weight based on the position of a candidate key in the probable key matrix could further enhance the strategy. However, chances of ghost peaks or spurious keys[22], whereby a wrong key will register higher DPA bias, discourage us from using a weighted system. In the next subsection we show an observation on how the probability of the correct key changes with number of traces.

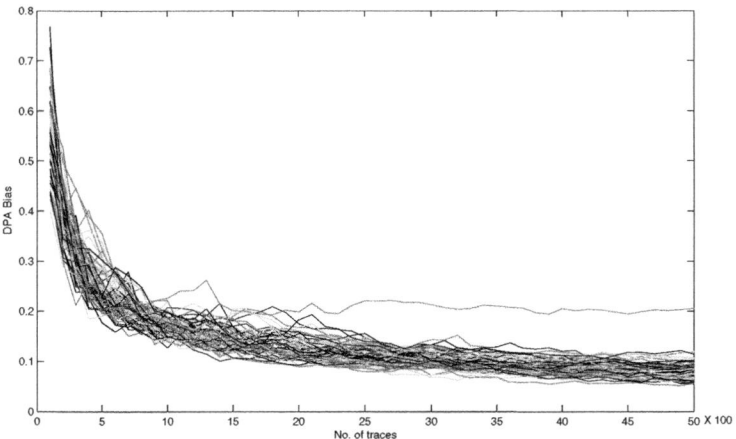

Fig. 1. Progression of DoM value with number of traces for Sbox-1, Bit-2 for DES

2.5 Probable Key Progression

In **Fig. 1** we plot the DoM values of the keys against the number of traces for the second bit of the first Sbox of DES. The plot shows us how the DoM value changes with the increase in the number of traces. As can be verified from the figure, the plot for one of the keys is distinctively different from the others. This plot corresponds to the DoM value of the correct key. The distinct progression of the actual key ensures that after a reasonable number of traces, the probability of the proposed scheme to return an unambiguous key becomes very high. In the next section, we try to develop a framework to give a theoretical bound on the most important parameter in PKDPA - the window-size.

3   Analytical Framework for Optimal Window-Size

We now give an approximate bound on the selection of window-size of probable key analysis. Let us consider a $(m \times n)$ Sbox. If we set the window-size to w, then we will have an $(w \times n)$ probable key matrix. We now define a random variable M_i such that

$$M_i = \left\{ \begin{array}{ll} 1 & \text{if any key occurs in the } i^{th} \text{ column of the probable key matrix} \\ 0 & \text{otherwise} \end{array} \right\}$$

The probability of occurrence of any randomly chosen key in a column of the probable key matrix is

$$Pr[M_i = 1] = \frac{w}{2^m}, \quad \text{where } 2^m \text{ is \# of keys guessed}$$

Under the assumption that $(M_1, M_2, \cdots, M_n)$ are independent random variables, we can write

$$M = \sum_{i=1}^{n} M_i$$

For, a random key $Pr[M < \frac{n}{2}]$ should be very high. Now, it can be noted that according to PKDPA if a key other than the correct key occurs more than or equal to $\lceil \frac{n}{2} \rceil$ times than it leads to an error. So, $Pr[M \geq \frac{n}{2}]$ gives the error probability of PKDPA. From the point of view of probable key analysis, we want to keep $Pr[M \geq \frac{n}{2}]$ for a randomly chosen key as low as possible. We then argue that with the error probability significantly low for a wrong key, if some key occurs more than or equal to $\lceil \frac{n}{2} \rceil$ times, then it must be the correct key with a very high probability. This is because $Pr[M \geq \frac{n}{2}]$ is high for a correct key as it will be correlated to all the target bits and hence tend to appear in all columns. Hence we can use this to devise a good distinguisher between the correct key and wrong keys. In the following paragraphs we try to draw a relation between the parameter window-size and error probability. We proceed by applying the theorem for multiplicative form of *Chernoff bound* [23] which is stated as follows:

Theorem 1. *[23] Let random variables* $(X_1, X_2, \cdots, X_n)$ *be independent random variables taking on values 0 or 1. Further, assume that* $Pr(X_i = 1) = p_i$. *Then, if we let* $X = \sum_{i=1}^{n} X_i$ *and* μ *be the expectation of* X, *for any* $\delta > 0$, *we have*

$$Pr[X \geq (1+\delta)\mu] \leq \left(\frac{e^\delta}{(1+\delta)^{(1+\delta)}} \right)^\mu \tag{7}$$

We are interested to find the Chernoff bound for the error probability $Pr[M \geq \frac{n}{2}]$ of a random (or wrong) key. For the random variable M defined earlier, the expectation $\mu = E(M)$ and $\delta > 0$ are as below.

$$E(M) = \left(\frac{w}{2^m} \right) n \tag{8}$$

$$\delta = \left\{ \frac{2^{(m-1)}}{w} \right\} - 1 \tag{9}$$

By Theorem 7, the error probability of a random key occurrence in the probable key matrix is bounded by

$$Pr\left[M \geq \frac{n}{2}\right] \leq \left[\frac{e^{\left\{ \frac{2^{(m-1)}}{w} - 1 \right\}}}{\left\{ \frac{2^{(m-1)}}{w} \right\}^{\left(\frac{2^{(m-1)}}{w} \right)}} \right]^{\left(\frac{w}{2^m} \right)n} \tag{10}$$

The R.H.S of **Equation** 10, thus gives an upper bound of the error probability of a randomly chosen (wrong) key to pass the test. From the condition that $\delta > 0$, we can easily derive the upper bound for the window-size.

$$\left\{ \frac{2^{(m-1)}}{w} \right\} - 1 > 0$$
$$\Rightarrow w < 2^{(m-1)}$$

We now give some theoretical results about the variation of error probability with the window-size. For AES, $m = n = 8$, while for DES and 3-DES, $m = 6, n = 4$. We plot the upper bound of the error probability given in **Equation 10** for all possible values of the window-size ($w < 2^{(m-1)}$) separately for AES and DES algorithms in **Fig. 2**. We also plot the value of confidence defined as (1 - error probability).

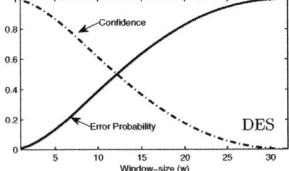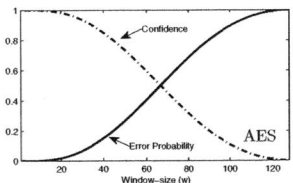

Fig. 2. Error probability and Confidence Vs Window-size for DES and AES

From **Fig. 2** one can infer that as we go on increasing the window-size the error probability of PKDPA also increases and attains very high value if w is close to $2^{(m-1)}$. In general we can say that the probability of error is directly proportional to the chosen window-size.

$$Pr\left[M \geq \frac{n}{2}\right]_{w_1} > Pr\left[M \geq \frac{n}{2}\right]_{w_2}, \quad \text{if} \quad w_1 > w_2$$

Tables 1 and **2** where we show the error probabilities for DES and AES w.r.t some discrete window-sizes. In the next paragraph we study the effect of window-size on the number of traces required to mount the attack.

Table 1. Error Probabilities(ϵ) for different window-sizes (DES)

w	1	3	5	10	15
ϵ	0.0068	0.0538	0.1320	0.3862	0.6358
$1 - \epsilon$	0.9932	0.9462	0.868	0.6138	0.3642

Table 2. Error Probabilities(ϵ) for different window-sizes (AES)

w	1	5	10	15	20	25	30	35
ϵ	1.9714e-007	1.0873e-004	0.0015	0.0064	0.0174	0.0364	0.0645	0.1022
$1-\epsilon$	0.99999980286	0.99989127	0.9985	0.9936	0.9826	0.9636	0.9355	0.8978

The estimates of error probability and confidence have been plotted in **Fig. 3** w.r.t the window-size. In addition to that we plot the number of traces that are required to attack AES Sbox - 15 against window-size (w). One can infer from the figure that ideal size of the window varies between 9 and 18. In this range least number of traces are required to mount PKDPA. If we increase w beyond 18 the trace count also increases rapidly. We also see that beyond a certain value of the window-size i.e., 34 there is a drastic increase in the number of required traces. Actually from our practical results we have found that for $w > 34$, the probability $Pr\left[M \geq \frac{n}{2}\right]$ for wrong keys becomes significantly high. It can be seen that the theoretical value of the error probability (ϵ) for $w = 35$ given in **Table 2** is also high. In case of DES, the optimal value of w varies from 4 to 10. We found that if we choose $w = 10$, we require 2,000 traces for a successful attack, while for $w = 5$, it reduces to 900. However if we choose $w = 1$ (classical DPA), number of traces required for a successful attack increases to 10,000. From our experimental results we have found that for all practical purposes $w = 10$ (AES) and $w = 5$ (DES, 3-DES) yields good results. In the next paragraph we compare classical DPA with PKDPA in relation to the above results.

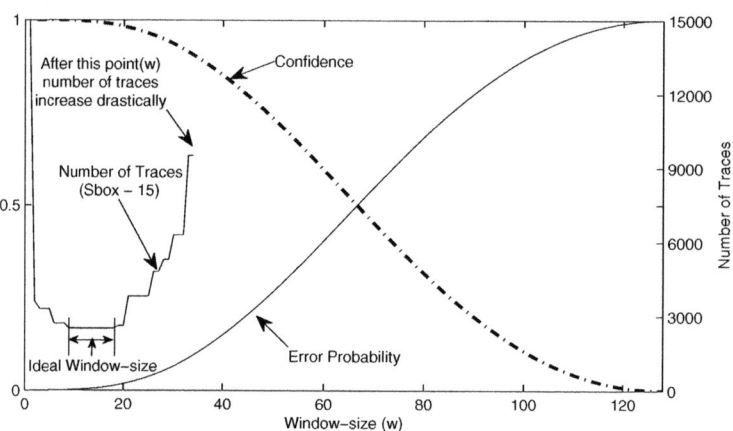

Fig. 3. Error probability, Confidence and # of Traces (Sbox- 15) Vs Window-size for AES

The plots in **Fig. 2** imply that the probability of error with window-size $w = 1$ is the least and steadily increases with $w > 1$. It is interesting to note that $w = 1$ actually represents classical DPA where we look only for the key with the highest DPA bias. From **Table 1** and **2** we can see that the probability of error (ϵ) is significantly low for $w = 1$. However, if we choose $w = 10$ (AES) or $w = 5$ (DES), then ϵ still remains very low. Conversely, we can say that the level of confidence ($1 - \epsilon$) in classical DPA is very high while for PKDPA it is still considerably high. In practice, we see that by slightly lowering the level of confidence we are able to considerably reduce the number of traces required for a successful attack. However, one should be careful in choosing the window size as choosing a very high window-size could lead to drastic increase in required number of traces. Before reporting the case-studies we illustrate properties of the Probable Key Matrix.

4   The Probable Key Matrix

For a better interpretation of the experimental results that are reported in the next section, it is vital to be able to understand a prime element of PKDPA - the probable key matrix. The probable key matrix comes into the picture after completion of the probable key extraction phase illustrated in subsection 2.3. It is the master set of all the probable key sets corresponding to respective target bits. A real-life probable key matrix is presented in **Fig. 4**. One can note that the first row of the matrix actually represents the keys which have registered highest DPA bias. In other words, these keys are the ones that a classical DPA attack would have returned as correct keys. Each column of the matrix denotes the probable key for the corresponding target bit. The window-size (w) is 10 and hence we have total of 10 probable keys in each column which are arranged in decreasing order of their DPA bias. In order to retrieve the correct key one has to perform a frequency analysis of the probable key matrix. The PKDPA attacks are reported in the next section.

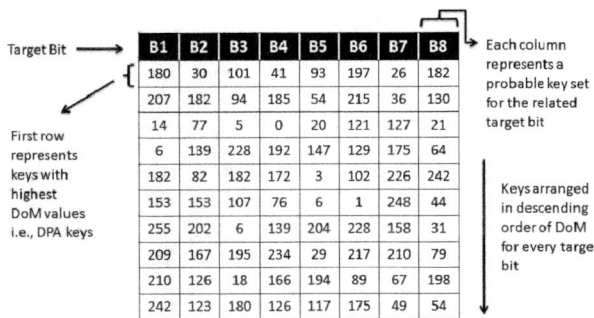

Fig. 4. Probable Key Matrix for AES Sbox 13, $w = 10$

5   Experimental Results

In this section, we report the experimental results that support our claims about PKDPA. Following is the test setup used for the experiment: **FPGA:** Spartan3 xc3s400-4pq208, **Oscilloscope:** Tektronix Digital Phosphor Oscilloscope TDS5032B, **Amplifier:** Tektronix TCPA300 Current Amplifier, **Probe:** Tektronix TCP312 30A DC Current Probe, **Sampling Rate:** 250 MegaSamples/Second, **Resolution:** 2000. We now present our case-studies on the unmasked and masked versions of AES.

5.1   Case-Study 1: AES

Attack Results. The attack has been mounted on a standard iterative implementation of AES. We first show that for a trace count of 2500, classical mono-bit DPA does not return any conclusive results. It is evident from the differential plots given in **Fig. 5** that the correct key, which is 12, cannot be distinctively identified except for the case of the 2nd target bit (**Fig. 5(a)**). In fact a precise look at the results show that for the 3rd and 6th target bits, DPA returns key 244 as the correct key. This could even trick an attacker into believing that 244 might be the correct key. Thus we see that for 2500 traces DPA leads to indecisive results. We deploy a probable key analysis using the same number of traces in the next paragraph.

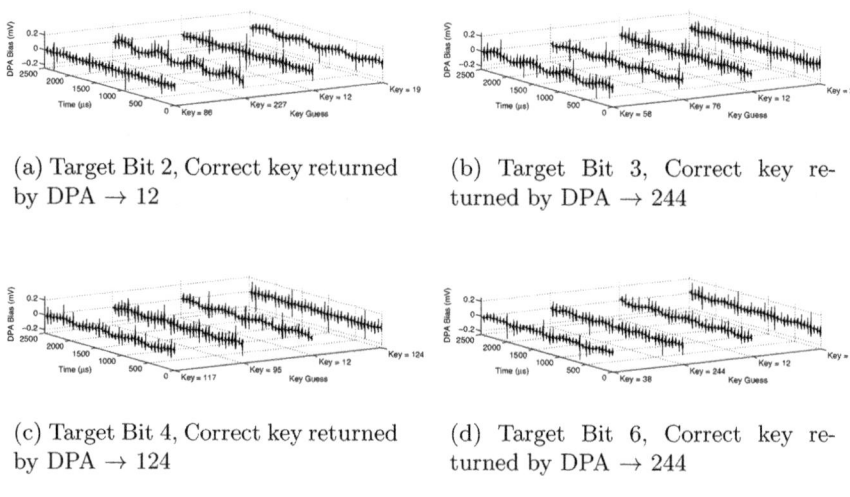

(a) Target Bit 2, Correct key returned by DPA → 12

(b) Target Bit 3, Correct key returned by DPA → 244

(c) Target Bit 4, Correct key returned by DPA → 124

(d) Target Bit 6, Correct key returned by DPA → 244

Fig. 5. AES Differential plots for some target bits of Sbox-15 with 2500 traces

The window-size is set to be 10. As the number of output bits of an AES Sbox is 8 so we get a (10×8) probable key matrix. **Table 3** shows the probable key matrix for $2,500$ traces. From the frequency analysis (**Table 4**), we infer that key 12 is the most frequent key with frequency of occurrence being 4. So by probable key analysis, key 12 is concluded to be the correct key. Similar results have been obtained when PKDPA was mounted on DES and Triple-DES.

5.2   Case-Study 2: Masked AES

There are several approaches to securing hardware implementations of block ciphers against differential power attacks. These approaches can be broadly classified under two types namely, hardware countermeasures and algorithmic countermeasures. Algorithmic approaches are generally preferred as they can be optimized for a specific cipher. As regards AES, various algorithmic countermeasures [24,25,26,27] have been reported. However, all of them hover around the

Table 3. AES Probable Key Matrix for Sbox-15 with 2500 Traces

Bit1	Bit2	Bit3	Bit4	Bit5	Bit6	Bit7	Bit8
11	12	244	124	86	244	242	147
23	86	76	95	64	38	143	107
61	227	58	244	69	67	128	42
133	19	217	12	210	17	44	88
197	161	164	142	127	124	174	137
35	38	69	139	60	103	41	12
22	191	52	117	218	91	36	122
220	164	123	74	193	196	68	125
238	105	12	73	18	82	133	159
178	26	193	147	89	78	202	96

Table 4. Frequency Matrix for unmasked AES Sbox-15 with 2500 Traces

Key	Freq.	Key	Freq.	Key	Freq.	Key	Freq.	Key	Freq.	Key	Freq.	Key	Freq.
12	4	96	1	41	1	73	1	105	1	142	1	210	1
244	3	11	1	42	1	74	1	107	1	143	1	217	1
38	2	17	1	44	1	76	1	117	1	159	1	218	1
69	2	18	1	52	1	78	1	122	1	161	1	220	1
86	2	19	1	58	1	82	1	123	1	174	1	227	1
124	2	22	1	60	1	88	1	125	1	178	1	238	1
133	2	23	1	61	1	89	1	127	1	191	1	242	1
147	2	26	1	64	1	91	1	128	1	196	1	Remaining	
164	2	35	1	67	1	95	1	137	1	197	1	keys do	
193	2	36	1	68	1	103	1	139	1	202	1	not occur	

principle of masking. A mask is defined as a random value which is added to the intermediate values of AES. The intention is to randomize the computations that use the secret key and hence make the information leakage unpredictable. While schemes like [24,26] have been shown to be vulnerable to to a certain type of differential side-channel attack called the zero-value attack, [25] is unrealistic from the perspective of hardware requirements. In [27], Oswald et. al. proposed a new masking strategy which overcame the limitations of the predecessors. They use a combination of additive and multiplicative masking and claim that their scheme is particularly suited for implementation in dedicated hardware. The next paragraph details the implementation of the masked AES that was attacked.

Implementation. The implementation of the masked AES is similar to the unmasked one. The only difference is that the look-up table based Sbox in the unmasked AES was replaced by a masked Sbox. The masking technique proposed in [27] was adopted to design the masked Sbox. As per the specifications in [27], one of the inputs of the masked Sbox was a random seed. In our design we provided the seed from outside. We now proceed with attack results.

Attack Results. Attacks on the protected architecture [27] implemented on an ASIC has been reported in [28]. The attacks reveal that simple power models like Hamming-weight or bit model and correlation-matrices fail to retrieve the key with even *1 million* traces. The authors conclude that the attacks mounted using the bit model have in general been unsuccessful. In order to attack such a masked scheme, the authors of [28] suggest using power models derived from simulations of back-annotated netlists. Using such power models, they still needed as many as **30,000** traces to break the masked implementation. However, they themselves admit that *"an attacker usually does not have easy access to the back-annotated netlist of a product"*. Thus, though the authors succeeded in attacking masked AES using simulated power models, the assumptions on the knowledge of the attacker about the implementation are too strong and hence, make such attacks less realistic.

Our results are not directly comparable with [28] due to the basic difference in platform namely ASIC and FPGA. Nevertheless, our approach relies on less stringent assumptions. We implemented the masking technique suggested in [27] on an FPGA implementation of AES. We mounted PKDPA with 13000 traces and the probable key matrix for the 15th Sbox is depicted in **Table 5** for a window-size of 15. The choice of the window-size is consistent with the theoretical results in **Table 2**. A closer look at **Table 5** shows us that the correct key is present in the probable key sets for 1st through 5th, and 7th bits for a total of 6 times. Being the most frequent key and conforming to PKDPA principle, key 12 is inferred as the correct one. The result aligns with actual key used. Thus PKDPA, though using the bit model, successfully retrieves the key for **13,000** power traces. Moreover, as is evident from **Table 5**, both mono-bit DPA and majority voting fail to come up with the right key.

Table 5. Masked AES Probable Key Matrix for Sbox-15 with 13000 Traces

Bit1	Bit2	Bit3	Bit4	Bit5	Bit6	Bit7	Bit8
111	163	196	**12**	129	42	130	125
12	215	239	147	**12**	76	16	0
216	54	62	169	113	218	139	143
94	22	249	200	13	37	98	126
139	181	86	87	244	58	10	2
172	50	113	67	30	22	176	225
160	226	169	97	218	97	116	73
57	194	240	250	183	117	146	178
11	137	76	81	145	230	244	200
110	42	137	201	35	173	158	90
44	162	102	180	182	138	44	83
233	72	176	0	216	227	245	22
237	**12**	55	142	67	56	**12**	145
208	253	**12**	94	93	155	236	13
103	73	19	84	203	191	147	111

6   A Comparative Study

With a multitude of attacks strategies and platforms documented in literature, it is difficult to compare them based on the results reported in the various texts. In order to compare different DPA attacks, various approaches have been suggested like type of assumptions about the leakage, number of measurements etc. However, from the perspective of a practical adversary the latter one is of particular significance. Below we try to highlight results reported by other proposed enhancement schemes. In order to emphasize the novelty of our work we then try to compare results of attacks mounted on the same architecture using conventional DPA, majority voting and PKDPA. We restate the fact that the metric for measuring the efficiency is the number of power traces required to mount the attack.

The authors of [29] report a successful classical DPA attack on DES using 10,000 traces. Classical DPA in our case reveals the key with 4,000 measurements. Though the implementations attacked in both works are comparable, difference lies in the FPGA platforms used. [29] uses *Virtex* while we use *Spartan-3* FPGAs as our target device. To validate the results, [3] uses simulated power traces and [2] uses a micro-controller based implementation. When we applied the proposed probable key analysis scheme to attack DES, we could reduce the number of traces required to 900. As regards 3-DES, there are no benchmarking results for classical DPA available in literature. So we report only our results. Classical DPA attack on 3-DES retrieves the key with 10,000 traces. PKDPA mounts a successful attack with 3,200 traces.

In [19] *Standaert et. al.* reports the first practical DPA on an FPGA based implementation of AES. As rightly pointed out by the authors, very few works exist which actually attack real and optimized AES designs. They present a detailed overview of power attacks on an AES core running on a Xilinx Virtex device. The attacks are based on correlation between a Global prediction matrix and a Global consumption matrix. The computation of the global prediction matrix requires detailed knowledge of the target registers whose transitions can be predicted and which leak information, thereby making the leakage model very demanding. This helps to reveal the AES key using 1000 power traces. The number of power traces required can be reduced to 300 by improving the measurements [20]. These results can be attributed to the observations made in [14] where the authors extensively study CPA, mono-bit DPA and linear regression attacks [15] in the light of a perfect leakage model as well as random leakage model. Their work reveals that CPA remains the best performer when the chosen leakage model closely matches the actual leakage. On the contrary, under the random leakage scenario, mono-bit DPA and linear regression attacks show higher efficiency. Additionally, *it might be noted that according to [21], the bit model of power analysis, which has been adapted in our work, makes the least assumptions about the attacked device.* In other words, the attacker requires much lesser knowledge about the architecture. To the best of our knowledge, the only published result which uses the bit model to attack an AES ASIC implementation is [21]. The authors in [21] show that with 1,000,000 traces they fail to mount a

Table 6. A Comparative Study

Cipher	Scheme	# of Traces	Validation	Attack Model / Knowledge Required
DES	[3]	Not reported	Simulation	
	[2]	2048	Micro-controller	
	Classical DPA [29]	10,000	Virtex XCV800	Bit Model
	Classical DPA (Ours)	4,000	Spartan-3 XC3S400	Bit Model
	PKDPA	**900**		
3-DES	Classical DPA(Ours) (No reported results)	10,000	Spartan-3 XC3S400	Bit Model
	PKDPA	**3,200**		
AES	Classical DPA [21]	6,532	ASIC	Hamming Distance Model
	Classical DPA (Ours)	15,000	Spartan-3 XC3S400	Bit Model
	[19]	1,000	Virtex XCV800	Correlation Matrix - Extensive Knowledge about target registers, architecture etc, required
	PKDPA	**2,500**	Spartan-3 XC3S400	Bit Model - Requires little knowledge about target device
Masked AES[27]	DPA [28]	Not possible with 1,000,000 traces	ASIC	Bit Model
		30,000	ASIC	Power model derived from simulation - Requires knowledge of target netlist
	PKDPA	**13,000**	Spartan-3 XC3S400	Bit Model

conclusive DPA attack on AES. They suggest that to mount a successful attack very high number of traces would be necessary. In our work, where an FPGA implementation is targeted, we show that by using $PKDPA$, with 2,500 traces we succeed to retrieve the entire AES key. In order to appreciate the power of $PKDPA$ it is worth to restate the fact that with 2,500 traces classical DPA fails to retrieve the correct key for any of the target bits of the 15th Sbox of AES (as already shown in the case study).

PKDPA has been also experimented on a masked AES implementation [27]. Attacks on the same protected architecture implemented on an ASIC has been reported in [28], which reveal that simple power models like bit model and correlation-matrices fail to retrieve the key with even *1 million* traces. In order to attack such a masked scheme, the authors of [28] have used power models derived from netlist simulations thus requiring extensive knowledge about the target netlist. On an FPGA implementation, PKDPA reveals the key at a much reduced number of power traces. The observations and comparisons presented here are summarized in **Table 6**.

A slightly different idea from conventional DPA may be found in [7]. The authors suggest to perform a majority voting among the results for all the target bits to infer the correct key. We observe in our experiments that the method in [7] finds the correct key with lesser number of traces than conventional DPA. A closer look at the majority voting reveals that it is a special case of PKDPA i.e, if in case of PKDPA the window-size is chosen to be one then it reduces to

Table 7. A comparative study with the same set of traces. These results are for three different DPA schemes mounted using the same set of traces. The values represent # of traces.

	DES	3-DES	AES
Classical DPA	4,000	10,000	15,000
Majority Voting [7]	1,100	3,500	3,600
PKDPA	900	3,200	2,500

the majority voting presented in [7]. This justifies why PKDPA further reduces the number of traces. **Table 8** enlists the keys with highest DPA biases for 13th Sbox of AES with 2, 500 power traces. Though the correct key, which is 182, appears for bit 8 but a majority voting is completely inconclusive at this stage. However a frequency analysis on the probable key matrix (**Table 9**) for the same number of traces reveals 182 to be the most frequent key. Hence by applying PKDPA we successfully retrieve the correct key when both classical DPA and majority voting [7] *fail*. We emphasize on the fact that PKDPA is able to capture the correct key at an earlier stage i.e., with a much lesser number of traces. Moreover, PKDPA is also resistant to ghost peaks [22]. Since we look at more than one key for each target bit and also on keys returned by multiple target bits, so the probability that we conclude a wrong key due to a ghost peak is considerably low. In order to present a more fair comparison we tally the results of some DPA schemes with PKDPA in **Table 7**, based on the same set of traces that we obtained.

Table 8. Keys returned for all target bits by classical DPA on Sbox-13 with 2500 Traces (A scenario where majority voting scheme [7] fails)

Bit1	Bit2	Bit3	Bit4	Bit5	Bit6	Bit7	Bit8
180	30	101	41	93	197	26	182

Table 9. Probable Key Matrix for Sbox-13 with 2500 Traces

Bit1	Bit2	Bit3	Bit4	Bit5	Bit6	Bit7	Bit8
180	30	101	41	93	197	26	182
207	182	94	185	54	215	36	130
14	77	5	0	20	121	127	21
6	139	228	192	147	129	175	64
182	82	182	172	3	102	226	242
153	153	107	76	6	1	248	44
255	202	6	139	204	228	158	31
209	167	195	234	29	217	210	79
210	126	18	166	194	89	67	198
242	123	180	126	117	175	49	54

7 Conclusion

An improved DPA analysis, PKDPA, is proposed which reveals the key of the targeted cipher using less number of power traces. Being based on the bit model of power analysis, less knowledge about the design architecture and leakage model is needed, thus leading to a practical form of attack. The error probability of the proposed method has been theoretically analyzed with respect to parameters, like window-size. The theoretical framework is supported with real life attacks on standard FPGA implementations of DES, Triple-DES and AES, which show that the attack makes a marked improvement over the mono-bit DPA attack using difference of means as the distinguisher and needs lesser number of traces to ascertain the key. The attack is shown to be capable of discovering the AES key, even for masked implementations using only $13,000$ power traces. Detailed comparative studies have been furnished to demonstrate the efficiency of PKDPA over conventional DPA and its other enhancements. In the light of recent works, where different standard DPA attacks have been shown to be equivalent, under reasonable assumptions, the enhancement of the popular mono-bit DPA presented in this work becomes relevant.

References

1. Kocher, P., Jaffe, J., Jun, B.: Differential Power Analysis. In: Wiener, M. (ed.) CRYPTO 1999. LNCS, vol. 1666, pp. 388–397. Springer, Heidelberg (1999)
2. Bévan, R., Knudsen, E.: Ways to Enhance Differential Power Analysis. In: Lee, P.J., Lim, C.H. (eds.) ICISC 2002. LNCS, vol. 2587, pp. 327–342. Springer, Heidelberg (2003)
3. Boracchi, G., Breveglieri, L.: A Study on the Efficiency of Differential Power Analysis on AES S-Box. Technical Report (January 15, 2007)
4. Mangard, S., Oswald, E., Standaert, F.-X.: All for one-one for all: Unifying univariate DPA attacks. IET Information Security 5(2), 100–110 (2011)
5. National Institute of Standards and Technology, Data Encryption Standard, in Federal Information Processing Standard 46-2,
   http://www.itl.nist.gov/fipspubs/fip46-2.htm
6. Messerges, T.S., Dabbish, E.A., Sloan, R.H.: Examining Smart-Card Security under the Threat of Power Analysis Attacks. IEEE Trans. Comput. 51(5), 541–552 (2002)
7. Yu, P., Schaumont, P.: Secure FPGA circuits using controlled placement and routing. In: Proceedings of International Conference on Hardware Software Codesign (CODES+ISSS), pp. 45–50. ACM (2007)
8. Brier, E., Clavier, C., Olivier, F.: Correlation Power Analysis with a Leakage Model. In: Joye, M., Quisquater, J.-J. (eds.) CHES 2004. LNCS, vol. 3156, pp. 16–29. Springer, Heidelberg (2004)
9. Le, T.-H., Clédière, J., Canovas, C., Robisson, B., Servière, C., Lacoume, J.-L.: A Proposition for Correlation Power Analysis Enhancement. In: Goubin, L., Matsui, M. (eds.) CHES 2006. LNCS, vol. 4249, pp. 174–186. Springer, Heidelberg (2006)
10. Katashita, T., Satoh, A., Sugawara, T., Homma, N., Aoki, T.: Enhanced Correlation Power Analysis Using Key Screening Technique. In: RECONFIG 2008: Proceedings of the 2008 International Conference on Reconfigurable Computing and FPGAs, pp. 403–408. IEEE Computer Society, Washington, DC, USA (2008)
11. Li, H., Wu, K., Peng, B., Zhang, Y., Zheng, X., Yu, F.: Enhanced Correlation Power Analysis Attack on Smart Card. In: International Conference for Young Computer Scientists, pp. 2143–2148 (2008)
12. Le, T. h., Nguyen-vuong, Q.t., Canovas, C., Clédière, J.: Novel Approaches for Improving the Power Consumption Models in Correlation Analysis,
    http://eprint.iacr.org/2007/306.pdf

13. Chari, S., Rao, J.R., Rohatgi, P.: Template Attacks. In: Kaliski Jr., B.S., Koç, Ç.K., Paar, C. (eds.) CHES 2002. LNCS, vol. 2523, pp. 13–28. Springer, Heidelberg (2003)
14. Doget, J., Prouff, E., Rivain, M., Standaert, F.-X.: Univariate side channel attacks and leakage modeling. J. Cryptographic Engineering 1(2), 123–144 (2011)
15. Schindler, W., Lemke, K., Paar, C.: A Stochastic Model for Differential Side Channel Cryptanalysis. In: Rao, J.R., Sunar, B. (eds.) CHES 2005. LNCS, vol. 3659, pp. 30–46. Springer, Heidelberg (2005)
16. Standaert, F.-X., Malkin, T.G., Yung, M.: A Unified Framework for the Analysis of Side-Channel Key Recovery Attacks. In: Joux, A. (ed.) EUROCRYPT 2009. LNCS, vol. 5479, pp. 443–461. Springer, Heidelberg (2009)
17. Standaert, F.-X., van Oldeneel tot Oldenzeel, L., Samyde, D., Quisquater, J.-J.: Differential Power Analysis of FPGAs: How Practical is the Attack? In: Y. K. Cheung, P., Constantinides, G.A. (eds.) FPL 2003. LNCS, vol. 2778, Springer, Heidelberg (2003)
18. Örs, S.B., Oswald, E., Preneel, B.: Power-Analysis Attacks on an FPGA First Experimental Results. In: Walter, C.D., Koç, Ç.K., Paar, C. (eds.) CHES 2003. LNCS, vol. 2779, pp. 35–50. Springer, Heidelberg (2003)
19. Standaert, F.-X., Örs, S.B., Preneel, B.: Power Analysis of an FPGA: Implementation of Rijndael: Is Pipelining a DPA Countermeasure? In: Joye, M., Quisquater, J.-J. (eds.) CHES 2004. LNCS, vol. 3156, pp. 30–44. Springer, Heidelberg (2004)
20. Standaert, F.-X., Mace, F., Peeters, E., Quisquater, J.-J.: Updates on the security of fPGAs against power analysis attacks. In: Bertels, K., Cardoso, J.M.P., Vassiliadis, S. (eds.) ARC 2006. LNCS, vol. 3985, pp. 335–346. Springer, Heidelberg (2006)
21. Mangard, S., Oswald, E., Popp, T.: Power Analysis Attacks: Revealing the Secrets of Smart Cards (Advances in Information Security). Springer-Verlag New York, Inc., Secaucus (2007)
22. Canovas, C., Clédière, J.: What do S-boxes Say in Differential Side Channel Attacks? Cryptology ePrint Archive. Tech. Rep. (2005)
23. Hagerup, T., Rüb, C.: A guided tour of Chernoff bounds. Information Processing Letters 33(6), 305–308 (1990)
24. Akkar, M.-L., Giraud, C.: An Implementation of DES and AES, Secure against Some Attacks. In: Koç, Ç.K., Naccache, D., Paar, C. (eds.) CHES 2001. LNCS, vol. 2162, pp. 309–318. Springer, Heidelberg (2001)
25. Golic, J.D., Tymen, C.: Multiplicative Masking and Power Analysis of AES. In: Kaliski Jr., B.S., Koç, Ç.K., Paar, C. (eds.) CHES 2002. LNCS, vol. 2523, pp. 198–212. Springer, Heidelberg (2003)
26. Trichina, E., Seta, D.D., Germani, L.: Simplified Adaptive Multiplicative Masking for AES. In: Kaliski Jr., B.S., Koç, Ç.K., Paar, C. (eds.) CHES 2002. LNCS, vol. 2523, pp. 187–197. Springer, Heidelberg (2003)
27. Oswald, E., Mangard, S., Pramstaller, N., Rijmen, V.: A Side-Channel Analysis Resistant Description of the AES S-Box. In: Gilbert, H., Handschuh, H. (eds.) FSE 2005. LNCS, vol. 3557, pp. 413–423. Springer, Heidelberg (2005)
28. Mangard, S., Pramstaller, N., Oswald, E.: Successfully Attacking Masked AES Hardware Implementations. In: Rao, J.R., Sunar, B. (eds.) CHES 2005. LNCS, vol. 3659, pp. 157–171. Springer, Heidelberg (2005)
29. Standaert, F.-X., Örs, S.B., Quisquater, J.-J., Preneel, B.: Power Analysis Attacks Against FPGA Implementations of the DES. In: Becker, J., Platzner, M., Vernalde, S. (eds.) FPL 2004. LNCS, vol. 3203, pp. 84–94. Springer, Heidelberg (2004)

Formal Analysis of the Entropy / Security Trade-off in First-Order Masking Countermeasures against Side-Channel Attacks

Maxime Nassar[1,2], Sylvain Guilley[2], and Jean-Luc Danger[2]

[1] Bull TrustWay, Rue Jean Jaurès, B.P. 68,
78 340 Les Clayes-sous-Bois, France
[2] Institut TELECOM / TELECOM ParisTech, CNRS LTCI (UMR 5141),
46 rue Barrault, 75 634 Paris Cedex, France
{nassar,guilley,danger}@TELECOM-ParisTech.fr

Abstract. Several types of countermeasures against side-channel attacks are known. The one called masking is of great interest since it can be applied to any protocol and/or algorithm, without nonetheless requiring special care at the implementation level. Masking countermeasures are usually studied with the maximal possible entropy for the masks. However, in practice, this requirement can be viewed as too costly. It is thus relevant to study how the security evolves when the number of mask values decreases.

In this article, we study a first-order masking scheme, that makes use of one n-bit mask taking values in a strict subset of $\mathbb{F}_2^n$. For a given entropy budget, we show that the security does depend on the choice of the mask values. More specifically, we explore the space of mask sets that resist first and second-order correlation analysis (CPA and 2O-CPA), using exhaustive search for word size $n \leqslant 5$ bit and a SAT-solver for n up to 8 bit. We notably show that it is possible to protect algorithms against both CPA and 2O-CPA such as AES with only 12 mask values. If the general trend is that more entropy means less leakage, some particular mask subsets can leak less (or on the contrary leak remarkably more). Additionally, we exhibit such mask subsets that allows a minimal leakage.

Keywords: side-channel attacks (SCAs), masking countermeasure, non-injective leakage function, correlation power analysis (CPA), second-order CPA (2O-CPA), mutual information analysis (MIA), entropy *vs* security tradeoff, SAT-solvers.

1 Introduction

Implementations of cryptographic algorithms are vulnerable to so-called side-channel attacks. They consist in analysing the leakage of the device during its operation, in a view to relate it to the internal data it processes. The prerequisite of the attack is a physical access to the targeted device. The attacker thus measures some analogue quantity, such as the power [10] or the radiated field [7]. Several ways to resist side-channel have been suggested. They are often

D.J. Bernstein and S. Chatterjee (Eds.): INDOCRYPT 2011, LNCS 7107, pp. 22–39, 2011.

referred to as "countermeasures". High level countermeasures intend to deny the exploitation of the leakage by updating the secrets on a regular basis. It results in leakage-resilient protocols. They are nice as they indeed manage to thwart any kind of side-channel attacks, but require that the user adopts a new protocol. Therefore, other countermeasures have been devised that operate at a lower level, without altering the protocol. Typically, hiding strategies aim at leaking a constant side-channel. Although relevant from a theoretical perspective, this approach nonetheless requires physical hypotheses about resources indiscernibility that are not trivial to meet. Masking is another option, that is transparent to the user and does not demand any special backend balance. We therefore focus on this countermeasure. It consists in computing on data whose representation is randomized. The more entropy is used, the more secure the countermeasure can be (if the entropy is used intelligently). In this paper, we rather investigate the effect of the reduction of the entropy on the security. Moreover, we concentrate on a first-order masking scheme, $i.e.$ that uses only one mask, that takes a restricted number of values.

The rest of the article is structured as follows. The studied countermeasure, called the rotating tables, is described in Sec. 2. This section introduces the leakage model considered in the sequel, and defines the notion of leakage and security metrics. The rotating tables countermeasure is then evaluated in the formal framework presented in [21]. Namely, its leakage is characterized in Sec. 3 and its resistance against CPA and 2O-CPA is quantified in Sec. 4. It is shown in the section that it is possible to reduce the leakage at a constant budget for masks of $n = 5$ bits. Masks of larger bitwidth, such as $n = 8$, are studied in Sec. 5. The exploration is conducted with the help of a SAT-solver. Conclusions and perspectives are in Sec. 6.

2   Description of the Rotating Tables Countermeasure

The goal of this section is to introduce the leakage model that will be studied next, and to explain why the cost of the countermeasure can be greatly reduced by limiting the mask values. We first give in subsection 2.1 a brief overview of a masking countermeasure with randomly selected precomputed tables. Then, in subsection 2.2, the leakage of this countermeasure is derived.

2.1   Rationale

Unprotected implementations are vulnerable to SCAs because they manipulate sensitive variables, that leak some physical quantities that depend somehow on them. Therefore, in a Boolean masking scheme, they are replaced by the exclusive-or (XOR) with random variables. Let us take the example of a first-order masking scheme, where one mask m goes along with one the sensitive variable z. The bitvectors z and m have the same size, namely n bits. We call $\mathcal{S}_0 \doteq z \oplus m$ and $\mathcal{S}_1 \doteq m$ the two shares. The preconditions on the shares is that the sensitive variable can be recovered by XORing them: $Z = \mathcal{S}_0 \oplus \mathcal{S}_1$.

The linear operations with respect to the XOR are straightforward. Indeed, to compute a linear operation S on z using the shares, it suffices to apply S on each share. As a matter of fact, it is trivial to check the following post-condition: $S(z) = S(\mathcal{S}_0) \oplus S(\mathcal{S}_1)$. Nonetheless, if S is a non-linear operation, this equality does not hold, and it is necessary to use judiciously both shares to be able to compute $S(z)$. This operation is costly in general [23] (unless some algebraic properties of the non-linear function S can be taken advantage of [16]) and error-prone [11].

Therefore, it is sometimes relevant to compute on only one share, namely $\mathcal{S}_0$. This share traverses the linear parts of the algorithm, and is all-in-one:

1. Demasked at the entrance of a non-linear function S,
2. Applied S, and
3. Remasked so as to propagate through the next linear part.

For sure, the demasking and remasking operations are very sensitive. Nonetheless, the composition of the three operations can be tabulated: a table, such as a ROM block, conceals the intermediate variables (as in whitebox cryptography). Indeed, in cryptography, the non-linear function S will typically be a substitution box (*aka* sbox), that is hard to compute analytically, thus better saved in memory provided there are enough resources to store it. In this case, the intermediate variables never appear. For more details on the implementation of this table, we refer the interested reader to [14, Sec. 2], and more specifically to the paragraphs that concern the "sbox secure calculation".

In a platform that embarks an operating system, a task can be scheduled to recompute the masked sboxes $z \mapsto m_{\text{out}} \oplus S(z \oplus m_{\text{in}})$ periodically. Nonetheless, some embedded systems cannot afford a supervision for the masks update. Also, this process of mask refresh is itself sensitive, and should be protected adequately. In a view to relieve this constraint, one can get rid off the recomputation, and use masked sboxes that had been entered initially. This option is especially favorable for the cryptosystem that reuses several times the same sbox in each round (such as AES). The goal is not to create a security by obscurity solution. Indeed, the masks m_{in} and m_{out} can be disclosed without compromising the countermeasure. The randomness that characterizes the masking scheme will result from the choice of the sbox for each computation. Let us take the example of a hardware implementation of AES that computes one round per clock cycle. Sixteen masked $S[i]$ sboxes, $i \in [\![0, 15]\!]$, must be available in parallel. We assume that the masks $m_{\text{in}}[i]$ and $m_{\text{out}}[i]$ satisfy this chaining relationship: $\forall i \in [\![0, 15]\!], m_{\text{out}}[i] = m_{\text{in}}[i+1 \mod 16]$. Then the computation of an AES-128 can start by drawing a random number $r \in [\![0, 15]\!]$; the algorithm then invokes $S[j + k \mod 16]$ to compute the sbox of byte $j \in [\![0, 16]\!]$ of the state at round $k \in [\![0, 9]\!]$. Because of the chaining property, the linear parts of AES in-between the sboxes are consistently masked and demasked with the same mask. This ensures the correctness of the AES encryption implemented with the rotating sboxes countermeasures.

The overhead of the countermeasure is directly linked to the number of masks[1]. Indeed, more masks mean more memory to store the masked tables. Also, the more tables, the more multiplexing logic to access them, which increases the critical path in a hardware implementation. Thus, in the sequel, we endeavour to reduce the number of masks, while nonetheless keeping an acceptable security level.

2.2   Modelization

Hardware implementations of AES are preferably attacked on the last round. Indeed, it is possible to guess one byte, noted y of the round 9 from one byte of the ciphertext x simply by guessing one byte of the last round key, because there is no MixColumns operation in the last round. The leakage is a function of the distance between x and y, *i.e.* $x \oplus y$ [22]. Now, when the rotating tables countermeasure is applied, the value y is actually replaced by $y \oplus m$, where m is one of the 16 mask values. The sensitive variable is the value $x \oplus y$, noted z. In a view to introduce statistical notions, we denote by capital letters (Z and M) the random variables and by small letters (z and m) their realizations. The leakage function thus has the form:

$$\mathcal{L}(Z, M) = \mathscr{L}(Z \oplus M) . \tag{1}$$

In this expression, Z and M are n-bit vectors, *i.e.* live in $\mathbb{F}_2^n$. The leakage function $\mathscr{L} : \mathbb{F}_2^n \rightarrow \mathbb{R}$ depends on the hardware. In a conservative perspective, $\mathscr{L}$ is assumed to be bijective. This choice is the most favorable to the attacker, and is thus considered in the leakage estimation. Now, in practice, the leakage functions are not bijective. The canonical example is that of the Hamming weight leakage, where each bit of $Z \oplus M$ dissipate the same. Let us denote by x_i the component $i \in [\![1, n]\!]$ of $x \in \mathbb{F}_2^n$. The Hamming weight of x is expressed as $\mathsf{HW}[x] = \sum_{i=1}^n x_i$.

We underline that this section was not meant to introduce a new countermeasure (the rotating sboxes). Indeed, this pragmatic countermeasure is already well known and adopted in the industry. We simply wished to provide the reader with a pedagogical introduction to the leakage function of Eqn. (1). This function will now be studied formally, as per the guidelines presented in [21]. More precisely, we employ:

- The mutual information between the $\mathcal{L}(Z, M)$ and the sensitive variable Z with $\mathscr{L}$ bijective as a leakage metric. This quantity is noted $\mathsf{I}[\mathcal{L}(Z, M); Z]$ — basic definitions of information theory applied to SCAs can be found in [21] — and referred to as "mutual information as a metric" (MIM [24]). We recall that a leakage metric points out vulnerabilities, that could in practice not be exploited by an attacker.

---

[1] Notice that in the rest of the article, we have only <u>one</u> masking variable, that takes <u>few</u> values. We sometimes refer to them as the "number of masks"; we attract the reader's attention on the fact this expression shall not be confused with "multi-masks" countermeasures, also known as "high-order" masking schemes.

– Security metrics to quantify the easiness to actually turn a leakage into a successful attack. In this case, we will focus on $\mathscr{L} = $ HW. First of all, the optimal correlation between HW$[Z \oplus M]$ and Z is considered a metric. It is traditionally called the (first-order) correlation power analysis, or CPA [3]. But CPA can be defeated easily with only two mask values. Therefore it is important to consider higher-order CPA (HO-CPA), and notably the second-order CPA, also abridged 2O-CPA [26]. However, CPA and 2O-CPA exploit only the first two moments of the distribution of $\mathcal{L}(Z, M)$. Therefore, we also use a second security metric, namely the mutual information. It is known in the literature as MIA [2]. Security-wise, our goal is to minimize the first and second-order correlation coefficients and the MIA.

3   Information Theoretic Evaluation of the Countermeasure

The specificity of this study is to consider masks M that are not completely entropic. Thus, the probability $\mathsf{P}[M = m]$ depends on m. Our target is to restrict to a relevant subset of the masks uniformly, that is every mask is used with the same probability. We call $J \subseteq \mathbb{F}_2^n$ the set of masks actually used. Thus:

$$\mathsf{P}[M = m] = \begin{cases} 1/\mathsf{Card}[J] & \text{if } m \in J, \text{ and} \\ 0 & \text{otherwise.} \end{cases}$$

We also write this probability law $M \sim \mathcal{U}(J)$. From an information theoretic point of view, we can characterize the entropy of M. By definition, $\mathsf{H}[M] = -\sum_{m \in J} \frac{1}{\mathsf{Card}[J]} \log_2 \frac{1}{\mathsf{Card}[J]} = \log_2 \mathsf{Card}[J]$ bit. The minimal number of masks is 1, which corresponds to the absence of countermeasure (take $M = 0$ in Eqn. (1)). At the opposite, when all the 2^n masks are used, the countermeasure is optimal.

Eventually, we assume that the attacker does not conduct a chosen message attack, *i.e.* $Z \sim \mathcal{U}(\mathbb{F}_2^n)$. We notice that even if the attacker cannot actually choose the messages, she has nonetheless the possibility to discard some messages so as to artificially bias the side-channel attack. But a priori, the attacker does not know which plaintext Z to favor. A biased side-channel attack has been detailed in [9,25]. However, this attack is adaptative, and thus requires that a breach be already found. Nonetheless, in our context, we target the protection of the secret at the early stages of the attack; the attacker still does not have any clue about the most likely hypotheses for the secret. This hypothesis is called the *non-adaptive known plaintext model* in [21].

Whatever the actual leakage function $\mathscr{L}$, $\mathsf{I}[\mathscr{L}(Z \oplus M); Z] = 0$ if $\mathsf{H}[M] = n$ bit (or equivalently, if $M \sim \mathcal{U}(\mathbb{F}_2^n)$). So with all the masks, the countermeasure is perfect.

If $\mathscr{L}$ is bijective (*e.g.* $\mathscr{L} = \mathsf{Id}$), then $\mathsf{I}[\mathscr{L}(Z \oplus M); Z] = n - \mathsf{H}[M]$. This results directly from the observation that:

– $\mathsf{H}[\mathscr{L}(Z \oplus M)] = \mathsf{H}[\mathscr{L}(Z)] = n$ bit, since $Z \sim \mathcal{U}(\mathbb{F}_2^n)$, and
– $\mathsf{H}[\mathscr{L}(Z \oplus M) \mid Z] = \mathsf{H}[M]$ bit because Z and M are independent.

We notice that this quantity is independent of the exact J, provided $\text{Card}[J]$ is fixed. This means that degrading the countermeasure (*i.e.* choosing $\text{Card}[J] < 2^n$) introduces a vulnerability, while decreasing the cost.

Now, it can checked to which extent this vulnerability is exploitable, considering a realistic leakage function. Specifically, it can be shown that if $\mathscr{L}$ is not injective, then the MIA metric $\mathsf{I}[\mathscr{L}(Z \oplus M); Z]$ depends on J. More precisely, when J as two (complementary) elements, then the MIA is independent of J. But when J is made up of strictly more than two masks, the MIA depends on J. For example, on $n = 8$ bits,

- $\mathsf{I}[\mathscr{L}(Z \oplus M); Z] = 1.42701$ bit if $J = \{\texttt{0x00}, \texttt{0x0f}, \texttt{0xf0}, \texttt{0xff}\}$, but
- $\mathsf{I}[\mathscr{L}(Z \oplus M); Z] = 0.73733$ bit if $J = \{\texttt{0x00}, \texttt{0x01}, \texttt{0xfe}, \texttt{0xff}\}$.

Thus, it is relevant to search for mask sets, at a constant budget (*i.e.* for a given $\text{Card}[J]$), that minimize the mutual information $\mathsf{I}[\text{HW}[Z \oplus M]; Z]$. Nonetheless, without a method, it is not obvious to conduct a reasoned search. Indeed, the default solution is to draw at random one mask set J and to compute $\mathsf{I}[\text{HW}[Z \oplus M]; Z]$. It is immediate to see that such method will indeed provide solutions harder to attack using MIA than the others, but that will maybe fail in front of other less sophisticated attacks. Typically, J sets only constrained by their cardinality are likely to yield functions trivially attackable by CPA. We therefore propose the following method:

- First mask sets J that resist first and second order correlation attacks (*i.e.* CPA and 2O-CPA, the easiest attacks against single-masked countermeasures) are found. This is the topic of Sec. 4.
- Then, amongst these solutions, those minimizing the risk of MIA are selected. Section 5 specifically analyses this point (already quickly discussed in Sec. 4.5).

Another argument to focus primarily on CPA and 2O-CPA is that they require in practice less side-channel measurements to succeed the attack than MIA. Indeed, MIA, as well all other information theoretic-based attacks (*e.g.* template attacks [5] and stochastic attacks [17]), need to estimate conditional probability functions, which needs many traces [8]. Also, from the certification standpoint, the common criteria [1] demand that the implemented countermeasures resist "state-of-the-art" attacks [6]. Now, CPA and 2O-CPA are much more studied in the information technology security evaluation facilities (ITSEFs) than information theoretic attacks.

4   Security against CPA and 2O-CPA

The average of the leakage function given in Eqn. (1) depends on $\mathscr{L} : \mathbb{F}_2^n \to \mathbb{R}$. As already mentioned, to conduct exact computations and to match with realistic leakage functions observed in practice, we opt for the Hamming weight ($\mathscr{L} = \text{HW}$). Thus the average of leakage function, noted $\mathsf{E}\mathcal{L}(Z, M)$, is equal to:

$$\mathsf{E}\,\text{HW}[Z \oplus M] = \frac{1}{\text{Card}[J]} \sum_{m \in J} \frac{1}{2^n} \sum_{z \in \mathbb{F}_2^n} \text{HW}[z \oplus m] = \frac{1}{\text{Card}[J]} \sum_{m \in J} \frac{n}{2} = \frac{n}{2}. \quad (2)$$

Against HO-CPA of order $d \geqslant 1$, the most powerful attacker correlates her guesses about the sensitive variable with the optimal function [15] defined as:

$$f_{\text{opt}}^{(d)}(z) \doteq \mathsf{E}\left((\mathcal{L}(Z,M) - \mathsf{E}\mathcal{L}(Z,M))^d \mid Z = z \right)$$

$$= \mathsf{E}\left(\left(\mathsf{HW}[Z \oplus M] - \frac{n}{2} \right)^d \mid Z = z \right)$$

$$= \frac{1}{\mathsf{Card}[J]} \sum_{m \in J} \left(\frac{-1}{2} \sum_{i=1}^{n} (-1)^{(z \oplus m)_i} \right)^d , \tag{3}$$

because if $b \in \{0,1\}$, then $b - \frac{1}{2} = -\frac{1}{2}(-1)^b$. Recall that the rotating tables countermeasure uses only one mask variable M, and thus leaks at only one date (*i.e.* for a given timing sample). In this context, HO-CPA consists in studying the linear dependency between the d-th moments of the leakage classes and the optimal function $f_{\text{opt}}^{(d)}(z)$ of the sensitive variable z.

For the designer of the countermeasure, the objective is to make Eqn. (3) independent of z. There is always a solution that consists in choosing $J = \mathbb{F}_2^n$. Nonetheless, with $\mathsf{Card}[J] < 2^n$, the existence of solutions is a priori not trivial. In this case, if is impossible to find masks that keep $f_{\text{opt}}^{(d)}(z)$ (defined in Eqn. (3)) independent from z, the secondary goal is to minimize the correlation coefficient:

$$\rho_{\text{opt}}^{(d)} \doteq \frac{\mathsf{Var}\left(f_{\text{opt}}^{(d)}(Z) \right)}{\mathsf{Var}\left((\mathcal{L}(Z,M) - \mathsf{E}\mathcal{L}(Z,M))^d \right)} = \frac{\mathsf{Var}\left(\mathsf{E}\left((\mathsf{HW}[Z \oplus M] - \frac{n}{2})^d \mid Z \right) \right)}{\mathsf{Var}\left((\mathsf{HW}[Z \oplus M] - \frac{n}{2})^d \right)} . \tag{4}$$

In this equation, Var represents the variance, defined on a random variable X as $\mathsf{Var}(X) \doteq \mathsf{E}\left(X - \mathsf{E}X\right)^2$.

In the two next subsections 4.1 and 4.2, the analytical expression of Eqn. (4) is derived. Then these expressions are unified in subsection 4.3 by replacing the notion of subset J by an indicator function f. The sets of masks that completely allow to deny CPA and 2O-CPA are given exhaustively in subsection 4.4 for $n = 4$ and in subsection 4.5 for $n = 5$.

4.1   Resistance against First-Order Correlation Attacks

When $d = 1$, Eqn. (4) is equal to:

$$\rho_{\text{opt}}^{(1)} = \frac{1}{n} \sum_{i=1}^{n} \left(\frac{1}{\mathsf{Card}[J]} \sum_{m \in J} (-1)^{m_i} \right)^2 . \tag{5}$$

This correlation $\rho_{\text{opt}}^{(1)}$ can be equal to zero if and only if (iff), for all $i \in [\![1,n]\!]$, $\mathsf{E}M_i = 1/2$. This means that the masks are balanced. It is possible to find such masks iff $\mathsf{Card}[J]$ is a multiple of two. A construction consists in building a set of masks by adding a new mask and its complement. Conversely, in a set containing

Table 1. Mask sets J that make the masking countermeasure immune to first order CPA. The masks go by pair, symmetrically with the middle of the table.

	Card$[J] = 2^4$	Card$[J] = 2^3$	Card$[J] = 2^2$	Card$[J] = 2^1$
	0000	0000	0000	0000
	0001			
	0010			
	0011	0011	0011	
	0100	0100		
	0101			
	0110			
J	0111	0111		
	1000	1000		
	1001			
	1010			
	1011	1011		
	1100	1100	1100	
	1101			
	1110			
	1111	1111	1111	1111

an odd number of different masks, it is impossible to as many ones as zeros for any component. For instance, we illustrate how to generate balanced sets of masks in the case $n = 4$ in Tab. 1.

A trivial example consists in taking two masks, m and $\neg m$ (such as 0x00 and 0xff on $n = 8$ bits). This is sufficient to thwart first-order attacks. At the opposite, without mask (J is equal to the singleton $\{0\text{x}00\}$) or with a single mask ($J = \{m\}$, whatever $m \in \mathbb{F}_2^n$), the correlation coefficient reaches its maximum (*i.e.* $+1$, because Eqn. (4) considers a correlation in absolute value).

4.2   Resistance against Second-Order Correlation Attacks

When $d = 2$, Eqn. (4) is equal to:

$$\rho_{\text{opt}}^{(2)} = \frac{1}{n(n-1)} \left(\frac{1}{\text{Card}[J]^2} \sum_{(m,m') \in J^2} \left(\sum_{i=1}^{n} (-1)^{(m \oplus m')_i} \right)^2 - n \right). \quad (6)$$

As an illustration, we show in Tab. 2 the optimal correlation coefficients of order 1 and 2 for the masks sets of Tab. 1 ($n = 4$ bit). We have added a column (the last one), for $\mathscr{L} = \text{Id}$; also, in the last row, we have included a constant masking (unprotected implementation), which serves as a reference.

Table 2. Security metrics for the masks sets of Tab. 1 and the singleton

Card[J]	H[M]	$\rho_{\text{opt}}^{(1)}$	$\rho_{\text{opt}}^{(2)}$	I[HW[$Z \oplus M$]; Z]	I[$Z \oplus M$; Z]
2^4	4	0	0	0	0
2^3	3	0	0.166667	0.15564	1
2^2	2	0	0.333333	1.15564	2
2^1	1	0	1	1.40564	3
2^0	0	1	1	2.03064	4

4.3   Expression of $\rho_{\text{opt}}^{(1,2)}$ as a Function of an Indicator f

The expressions of $\rho_{\text{opt}}^{(1)}$ and $\rho_{\text{opt}}^{(2)}$ (altogether referred to as $\rho_{\text{opt}}^{(1,2)}$) defined in Eqn. (5) and (6) lay a mathematical ground to search for suitable J. Nonetheless, these equations remain at the set-theory level. To simplify the problem, we introduce the Boolean function $f : \mathbb{F}_2^n \to \mathbb{F}_2$, defined as: $\forall m \in \mathbb{F}_2^n, f(m) = 1 \Leftrightarrow m \in J$. Then, we can simply replace "$\sum_{m \in J}$" by "$\sum_{m \in \mathbb{F}_2^n} f(m)$" in the equations previously established.

The Fourier transform $\hat{f} : \mathbb{F}_2^n \to \mathbb{Z}$ of the Boolean function $f : \mathbb{F}_2^n \to \mathbb{F}_2$ is defined as $\forall a \in \mathbb{F}_2^n, \hat{f}(a) \doteq \sum_{m \in \mathbb{F}_2^n} f(m)(-1)^{a \cdot m}$. It allows for instance to write $\text{Card}[J] = \sum_{m \in J} 1 = \sum_{m \in \mathbb{F}_2^n} f(m) = \hat{f}(0)$. Recall $\text{Card}[J] \in [\![1, 2^n]\!]$, hence $\hat{f}(0) > 0$.

Then Eqn. (5) rewrites:

$$\rho_{\text{opt}}^{(1)} = \frac{1}{n} \sum_{i=1}^{n} \left(\frac{\hat{f}(e_i)}{\hat{f}(0)} \right)^2, \tag{7}$$

where e_i are the canonical basis vectors $(0, \cdots, 0, 1, 0, \cdots, 0)$, the unique 1 laying at position i.

Also, Eqn. (6) rewrites:

$$\rho_{\text{opt}}^{(2)} = \frac{1}{n(n-1)} \sum_{(i,i') \in [\![1,n]\!]^2} \left(\left(\frac{\hat{f}(e_i \oplus e_{i'})}{\hat{f}(0)} \right)^2 - n \right)$$

$$= \frac{1}{n(n-1)} \sum_{\substack{(i,i') \in [\![1,n]\!]^2 \\ i \neq i'}} \left(\frac{\hat{f}(e_i \oplus e_{i'})}{\hat{f}(0)} \right)^2. \tag{8}$$

Thus, the rotating tables countermeasure resists:

1. First-order attacks iff $\forall a$, HW[a] = 1 $\Rightarrow \hat{f}(a) = 0$;
2. First and second-order attacks iff $\forall a$, $1 \leqslant$ HW[a] $\leqslant 2 \Rightarrow \hat{f}(a) = 0$.

As a sanity check, we can verify that these properties hold when all the 2^n masks are used, *i.e.* when f is constant (and furthermore equal to 1). Indeed, in this case, $\hat{f}(a) = \sum_m f(m)(-1)^{a \cdot m} = \sum_m (-1)^{a \cdot m} = 2^n \delta(a)$, where δ is the Kronecker symbol.

Now, we notice that for Boolean functions, the notions of Fourier and Walsh transforms are very alike. Indeed,

$$\forall a \neq 0, \hat{f}(a) = \sum_m f(a)(-1)^{a \cdot m} = \sum_m (-1)^{a \cdot m} \tfrac{1}{2}\left(1 - (-1)^{f(m)}\right) = -\tfrac{1}{2}\widehat{(-1)^f}(a).$$

Therefore, the previous conditions are equivalent to saying the following: the countermeasure resists $d \in \{1, 2\}$ order CPA iff $\forall a$, $\mathsf{HW}[a] \leqslant d \Rightarrow \widehat{(-1)^f}(a) = 0$.

We insist that this characterization is <u>not</u> equivalent to saying that f is d-resilient (defined in [4, page 45]). Indeed, a resilient function is balanced, which is explicitly not the case of f. Therefore, we study in the sequel a new kind of Boolean functions, that have everything in common with resilient functions but the balancedness of the plain function. The corollary is that, to the authors' best knowledge, no known construction method exists for this type of functions. Nonetheless, it is interesting to get an intuition about what characterizes a good resilient function. In [4, §7.1, page 95], it is explained that the highest degree of resiliency of a $f : \mathbb{F}_2^n \to \mathbb{F}_2$ is $n - 2$. This maximum is reached by affine functions (functions of unitary algebraic degree). Nonetheless, in our case, affine functions are not the best choice, because they are balanced. This means that the cardinality of their support (i.e. $\mathsf{Card}[J]$) is 2^{n-1}, which is large. Therefore, we will be interested, whenever possible, by non-affine functions f of algebraic degree strictly greater than one (noted $d_{\mathrm{alg}}^{\circ}(f) > 1$).

4.4   Functions $f : \mathbb{F}_2^4 \to \mathbb{F}_2$ That Cancel $\rho_{\mathrm{opt}}^{(1,2)}$

For $n = 4$, all the sets J can be tested. The table 3 reports all the functions f that cancel $\rho_{\mathrm{opt}}^{(1)}$ and $\rho_{\mathrm{opt}}^{(2)}$. In this table, the truth-table of f, given in the first column, is encoded in hexadecimal. We note $\mathsf{HW}[f]$ the number of ones in the truth-table, and recall that $\mathsf{HW}[f] = \mathsf{Card}[J]$. Columns 4, 5 and 6 are security metrics, whereas column 7 is the leakage metric (MIM). There are non-trivial solutions only for $\mathsf{Card}[J]$ equal to half of the complete mask set cardinal. The MIA (column 6) shows two values: 0.219361 and 1 bit. Those values shall be contrasted with the MIA:

- Without countermeasure ($\mathsf{Card}[J] = 1$): MIA = 2.19819 bit and
- With two complementary masks ($\mathsf{Card}[J] = 2$, which thwarts CPA but not 2O-CPA): MIA = 1.1981 bit.

Thus the countermeasure resists better correlation and information theoretic attacks, at the expense of more masks. Indeed, apart from $f = 1$, all the solutions are affine ($d_{\mathrm{alg}}^{\circ}(f) = 1$), and thus have a Hamming weight of $2^{n-1} = 8 \gg 2$.

In this table, some functions belong to equivalent classes. Namely, two of them can be identified:

- The permutations of the bits (because the summations over i in Eqn. (7) or i, i' in Eqn. (8) is invariant in any change of the bits order), and

Table 3. All the functions $f : \mathbb{F}_2^4 \to \mathbb{F}_2$ that cancel $\rho_{\mathrm{opt}}^{(1,2)}$

f	HW$[f]$	H$[M]$	$\rho_{\mathrm{opt}}^{(1)}$	$\rho_{\mathrm{opt}}^{(2)}$	I$[$HW$[Z \oplus M]; Z]$	I$[Z \oplus M; Z]$	$d_{\mathrm{alg}}^{\circ}(f)$
0x3cc3	8	3	0	0	0.219361	1	1
0x5aa5	8	3	0	0	0.219361	1	1
0x6699	8	3	0	0	0.219361	1	1
0x6969	8	3	0	0	0.219361	1	1
0x6996	8	3	0	0	1	1	1
0x9669	8	3	0	0	1	1	1
0x9696	8	3	0	0	0.219361	1	1
0x9966	8	3	0	0	0.219361	1	1
0xa55a	8	3	0	0	0.219361	1	1
0xc33c	8	3	0	0	0.219361	1	1
0xffff	16	4	0	0	0	0	0

- The complementation. Indeed, $\widehat{\neg f}(a) = \sum_{m \in \mathbb{F}_2^n} \neg f(m)(-1)^{a \cdot m} = \sum_{m \in \mathbb{F}_2^n}(1 - f(m))(-1)^{a \cdot m} = 2^n \delta(a) - \hat{f}(a)$. Now, in Eqn. (7) and (8), $a \neq 0$ and $\hat{f}$ is involved squared. Thus $\rho_{\mathrm{opt}}^{(1,2)}(\neg f) = \rho_{\mathrm{opt}}^{(1,2)}(f)$.

The same can be said for the mutual information. This lemma is useful:

Lemma 1. *Let A and B be two random variables and ϕ a bijection; then* $\mathsf{I}[A; \phi(B)] = \mathsf{I}[A; B]$.

This equality is obtained simply by writing the definition of the mutual information as a function of the probabilities, and by doing a variable change. Then:

- Let us call σ a permutation of $[\![1, n]\!]$. This function is a bijection, and its inverse is also a permutation. The Hamming weight is invariant if σ is applied on its input (*i.e.* HW $=$ HW $\circ \sigma$). Hence HW$[Z \oplus \sigma(M)] =$ HW$[\sigma^{-1}(Z \oplus \sigma(M))] =$ HW$[\sigma^{-1}(Z) \oplus M]$ (because σ is furthermore linear with respect to the addition). Let us note $Z' = \sigma^{-1}(Z)$, a random variable that is also uniform. Thus, I$[$HW$[Z \oplus \sigma(M)]; Z] =$ I$[$HW$[Z' \oplus M]; \sigma(Z')]$. By considering $\phi = \sigma$, we prove that I$[$HW$[Z \oplus \sigma(M)]; Z] =$ I$[$HW$[Z' \oplus M]; Z'] =$ I$[$HW$[Z \oplus M]; Z]$, because Z and Z' have the same probability density function.
- Regarding the complementation, it is straightforward to note that HW$[Z \oplus \neg M] =$ HW$[\neg(Z \oplus M)] = n -$ HW$[Z \oplus M]$. By considering $\phi : x \mapsto n - x$, we also have the invariance of the mutual information by the complementation of the mask.

So, there are eventually only three classes of functions listed in Tab. 3, modulo the two abovementioned equivalence classes. They are summarized below:

1. $f(x_1, x_2, x_3, x_4) = \bigoplus_{\substack{i \in I \subseteq [\![1,4]\!] \\ \mathrm{Card}[I]=3}} x_i$, (*aka* 0x3cc3, 0x5aa5, 0x6699, 0x6969) or complemented (*aka* 0x9696, 0x9966, 0xa55a, 0xc33c); According to the

criteria stated at the end of Sec. 3, those functions are the best solutions for $n = 4$.

2. $f(x_1, x_2, x_3, x_4) = \bigoplus_{i=1}^{4} x_i$ (*aka* 0x6996) or $f(x_1, x_2, x_3, x_4) = 1 \oplus \bigoplus_{i=1}^{4} x_i$ (*aka* 0x9669), that has no advantage of the previous solutions;
3. The constant function $f = 1$ (*aka* 0xffff).

To resist first-order attacks, the masks set can be partitioned in two complementary sets; this means that there exists $\tilde{J}$, a subset of J, such that: $J = \tilde{J} \cup \neg\tilde{J}$, where $\neg\tilde{J} \doteq \{\neg m, m \in J\}$. Incidentally, we notice that this is not a mandatory property. Typically, this property is not verified any longer at order 2. For instance, in the solution $f = $ 0x3cc3, 0x0 $\in J$ but $\neg$0x0 $= $ 0xf $\notin J$.

In conclusion, when $n = 4$ and the designer cannot afford using all the 16 masks, then with 8 masks, the rotating tables countermeasure is able to resist CPA, 2O-CPA and leak the minimal value of 0.219361 bit (about ten times less than the unprotected implementation, for which the MIA is 2.19819 bit).

4.5   Functions $f : \mathbb{F}_2^5 \to \mathbb{F}_2$ That Cancel $\rho_{\text{opt}}^{(1,2)}$

For $n = 5$, all the subsets J of $\mathbb{F}_2^5$ (2^{32} of them, it is the maximum achievable on a personal computer, as precised in [4, page 6]) have been tested. There are 1057 functions that cancel $\rho_{\text{opt}}^{(1,2)}$. The lowest value for $\text{HW}[f]$ is 8. There are 60 functions of weight 8, but only three classes modulo the invariants. The functions, sorted regarding their properties, are shown in Tab. 4. As opposed to the case $n = 4$, there are non-affine solutions. In this table, only the number of equivalent classes is given. For a list of all functions, refer to the ePrint extended version of the paper [13].

Table 4. Summary of the security metrics of $f : \mathbb{F}_2^5 \to \mathbb{F}_2$ that cancel $\rho_{\text{opt}}^{(1,2)}$

Nb. classes	HW[f]	H[M]	$\rho_{\text{opt}}^{(1)}$	$\rho_{\text{opt}}^{(2)}$	I[HW[$Z \oplus M$]; Z]	I[$Z \oplus M$; Z]	$d_{\text{alg}}^{\circ}(f)$
3	8	3	0	0	0.32319	2	2
4	12	3.58496	0	0	0.18595	1.41504	3
2	16	4	0	0	0.08973	1	1
2	16	4	0	0	0.08973	1	2
4	16	4	0	0	0.12864	1	2
2	16	4	0	0	0.16755	1	1
4	16	4	0	0	0.26855	1	2
6	16	4	0	0	0.32495	1	2
1	16	4	0	0	1	1	1
4	20	4.32193	0	0	0.07349	0.67807	3
3	24	4.58496	0	0	0.04300	0.41504	2
1	32	5	0	0	0	0	0

The greater $\mathsf{H}[M]$, the smaller the mutual information with $\mathscr{L} = \mathsf{HW}$ in general, but for some remarkable solutions (*e.g.* the one MIA $= \mathsf{I}[\mathsf{HW}[Z \oplus M]; Z] = 1$ of algebraic degree 1 for $\mathsf{HW}[f] = 16$). Also, it is worth noting that for a given budget (*e.g.* 16 masks) and security requirement (resistance against CPA and 2O-CPA), some solutions are better than the others against MIA. Indeed, the leaked information in Hamming weight model spans from 0.0897338 bit to 1 bit.

5   Exploring More Solutions Using SAT-Solvers

In order to explore problems of greater complexity, SAT-solver are indicated tools. We model f as a set of 2^n Boolean unknowns. The problem consists in finding f such that $\forall a, 1 \leqslant \mathsf{HW}[a] \leqslant 2, \hat{f}(a) = 0$, for a given $\mathsf{Card}[J] = \hat{f}(0)$. A SAT-solver either:

- Proves that there is no solution, or
- Proves that a solution exists, and provides for (at least) one.

We notice that a SAT-solver may not terminate on certain instances of large exploration space; this has not been an issue in the work we report here. In this section, we first explain how our problem can be fed into a SAT-solver. Then, we use a SAT-solver in the case $n = 8$, relevant for AES. We look for low $\mathsf{Card}[J]$ solutions, and for a given $\mathsf{Card}[J]$, for the solutions of minimal MIA.

5.1   Mapping of the Problem into a SAT-Solver

Knowing that $\mathsf{Card}[J] = \hat{f}(0)$, the problem $\rho_{\mathrm{opt}}^{(1,2)}(f) = 0$ rewrites:

$$\forall a, 1 \leqslant \mathsf{HW}[a] \leqslant 2, \quad \sum_x f(x)(-1)^{a \cdot x} = 0 \quad \Leftrightarrow$$

$$\forall a, 1 \leqslant \mathsf{HW}[a] \leqslant 2, \quad \sum_x f(x) \wedge (a \cdot x) = \frac{1}{2} \sum_x f(x) = \frac{1}{2}\mathsf{Card}[J] \; . \quad (9)$$

A SAT-solver verifies the validity of clauses, usually expressed in conjunctive normal form (CNF). It is known that cardinality constraints can be formulated compactly thanks to Boolean clauses. More precisely, any condition "$\leqslant k(x_1, \cdots, x_n)$", for $0 \leqslant k \leqslant n$, can be expressed in terms of CNF clauses [18]. We note that:

$$\mathsf{HW}[x] \leqslant k \quad \Leftrightarrow \quad n - \mathsf{HW}[\neg x] \leqslant k \quad \Leftrightarrow \quad \mathsf{HW}[\neg x] \geqslant n - k \; .$$

Hence, satisfying $\geqslant k(x_1, \cdots, x_n)$ is equivalent to satisfying $\leqslant n - k(\neg x_1, \cdots, \neg x_n)$. Thus, testing the equality of a Hamming to $\frac{1}{2}\mathsf{Card}[J]$ can be achieved by the conjunction of two clauses: $\leqslant \frac{1}{2}\mathsf{Card}[J](x_1, \cdots, x_n)$ and $\leqslant n - \frac{1}{2}\mathsf{Card}[J](\neg x_1, \cdots, \neg x_n)$.

The $n = 8$, the number of useful literals, $\{f(x), x \in \mathbb{F}_2^n\}$, is 2^8. However, the constraints $\mathsf{Card}[J] = \hat{f}(0)$ and $\rho_{\mathrm{opt}}^{(1,2)}(f) = 0$ (see Eqn. (9)) introduce 1,105,664 auxiliary variables and translate into 2,219,646 clauses, irrespective of $\mathsf{Card}[J] \in \mathbb{N}^*$.

5.2   Existence of Low Hamming Weight Solutions for $n = 8$

The software `cryptominisat` [19,20] is used to search for solutions. The problem is tested for all the Card[J] from 2 to 2^n, by steps of 2, as independent problems. Each problem requires a few hours to be solved. Impressively low Hamming weight solutions are found. The table 5 represents some of them. There are solutions only for Card[J] $\in \{4 \times \kappa, \kappa \in [\![3, 61]\!] \cup \{64\}\}$. Also, the mutual information with a Hamming weight leakage as a function of H[M] is plotted in Fig. 1. These values are low when compared to:

  – MIA = 2.5442 bit without masking (Card[J] = 1) and
  – MIA = 1.8176 bit with a mask that takes two complementary values (Card[J] = 2).

Those MIA figures concern countermeasures that do not protect against 2O-DPA. The table 5 basically indicates that the margin gain in MIA resistance decreases when the cost of the countermeasures, proportional to HW[f], increases.

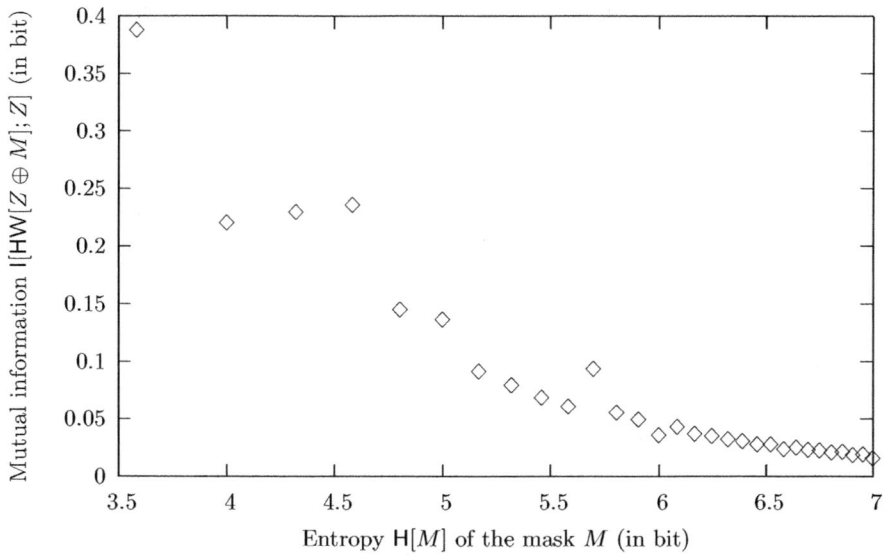

Fig. 1. Mutual information of the leakage in Hamming weight with the sensitive variable Z, for one solution that cancels $\rho_{\mathrm{opt}}^{(1,2)}$ found by the SAT-solver

5.3   Exploration of Solutions for $n = 8$ and a Fixed Card[J]

There are nonequivalent solutions for a same Card[J]. Various seeds of the SAT-solver are needed to discover these solutions. All the solutions found by the SAT-solver for Card[J] = 12 have the same MIA value: 0.387582 bit. It can also be shown that for Card[J] = 16, various MIA values exist. The SAT-solver has

Table 5. Metrics for one $f : \mathbb{F}_2^8 \to \mathbb{F}_2$ that cancel $\rho_{\text{opt}}^{(1,2)}$, found by a SAT-solver

HW[f]	H[M]	$\rho_{\text{opt}}^{(1)}$	$\rho_{\text{opt}}^{(2)}$	I[HW[$Z \oplus M$]; Z]	I[$Z \oplus M$; Z]	$d_{\text{alg}}^{\circ}(f)$
12	3.58496	0	0	0.387582	4.41504	6
16	4	0	0	0.219567	4	5
20	4.32193	0	0	0.228925	3.67807	6
24	4.58496	0	0	0.235559	3.41504	5
28	4.80735	0	0	0.144147	3.19265	6
32	5	0	0	0.135458	3	5
36	5.16993	0	0	0.090575	2.83007	6
40	5.32193	0	0	0.078709	2.67807	5
44	5.45943	0	0	0.067960	2.54057	6
48	5.58496	0	0	0.060515	2.41504	5
52	5.70044	0	0	0.092676	2.29956	6
56	5.80735	0	0	0.054936	2.19265	5
60	5.90689	0	0	0.049069	2.09311	6
64	6	0	0	0.035394	2	2
68	6.08746	0	0	0.042374	1.91254	6
72	6.16993	0	0	0.036133	1.83007	5
76	6.24793	0	0	0.034194	1.75207	6
80	6.32193	0	0	0.031568	1.67807	5
84	6.39232	0	0	0.030072	1.60768	6
88	6.45943	0	0	0.026941	1.54057	5
92	6.52356	0	0	0.027042	1.47644	6
96	6.58496	0	0	0.022992	1.41504	5
100	6.64386	0	0	0.024316	1.35614	6
104	6.70044	0	0	0.022257	1.29956	5
108	6.75489	0	0	0.021458	1.24511	6
112	6.80735	0	0	0.019972	1.19265	4
116	6.85798	0	0	0.020481	1.14202	6
120	6.90689	0	0	0.018051	1.09311	5
124	6.9542	0	0	0.018397	1.0458	6
128	7	0	0	0.015095	1	1
⋮	⋮	⋮	⋮	⋮	⋮	⋮

notably came across, from best to worst: 0.181675, 0.213996, 0.215616, 0.216782, 0.219567, 0.220733, 0.246318, 0.249556, 0.251888, 0.253508, 0.254674, 0.257459, 0.388196, 0.434113, 1.074880 and 1.074950. We insist that with the SAT-solver, we find some solutions, but we cannot easily classify them. Thus we are unsure we have indeed found the best one. Nonetheless, it is already of great practical importance to exhibit some solutions.

6   Conclusions and Perspectives

Masking is a pro-active countermeasure against side-channel attacks. It implies adequately extra random variables amidst the computation in order to remove dependencies between the leakage of computation and guesses of internal sensitive values by a prospective attacker. Based on a representative first-order leakage model, this article explores the connections between the mask entropy and the best achievable security. If the implementation leaks its data values, then the leakage increases in proportion of the mask entropy reduction. Nonetheless, in practice, the implementation leaks a non-bijective value of its internal variables, such as the sum of their n bits. In this case, we show that the leakage is never null when limiting to a subset of few mask values amongst the 2^n possible. Furthermore, higher-order attacks can defeat this protection even if the mask losses as little as 1 single bit of entropy. Thus, we explore other mask entropy *vs* security tradeoffs. Our methodology is to demand resistance against CPA and 2O-CPA, and to minimize the leakage.

The criteria for masks selection has been formalized as a condition on the Walsh transform of an indicator function. This criteria has been used heuristically in a SAT-solver, but we expect that constructive methods based on the Boolean theory, for all n, can be invented. We exhibit the best solutions for $n = 4$ and $n = 5$, and prove the existence of varied values of mutual information for some masks cardinality for $n = 8$ (thanks to the SAT-solver). We notably show that amongst the masks subsets that allow for a resistance at orders 1 and 2 against CPA, some are less sensitive to MIA than others, especially for Card$[J] = 16$. Therefore, there is a real opportunity for the designer to reduce the cost of the countermeasure in a reasoned way. We insist that, at first sight, it can seem very audacious to mask an eight bit sensitive data with only four bits of mask. But it is indeed possible due to the high non-injectivity of the HW function, that maps 256 values into only 9.

Controlling the overhead in terms of resources is an enabler for masking technologies. Some countermeasures are expensive and our proposed tradeoff definitely shows that it is possible to quantify the security loss when one downgrades a countermeasure. As a perspective, we note that to further save area and speed, instead of storing the sboxes in RAM and selecting them randomly, we could take advantage of the dynamic partial reconfiguration of modern FPGAs to do so [12]. The idea is that even if computed at full throughput, the attacker does not have enough time to collect enough traces with a consistent set of sboxes to succeed an attack. This assumption is the same as those used for the resilience "leakage-proof" countermeasures.

References

1. Common Criteria (aka CC) for Information Technology Security Evaluation (ISO/IEC 15408), http://www.commoncriteriaportal.org/
2. Batina, L., Gierlichs, B., Prouff, E., Rivain, M., Standaert, F.X., Veyrat-Charvillon, N.: Mutual Information Analysis: a Comprehensive Study. J. Cryptology 24(2), 269–291 (2011)
3. Brier, É., Clavier, C., Olivier, F.: Correlation Power Analysis with a Leakage Model. In: Joye, M., Quisquater, J.-J. (eds.) CHES 2004. LNCS, vol. 3156, pp. 16–29. Springer, Heidelberg (2004)
4. Carlet, C.: Boolean Functions for Cryptography and Error Correcting Codes. In: Crama, Y., Hammer, P. (eds.) Boolean Methods and Models. Cambridge University Press (2008)
5. Chari, S., Rao, J.R., Rohatgi, P.: Template Attacks. In: Kaliski Jr., B.S., Koç, Ç.K., Paar, C. (eds.) CHES 2002. LNCS, vol. 2523, pp. 13–28. Springer, Heidelberg (2003)
6. Criteria, C.: Application of Attack Potential to Smartcards, Mandatory Technical Document, Version 2.7, Revision 1, CCDB-2009-03-001 (March 2009), http://www.commoncriteriaportal.org/files/supdocs/CCDB-2009-03-001.pdf
7. Gandolfi, K., Mourtel, C., Olivier, F.: Electromagnetic Analysis: Concrete Results. In: Koç, Ç.K., Naccache, D., Paar, C. (eds.) CHES 2001. LNCS, vol. 2162, pp. 251–261. Springer, Heidelberg (2001)
8. Gierlichs, B., Batina, L., Tuyls, P., Preneel, B.: Mutual Information Analysis. In: Oswald, E., Rohatgi, P. (eds.) CHES 2008. LNCS, vol. 5154, pp. 426–442. Springer, Heidelberg (2008)
9. Köpf, B., Basin, D.: An information-theoretic model for adaptive side-channel attacks. In: CCS 2007: Proceedings of the 14th ACM Conference on Computer and Communications Security, pp. 286–296. ACM, New York (2007)
10. Mangard, S., Oswald, E., Popp, T.: Power Analysis Attacks: Revealing the Secrets of Smart Cards (December 2006) ISBN 0-387-30857-1, http://www.springer.com/, http://www.dpabook.org/
11. Mangard, S., Schramm, K.: Pinpointing the Side-Channel Leakage of Masked AES Hardware Implementations. In: Goubin, L., Matsui, M. (eds.) CHES 2006. LNCS, vol. 4249, pp. 76–90. Springer, Heidelberg (2006)
12. Mentens, N., Gierlichs, B., Verbauwhede, I.: Power and Fault Analysis Resistance in Hardware Through Dynamic Reconfiguration. In: Oswald, E., Rohatgi, P. (eds.) CHES 2008. LNCS, vol. 5154, pp. 346–362. Springer, Heidelberg (2008)
13. Nassar, M., Guilley, S., Danger, J.L.: Formal Analysis of the Entropy / Security Trade-off in First-Order Masking Countermeasures against Side-Channel Attacks — Complete version. Cryptology ePrint Archive, Report 2011/534 (September 2011), http://eprint.iacr.org/2011/534
14. Prouff, E., Rivain, M.: A Generic Method for Secure SBox Implementation. In: Kim, S., Yung, M., Lee, H.-W. (eds.) WISA 2007. LNCS, vol. 4867, pp. 227–244. Springer, Heidelberg (2008)
15. Prouff, E., Rivain, M., Bevan, R.: Statistical Analysis of Second Order Differential Power Analysis. IEEE Trans. Computers 58(6), 799–811 (2009)
16. Rivain, M., Prouff, E.: Provably Secure Higher-Order Masking of AES. In: Mangard, S., Standaert, F.-X. (eds.) CHES 2010. LNCS, vol. 6225, pp. 413–427. Springer, Heidelberg (2010)

17. Schindler, W.: Advanced stochastic methods in side channel analysis on block ciphers in the presence of masking. Journal of Mathematical Cryptology 2(3), 291–310 (2008) ISSN (Online) 1862-2984, ISSN (Print) 1862-2976, doi:10.1515/JMC.2008.013
18. Sinz, C.: Towards an Optimal CNF Encoding of Boolean Cardinality Constraints. In: van Beek, P. (ed.) CP 2005. LNCS, vol. 3709, pp. 827–831. Springer, Heidelberg (2005)
19. Soos, M.: SAT-solver "cryptominisat", Version 2.9.0 (January 20, 2011), https://gforge.inria.fr/projects/cryptominisat
20. Soos, M., Nohl, K., Castelluccia, C.: Extending SAT Solvers to Cryptographic Problems. In: Kullmann, O. (ed.) SAT 2009. LNCS, vol. 5584, pp. 244–257. Springer, Heidelberg (2009)
21. Standaert, F.-X., Malkin, T.G., Yung, M.: A Unified Framework for the Analysis of Side-Channel Key Recovery Attacks. In: Joux, A. (ed.) EUROCRYPT 2009. LNCS, vol. 5479, pp. 443–461. Springer, Heidelberg (2009)
22. Standaert, F.X., Peeters, É., Macé, F., Quisquater, J.J.: Updates on the Security of FPGAs Against Power Analysis Attacks. In: Bertels, K., Cardoso, J.M.P., Vassiliadis, S. (eds.) ARC 2006. LNCS, vol. 3985, pp. 335–346. Springer, Heidelberg (2006)
23. Standaert, F.X., Rouvroy, G., Quisquater, J.J.: FPGA Implementations of the DES and Triple-DES Masked Against Power Analysis Attacks. In: Proceedings of FPL 2006, Madrid, Spain. IEEE (2006)
24. Veyrat-Charvillon, N., Standaert, F.-X.: Mutual Information Analysis: How, When and Why? In: Clavier, C., Gaj, K. (eds.) CHES 2009. LNCS, vol. 5747, pp. 429–443. Springer, Heidelberg (2009)
25. Veyrat-Charvillon, N., Standaert, F.-X.: Adaptive Chosen-Message Side-Channel Attacks. In: Zhou, J., Yung, M. (eds.) ACNS 2010. LNCS, vol. 6123, pp. 186–199. Springer, Heidelberg (2010)
26. Waddle, J., Wagner, D.: Towards Efficient Second-Order Power Analysis. In: Joye, M., Quisquater, J.-J. (eds.) CHES 2004. LNCS, vol. 3156, pp. 1–15. Springer, Heidelberg (2004)

Square Always Exponentiation

Christophe Clavier[1], Benoit Feix[1,2], Georges Gagnerot[1,2], Mylène Roussellet[2], and Vincent Verneuil[2,3]

[1] XLIM-CNRS, Université de Limoges, France
firstname.familyname@unilim.fr
[2] INSIDE Secure, Aix-en-Provence, France
firstname-first-letterfamilyname@insidefr.com
[3] Institut de Mathématiques de Bordeaux, Talence, France

Abstract. Embedded exponentiation techniques have become a key concern for security and efficiency in hardware devices using public key cryptography. An exponentiation is basically a sequence of multiplications and squarings, but this sequence may reveal exponent bits to an attacker on an unprotected implementation. Although this subject has been covered for years, we present in this paper new exponentiation algorithms based on trading multiplications for squarings. Our method circumvents attacks aimed at distinguishing squarings from multiplications at a lower cost than previous techniques. Last but not least, we present new algorithms using two parallel squaring blocks which provide the fastest exponentiation to our knowledge.

Keywords: Public key cryptography, exponentiation, long integer arithmetic, side-channel analysis, atomicity.

1 Introduction

Nowadays most embedded devices implementing public key cryptography use RSA [16] for encryption and signature schemes, or cryptographic primitives over $(\mathbb{F}_p, \times)$ such as DSA [7] and the Diffie-Hellman key agreement protocol [6]. All these algorithms require the computation of modular exponentiations. Since the emergence of the so-called *side-channel analysis*, embedded devices implementing these cryptographic algorithms must be protected against a wider and wider class of attacks.

Moreover, the cost and timing constraints are crucial in many applications of embedded devices (e.g. banking, transport, etc.). This often requires cryptographic implementors to choose the best compromise between security and speed. Improving the efficiency of algorithms or countermeasures generates thus a lot of interest in the industry.

An exponentiation is generally processed using a sequence of multiplications, some of them having different operands and some of them being squarings. In [2], Amiel et al. showed that this distinction can provide exploitable side-channel leakages to an attacker. Classical countermeasures consist of using exponentiation algorithms where the sequence of multiplications and squarings does not depend on the secret exponent.

D.J. Bernstein and S. Chatterjee (Eds.): INDOCRYPT 2011, LNCS 7107, pp. 40–57, 2011.

Our contribution is to propose a new exponentiation scheme using squarings only, which is faster than the classical countermeasures. Also, we introduce new algorithms having a particularly low cost when two squarings can be parallelized.

This paper is organized as follow: in Section 2 we recall classical exponentiation algorithms and present some well-known side-channel attacks and countermeasures. Then we propose our new countermeasure in Section 3 and study its efficiency from the parallelization point of view in Section 4. Finally we present some practical results in Section 5 and we conclude in Section 6.

2   Background on Exponentiation on Embedded Devices

We recall in this section some classical exponentiation algorithms. First we present the *square-and-multiply* algorithms upon which are based most of the exponentiation methods. Then we introduce the *side-channel analysis* and in particular the *simple power analysis* (SPA). We present some algorithms immune to this attack, and we finally recall a particular side-channel attack aimed at distinguishing squarings from multiplications in an exponentiation operation.

2.1   Square-and-Multiply Algorithms

Many exponentiation algorithms have been proposed in the literature. Among the numerous references an interested reader can refer for instance to [14] for details. Alg. 2.1 and Alg. 2.2 are two variants of the classical square-and-multiply algorithm which is the simplest approach to compute an RSA exponentiation.

Alg. 2.1 Left-to-Right Square-and-Multiply Exponentiation

Input: $m, n \in \mathbb{N}$, $m < n$, $d = (d_{k-1}d_{k-2} \ldots d_0)_2$
Output: $m^d \mod n$
  1: $a \leftarrow 1$
  2: **for** $i = k - 1$ **to** 0 **do**
  3:     $a \leftarrow a^2 \mod n$
  4:     **if** $d_i = 1$ **then**
  5:         $a \leftarrow a \times m \mod n$
  6: **return** a

Considering a balanced exponent d, these algorithms require on average $1S + 0.5M$ per bit of exponent to perform the exponentiation – S being the cost of a modular squaring and M the cost of a modular multiplication. It is generally considered in the literature – and corroborated by our experiments – that on cryptographic coprocessors $S \approx 0.8M$.

These algorithms are no longer used in embedded devices for security applications since the emergence of the side-channel analysis.

Alg. 2.2 Right-to-Left Square-and-Multiply Exponentiation

Input: $m, n \in \mathbb{N}$, $m < n$, $d = (d_{k-1} d_{k-2} \ldots d_0)_2$
Output: $m^d \mod n$
1: $a \leftarrow 1$; $b \leftarrow m$
2: **for** $i = 0$ **to** $k-1$ **do**
3:    **if** $d_i = 1$ **then**
4:       $a \leftarrow a \times b \mod n$
5:    $b \leftarrow b^2 \mod n$
6: **return** a

2.2   Side-Channel Analysis on Exponentiation

Side-channel analysis was introduced in 1996 by Kocher in [12] and completed in [13]. Many attacks have been derived in the following years.

On one hand, *passive* attacks rely on the following physical property: a microprocessor is physically made of thousands of logical gates switching differently depending on the executed operations and on the manipulated data. Therefore the power consumption and the electromagnetic radiation, which depend on those gates switchings, reflect and may leak information on the executed instructions and the manipulated data. Consequently, by monitoring such side-channels of a device performing cryptographic operations, an observer may infer information on the implementation of the program executed and on the – potentially secret – data involved.

On the other hand, *active* attacks intend to physically tamper with computations and/or stored values in memories. Such effects are generally obtained using clock or power glitches, laser beam, etc.

Finally some works [1] have highlighted the fact that passive and active attacks may be combined to threaten implementations applying countermeasures against both of them but not against their simultaneous use.

In the remainder of this section we focus on two passive attacks : the SPA presented hereafter with classical countermeasures, and a particular analysis from [2] discussed in Section 2.3.

Simple Power Analysis. Simple side-channel analysis [11] consists in observing a difference of behavior depending on the value of the secret key on the component performing cryptographic operations by using a single measurement.

In the case of an exponentiation, the original SPA is based on the fact that, if the squaring operation has a different pattern than a multiplication, the secret exponent can be directly read on the curve. For instance, in Alg. 2.1, a 0 exponent bit implies a squaring to be followed by another squaring, while a 1 bit causes a multiplication to follow a squaring. Classical countermeasures consist of using *regular* algorithms or applying the *atomicity* principle, as detailed in the following.

Regular Algorithms. These algorithms include the well known *square-and-multiply always* and Montgomery ladder algorithms [15,10]. The latter is

Alg. 2.3 Montgomery Ladder Exponentiation

Input: $m, n \in \mathbb{N}$, $m < n$, $d = (d_{k-1}d_{k-2}\ldots d_0)_2$
Output: $m^d \mod n$
1: $R_0 \leftarrow 1$; $R_1 \leftarrow m$
2: **for** $i = k - 1$ **to** 0 **do**
3:      $R_{1-d_i} \leftarrow R_0 \times R_1 \mod n$
4:      $R_{d_i} \leftarrow R_{d_i}^2 \mod n$
5: **return** R_0

Alg. 2.4 Left-to-Right Multiply Always Exponentiation

Input: $m, n \in \mathbb{N}$, $m < n$, $d = (d_{k-1}d_{k-2}\ldots d_0)_2$
Output: $m^d \mod n$
1: $R_0 \leftarrow 1$; $R_1 \leftarrow m$; $i \leftarrow k - 1$; $t \leftarrow 0$
2: **while** $i \geq 0$ **do**
3:      $R_0 \leftarrow R_0 \times R_t \mod n$
4:      $t \leftarrow t \oplus d_i$; $i \leftarrow i - 1 + t$ [$\oplus$ is bitwise XOR]
5: **return** R_0

presented hereafter in Alg. 2.3. It is generally preferred over the square-and-multiply always method since it does not involves dummy multiplications which makes it naturally immune to the C safe-error attacks [18,10].

Such regular algorithms perform one squaring and one multiplication at every iteration and thus require $1M + 1S$ per exponent bit.

Atomicity Principle. This method, presented by Chevallier-Mames et al. in [4], can be applied to protect the square-and-multiply algorithm against the SPA. It yields the so-called *multiply always* algorithm, since all squarings are performed as classical multiplications. We present a left-to-right multiply always algorithm in Alg. 2.4.

The interest of the multiply always algorithm is its better performances compared to the regular ones. Indeed it performs an exponentiation using on average $1.5M$ per exponent bit.

2.3   Distinguishing Squarings from Multiplications

Amiel et al. showed in [2] that the average Hamming weight of the output of a multiplication $x \times y$ has a different distribution whether:

 - The operation is a squaring performed using the multiplication routine, i.e. $x = y$, x uniformly distributed in $[0, 2^k - 1]$,
 - Or the operation is an "actual" multiplication, i.e. x and y independent and uniformly distributed in $[0, 2^k - 1]$.

This attack can thus target an atomic implementation such as Alg. 2.4 where the same multiplication operation is used to perform $x \times x$ and $x \times y$.

First, many exponentiation curves using a fixed exponent but variable data have to be acquired and averaged. Then, considering the average curve, the aim of the attack is to reveal if two consecutive operations are identical – i.e. two squarings – or different – i.e. a squaring and a multiplication. As in the classical SPA, two consecutive squarings reveal that a 0 bit has been manipulated whereas a squaring followed by a multiplication reveals a 1 bit. This information is obtained using the above-mentioned leakage by subtracting the parts of the average curve corresponding to two consecutive operations: peaks occur if one is a squaring and the other is a multiplication while subtracting two squarings should produce only noise. It is worth noticing that no particular knowledge on the underlying hardware implementation is needed which in practice increases the strength of this analysis.

A classical countermeasure against this attack is the randomization of the exponent[1], i.e. $d^* \leftarrow d + r\varphi(n)$, r being a random value. The result is obtained as $m^d \bmod n = m^{d^*} \bmod n$.

In spite of the possibility to apply the exponent randomization, this attack brings into light an intrinsic flaw of the multiply always algorithm: the fact that at some instant a multiplication performs a squaring $(x \times x)$ or not $(x \times y)$ depending on the exponent. In the rest of this paper we propose new atomic algorithms that are exempt from this weakness.

3   Square Always Countermeasure

We present in this section new exponentiation algorithms which simultaneously benefit from efficiency of the atomicity principle and immunity against the afore-mentioned weakness of the multiply always method.

3.1   Principle

It is well known that a multiplication can be performed using squarings only. Therefore we propose the following countermeasure which consists in using either expression (1) or (2) to perform all the multiplications in the exponentiation. Combined with the atomicity principle, this countermeasure completely prevents the attack described in Section 2.3 since only squarings are performed.

$$x \times y = \frac{(x+y)^2 - x^2 - y^2}{2} \tag{1}$$

$$x \times y = \left(\frac{x+y}{2}\right)^2 - \left(\frac{x-y}{2}\right)^2 \tag{2}$$

At the first glance, (1) requires three squarings to perform a multiplication whereas (2) requires only two. Further analysis reveals however that using (1) or (2) in Alg. 2.1 and 2.2 has always the cost of replacing multiplications by twice more squarings. Indeed, notice that in the multiplication $a \leftarrow a \times m$ of Alg. 2.1 m is a constant operand. Therefore implementing $a \times m$ using (1) yields

---

[1] Notice however that the randomization of the message has no effect on this attack, or even makes it easier by providing the required data variability.

$y = m$, thus m^2 mod n can be computed only once at the beginning of the exponentiation. The cost of computing y^2 can then be neglected.

This trick does not apply to Alg. 2.2 since no operand is constant in step 4. However $b \leftarrow b^2$ is the following operation. Using equation (1) in Alg. 2.2 then yields to store $t \leftarrow y^2$ and save the following squaring: $b \leftarrow t$. The resulting cost is thus equivalent as trading one multiplication for two squarings.

Remark In our context, (1) or (2) refer to operations modulo n. Notice however that divisions by 2 in these equations require neither inversion nor multiplication. For example, we recommend computing $z/2$ mod n in the following atomic way:

$$t_0 \leftarrow z$$
$$t_1 \leftarrow z + n$$
$$\alpha \leftarrow z \ \text{mod} \ 2$$
$$\textbf{return} \ t_\alpha/2$$

3.2   Atomic Algorithms

Trading multiplications for squarings in Alg. 2.1 and 2.2 just requires to apply formula (1) or (2) at step 5 in Alg. 2.1 or step 4 in Alg. 2.2. However the resulting algorithms would still present a leakage since different operations would be performed when processing a 0 or 1 bit. Hence it is necessary to apply the atomicity principle on these algorithms.

This step is achieved by identifying a minimal pattern of operations to be performed on each loop iteration and rewrite the algorithms using this pattern. For the considered algorithms, the minimal pattern should obviously contain a single squaring since it is the only operation required by the processing of a 0 bit and performing dummy squarings would lessen the performances of the algorithm. An addition, subtraction and division by 2 should also be present to compute formulas (1) or (2). Finally some more operations are required to manage the loop counter and the pointer on exponent bits.

Algorithm 3.1 presented hereafter details how to implement atomically the square always method in a left-to-right exponentiation using equation (1).

As in [4] we use a matrix for a more readable and efficient implementation:

$$M = \begin{pmatrix} 1\,1\,1\,0\,2\,1\,1\,1\,2\,1 \\ 2\,0\,1\,2\,2\,2\,2\,2\,3\,0 \\ 1\,1\,3\,0\,0\,0\,0\,2\,0\,0 \\ 3\,3\,3\,0\,3\,3\,1\,1\,3\,1 \end{pmatrix}$$

The main loop of Alg. 3.1 can be viewed as a four state machine where each row j of M define the operands of the atomic pattern. The atomic pattern itself is given by the content of the loop, i.e. steps 4 to 9. An exponent bit d_i is processed by the state $j = 0$ (resp. $j = 3$) if the previous bit d_{i+1} is a 0 (resp. a 1). This state is followed by the processing of the next bit if $d_i = 0$, or by the states $j = 1$ and $j = 2$ if $d_i = 1$. For more clarity, we present below the four sequences of operations corresponding to each state. The dummy operations are identified by a $\star$.

Alg. 3.1 Left-to-Right Square Always Exponentiation with (1)

Input: $m, n \in \mathbb{N}$, $m < n$, $d = (d_{k-1}d_{k-2}\ldots d_0)_2$
Output: $m^d \bmod n$
1: $R_0 \leftarrow 1$; $R_1 \leftarrow m$; $R_2 \leftarrow 1$; $R_3 \leftarrow m^2/2 \bmod n$
2: $j \leftarrow 0$; $i \leftarrow k-1$
3: **while** $i \geq 0$ **do**
4:    $R_{M_{j,0}} \leftarrow R_{M_{j,1}} + R_{M_{j,2}} \bmod n$
5:    $R_{M_{j,3}} \leftarrow R_{M_{j,3}}^2 \bmod n$
6:    $R_{M_{j,4}} \leftarrow R_{M_{j,5}}/2 \bmod n$
7:    $R_{M_{j,6}} \leftarrow R_{M_{j,7}} - R_{M_{j,8}} \bmod n$
8:    $j \leftarrow d_i(1 + (j \bmod 3))$
9:    $i \leftarrow i - M_{j,9}$
10: **return** R_0

$j = 0$	
($d_i = 0$ or 1)	
$R_1 \leftarrow R_1 + R_1 \bmod n$	$\star$
$R_0 \leftarrow R_0^2 \bmod n$	
$R_2 \leftarrow R_1/2 \bmod n$	$\star$
$R_1 \leftarrow R_1 - R_2 \bmod n$	$\star$
$j \leftarrow d_i$	[$\star$ if $d_i = 0$]
$i \leftarrow i - (1 - d_i)$	[$\star$ if $d_i = 1$]

$j = 2$	
($d_i = 1$)	
$R_1 \leftarrow R_1 + R_3 \bmod n$	$\star$
$R_0 \leftarrow R_0^2 \bmod n$	
$R_0 \leftarrow R_0/2 \bmod n$	
$R_0 \leftarrow R_2 - R_0 \bmod n$	
$j \leftarrow 3$	
$i \leftarrow i - 1$	

$j = 1$	
($d_i = 1$)	
$R_2 \leftarrow R_0 + R_1 \bmod n$	
$R_2 \leftarrow R_2^2 \bmod n$	
$R_2 \leftarrow R_2/2 \bmod n$	
$R_2 \leftarrow R_2 - R_3 \bmod n$	
$j \leftarrow 2$	
$i \leftarrow i$	$\star$

$j = 3$	
($d_i = 0$ or 1)	
$R_3 \leftarrow R_3 + R_3 \bmod n$	$\star$
$R_0 \leftarrow R_0^2 \bmod n$	
$R_3 \leftarrow R_3/2 \bmod n$	$\star$
$R_1 \leftarrow R_1 - R_3 \bmod n$	$\star$
$j \leftarrow d_i$	
$i \leftarrow i - (1 - d_i)$	[$\star$ if $d_i = 1$]

We also present in Alg. 3.2 a right-to-left variant of the square always exponentiation using equation (2). This algorithm requires the following matrix:

$$M = \begin{pmatrix} 0\,0\,2\,0\,0\,0\,2\,1 \\ 2\,1\,2\,2\,1\,0\,1\,0 \\ 0\,2\,1\,1\,0\,0\,2\,0 \\ 0\,0\,0\,0\,1\,2\,1\,1 \end{pmatrix}$$

As for the previous algorithm, the main loop of Alg. 3.2 has four states. Here, the state $j = 0$ corresponds to the processing a 0 bit and the sequence $j = 1$, $j = 2$, and $j = 3$ corresponds to the processing of a 1 bit, as detailed below.

Alg. 3.2 Right-to-Left Square Always Exponentiation with (2)

Input: $m, n \in \mathbb{N}$, $m < n$, $d = (d_{k-1}d_{k-2} \ldots d_0)_2$
Output: $m^d \mod n$
1: $R_0 \leftarrow m$; $R_1 \leftarrow 1$; $R_2 \leftarrow 1$
2: $i \leftarrow 0$; $j \leftarrow 0$
3: **while** $i \leq k - 1$ **do**
4:    $j \leftarrow d_i(1 + (j \mod 3))$
5:    $R_{M_{j,0}} \leftarrow R_{M_{j,1}} + R_0 \mod n$
6:    $R_{M_{j,2}} \leftarrow R_{M_{j,3}}/2 \mod n$
7:    $R_{M_{j,4}} \leftarrow R_{M_{j,5}} - R_{M_{j,6}} \mod n$
8:    $R_{M_{j,3}} \leftarrow R_{M_{j,3}}{}^2 \mod n$
9:    $i \leftarrow i + M_{j,7}$
10: **return** R_1

$j = 0$		$j = 2$	
$(d_i = 0)$		$(d_i = 1)$	
$j \leftarrow 0$	[$\star$ if j was 0]	$j \leftarrow 2$	
$R_0 \leftarrow R_0 + R_0 \mod n$	$\star$	$R_0 \leftarrow R_2 + R_0 \mod n$	$\star$
$R_2 \leftarrow R_0/2 \mod n$	$\star$	$R_1 \leftarrow R_1/2 \mod n$	
$R_0 \leftarrow R_0 - R_2 \mod n$	$\star$	$R_0 \leftarrow R_0 - R_2 \mod n$	$\star$
$R_0 \leftarrow R_0{}^2 \mod n$		$R_1 \leftarrow R_1{}^2 \mod n$	
$i \leftarrow i + 1$		$i \leftarrow i$	$\star$

$j = 1$		$j = 3$	
$(d_i = 1)$		$(d_i = 1)$	
$j \leftarrow 1$		$j \leftarrow 3$	
$R_2 \leftarrow R_1 + R_0 \mod n$		$R_0 \leftarrow R_0 + R_0 \mod n$	$\star$
$R_2 \leftarrow R_2/2 \mod n$		$R_0 \leftarrow R_0/2 \mod n$	$\star$
$R_1 \leftarrow R_0 - R_1 \mod n$		$R_1 \leftarrow R_2 - R_1 \mod n$	
$R_2 \leftarrow R_2{}^2 \mod n$		$R_0 \leftarrow R_0{}^2 \mod n$	
$i \leftarrow i$	$\star$	$i \leftarrow i + 1$	

3.3   Performance Analysis

Algorithms 3.1 and 3.2 are mostly equivalent in terms of operations realized in a single loop. The number of dummy operations (additions, subtractions and halvings) introduced to fill the atomic blocks are the same in the two versions – it is generally considered that the cost of these operations is negligible compared to multiplications and squarings. Both algorithms require $2S$ per exponent bit on average or $1.6M$ if $S/M = 0.8$ which represents a theoretical 11.1% speed-up over Alg. 2.3 which is the fastest known regular algorithm immune to the attack from [2]. Table 1 compares the efficiency of the multiply always, Montgomery ladder, and square always algorithms when $S = M$ and $S/M = 0.8$.

In addition, our algorithms can be enhanced using the sliding window or m-ary exponentiation techniques [14,9] while the Montgomery ladder cannot. These techniques are known to provide a substantial speed-up on Alg. 2.4 when extra

Table 1. Comparison of the expected cost of SPA protected exponentiation algorithms (including the multiply always which is not immune to the attack from [2])

Algorithm	General cost	$S/M = 1$	$S/M = 0.8$	# registers
Multiply always (Alg. 2.4)	$1.5M$	$1.5M$	$1.5M$	2
Montgomery ladder (Alg. 2.3)	$1M + 1S$	$2M$	$1.8M$	2
L.-to-r. Square always (Alg. 3.1)	$2S$	$2M$	**$1.6M$**	4
R.-to-l. Square always (Alg. 3.2)	$2S$	$2M$	**$1.6M$**	3

memory is available. Though we did not investigate this path, we believe that a comparable trade-off between space and time can be expected.

3.4   Security Considerations

Our algorithms are protected against the SPA by the implementation of the atomicity principle. The analysis from [2] cannot apply either since only squarings are involved. As a matter of comparison, notice that the exponent blinding countermeasure does not fundamentally remove the source of the leakage but only renders this attack practically infeasible. Embedded implementations should also be protected against the *differential power analysis* (DPA) which we do not detail in this study. However it is worth noticing that classical DPA countermeasures, like exponent or modulus randomization, can be applied as well. The interested reader may refer to [13,5].

We recommend implementing Alg. 3.2 instead of Alg. 3.1 since left-to-right algorithms are vulnerable to the chosen message SPA and *doubling attack* [8], and more subject to combined attacks [1]. Besides, Alg. 3.2 requires one less register than Alg. 3.1.

It is well-known that algorithms using dummy operations generally succumb to safe-error attacks. Immunity to C and M safe-errors can be easily obtained by applying the exponent randomization technique, which also prevent the DPA. Nevertheless, special care has been taken in our algorithms to ensure that inducing a fault in any of the dummy operations would produce an erroneous result. For instance, in the following sequence of dummy operations in Alg. 3.2 ($j = 0$), no operation can be tampered with without corrupting R_0 and thus the result of the exponentiation:

$$R_0 \leftarrow R_0 + R_0 \mod n$$
$$R_2 \leftarrow R_0/2 \mod n$$
$$R_0 \leftarrow R_0 - R_2 \mod n$$

Only operations $i \leftarrow i$ and $j \leftarrow 0$, appearing in some instances of Alg. 3.1 and 3.2 patterns, have not been protected for readability reasons. It is easy to fix these points: perform $i \leftarrow i \pm M_{j,\cdot} + \alpha$ instead of $i \leftarrow i \pm M_{j,\cdot}$ in Alg. 3.1 and 3.2 and add a step $i \leftarrow i - \alpha$ in the loop. The $j \leftarrow d_i(1 + \dots)$ operation should be protected in the same manner. In the end, our algorithms are immune to C safe-error attacks.

Further work may focus on implementing on our algorithms the *infective computation* strategy presented by Schmidt et al. in [17] in order to counterfeit the combined attacks.

4   Parallelization

It is well known that the Montgomery ladder algorithm is well suited for parallelization. It is thus natural to ask if the square always algorithms have the same property. For example the two squarings needed to perform a classical multiplication using equation (2) are independent and can therefore be performed simultaneously. The same strategy applies for equation (1).

We believe that the interest of this section extends beyond the scope of embedded systems. Nowadays most of computers are provided with several processors which enables using parallelized algorithms to speed-up computations.

4.1   Parallelized Algorithms

We noticed that right-to-left exponentiations are more suited for parallelization than their left-to-right counterpart since more operations are independent. For example in Alg. 2.2 one can first perform all squarings (step 5), store all values corresponding to a $d_i = 1$, and then perform the remaining multiplications. We present in Alg. 4.1 a right-to-left square always algorithm using (2) and two parallel squaring blocks (i.e. two 1-operand multipliers). For a better readability Alg. 4.1 is not atomic and two operations o_1 and o_2 performed simultaneously are denoted $o_1 \parallel o_2$.

Algorithm 4.2 is an atomic variant of Alg. 4.1. It requires two extra registers compared to the non atomic version and the following matrices:

$$M = \begin{pmatrix} 1 & 1 & 5 & 6 & 5 & 5 & 5 & 0 & 1 \\ 0 & 6 & 4 & 3 & 0 & 1 & 3 & 1 & 1 \\ 2 & 5 & 3 & 1 & 5 & 5 & 5 & 0 & 0 \\ 2 & 5 & 0 & 6 & 0 & 1 & 5 & 0 & 1 \end{pmatrix} \quad N = \begin{pmatrix} 1 & 1 & 0 \\ 5 & 2 & 2 \end{pmatrix}$$

It is possible to further enhance the efficiency of Alg. 4.1 if more memory is available by storing more free squarings when 1's sequences are processed. This observation yields Alg. 4.3 which allows the storage of *extramax* simultaneous precomputed squarings using as many registers $R_3, R_4, \ldots R_{extramax+2}$. Alg. 4.3 with *extramax* = 1 is thus equivalent to algorithms 4.1 and 4.2. Though Alg. 4.3 is not atomic for readability reasons and because of the difficulty to write an atomic algorithm depending on a variable (here *extramax*), it should be possible to write an atomic version for each *extramax* value in the same way than we processed with Alg. 4.1.

Alg. 4.1 Right-to-Left Parallel Square Always Exponentiation with (2)

Input: $m, n \in \mathbb{N}$, $m < n$, $d = (d_{k-1} \ldots d_0)_2$, require 5 k-bit registers a, b, R_0, R_1, R_2
Output: $m^d \mod n$
1: $a \leftarrow 1$; $b \leftarrow m$; extra $\leftarrow 0$
2: **for** $i = 0$ **to** $k - 1$ **do**
3:    **if** $d_i = 1$ **then**
4:       **if** extra $= 0$ **then**
5:          $R_0 \leftarrow (a - b)^2 \mod n$ $||$ $R_1 \leftarrow b^2 \mod n$
6:          $a \leftarrow (a + b)^2 \mod n$ $||$ $R_2 \leftarrow R_1{}^2 \mod n$
7:          $a \leftarrow (a - R_0)/4 \mod n$
8:          $b \leftarrow R_1$
9:          $R_1 \leftarrow R_2$
10:          extra $\leftarrow 1$
11:      **else**
12:          $R_0 \leftarrow (a - b)^2 \mod n$ $||$ $a \leftarrow (a + b)^2 \mod n$
13:          $a \leftarrow (a - R_0)/4 \mod n$
14:          $b \leftarrow R_1$
15:          extra $\leftarrow 0$
16:   **else**
17:      **if** extra $= 0$ **then**
18:          $b \leftarrow b^2 \mod n$
19:      **else**
20:          $b \leftarrow R_1$
21:          extra $\leftarrow 0$
22: **return** a

Alg. 4.2 Right-to-Left Atomic Parallel Square Always Exp. with (2)

Input: $m, n \in \mathbb{N}$, $m < n$, $d = (d_{k-1}d_{k-2} \ldots d_0)_2$, require 7 k-bit registers R_0 to R_6
Output: $m^d \mod n$
1: $R_0 \leftarrow 1$; $R_1 \leftarrow m$; $v \leftarrow (0,0,0)$; $u \leftarrow 1$ [v_0 is i and v_1 is extra from Alg. 4.1]
2: **while** $v_0 \leq k - 1$ **do**
3:    $j \leftarrow d_{v_0}(v_1 + u + 1)$
4:    $R_5 \leftarrow (R_0 - R_1)/2 \mod n$
5:    $R_6 \leftarrow (R_0 + R_1)/2 \mod n$
6:    $R_{M_{j,0}} \leftarrow R_{M_{j,1}}{}^2 \mod n$ $||$ $R_{M_{j,2}} \leftarrow R_{M_{j,3}}{}^2 \mod n$
7:    $R_{M_{j,4}} \leftarrow R_0 - R_2 \mod n$
8:    $R_{M_{j,5}} \leftarrow R_3$
9:    $R_{M_{j,6}} \leftarrow R_4$
10:    $v_1 \leftarrow M_{j,7}$
11:    $u \leftarrow M_{j,8}$
12:    $t \leftarrow 1 - v_1(1 - d_{v_0+1})$
13:    $R_{N_{t,0}} \leftarrow R_3$
14:    $v_{N_{t,1}} \leftarrow 0$
15:    $v_{N_{t,2}} \leftarrow v_{N_{t,2}} + 1$
16:    $v_0 \leftarrow v_0 + u$
17: **return** R_0

Alg. 4.3 Right-to-Left Generalized Parallel Square Always Exp. with (2)

Input: $m, n \in \mathbb{N}$, $m < n$, $d = (d_{k-1}d_{k-2}\ldots d_0)_2$, extramax $\in \mathbb{N}^*$, require extramax $+4$
   k-bit registers a, R_0, R_1, $\ldots R_{extramax+2}$

Output: $m^d \bmod n$

1: $a \leftarrow 1$; $R_1 \leftarrow m$; extra $\leftarrow 0$
2: **for** $i = 0$ **to** $k - 1$ **do**
3:   **if** $d_i = 1$ **then**
4:     **if** extra $<$ extramax **then**
5:       $R_0 \leftarrow (a - R_1)^2 \bmod n$ $||$ $R_{extra+2} \leftarrow R_{extra+1}{}^2 \bmod n$
6:       $a \leftarrow (a + R_1)^2 \bmod n$ $||$ $R_{extra+3} \leftarrow R_{extra+2}{}^2 \bmod n$
7:       $a \leftarrow (a - R_0)/4 \bmod n$
8:       $(R_1, R_2, \ldots R_{extramax+1}) \leftarrow (R_2, R_3, \ldots R_{extramax+2})$
9:       extra $\leftarrow$ extra $+1$
10:    **else**
11:      $R_0 \leftarrow (a - R_1)^2 \bmod n$ $||$ $a \leftarrow (a + R_1)^2 \bmod n$
12:      $a \leftarrow (a - R_0)/4 \bmod n$
13:      $(R_1, R_2, \ldots R_{extramax+1}) \leftarrow (R_2, R_3, \ldots R_{extramax+2})$
14:      extra $\leftarrow$ extra -1
15: **else**
16:    **if** extra $= 0$ **then**
17:      $R_1 \leftarrow R_1{}^2 \bmod n$
18:    **else**
19:      $(R_1, R_2, \ldots R_{extramax+1}) \leftarrow (R_2, R_3, \ldots R_{extramax+2})$
20:      extra $\leftarrow$ extra -1
21: **return** a

Remark Notice that multiple assignments of steps 8, 13, and 19 may be traded for a cheap index increment if registers R_1, R_2, $\ldots$, $R_{extramax+2}$ are managed as a cyclic buffer.

4.2   Cost of Parallelized Algorithms

We demonstrate in Appendix A that, as the length of the exponent tends to infinity, the cost per exponent bit of Alg. 4.3 tends to:

$$\left(1 + \frac{1}{4\,extramax+2}\right) S$$

It yields a cost of $7S/6$ for Alg. 4.1, 4.2, and 4.3 with extramax $= 1$, $11S/10$ for extramax $= 2$, $15S/14$ for extramax $= 3$, etc. The difference between this limit and costs actually observed in our simulations is negligible for 1024-bit or longer exponents.

It is remarkable that if $S/M = 0.8$ these costs become respectively $0.93M$, $0.88M$, $0.86M$, etc. per exponent bit. We believe that such performances cannot be achieved by binary algorithms using two parallelized 2-operands multiplication blocks. Indeed at least k multiplications have to be performed sequentially in Alg. 2.1 and 2.2, which requires at least $1M$ per exponent bit. Moreover when

Table 2. Comparison of the expected cost of parallelized exponentiation algorithms

Algorithm	General cost	$S/M = 1$	$S/M = 0.8$
Parallelized Montgomery ladder	$1M$	$1M$	$1M$
Alg. 4.1, Alg. 4.2, Alg. 4.3 with $extramax = 1$	$7S/6$	$1.17M$	**$0.93M$**
Alg. 4.3 with $extramax = 2$	$11S/10$	$1.10M$	**$0.88M$**
Alg. 4.3 with $extramax = 3$	$15S/14$	$1.07M$	**$0.86M$**
$\vdots$	$\vdots$	$\vdots$	$\vdots$
Alg. 4.3 with $extramax \to \infty$	$1S$	$1M$	**$0.8M$**

$extramax$ tends to infinity, the cost of Alg. 4.3 tends to $1S$, which we believe to be the optimal cost of an exponentiation algorithm based on the binary decomposition of the exponent since k squarings at least have to be performed sequentially.

Table 2 summarizes the theoretical cost of parallelized algorithms cited in this study.

5   Practical Results

In this section, we briefly present practical implementation results of the non-parallelized square always algorithm. As discussed in Section 3.4 we focused the right-to-left version.

We implemented this algorithm and the Montgomery ladder on an Atmel AT90SC smart card chip. This component is provided with an 8-bit AVR core and the AdvX coprocessor dedicated to long integer arithmetic. We used the Barrett reduction [3] to implement modular arithmetic.

We present in Table 3 the memory (code and RAM) and timing figures obtained with the chip and the AdvX running at 30 MHz. The observed speed-up of the square always algorithm over the Montgomery ladder is 5% on average. This is less than the predicted 11% but the difference can be explained by the neglected operations of the atomic pattern. Keep in mind that such results highly depend on the considered device and its hardware capabilities.

Table 3. On chip comparison of the Montgomery ladder and square always algorithms

Algorithm	Key Length (b)	Code Size (B)	RAM used (B)	Timings (ms)
Montgomery ladder (Alg. 2.3)	512	360	128	30
	1024	360	256	200
	2048	360	512	1840
Square Always (Alg. 3.2)	512	510	192	28
	1024	510	384	190
	2048	510	768	1740

We performed careful SPA on both implementations and observed no leakage on power traces.

6   Conclusion

In this paper we show that trading multiplications for squarings in an exponentiation scheme together with the atomicity principle provides a new countermeasure against side-channel attacks aimed at distinguishing squarings from multiplications. Moreover, this countermeasure is intrinsically more secure against such analysis than the classical multiply always atomic algorithm with exponent blinding, and provides better performances and flexibility towards space/time trade-offs than regular algorithms such as the Montgomery ladder or the square-and-multiply always.

As a complementary work, we present new algorithms using two parallel squaring blocks, and show how to write them atomically. We point out that, as far as we know, it leads to the fastest results in terms of speed. On the hardware side, an interesting conclusion is that two parallel squaring blocks enable faster exponentiation algorithms than two parallel multiplication blocks.

We believe that these observations are of great interest for the embedded devices industry and for everyone looking for fast exponentiation.

Acknowledgments. The authors would like to thank Sean Commercial and Alexandre Venelli for their valuable comments and advice on this manuscript. We would also like to thank the anonymous reviewers of this paper for their fruitful comments and advice.

References

1. Amiel, F., Feix, B., Marcel, L., Villegas, K.: Passive and Active Combined Attacks. In: Workshop on Fault Detection and Tolerance in Cryptography - FDTC 2007, IEEE Computer Society Press, Los Alamitos (2007)
2. Amiel, F., Feix, B., Tunstall, M., Whelan, C., Marnane, W.P.: Distinguishing Multiplications from Squaring Operations. In: Avanzi, R.M., Keliher, L., Sica, F. (eds.) SAC 2008. LNCS, vol. 5381, pp. 346–360. Springer, Heidelberg (2009)
3. Barrett, P.: Implementing the Rivest Shamir and Adleman Public Key Encryption Algorithm on a Standard Digital Signal Processor. In: Odlyzko, A.M. (ed.) CRYPTO 1986. LNCS, vol. 263, pp. 311–323. Springer, Heidelberg (1987)
4. Chevallier-Mames, B., Ciet, M., Joye, M.: Low-Cost Solutions for Preventing Simple Side-Channel Analysis: Side-Channel Atomicity. IEEE Transactions on Computers 53(6), 760–768 (2004)
5. Coron, J.-S.: Resistance Against Differential Power Analysis for Elliptic Curve Cryptosystems. In: Koç, Ç.K., Paar, C. (eds.) CHES 1999. LNCS, vol. 1717, pp. 292–302. Springer, Heidelberg (1999)
6. Diffie, W., Hellman, M.E.: New Directions in Cryptography. IEEE Transactions on Information Theory 22(6), 644–654 (1976)
7. FIPS PUB 186-3. Digital Signature Standard. National Institute of Standards and Technology (October 2009)
8. Fouque, P.-A., Valette, F.: The Doubling Attack – Why Upwards is Better Than Downwards. In: Walter, C.D., Koç, Ç.K., Paar, C. (eds.) CHES 2003. LNCS, vol. 2779, pp. 269–280. Springer, Heidelberg (2003)

9. Hankerson, D., Menezes, A.J., Vanstone, S.: Guide to Elliptic Curve Cryptography. Springer Professional Computing Series (January 2003)
10. Joye, M., Yen, S.-M.: The Montgomery Powering Ladder. In: Kaliski Jr., B.S., Koç, Ç.K., Paar, C. (eds.) CHES 2002. LNCS, vol. 2523, pp. 291–302. Springer, Heidelberg (2003)
11. Kocher, P., Jaffe, J., Jun, B.: Introduction to Differential Power Analysis and Related Attacks (1998)
12. Kocher, P.C.: Timing Attacks on Implementations of Diffie-Hellman, RSA, DSS, and Other Systems. In: Koblitz, N. (ed.) CRYPTO 1996. LNCS, vol. 1109, pp. 104–113. Springer, Heidelberg (1996)
13. Kocher, P.C., Jaffe, J., Jun, B.: Differential Power Analysis. In: Wiener, M. (ed.) CRYPTO 1999. LNCS, vol. 1666, pp. 388–397. Springer, Heidelberg (1999)
14. Menezes, A., van Oorschot, P.C., Vanstone, S.A.: Handbook of Applied Cryptography. CRC Press (1996)
15. Montgomery, P.L.: Speeding the Pollard and Elliptic Curve Methods of Factorization. MC 48, 243–264 (1987)
16. Rivest, R.L., Shamir, A., Adleman, L.: A Method for Obtaining Digital Signatures and Public-Key Cryptosystems. Communications of the ACM 21, 120–126 (1978)
17. Schmidt, J.-M., Tunstall, M., Avanzi, R., Kizhvatov, I., Kasper, T., Oswald, D.: Combined Implementation Attack Resistant Exponentiation. In: Abdalla, M., Barreto, P.S.L.M. (eds.) LATINCRYPT 2010. LNCS, vol. 6212, pp. 305–322. Springer, Heidelberg (2010)
18. Yen, S.-M., Joye, M.: Checking Before Output Not Be Enough Against Fault-Based Cryptanalysis. IEEE Trans. Computers 49(9), 967–970 (2000)

A   Cost of Algorithm 4.3

We present hereafter a demonstration of the claimed asymptotic cost of Alg. 4.3.

We first recall the principle of this algorithm: since 3 squarings are required to process a 1 bit, a fourth squaring slot is available at the same cost ($2S$). Thus, the algorithm scans the exponent from the right to the left and computes one squaring in advance at each 1 bit ($\nearrow$ in the following), within the limit of $extramax$. Then, as 0's are processed, the free squarings are consumed ($\searrow$) at null cost ($0S$). Two other cases may happen: first, a 1 bit can be processed but $extramax$ squarings are already stored in registers, then one free squaring is consumed ($\searrow$) and $1S$ is enough to perform the two other squarings. Second, a 0 bit can be processed with no free squaring in registers ($extra = 0$). Only in this latter case one squaring is performed at the cost of $1S$ and the parallel squaring slot is wasted ($\rightarrow$).

We can consider the evolution of $extra$ as exponent bits are processed using a diagram as below. For example, we have represented here the evolution of $extra$ for the 5 first bits of an exponent $d = (d_{k-1} \ldots 00110)_2$ with $extramax \geq 2$. The cost of the first 0 bit is $1S$ since $extra = 0$ at the beginning of the exponentiation, the cost of two next 1 bits is $2S$ each and $extra$ is incremented, finally the two last 0 bits have cost $0S$ and $extra$ is decremented. The total cost of the 5 bits is $5S$.

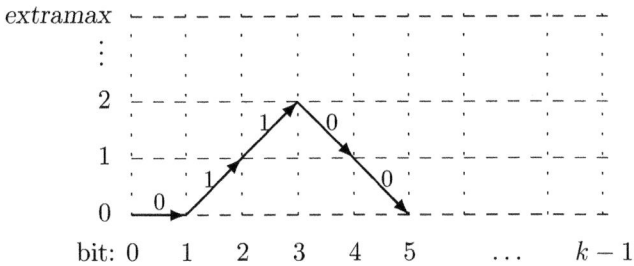

Observe now that the same bits have a higher cost if $extramax = 1$: as previously the two first bits 01 cost $1S$ and $2S$ respectively. However, the next 1 bit cannot lead to the computation of a second free squaring since $extramax = 1$. So the bit is processed at the cost of $1S$ and the free squaring is lost. Finally, the two last 0's cost $1S$ each since no free squaring is stored anymore. The cost of the sequence is $6S$.

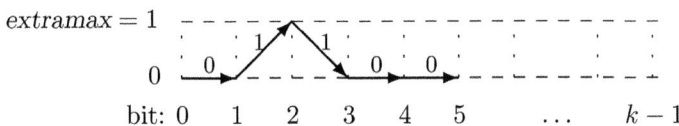

For a given exponent and $extramax$, let's call a c-cycle a sequence of bits starting with $extra = c$, ending with $extra = c$, and inside which $extra > c$. In particular, we can decompose any exponent as a sequence of 0-cycles, except that the last one may be unterminated with $extra > 0$.

Then, let B_c^e stand for the expected number of bits of a c-cycle when $extramax = e$ and C_c^e its expected cost.

extramax = 1 For a random exponent and $extramax = 1$, a 0-cycle is "0" with probability $1/2$ and "$1x$", $x \in \{0, 1\}$ otherwise. The cost of a 0-cycle "0" is $1S$ and the cost of a 0-cycle "$1x$" is $2S$ if $x = 0$ which happens with probability $1/2$, or $3S$ if $x = 1$.

$$B_0^1 = 1/2 \times 1 + 1/2 \times 2 = 3/2$$

$$C_0^1 = 1/2 \times 1S + 1/2 \times (1/2 \times 2S + 1/2 \times 3S) = 7S/4$$

The expected cost of a 0-cycle with $extramax = 1$ is then $C_0^1/B_0^1 = 7S/6$ per bit. As the length of the exponent tends to infinity, the contribution of the possibly unterminated last 0-cycle becomes negligible. Therefore the cost per bit of a random exponent tends to the cost per bit of a 0-cycle as its length tends to infinity. So we can approximate the cost of algorithms 4.1, 4.2 and 4.3 to $7S/6$ for exponents of thousands of bits.

extramax = e A 0-cycle starts with a 0 with probability $1/2$ and with a 1 otherwise. In the first case its cost is $1S$ as previously. Let $\tilde{B}_c^e$, respectively $\tilde{C}_c^e$, denote the expected length, respectively the expected cost, of a c-cycle starting with a 1 bit when $extramax = e$.

$$B_0^e = 1/2 \times 1 + 1/2 \times \tilde{B}_0^e \tag{3}$$

$$C_0^e = 1/2 \times 1S + 1/2 \times \tilde{C}_0^e \tag{4}$$

First we demonstrate that $\tilde{B}_0^e = 2e$. As depicted below, one can observe that $\tilde{B}_0^e = \tilde{B}_1^{e+1}$.

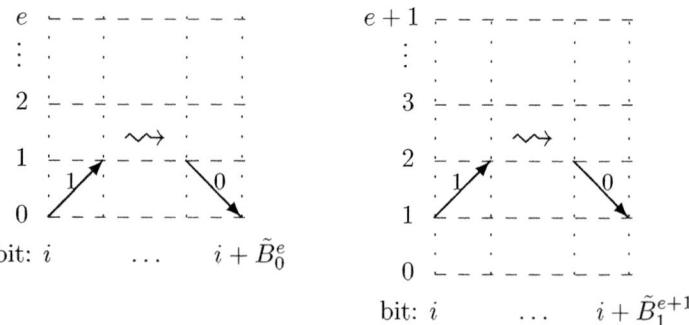

As depicted hereafter, the length $\tilde{B}_0^{e+1}$ of a 0-cycle with $extramax = e+1$ and starting by a 1 bit is $s\tilde{B}_1^{e+1} + 2$ where s is the number of inner 1-cycles starting by a 1 bit. Notice also that $s = i$ with probability $2^{-(i+1)}$, which gives:

$$\tilde{B}_0^{e+1} = 2 + \sum_{i=0}^{\infty} \frac{i\tilde{B}_1^{e+1}}{2^{(i+1)}} = 2 + \tilde{B}_1^{e+1} = 2 + \tilde{B}_0^e$$

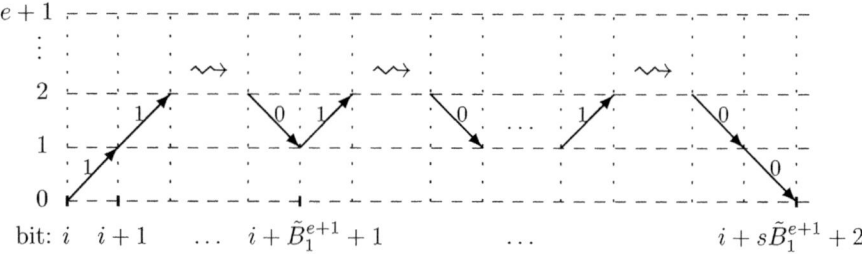

$(\tilde{B}_0^e)_{e \geq 1}$ is thus an arithmetic progression with common difference 2 and $\tilde{B}_0^1 = 2$. This yields $\tilde{B}_0^e = 2e$.

In a same manner, we can observe that:

$$\tilde{C}_0^{e+1} = 2S + \sum_{i=0}^{\infty} \frac{i\tilde{C}_1^{e+1}}{2^{(i+1)}} = 2S + \tilde{C}_1^{e+1} = 2S + \tilde{C}_0^e$$

Since $\tilde{C}_0^1 = 5S/2$ we obtain that $\tilde{C}_0^e = (1/2 + 2e)S$.

Using the above results in equations (3) and (4), we obtain finally:

$$B_0^e = 1/2 \times 1 + 1/2 \times 2e = 1/2 + e$$

$$\text{and} \quad C_0^e = 1/2 \times 1S + 1/2 \times (1/2 + 2e)S = (3/4 + e)S$$

The expectation of the cost per bit of a 0-cycle is then:

$$\frac{C_0^e}{B_0^e} = \left(\frac{3/4 + e}{1/2 + e}\right)S = \left(1 + \frac{1}{4e + 2}\right)S$$

Therefore the expectation of the cost of Alg. 4.3 with *extramax* $= e$ tends then to $(1 + \frac{1}{4e+2})S$ as the length of the exponent tends to infinity.

An Enhanced Differential Cache Attack on CLEFIA for Large Cache Lines

Chester Rebeiro, Rishabh Poddar, Amit Datta, and Debdeep Mukhopadhyay

Department of Computer Science and Engineering
Indian Institute of Technology Kharagpur, India
{chester,rishavp,adatta,debdeep}@cse.iitkgp.ernet.in

Abstract. Reported results on cache trace attacks on CLEFIA do not work with increased cache line size. In this paper we present an enhanced cache trace attack on CLEFIA using the differential property of the s-boxes of the cipher and the diffusion properties of the linear transformations of the underlying Feistel structures. The attack requires 3 round keys, which are obtained by monitoring cache access patterns of 4 rounds of the cipher. A theoretical analysis is made on the complexity of the attack, while experimental results are presented to show the effectiveness of power and timing side-channels in deducing cache access patterns. The efficacy of the attack is theoretically justified by showing the effect of cache line size on the time and space complexity of the attack. Finally countermeasures that guarantee security against cache-attacks are compared for their efficiency on large cache lines.

1 Introduction

In 2000, John Kelsey, Bruce Schneier, David Wagner, and Chris Hall prophesied that cryptographic ciphers implemented on systems with cache memory are vulnerable to side-channel attacks [9]. These attacks, which came to be known as *cache attacks*, exploited the non-uniform behavior between a cache hit and a cache miss. The differential behavior between a cache hit and a miss is manifested through side-channels such as timing, power, and electro-magnetic radiation. Over the decade, several forms of cache attacks have been discovered [1,4,5,7,10,11,12,18,17], which provides different strategies to extract information from the side-channels. Cache attacks have been found to be a serious threat to the security of modern crypto-systems due to the small number of encryptions that need to be monitored and the possibility of remote attacks [2,8].

Most of the cache attacks developed target AES and all of them follow a *divide-and-conquer* approach. Some of these attacks are [1,4,5,7,10]. The attacks on AES split the 128 bit key into 16 bytes and then recovers each byte independently. Obtaining all 16 bytes can easily be done by targeting just the first round of the cipher. In Feistel ciphers such as CAMELLIA [3] and CLEFIA [14] however, the first round provides only 64 bits of the key. Obtaining the remaining 64 bits requires more rounds to be attacked. This is difficult due to two reasons. First, for inner rounds it becomes more difficult to have control of the inputs to that

D.J. Bernstein and S. Chatterjee (Eds.): INDOCRYPT 2011, LNCS 7107, pp. 58–75, 2011.
© Springer-Verlag Berlin Heidelberg 2011

round; an essential requirement in cache-attacks (except for a variant in [16], which requires knowledge of neither the plaintext nor ciphertext and depends on a spy process to reveal the cache access patterns). Second, the correctness of the round key recovered depends on the correctness of the previous round keys obtained, this adds to the unpredictability.

In [12] and [19], a cache attack on CLEFIA was described, which required attacking 3 rounds of the cipher to completely recover the 128 bit secret key. However, the attack in [19] by Zhao and Wang used a strong assumption of misaligned tables. Misalignment of tables does not happen unless forced by the programmer; for example to optimize for space. The attack in [12] used a combination of differential techniques with cache access patterns to attack CLEFIA in 2^{14} encryptions. The attack however is restricted to cache memories with a cache line size of 32 bytes. This is not the conventional cache line size on modern day microprocessors. In this paper we present and critically analyze a new attack on CLEFIA that can be mounted on standard microprocessors. The contributions of this paper are as follows:

- We enhance the attack on CLEFIA in [12] to eliminate the constraint of the cache line size. On a microprocessor with a 64-byte cache line, the attack has a complexity of 2^{11} encryptions, while on a system with a 32-byte cache line, the attack has an estimated complexity of 2^{12} encryptions. This is lesser than the 2^{14} encryptions reported in [12].
- We critically analyze the difficulty in attacking more than one round of a Feistel cipher using the proposed attack on CLEFIA.
- A study is made on the effect of cache-line size on the complexity of the attack. Based on this analysis we show a peculiar property that the attack is better suited for microprocessors with large cache lines, provided the entire table does not completely fit into a single line in the cache.
- A discussion on countermeasures is done and their overhead on the performance compared.

The outline of the paper is as follows: Section 2 provides a brief introduction to CLEFIA and the differential cache attack proposed in [12]. The drawbacks of the existing attack and the new attack's principle are presented in Section 3. This section also presents a theoretical analysis on the complexity of attacking different rounds of a cipher. Section 4 presents the complete attack on CLEFIA. Section 5 compares the capabilities of a power adversary and a timing adversary with respect to the number of measurements required to distinguish between a cache hit and a miss. The effect of the cache line size and number of encryptions required is analyzed in Section 6, while countermeasures for the attack are discussed in Section 7. The final section has the conclusion of the paper.

2 Preliminaries

In this section we first give a brief description of the CLEFIA structure, then present the attack in [12].

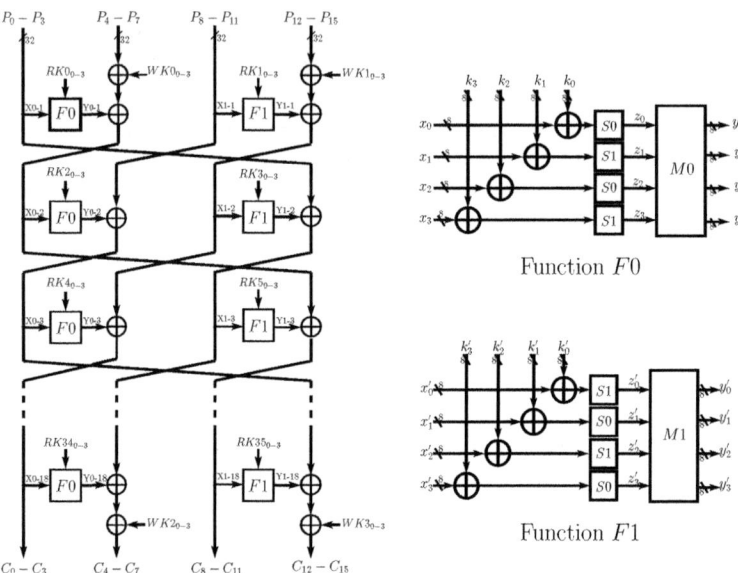

Fig. 1. CLEFIA Block Diagram

2.1 The CLEFIA Structure

The 128-bit block cipher CLEFIA [15] has a type-2 generalized Feistel structure [20] as shown in Figure 1. The 16 bytes plaintext input P_0 to P_{15} is grouped in 4 byte words. There are 18 rounds and in each round, the key addition, substitution and diffusion is provided by the application of two F functions, $F0$ and $F1$. The substitution in the F functions are done by two 256 element s-boxes $S0$ and $S1$, while the diffusion is done by the self-inverting matrices $M0$ and $M1$, which are defined as follows.

$$M0 = \begin{pmatrix} 1 & 2 & 4 & 6 \\ 2 & 1 & 6 & 4 \\ 4 & 6 & 1 & 2 \\ 6 & 4 & 2 & 1 \end{pmatrix} \qquad M1 = \begin{pmatrix} 1 & 8 & 2 & A \\ 8 & 1 & A & 2 \\ 2 & A & 1 & 8 \\ A & 2 & 8 & 1 \end{pmatrix} \qquad (1)$$

Each round has an addition of round keys. The ith round uses the round keys RKi and $RKi+1$. Each of these round keys are of 32 bits. Additionally, whitening keys $WK0$ to $WK3$ are applied at the start and end of encryption as seen in Figure 1.

2.2 Cache Attacks on CLEFIA

All cache attacks target structures in the block cipher such as in Figure 2. The figure shows two accesses to table S with indices $(x \oplus k1)$ and $(y \oplus k2)$. When

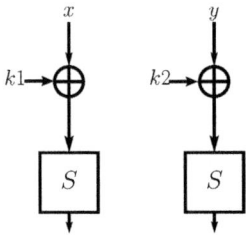

Fig. 2. Look-up Structure Targeted in Cache Attacks

a cache hit occurs the following relation holds leading to leakage of information about the ex-or of the keys; assuming that the inputs x and y are known.

$$\langle k1 \oplus k2 \rangle = \langle x \oplus y \rangle \tag{2}$$

We note that due to the effects of the cache line, only the most significant bits can be equated, therefore $\langle \cdot \rangle$ refers to only these most significant bits. If the size of $k1$ and $k2$ is l bits, and there are 2^δ elements that share a cache line, then only the most significant $b = l - \delta$ bits satisfy the above equation. Similarly, when a cache miss occurs, the following inequality holds.

$$\langle k1 \oplus k2 \rangle \neq \langle x \oplus y \rangle \tag{3}$$

Given any set of 4 round keys $(RK4i, RK4i+1, RK4i+2, RK4i+3)$, where $i \mod 2 = 0$, CLEFIA's key expansion algorithm can be reversed to obtain 121 out of the 128 bit secret key. For cache-attacks, determining the first set of round keys $RK0$, $RK1$, $RK2$, and $RK3$ is most suited. However, the presence of the whitening keys $WK0$ and $WK1$ makes the determination of these round keys not straight forward. In [12] this difficulty is circumvented by having 3 stages in the attack. First $RK0$ and $RK1$ are discovered, then $WK0 \oplus RK2$ and $WK1 \oplus RK3$, and finally $RK4$ and $RK5$. Determining each of these keys requires knowing the inputs to the respective F functions. For example, determining $WK0 \oplus RK2$ requires knowledge of $X0\text{-}2$ (see Figure 1), which in turn can be only computed if $RK0$ is known. Thus, an error in determining $RK0$ would also cause an error in $WK0 \oplus RK2$.

3   Enhancing the Differential Cache Attack

In the first part of this section, we discuss the reason why the attack proposed in [12] fails on systems with large cache lines. We then present a general technique to enhance the attack that would work with larger cache lines.

3.1   Why the Attack in [12] Fails for Large Cache Lines?

Figure 3 shows the application of three rounds of an F function (either $F0$ or $F1$). The differential attack in [12], used the fact that if at-least 3 bits of the

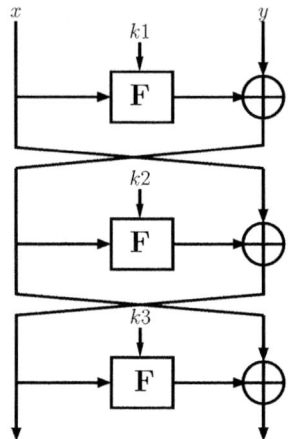

Fig. 3. Three Round Feistel Structure

difference of each output byte of F is known, then properties of the matrix ($M0$ for $F0$ and $M1$ for $F1$) can be used to obtain few output bits of the s-boxes. The difference distribution tables of the s-boxes can then be used to obtain a set of candidate keys. The crucial part of obtaining 3 bits of the F function output is done by forcing cache hits in the s-box tables of the second round. Since the CLEFIA implementation[1] attacked used 256 byte tables, and elements within a cache line cannot be distinguished, the table should occupy at-least 8 cache lines in order to obtain the required 3 bits of information. This means that each cache line can be of at-most $\frac{256}{8} = 32$ bytes. If the cache has the more conventional 64 byte cache lines, then only 2 bits of information can be obtained. This is insufficient to apply the properties of the matrices in order to derive the output difference of the s-boxes.

3.2 The Proposed Differential Cache Attack

CLEFIA is based on the generalized Feistel structure therefore we use Figure 3 to explain the principle of the new attack. The input x consists of 4 concatenated bytes $(x_0|x_1|x_2|x_3)$ and is known as the *differential introducing input*, while y comprising of the bytes $(y_0|y_1|y_2|y_3)$ is the *restoring input*. The F in the figure is either CLEFIA's $F0$ or $F1$ function (see Figure 1). For any fixed value of x, each byte of y is varied until cache hits are obtained in all s-box tables in the second round F function. This is called the *collision setup* phase. In this state, the following equation holds for $0 \le i \le 3$ if a cache hit is obtained.

$$\langle x_i \oplus k1_i \rangle = \langle y_i \oplus k2_i \oplus F(x, k1)_i \rangle \tag{4}$$

---

[1] http://www.sony.net/Products/cryptography/clefia/download/data/clefia_ref.c

Similarly, the following inequality holds if a cache miss is obtained.

$$\langle x_i \oplus k1_i \rangle \neq \langle y_i \oplus k2_i \oplus F(x, k1)_i \rangle \tag{5}$$

Now the input byte x_0 is displaced by $d_{x_0} \neq 0$, while all other bytes of x are unchanged. After the s-box operation, the displacement is diffused to all output bytes of the F function (since the branch number for the F functions in CLEFIA are 5 [14]), due to which, some of the cache hits in the second round are lost. The cache hits that remain are due to the tables in the second round being accessed in the same cache lines as before. The collision state is restored by reinforcing cache hits in the second round by modifying y. This is called the *restoring phase*. Let $y' = (y'_0|y'_1|y'_2|y'_3)$ be the new value of y. The differences at the output of the F function is $d_{y_i} = y_i \oplus y'_i$.

From the difference distribution table for the s-box, one can derive the set of possible output differentials corresponding to the input differential d_{x_0}. Let this set be called $\mathcal{D}$. For every differential $d_o \in \mathcal{D}$, the matrix product $M \cdot (d_o, 0, 0, 0)^T$ is computed to obtain the differentials $(d_{z_0}, d_{z_1}, d_{z_2}, d_{z_3})^T$ (M is either $M0$ or $M1$ depending on the F function used). The property that is exploited in the attack is that for the correct s-box output differential d_o, $\langle d_{z_i} \rangle = \langle d_{y_i} \rangle$ for $0 \leq i \leq 3$. Each equality leads to some information about the key $k1_0$. In both $M0$ and $M1$ matrices, 3 out of the 4 equalities reveal unique information, therefore it is sufficient to have only 3 equality tests instead of 4. In a similar way, displacements introduced at x_1, x_2, and x_3 would lead to leakages in $k1_1, k1_2$, and $k1_3$ respectively. The collision set up phase is common for all keys. However the restoring phases differ. Algorithm 1 shows the technique used to generate the candidate keys given that the collisions have been set up.

Algorithm 1. find: Finding Key Byte $k1_i$ assuming collisions have been setup

---

    Input: $i \in \{0, 1, 2, 3\}$, the differential introducing input x and restoring input y
    Output: $S^1_{k1_i}$: Candidate Key Set for $k1_i$

1 **begin**
2       $S^1_{k1_i} \leftarrow \{\}$
3       $x_i \leftarrow x_i \oplus d_{x_i}$
4   *Restore collisions*: Find y'_0, y'_1, y'_2 which causes collision in the first three accesses of
        the 2nd round
5       $\mathcal{D} \leftarrow$ output difference set corresponding to the input difference d_{x_i}
6   **foreach** $d_o \in D$ **do**
7          $(d_{z0}, d_{z1}, d_{z2}, d_{z3})^T \leftarrow (d_o, 0, 0, 0)^T \cdot M0$
8   **if** $(\langle d_{z0} \rangle = \langle d_{y_0} \rangle$ and $\langle d_{z1} \rangle = \langle d_{y_1} \rangle$ and $\langle d_{z2} \rangle = \langle d_{y_2} \rangle)$ **then**
9             $S^1_{k1_i} \leftarrow S^1_{k1_i} \cup \{d_o\}$
10   **end**
11   **end**
12 **end**

The Amount of Leakage: Let the number of bits revealed due to Equation 4 be b. This means that b bits in each of d_{y_i} (where $0 \leq i \leq 3$) is revealed. Since each

Table 1. Expected Number of Candidate Key Bytes with Varying Cache Line sizes

CacheLine Size →	128	64	32	16	8
Matrix, S-box ↓	$(b = 1)$	$(b = 2)$	$(b = 3)$	$(b = 4)$	$(b = 5)$
$M0, S0$	81.9	40.9	20.4	10.2	5.1
$M0, S1$	64.4	32.2	16.1	8.0	4.0
$M1, S0$	81.9	20.4	10.2	5.1	2.5
$M1, S1$	64.4	16.1	8.0	4.0	2.0

d_{y_i} is a linear transformation of the difference output of the s-box, a few bits of this output difference is obtained. Further, a valid input-output difference pair in $S0$ has on average 1.28 input possibilities, while $S1$ has 1.007. Table 4 shows the expected number of candidate keys obtained for a key byte (the size of the set $S_{k1_i}^1$ in Algorithm 1) for different cache line sizes (assuming each element in the table occupies one byte). As seen in the table, this depends on the matrix as well as the s-box. The number of candidate keys can be reduced by repeating the attack several times and taking the intersection between the sets. If q repetitions are done then the expected number of candidate keys for $k1_i$ at the end of the q repetitions is:

$$|S_{k1_i}^q| = \frac{(|S_{k1_i}^1|)^q}{256^{q-1}} \tag{6}$$

3.3   Attacking a Feistel Structure from Cache Traces

Attack on Feistel ciphers such as CLEFIA requires keys from more than one round to be obtained. However two problems are likely to arise while recovering keys for rounds other than the first round. This part of the section discusses these problems.

Increase in the Number of Candidate Keys: For a key to be recovered, the input and output difference of the F function is required. Consider the task of recovering $k2$ in Figure 3. The input difference to the second round F function is $(y \oplus F(x, k1)) \oplus (y' \oplus F(x', k1))$, while the output difference obtained from the cache access patterns is $x \oplus x'$. Thus it is seen that determining any $k2_i$, for $0 \le i \le 3$, depends on the value of $k1$. Let the number of candidate keys for each byte of $k1$ be n_1, then the number of possible values for an output byte of $F(x, k1)$ has an upper bound of n_1^4. When $n_1 = 1$ and after q repetitions, each key byte $k2_i$ would have the same number of candidates as $k1_i$, (ie. $|S_{k1}^q|$). For $n_1 > 1$, each output byte of $F(x, k1)$ produces a different set of keys for $k2_i$. The union of these sets should be considered while determining $k2_i$. The estimated size of this union is less than $256(1 - (1 - \frac{|S_{k1_i}^q|}{256})^{n_1^4})$. In general, for round $r > 1$,

$$|S_{kr_i}^q| \le 256\left(1 - \left(1 - \frac{|S_{k1_i}^q|}{256}\right)^{n_{r-1}^4}\right) \tag{7}$$

We see that the expected number of candidates increases exponentially with the round number. There are two ways to reduce this set.

- Increasing q, the number of repetitions in r^{th} round, would reduce the size of the set $|S^q_{k1_1}|$, thus reducing the number of candidates.
- A more effective approach is to increase the number of repetitions of the previous rounds, thus reducing the value of n_{r-1}. We see that the number of keys in the first round most influences the size of the candidate key set for kr_i and this influence reduces as the round number increases. The best strategy would therefore be to have q for a round much larger than the q for the following rounds. This would result in $n_1 \ll n_2 \cdots \ll n_{r-1}$.

Control of the Round Inputs: Determining the key bytes for round r requires collisions between the s-box accesses in rounds r and $r + 1$. For the Equations (4) and (5) to be used we need to know the inputs (ie. x and y) to round r. This becomes increasingly difficult as r increases. Secondly, even if we were able to control these inputs, a cache hit in a table access in round $r + 1$ could be due to a collision with any of the accesses in the previous rounds, not necessarily with the access in round r. Methods to improve controllability and reduce ambiguity about collisions are influenced by the cipher's structure. In the next section we present several such strategies for CLEFIA.

4   The New Differential Cache Attack against CLEFIA

Just as in [12], the proposed attack first finds $RK0$ and $RK1$, then $RK2 \oplus WK0$ and $RK3 \oplus WK1$, and finally $RK4$ and $RK5$. With this information, the key expansion algorithm can be exploited (shown in [12]) to determine 121 bits out of 128 bits of the secret key.

4.1   Determining $RK0$ and $RK1$

To find $RK0$, the memory accesses in the first and second round $F0$ functions (Figure 1) are considered. This is similar to a two round Feistel structure (Figure 3) with the differential introducing input being $P_0 \cdots P_3$, while the restoring input being $P_4 \cdots P_7$. Since in CLEFIA, each s-box is used 4 times per round, a resulting cache hit in the second round would be due to collisions with any of these four accesses. Let $I\alpha^i_{S\beta}$ be the index to the ith access to table $S\beta$ in round α. Thus a collision in $I2^1_{S0}$ could be with one or more of the accesses $I1^1_{S0}$, $I1^2_{S0}$, $I1^3_{S0}$, and $I1^4_{S0}$. In order to apply Equations 4 and 5, we need to know which of the four accesses has caused the collision. To prevent this ambiguity, we ensure that all accesses in the first round are themselves colliding. That is for $S0$, $\langle I1^1_{S0} \rangle = \langle I1^2_{S0} \rangle = \langle I1^3_{S0} \rangle = \langle I1^4_{S0} \rangle$ and for $S1$, $\langle I1^1_{S1} \rangle = \langle I1^2_{S1} \rangle = \langle I1^3_{S1} \rangle = \langle I1^4_{S1} \rangle$. We call such a state the 1-*round collision state*. Obtaining the 1-round collision state requires finding the appropriate values for P_2, P_3, P_8, P_9, P_{10}, and P_{11}.

To find $RK1$, $F1$ of the first two rounds are considered with differential introducing inputs being $P_8 \cdots P_{11}$ and restoring inputs $P_{12} \cdots P_{15}$. Algorithm 2 uses the *find* procedure described in Algorithm 1 to determine $RK0$ and $RK1$.

Algorithm 2. Finding Candidate Keys for $RK0$ and $RK1$

Output: S_{RK0_i} and S_{RK1_i}: Respective candidate key sets for $RK0$ and $RK1$, where
$\quad\quad i \in \{0,1,2,3\}$

1 **begin**
2 $\quad$ Randomly select P_0 and P_1
3 $\quad$ $(P_2, P_3, P_8 \cdots P_{11}) \leftarrow$ One Round Collision State
4 $\quad$ *Set up collisions:* Find P_4, P_5, P_6 causing collisions in $I2^1_{S0}, I2^1_{S1}, I2^2_{S0}$ respectively
5 $\quad$ $S_{RK0_0} \leftarrow find(0, P_0 \cdots P_3, P_4 \cdots P_7)$
6 $\quad$ $S_{RK0_1} \leftarrow find(1, P_0 \cdots P_3, P_4 \cdots P_7)$
7 $\quad$ $S_{RK0_2} \leftarrow find(2, P_0 \cdots P_3, P_4 \cdots P_7)$
8 $\quad$ $S_{RK0_3} \leftarrow find(3, P_0 \cdots P_3, P_4 \cdots P_7)$
9 $\quad$ Find collision in $I2^2_{S1}$ using P_7
10 $\quad$ *Set up collisions:* Find P_{12}, P_{13}, P_{15} causing collisions in $I2^3_{S1}, I2^3_{S0}, I2^4_{S1}$ respectively
11 $\quad$ $S_{RK1_0} \leftarrow find(0, P_8 \cdots P_{11}, P_{12} \cdots P_{15})$
12 $\quad$ $S_{RK1_1} \leftarrow find(1, P_8 \cdots P_{11}, P_{12} \cdots P_{15})$
13 $\quad$ $S_{RK1_2} \leftarrow find(2, P_8 \cdots P_{11}, P_{12} \cdots P_{15})$
14 $\quad$ $S_{RK1_3} \leftarrow find(3, P_8 \cdots P_{11}, P_{12} \cdots P_{15})$
15 **end**

Analysis: Let ρ be the number of encryptions required to determine (from the side-channel information) if a memory access resulted in a cache hit or a cache miss. Let b be the number of bits that are revealed in Equation 4 or 5. Obtaining a collision requires $\rho(2^b - 1)$ encryptions, because b bits can have 2^b possibilities, and if $(2^b - 1)$ choices fail to give a collision, then the final choice is the correct one. Thus obtaining a 1-round collision state requires $6\rho(2^b - 1)$ encryptions. Algorithm 2 requires $3\rho(2^b - 1)$ encryptions each to set up collisions in lines 4 and 10. Moreover, each call to $find$ requires $3\rho(2^b - 1)$ encryptions and line 9 requires $\rho(2^b - 1)$ encryptions. Thus in total, finding both $RK0$ and $RK1$ requires $37\rho(2^b - 1)$ encryptions.

4.2   Determining $RK2 \oplus WK0$ and $RK3 \oplus WK1$

To obtain the candidate keys of $RK2 \oplus WK0$, the $F0$ functions in the second and third rounds are considered with $P_4 \cdots P_7$ used as differential introducing inputs, while $P_8 \cdots P_{11}$ used as restoring inputs. Just as in the first stage of the attack, ambiguities about collisions may arise when cache hits are forced in the tables in the third round $F0$. Therefore, before forcing hits in the third round, the cipher is put in a *2-round colliding state*. Besides the first access to each table, a 2-round colliding state has all remaining accesses in collision.

$$\langle I1^1_{S0}\rangle = \langle I1^2_{S0}\rangle = \langle I1^3_{S0}\rangle = \langle I1^4_{S0}\rangle = \langle I2^1_{S0}\rangle = \langle I2^2_{S0}\rangle = \langle I2^3_{S0}\rangle = \langle I2^4_{S0}\rangle$$

for $S0$ and for $S1$,

$$\langle I1^1_{S1}\rangle = \langle I1^2_{S1}\rangle = \langle I1^3_{S1}\rangle = \langle I1^4_{S1}\rangle = \langle I2^1_{S1}\rangle = \langle I2^2_{S1}\rangle = \langle I2^3_{S1}\rangle = \langle I2^4_{S1}\rangle$$

In spite of the 2-round colliding state, ambiguities about the collisions still occur in the third round $F0$ memory accesses. For example, a cache hit in $I3^1_{S0}$ can be forced by the plaintext byte P_8. However, changing P_8, may loose the cache

---

Algorithm 3. Elimination Method

Input: I_a: memory access where a collision is required, I_{b1} and I_{b2}: accesses which cause
           undesirable collisions, C_1 and C_2: controlling plaintexts for I_{b1} and I_{b2} respectively
Output: Collision at I_a is desirable or undesirable

```
 1  begin
 2      c_1 ← C_1
 3      C_1 ← (C_1 + cache line size)mod table size
 4      if (I_a not in collision) then
 5          return "Undesirable Collision"
 6      end
 7      C_1 ← c_1
 8      C_2 ← (C_2 + cache line size)mod table size
 9      if (I_a not in collision) then
10          return "Undesirable Collision"
11      end
12      C_1 ← (C_1 + cache line size)mod table size
13      if (I_a not in collision) then
14          return "Undesirable Collision"
15      end
16      return "Desirable Collision"
17  end
```

---

hits in the two $S0$ accesses in $F1$ of the second round (Note that the cache hits in the first round are not altered). Thus a collision in $I3_{S0}^1$ may be due to three reasons: the desirable cache hit with $I2_{S0}^1$ and undesirable cache hits due to collisions with $I2_{S0}^3$ and $I2_{S0}^4$. There are two ways to identify the ambiguous cache hits:

- *Elimination Method:* This method uses a set of controlling plaintext bytes C_1 and C_2. These inputs can control locations causing ambiguous cache hits ($I2_{S0}^3$ or/and $I2_{S0}^4$). In the elimination method C_1 is first varied and checked if a cache hit persists. Then C_2 varied and finally both C_1 and C_2 are varied. A cache hit is desirable only if all three tests result in cache hits. Algorithm 3 presents this method. For the example above, I_a is $I3_{S0}^1$, I_{b1} is $I2_{S0}^3$ and I_{b2} is $I2_{S0}^4$. The controlling plaintexts C_1 and C_2 are P_{12} and P_{14} respectively.
- *Probabilistic Method:* In the example described above, assume that P_8 is ex-ored by a non-zero value d, which is less than the size of the cache line. The small displacement of P_8 ensures that if $I3_{S0}^1$ was the desirable collision, then the collision would remain with probability 1. However the small displacement of P_8 would become a random change after the s-box, affecting all four outputs of the first round $F1$. Thus the accesses to $I2_{S0}^3$ and $I2_{S0}^4$ would also change randomly. Therefore, if the collision at $I3_{S0}^1$ is undesirable, the collision would remain with probability $(1 - (1 - \frac{1}{2^b})^2)$. Sufficient confidence in the correctness of the collision is obtained if this test is repeated $1/(1 - \frac{1}{2^b})^2$ times as shown in Algorithm 4.

The elimination method, which gives a deterministic result, is the suited technique. However application of this method is not always feasible (as in the case of determining $RK3 \oplus WK1$), in which case the probabilistic method will need to be applied. Algorithm 5 gives the procedure for determining the byte $(RK2 \oplus WK0)$.

Algorithm 4. Probabilistic Method

Input: I_a: access where a cache hit is required, P_a: plaintext byte used to force cache hits in I_a

Output: Cache hit at I_a is desirable or not

1 **begin**
2   **for** $i \in \{0, \cdots, \frac{2^{2b}}{(2^b-1)^2}\}$ **do**
3         $P_a \leftarrow P_a \oplus i$
4   **if** I_a *not in collision* **then**
5   **return** "Undesirable Collision"
6   **end**
7   **end**
8   **return** "Desirable Collision";
9 **end**

Algorithm 5. Finding Candidate Keys for $RK2 \oplus WK0$

Output: $S_{(RK0 \oplus WK0)_i}$: Candidate Key Set for $(RK0 \oplus WK0)_i$, where $i \in \{0, 1, 2, 3\}$

1 **begin**
2   Put cipher in Two round colliding state
3   *Set up collisions:* Find P_8, P_9, P_{10} causing collisions in $I3^1_{S0}, I3^1_{S1}, I3^2_{S0}$ respectively (using elimination method)
4     $S(RK0 \oplus WK0)_0 \leftarrow find(0, P_4 \cdots P_7, P_8 \cdots P_{11})$
5     $S(RK0 \oplus WK0)_1 \leftarrow find(1, P_4 \cdots P_7, P_8 \cdots P_{11})$
6     $S(RK0 \oplus WK0)_2 \leftarrow find(2, P_4 \cdots P_7, P_8 \cdots P_{11})$
7     $S(RK0 \oplus WK0)_3 \leftarrow find(3, P_4 \cdots P_7, P_8 \cdots P_{11})$
8 **end**

To determine $RK3 \oplus WK1$, the second and third round $F1$ functions are considered with $P_{12} \cdots P_{15}$ as the difference introducing inputs, while $P_0 \cdots P_3$ the restoring inputs. Starting from the 2-round colliding state, ambiguities in collisions in the third round $F1$ are due to collisions with the tables of $F0$ of the third round. Using the elimination method would imply P_8 to P_{15} be used as controlling inputs. But changing P_8 would also alter the cache hit in the third round $F1$. Therefore, the elimination method is not feasible and probabilistic method needs to be applied.

Analysis: For the elimination method, in addition to the $\rho(2^b - 1)$ encryptions that are required to find a collision, 3ρ additional encryptions are necessary to eliminate each of the undesirable collisions. Therefore in all $\rho(2^b + 5)$ encryptions are required to correctly identify a collision. In Algorithm 2, setting up collisions will therefore require $3\rho(2^b + 5)$ and each execution of *find* would need an additional $3\rho(2^b + 5)$. In total, there would be $15\rho(2^b + 5)$ encryptions required to find $RK0 \oplus WK0$.

For $RK3 \oplus WK1$, the probabilistic method is used. To determine the correct collision with significant probability, $\rho\frac{2^{b+1}}{(2^b-1)^2}$ encryptions are required in addition to the $\rho(2^b - 1)$ encryptions. Thus obtaining the whole of $RK3 \oplus WK1$ can be done in $16\rho((2^b - 1) + \frac{2^{b+1}}{(2^b-1)^2})$.

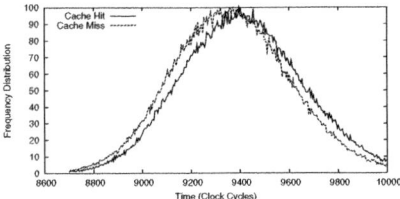

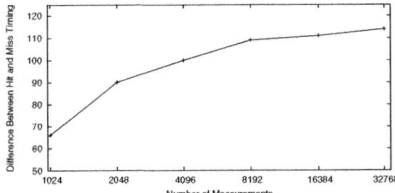

(a) Timing Distributions for $I1^2_{S0}$ for a Cache Miss and Cache Hit

(b) Difference between Expected Values for a Cache Miss and Cache Hit Distributions

Fig. 4. Analysis of Timing Side Channel on Intel Xeon (E5606)

4.3 Determining $RK4$ and $RK5$

To compute the candidate keys for $RK4$, we use the 3rd and 4th round $F0$ functions with $P_8 \cdots P_{11}$ as the difference introducing inputs and $P_{12} \cdots P_{15}$ the restoring inputs. The initial state is a 2-round colliding state and all tables in $F0$ of the third round are also in collision. The undesirable collisions occurring with $F1$ of round three can be detected by the probabilistic method. The number of encryptions required for $RK4$ is be found to be $16\rho((2^b-1)+\frac{2^{2b+1}}{(2^b-1)^2})$. Candidate keys for $RK5$ is similarly determined from the 3rd and 4th round $F1$ functions with $P_0 \cdots P_3$ as the difference introducing inputs and $P_4 \cdots P_7$ the restoring inputs. In this case however, more undesirable collisions can occur due to $F0$ in rounds three and four. Due to this, the number of encryptions required is increased to $16\rho((2^b - 1) + \frac{2^{4b+2}}{(2^b-1)^4})$.

5 Distinguishing between a Cache Hit and Miss

The attack on CLEFIA presented in the previous section relies on distinguishing between a cache hit and a miss. The number of encryptions (ρ) required to make this distinction depends on the form of side-channel used. In this section we determine ρ for two side-channels namely power consumption and timing.

Cache Access Patterns from Power Consumption: In [12], the attack was mounted on an embedded PowerPC in the Xilinx XC2VP30 FPGA present on the SASEBO side channel attack evaluation board [13]. Cache access patterns were identified by correlating the power trace obtained with templates. For a match, a correlation value of 0.997 was obtained, compared to 0.8 for a mismatched power trace. Thus a single trace ($\rho = 1$) is sufficient to obtain cache access patterns.

Cache Access Patterns from Timing: Obtaining side channel distributions from timing requires several encryptions to be done. The distinguisher is a shift in

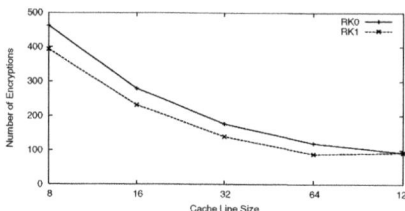
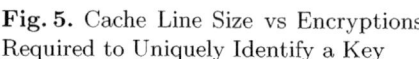
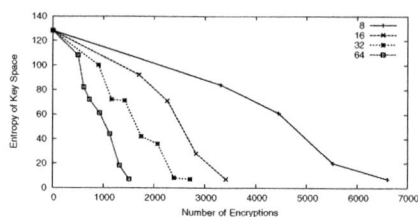

Fig. 5. Cache Line Size vs Encryptions Required to Uniquely Identify a Key

Fig. 6. Encryptions vs Key Space Remaining for Various Cache Lines

the timing distribution for a cache hit compared to the cache miss distribution. For example, the timing distributions for a hit and miss at $I1_{S0}^{2}$ is shown in Figure 4(a). The distributions are Gaussian and a noticeable difference between the distributions is observable. On an average, when there is a cache hit in $I1_{S0}^{2}$, CLEFIA takes around 100 clock cycles longer than when a miss is present in $I1_{S0}^{2}$, assuming that all encryptions start with a clean cache. The distance between the cache hit and miss distributions represents our confidence in the measurement. A small distance implies low confidence about the correctness of the measurements (and vice-versa). Figure 4(b) shows the distance between a cache hit and a cache miss distribution as the number of measurements increase. We find that the difference is around 60 clock cycles when 2^{10} encryptions are made. This difference increases as the number of measurements increase until it saturates at around 114 clock cycles. A ρ of 2^{10} ensures with certainly the correct detection of the cache hit.

6   Effect of Cache Line Size on the Number of Encryptions

In the cipher's implementation, if a table occupies 2^{b} cache lines, then $2^{b} - 1$ encryptions are required to identify a collision. Further, if 2^{δ} elements share a common cache line then there still remains an uncertainty of δ bits.

With increase in the size of the cache line, the table (which we assume to have the same size) occupies fewer lines in memory (b reduces), consequently fewer number of encryptions are required to identify a collision. On the other hand, an increase in the cache line size would increase δ causing an increase in the size of the candidate key set (as seen in Table 4). The attack reduces this uncertainty by repeating the experiment and taking intersections between the candidate key sets obtained. Thus we see that the cache line size has opposite effects on the number of encryptions required to find a collision and the number of encryptions required to reduce the candidate key set. Figure 5 shows the effect of cache line size on the number of encryptions required on average to identify $RK0$ and $RK1$ uniquely. This is assuming a power side-channel adversary where $\rho = 1$. As seen from the graph large cache lines are more prone to being attacked as they require fewer encryptions to be monitored. This can also be seen in Figure 6, which estimates the entropy of the remaining key space with increasing number of encryptions for the entire attack (assuming $\rho = 1$). This being said, if the cache line was large enough to fit the entire table then every access to the table (except the first) would result in a cache hit, and therefore the attack would fail.

7   Countermeasures Suited for Large Cache Lines

Although there have been several countermeasures proposed against cache attacks (such as in [6,11,16]), not all of them guarantee protection. Countermeasures that do guarantee protection either disable the cache, do not use tables, or implement the cipher to fit tables within a single cache line. Disabling the cache is known to slow down an execution 100 times [16] and affects the entire system, therefore is not an option. The other countermeasures too have significant performance overheads. In this section we compare these overheads and also suggest splitting of tables as a countermeasure for large cache lines.

Using Single Cache Lines: In [12], techniques were presented to implement CLEFIA's s-boxes using small table of 32 bytes. This table is small enough to fit into a single cache line, thus except for the first access to the table, all other memory accesses results in cache hits. However, the small tables implied substantial number of logical operations were required in addition to the table lookups; adding to the overhead. Naturally, when a cache line of larger dimensions is available, one would attempt to increase the number of tables to fit in the entire cache line. To the countermeasure in [12], we have added more tables, to adapt it for a 64 byte cache line (See Appendix-D).

Split Tables: Consider a table T having t elements with each element occupying u bits. Let the size of the cache line be c bytes. Then the table occupies $s = \frac{tu}{8c}$ lines in cache. Each of the t elements in the table is split into s parts and each part is put into a different table. There are therefore s tables, each of c bytes. To access an s-box element, all tables require to be read in order to obtain the element, therefore memory accesses are no longer dependent on the key protecting the implementation against cache attacks.

We found that on standard Intel platforms, the number of cache misses has a dominating influence on the execution time of a block cipher. In fact on an Intel Core 2 Duo platform, a cache miss has an average overhead of 10 clock cycles compared to a cache hit. If the size of the cache line (c) is large with respect to the table size (that is, the table occupies few cache lines), then with high probability the entire table is loaded into memory during the execution.

In the countermeasure proposed, the cumulative sum of the sizes of the s tables equals that of T. If we assume a large cache line exists, then the countermeasure has the same number of cache misses as the non-protected implementation. The only overhead of the countermeasure is in the non cache miss executions. This is also kept minimum if s is small, as is the case.

Table 7 shows the overheads of various countermeasures applied on CLEFIA implementation[2] for a cache memory with 64 byte cache line.

---

[2] http://www.sony.net/Products/cryptography/clefia/download/data/clefia_ref.c

Table 2. Execution Time for Different Implementations of CLEFIA on an Intel Core 2 Duo Processor

Implementation	Clock Cycles	Overhead
Reference Code (RC)	13452	-
RC without tables	47032	3.5
RC with countermeasure from [12] for 32 byte cache lines	24446	1.81
RC with countermeasure from [12] for 64 byte cache line	22356	1.66
Split tables	16235	1.2

8   Conclusion

In this paper we presented an attack on CLEFIA, which is not restricted to systems having small cache lines. The work shows that while existing attacks fail on systems with large cache lines, better exploitation of the differential properties of the non-linear layers and the diffusion properties of the linear layer inside the rounds, still lead to attacks. The work analyzes the effect of cache line size on the attack complexity, showing that systems with large cache lines are more prone to being attacked. This is supported by theoretical analysis of the time and space complexity of the attack. Further, techniques are suggested for CLEFIA to reduce these difficulties. The techniques presented in the paper can easily be adopted to break other ciphers such as CAMELLIA that have Feistel structures. Finally countermeasures for the attack are analyzed and their overheads compared on a standard computing platform.

References

1. Acıiçmez, O., Koç, Ç.K.: Trace-Driven Cache Attacks on AES (Short Paper). In: Ning, P., Qing, S., Li, N. (eds.) ICICS 2006. LNCS, vol. 4307, pp. 112–121. Springer, Heidelberg (2006)
2. Acıiçmez, O., Schindler, W., Koç, Ç.K.: Cache Based Remote Timing Attack on the AES. In: Abe, M. (ed.) CT-RSA 2007. LNCS, vol. 4377, pp. 271–286. Springer, Heidelberg (2006)
3. Aoki, K., Ichikawa, T., Kanda, M., Matsui, M., Moriai, S., Nakajima, J., Tokita, T.: *Camellia*: A 128-Bit Block Cipher Suitable for Multiple Platforms - Design and Analysis. In: Stinson, D.R., Tavares, S. (eds.) SAC 2000. LNCS, vol. 2012, pp. 39–56. Springer, Heidelberg (2001)
4. Bernstein, D.J.: Cache-timing Attacks on AES. Tech. rep. (2005)
5. Bertoni, G., Zaccaria, V., Breveglieri, L., Monchiero, M., Palermo, G.: AES Power Attack Based on Induced Cache Miss and Countermeasure. In: ITCC (1), pp. 586–591. IEEE Computer Society (2005)
6. Brickell, E., Graunke, G., Neve, M., Seifert, J.P.: Software Mitigations to Hedge AES Against Cache-based Software Side Channel Vulnerabilities. Cryptology ePrint Archive, Report 2006/052 (2006), http://eprint.iacr.org/
7. Canteaut, A., Lauradoux, C., Seznec, A.: Understanding Cache Attacks. Research Report RR-5881, INRIA (2006), http://hal.inria.fr/inria-00071387/en/

8. Crosby, S.A., Wallach, D.S., Riedi, R.H.: Opportunities and Limits of Remote Timing Attacks. ACM Trans. Inf. Syst. Secur. 12(3) (2009)
9. Kelsey, J., Schneier, B., Wagner, D., Hall, C.: Side Channel Cryptanalysis of Product Ciphers. J. Comput. Secur. 8(2,3), 141–158 (2000)
10. Osvik, D.A., Shamir, A., Tromer, E.: Cache Attacks and Countermeasures: The Case of AES. In: Pointcheval, D. (ed.) CT-RSA 2006. LNCS, vol. 3860, pp. 1–20. Springer, Heidelberg (2006)
11. Page, D.: Theoretical Use of Cache Memory as a Cryptanalytic Side-Channel (2002)
12. Rebeiro, C., Mukhopadhyay, D.: Cryptanalysis of CLEFIA Using Differential Methods with Cache Trace Patterns. In: Kiayias, A. (ed.) CT-RSA 2011. LNCS, vol. 6558, pp. 89–103. Springer, Heidelberg (2011)
13. Research Center for Information Security National Institute of Advanced Industrial Science and Technology: Side-channel Attack Standard Evaluation Board Specification, Version 1.0 (2007)
14. Shirai, T., Shibutani, K., Akishita, T., Moriai, S., Iwata, T.: The 128-Bit Blockcipher CLEFIA (Extended Abstract). In: Biryukov, A. (ed.) FSE 2007. LNCS, vol. 4593, pp. 181–195. Springer, Heidelberg (2007)
15. Sony Corporation: The 128-bit Blockcipher CLEFIA: Algorithm Specification (2007)
16. Tromer, E., Osvik, D.A., Shamir, A.: Efficient Cache Attacks on AES, and Countermeasures. Journal of Cryptology 23(2), 37–71 (2010)
17. Tsunoo, Y., Saito, T., Suzaki, T., Shigeri, M., Miyauchi, H.: Cryptanalysis of DES Implemented on Computers with Cache. In: Walter, C.D., Koç, Ç.K., Paar, C. (eds.) CHES 2003. LNCS, vol. 2779, pp. 62–76. Springer, Heidelberg (2003)
18. Tsunoo, Y., Tsujihara, E., Minematsu, K., Miyauchi, H.: Cryptanalysis of Block Ciphers Implemented on Computers with Cache. In: International Symposium on Information Theory and Its Applications, pp. 803–806 (2002)
19. Zhao, X., Wang, T.: Improved Cache Trace Attack on AES and CLEFIA by Considering Cache Miss and S-box Misalignment. Cryptology ePrint Archive, Report 2010/056 (2010), http://eprint.iacr.org/
20. Zheng, Y., Matsumoto, T., Imai, H.: On the Construction of Block Ciphers Provably Secure and Not Relying on Any Unproved Hypotheses. In: Brassard, G. (ed.) CRYPTO 1989. LNCS, vol. 435, pp. 461–480. Springer, Heidelberg (1990)

Appendix A: Number of Encryptions for Key Recovery

Table 3 summarizes the number of encryptions required to perform one repetition for a key ($q = 1$). In the table b is the number of bits revealed from the cache access profiles and $l = 2^b - 1$.

Appendix B: Plotting Figure 5

After q repetitions, the expected size of $RK0_i$ for $0 \leq i \leq 3$ is $|S^q_{RK0_i}|/256^{q-1}$. Therefore for the entire $RK0$ the set of candidate keys is

$$\frac{|S^q_{RK0_0}|}{256^{q-1}} \times \frac{|S^q_{RK0_1}|}{256^{q-1}} \times \frac{|S^q_{RK0_2}|}{256^{q-1}} \times \frac{|S^q_{RK0_3}|}{256^{q-1}}$$

To obtain a unique $RK0$, q should be found so that the above expression reduces to 1. In a similar way the trend for $RK1$ is computed.

Table 3. Number of Encryption required for $q = 1$

Key	Encryptions	Key	Encryptions
$RK0$	$21\rho l$	$RK1$	$16\rho l$
$RK2 \oplus WK0$	$15\rho(2^b + 5)$	$RK2 \oplus WK1$	$16\rho\left(l + \frac{2^{2b+1}}{l^2}\right)$
$RK4$	$16\rho\left(l + \frac{2^{2b+1}}{l^2}\right)$	$RK5$	$16\rho\left(l + \frac{2^{4b+2}}{l^4}\right)$

Appendix C: Plotting Figure 6

Figure 6 shows the estimated number of keys remaining versus the number of encryptions required. This appendix presents how the graph is obtained.

Estimating the Key Space Reduction: Table 4 gives the average number of candidate keys per byte for the F functions when $q = 1$. For example, when $b = 2$, each output byte for $F0$ in the first round has $(40.9 \times 32.2)^2$ number of options. However each output byte for $F0$ can take only 256 values. Therefore in this case $n_1^4 = 256$ in Equation 7 resulting in no key space reduction for the following round. The key space can be reduced by increasing q, which would then reduce the value of n_1^4. In a similar way, the number of outputs for $F0$ in the second round is used to compute the possible keys for the third round. We therefore obtain the key sets S_{RK0}^q, $S_{RK2\oplus WK0}^q$, and S_{RK4}^q. Similarly $F1$ is analyzed to obtain S_{RK1}^q, $S_{RK3\oplus WK1}^q$, and S_{RK5}^q.

Every possible combination formed by the keys in S_{RK0}^q, S_{RK1}^q, S_{RK4}^q, and S_{RK5}^q produces a candidate CLEFIA key, therefore the product of these sets have to be considered.

Encryptions per Round : This appendix gives details about the break up of the number of encryptions per round and the corresponding number of keys remaining.

Appendix D: Adapting the Countermeasure in [12] for 64 byte Cache Lines

The sboxes in CLEFIA are designed differently. $S0$ is composed of four sboxes $SS0$, $SS1$, $SS2$, and $SS3$; each of 16 bytes. The output of $S0$ is given by:

$$\beta_l = SS2[SS0[\alpha_l] \oplus 2 \cdot SS1[\alpha_h]]$$
$$\beta_h = SS3[SS1[\alpha_h] \oplus 2 \cdot SS0[\alpha_l]] \tag{8}$$

,where $\beta = (\beta_h|\beta_l)$, $\alpha = (\alpha_h|\alpha_l)$, and $\beta = S0[\alpha]$. In [12], a pair of the 16 byte sboxes share a 16 byte table. So all 4 sboxes require 32 bytes. For 64 byte cache lines, the sharing is not required and each sbox can exclusively use 16 bytes of memory. This produces a small speedup.

For the sbox $S1$, the output corresponding to the input byte α is given by $g((f(\alpha))^{-1})$, where g and f are affine transforms and the inverse is found in the

Table 4. Expected Number of Candidate Keys with Varying Cache Line sizes

Cache Line Size	Number of Repetitions			Encryptions Required	Key Space Remaining
	Round 1	Round 2	Round 3		
8	1	1	1	3303	2^{84}
	2	1	1	4450	2^{61}
	2	2	1	5518	2^{20}
	2	2	2	6606	2^{7}
16	1	1	1	1703	2^{92}
	2	1	1	2258	2^{71}
	2	2	1	2830	2^{28}
	2	2	2	3406	2^{7}
32	1	1	1	903	2^{100}
	2	1	1	1162	2^{72}
	2	2	1	1486	2^{53}
	2	2	2	1806	2^{38}
	3	2	2	2065	2^{14}
	3	3	2	2389	2^{8}
	3	3	3	2709	2^{7}
64	1	1	1	503	2^{108}
	2	1	1	614	2^{82}
	3	1	1	725	2^{72}
	3	2	1	925	2^{61}
	3	3	1	1125	2^{44}
	3	3	2	1317	2^{18}
	3	3	3	1509	2^{7}

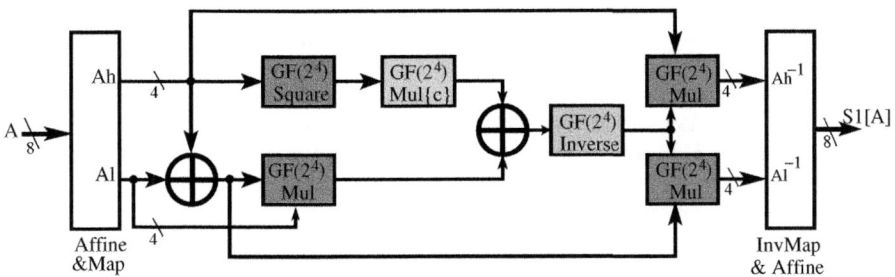

Fig. 7. Composite Field Implementation of $S1$ for 64 byte Tables

field $GF(2^8)$. To fit this into a 32 byte cache line, [12] used composite fields, and had some of the composite field operations such as inversion and multiplication by a constant stored in tables. However due to the 32 byte constraint, many operations were still performed by logical equations, resulting in significant overheads. Since in our case, the cache line is larger so more operations can be done using tables. Figure 7 shows our modified composite field implementation for $S1$. The lighter shaded boxes are tables which were also implemented in [12], while the darker boxes are new tables for 64 byte cache lines. The squaring is done using a table of 16 bytes and part of the $GF(2^4)$ multiplication is done using a single 32 byte table. This table stores the product of a 4 bit element and a 2 bit element. Thus a $GF(2^4)$ multiplication is done with only two table accesses.

Partial Key Exposure: Generalized Framework
to Attack RSA

Santanu Sarkar

Indian Statistical Institute, 203 B T Road, Kolkata 700 108, India
sarkar.santanu.bir@gmail.com

Abstract. In the domain of modern public key cryptography, RSA is the most popular system in use. Efficient factorization of the RSA modulus N, constituted as a product of two primes p, q of 'large' bitsize, is a challenging problem in RSA cryptanalysis. The solution to this factorization is aided if the attacker gains partial knowledge about the decryption exponent of RSA. This line of attack is called the Partial Key Exposure attack, and there exists an extensive literature in this direction.

In this paper, we study partial key exposure attacks on RSA where the number of unexposed blocks in the decryption exponent is more than one. The existing works have considered only one unexposed block and thus our work provides a generalization of the existing attacks. We propose lattice based approaches to factorize the RSA modulus $N = pq$ (for large primes p, q) when the number of unexposed blocks is $n \geq 1$. We also analyze the ISO/IEC 9796-2 standard signature scheme (based on CRT-RSA) with partially known messages.

Keywords: Factorization, ISO/IEC 9796-2 Signature, Lattice, Partial Key Exposure, RSA.

1 Introduction

RSA cryptosystem, publicly proposed in 1978 and named after its inventors Ron Rivest, Adi Shamir and Len Adleman, is the most popular Public Key Cryptosystem till date. One can describe the RSA scheme in a nutshell as follows [15].

Cryptosystem 1 (RSA). *Let us define $N = pq$ where p and q are primes. By definition of the Euler totient function, $\phi(N) = (p-1)(q-1)$.*

- KEYGEN: *Choose integer e co-prime to $\phi(N)$ and find $d = e^{-1} \bmod \phi(N)$.*
- KEYDIST: *Publish public key $\langle N, e \rangle$ and keep private key $\langle N, d \rangle$ secret.*
- ENCRYPT: *For message $M \in \mathbb{Z}_N$, ciphertext $C = M^e \bmod N$.*
- DECRYPT: *For ciphertext C, message $M = C^d \bmod N$.*

For encryption and decryption in RSA cryptosystem, one needs modular exponentiation. To reduce the cost of encryption, one can take a small e. In such a case, d becomes quite large, i.e., of the order of N, and the decryption process

D.J. Bernstein and S. Chatterjee (Eds.): INDOCRYPT 2011, LNCS 7107, pp. 76–92, 2011.
© Springer-Verlag Berlin Heidelberg 2011

will be much less efficient than the encryption. Consider that one likes to make the decryption process faster. Then the secret decryption exponent d has to be made small. However, Wiener [17] showed that when $d < \frac{1}{3}N^{0.25}$, one can factor N efficiently, thus making RSA insecure. This result has been improved by Boneh and Durfee [3] to provide the upper bound $d < N^{0.292}$.

Attacks Based on Partial Knowledge of d. Kocher [13] proposed a new attack on RSA to obtain the private exponent d. He showed that an attacker can get a few bits of d by timing characteristic of an RSA implementing device. In [2], it has been studied how many bits of d need to be known to mount an attack on RSA. The constraint in the work of [2] was the upper bound on e which is $\sqrt{N}$. The study attracted interest and the idea of [2] has been improved in [1] where the bound of e was increased upto $N^{0.725}$. Then the work of [7] improved the result for full size public exponent e. These attacks require knowledge of contiguous blocks of bits of the RSA secret keys. The results of Ernst et al. [7] are as follows:

Theorem 1. *Let $N = pq$ be an RSA modules with p, q are of same bit size. Let e be of full bit size and $d \leq N^{\delta}$. Then given $(\delta - \gamma) \log N$ Most Significant Bits of d, one can factor N in polynomial time if*

- $\gamma < \frac{5}{6} - \frac{1}{3}\sqrt{1 + 6\delta}$ *or*
- $\gamma < \frac{\beta}{3} + \frac{1}{2} - \frac{1}{3}\sqrt{4\beta^2 + 6\beta}$ *where $\beta = \max\{\gamma, \delta - \frac{1}{2}\}$*

Attacks on a CRT-RSA signature scheme. CRT-RSA is used to devise one of the most popular digital signature schemes. To sign a message m, one first needs to calculate

$$s_p = m^{d_p} \bmod p \quad \text{and} \quad s_q = m^{d_q} \bmod q.$$

Now the signature s for modulus $N = pq$ can be computed using CRT with s_p and s_q. Boneh et al. [4] showed that CRT-RSA implementations are vulnerable due to fault attacks. Suppose the attacker is able to induce a fault while s_q is being calculated, such that $s_q \neq m^{d_q} \bmod q$. Then using the faulty signature s, the attacker can factor N as $\gcd(s^e - m, N) = p$. The attack of Boneh et al. also works for any deterministic RSA encoding or any probabilistic signature scheme where some randomness is used to generate the signature and is also sent as a part of the signature. However, if some part of the message is unknown, the attack of Boneh et al. does not work any more. Recently Coron et al. [5,6] showed how to use the same idea in such a situation, and their attack was illustrated against ISO/IEC 9796-2 standards [11]. In ISO/IEC 9796-2, the encoded message is of the form

$$\mu(m) = 6A_{16} \parallel m[1] \parallel H(m) \parallel BC_{16},$$

where $m = m[1] \parallel m[2]$ is split into two parts. Note that $H(m)$, the hash value of m is unknown to the attacker. Coron et al. proved that when unknown part of $m[1]$ is small, one can factor N efficiently, thus making the scheme vulnerable.

1.1   Our Contribution

Motivated by the work of [8], we consider the situation where a few contiguous blocks of the RSA secret exponent d are unknown, and the encryption exponent e is of full bit size. In such a situation, the reconstruction idea of [9] will not work as e is of full bit size. We propose a lattice based approach to handle the situation. In [7], authors analyzed the case when number n of unknown contiguous blocks is one. We prove that when the number n of unknown blocks increases, one needs more bits to be known for polynomial time factorization of N.

We prove the apparently surprising result that even for an arbitrary n number of blocks, only $(\delta + \frac{1}{2}\sqrt{1 + 4\delta} - 1) \log N$ bits of $d(= N^\delta)$ are sufficient to reconstruct the complete d in the case when $d < N^{0.75}$. Note that as long as $\delta < 0.275$, $\delta + \frac{1}{2}\sqrt{1 + 4\delta} - 1$ is less than zero. Although the runtime of our reconstruction algorithm is polynomial in $\log N$, it is exponential in the number of unknown blocks n. So, in case $n = O(\text{poly}(\log \log N))$, our algorithm is a $\text{poly}(\log N)$ time algorithm.

Next, we consider the case when a large block MSBs of d is exposed. Using these known MSBs of d, we propose another lattice based approach. We prove that if attacker knows d_0 such that $|d - d_0| < N^{\delta - \frac{1}{2}}$, the knowledge of

$$\sqrt{2\left(\delta - \frac{1}{2}\right)^2 + \left(\delta - \frac{1}{2}\right)} \cdot \log N$$

many bits of d is sufficient for the factorization of N in time polynomial in $\log N$ but exponential in n.

Finally, we turn our attention to the CRT-RSA signature scheme, and consider the case where two faults occur for the two different primes individually. That is attacker inject a fault modulo p for one signature, and a fault modulo q for another signature. Let the sizes of the unknown blocks for each of the messages for two faulty signatures be $\delta_1 \log N, \cdots, \delta_n \log N$ and the size of hash value be $\gamma \log N$. We prove when

$$\delta_1 + \cdots + \delta_n + \gamma < \frac{n + 2}{2(2n + 3)},$$

one can factor N in time polynomial in $\log N$ but exponential in n. For the case $n = 1$, i.e., if the size of the unknown block of each of the messages is $\delta \log N$ and the size of the hash value is $\gamma \log N$, the approach of Coron et al. [5] works if $\delta + \gamma < 0.167$. Putting $n = 1$ in our bound, we obtain the upper bound as 0.30. The idea of using Coppersmith's method to improve on the bound 0.167, was already suggested in [5].

Organization. In summary, our work in this paper is stated and organized as follows.

- In Section 2, we study partial key exposure attacks when more than one block of d is unknown. In Section 3, we consider the case when a large block of MSBs of d is known.

– In Section 4, we present our attack on ISO/IEC 9796-2 standard signature scheme with partially known messages.

Our approaches are based on the technique of [12], which itself follows from the idea of [10]. For all the results that we obtain in this paper related to lattice based techniques, we need the following assumption. For brevity, we do not refer to this assumption in each of the results, though it should be considered implied in each of them.

Assumption 1. Let lattice reduction be executed using the idea of LLL [14] algorithm leading to polynomials in u variables. Consider that LLL outputs the polynomials $f_1, f_2, \ldots, f_i$, $i \geq u$, that have a common root. Then one can efficiently compute this root from $f_1, f_2, \ldots, f_i$ using techniques like calculation of resultants of these polynomials or finding a Gröbner basis.

We have implemented all the programs in SAGE 3.1.1 over Linux Ubuntu 8.04 on a laptop with Dual CORE Intel(R) Pentium(R) D CPU 1.83 GHz, 2 GB RAM and 2 MB Cache. In all cases except in Section 4 we take a 1024-bit RSA modulus N. In Section 4, we consider both 1024-bit and 2048-bit RSA moduli.

2   General Attacks Based on Partial Knowledge of d

In this section we consider the scenario when certain blocks of the decryption exponent d are known (exposed) and naturally the other parts are unknown (unexposed). Figure 1 provides a pictorial view of the attack model.

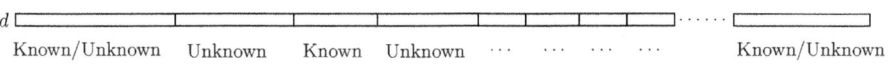

Known/Unknown   Unknown   Known   Unknown $\cdots$ $\cdots$ $\cdots$ $\cdots$ Known/Unknown

Fig. 1. Exposed and unexposed blocks of d

Before proceeding further, let us state the following technical result [16] that we will be using later.

Proposition 1. *For any fixed positive integer $r \geq 1$, and a large integer m,*

$$\sum_{t=1}^{m} t^r = \frac{m^{r+1}}{r+1} + o(m^{r+1}).$$

Theorem 2. *Let e be $O(N)$ and $d \leq N^\delta$. Suppose the bits of d are exposed except n many blocks, each of size $\gamma_i \log N$ bits for $1 \leq i \leq n$. Then one can factor N in polynomial in $\log N$ but exponential in n time if*

$$\sum_{i=1}^{n} \gamma_i < 1 - \frac{1}{2(n+2)} - \frac{n+1}{2(n+2)} \sqrt{4\delta + 1 + \frac{4\delta}{n+1}}.$$

Proof. Since d is unknown for n many blocks, one can write $d = a_0 + a_1 y_1 + \ldots + a_n y_n$, where $y_1, y_2, \ldots, y_n$ are unknown. Now from $ed = 1 + k(N + 1 - p - q)$, so we have $ea_0 + ea_1 y_1 + \ldots + ea_n y_n - 1 - k(N + 1 - p - q) = 0$. Hence, we are interested to find the root of the polynomial

$$f(x_1, \ldots, x_{n+1}, x_{n+2}) = ea_0 + ea_1 x_1 + \ldots + ea_n x_n - 1 + N x_{n+1} + x_{n+1} x_{n+2}.$$

Clearly, $f(y_1, \ldots, y_n, -k, 1 - p - q) = 0$. Let, $X_i = N^{\gamma_i}$ for $1 \le i \le n$, $X_{n+1} = N^\delta$, $X_{n+2} = N^{0.5}$. Clearly, $X_1, \ldots, X_{n+2}$ is the upper bound of absolute value of $y_1, \ldots, y_n, -k, 1 - p - q$, neglecting small constants. Using the extended strategy of [12, Section 2.2], we define

$$S = \bigcup_{0 \le j \le t} \{ x_1^{i_1} x_2^{i_2} \ldots x_{n+2}^{i_{n+2}+j} : x_1^{i_1} x_2^{i_2} \ldots x_{n+2}^{i_{n+2}} \text{ is a monomial of } f^m \}$$

$$M = \{ \text{monomials of } x_1^{i_1} x_2^{i_2} \ldots x_{n+2}^{i_{n+2}} f : x_1^{i_1} x_2^{i_2} \ldots x_{n+2}^{i_{n+2}} \in S \}$$

where m, t are non-negative integers. To use Coppersmith's method, it suffices to find at least $n + 1$ more polynomials $f_1, f_2, \ldots f_{n+1}$ that share the same root $(y_1, \ldots, y_n, -k, 1 - p - q)$ over the integers.

Considering a fixed n, we know from [12] that these polynomials can be found by LLL [14] algorithm in poly$(\log N)$ time if

$$X_1^{s_1} X_2^{s_2} \cdots X_{n+2}^{s_{n+2}} < W^s$$

for $s_j = \sum_{x_1^{i_1} \ldots x_{n+2}^{i_{n+2}} \in M \backslash S} i_j$ with $j = 1, \ldots, n + 2$, $s = |S|$, and

$$W = \| f(x_1 X_1, \ldots x_{n+2} X_{n+2}) \|_\infty \ge N X_{n+1} = N^{1+\delta}.$$

The calculation's of $s, s_1, \ldots, s_{n+2}$ are quite tedious and those are presented Appendix A. From the structure of the polynomial f, it is clear that $s_1, \ldots, s_n$ are equal. We get,

$$s \approx \frac{m^{n+2}}{(n+2)!} + \frac{t m^{n+1}}{(n+1)!},$$

$$s_1 = \cdots = s_n \approx \frac{m^{n+2}}{(n+2)!} + \frac{t m^{n+1}}{(n+1)!},$$

$$s_{n+1} \approx \frac{2 m^{n+2}}{(n+2)!} + t \frac{m^{n+1}}{(n+1)!},$$

$$s_{n+2} \approx \frac{m^{n+2}}{(n+2)!} + \frac{t m^{n+1}}{(n+1)!} + \frac{t^2 m^n}{2n!}$$

Consider $t = \tau m$, where $\tau \ge 0$ is a real number. Now putting values of $X_1, \ldots, X_{n+2}$ and $s_1, \ldots, s_{n+2}, s$, and the lower bound of W in

$$X_1^{s_1} X_2^{s_2} \ldots X_{n+2}^{s_{n+2}} < W^s,$$

we get

$$\left(\frac{1}{4}n^2 + \frac{3}{4}n + \frac{1}{2}\right)\tau^2 + \left(n\sum_{i=1}^{n}\gamma_i + 2\sum_{i=1}^{n}\gamma_i - \frac{1}{2}n - 1\right)\tau + \left(\sum_{i=1}^{n}\gamma_i + \delta - \frac{1}{2}\right) < 0.$$

$$(1)$$

The optimal value of τ is $\frac{1-2\sum_{i=1}^{n}\gamma_i}{n+1}$. Putting this optimal value of τ in Equation (1), we get

$$\sum_{i=1}^{n}\gamma_i < \frac{(2n^2 + 3n) - \sqrt{4n^4\delta + n^4 + 12n^3\delta + 2n^3 + 8n^2\delta + n^2}}{2(n^2 + 2n)}.$$

From which we have the required condition.

Using the strategy of [12, Section 2.2], one can construct a lattice L from S, M. The bit size of the entries of L is poly($\log N$), and

$$\dim(L) = |M| = \frac{(m+1)^{n+2}}{(n+2)!} + \frac{t(m+1)^{n+1}}{(n+1)!} + o((m+1)^{n+2}).$$

The running time of our algorithm is dominated by the LLL algorithm run on L, which takes time polynomial in the dimension of the lattice and in the bitsize of the entries. Since the lattice dimension in our case is exponential in n, so the total running time for this method is polynomial in $\log N$ but exponential in n. □

Note that when $n = 1$ i.e, number of unknown block is one, then the situation is analyzed in [7].

Now putting $n = 1$ in our Theorem 2, we get $\gamma_1 < \frac{5}{6} - \frac{1}{3}\sqrt{1 + 6\delta}$, same bound as in the first result of the Theorem 1. However, since we assume bits of d are known from any position, we can not get any partial information of k. This is the reason we can not obtain other result of Theorem 1. Also note that total number of unknown bits of d is $\sum_{i=1}^{n}\gamma_i \log N$. Now from Theorem 2

$$\sum_{i=1}^{n}\gamma_i < 1 - \frac{1}{2(n+2)} - \frac{n+1}{2(n+2)}\sqrt{4\delta + 1 + \frac{4\delta}{n+1}}.$$

Also when n increases, then upper bound of $\sum_{i=1}^{n}\gamma_i$ decreases, i.e, one needs much number of bits $\left(\delta - \sum_{i=1}^{n}\gamma_i\right)\log N$ to be known. In Table 1, we present few numerical values of the upper bound of unknown bits.

Now we have,

$$\lim_{n\to\infty} 1 - \frac{1}{2(n+2)} - \frac{n+1}{2(n+2)}\sqrt{4\delta + 1 + \frac{4\delta}{n+1}} = 1 - \frac{\sqrt{1 + 4\delta}}{2}.$$

Table 1. Numerical upper bound of unknown bits of d for different n

δ	n = 1	n = 2	n = 3	n = 4
0.30	0.275	0.270	0.267	0.266
0.35	0.246	0.240	0.237	0.234
0.40	0.219	0.211	0.207	0.205
0.45	0.192	0.183	0.179	0.176
0.50	0.167	0.157	0.152	0.148
0.55	0.142	0.131	0.125	0.122
0.60	0.118	0.106	0.100	0.096
0.65	0.095	0.082	0.075	0.071
0.70	0.073	0.059	0.051	0.047
0.75	0.051	0.036	0.028	0.023
0.80	0.030	0.014	0.005	0.000

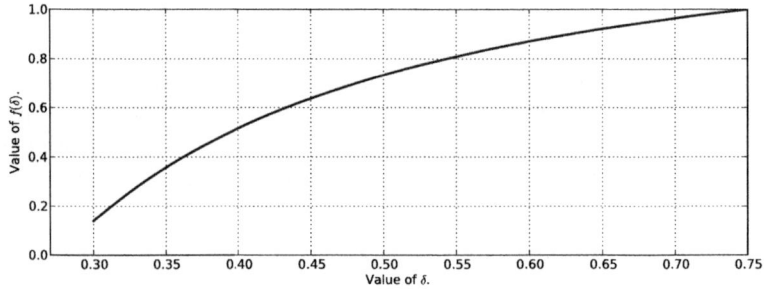

Fig. 2. Partial Key Exposure Attack for d. Plot of $f(\delta) = 1 + \frac{\sqrt{1+4\delta}}{2\delta} - \frac{1}{\delta}$ vs. values of δ.

So, when $\sqrt{1 + 4\delta} < 2$, i.e., when $\delta < 0.75$, knowledge of $\left(\delta + \frac{\sqrt{1+4\delta}}{2} - 1\right) \log N$ many bits of d is sufficient to factor N irrespective of their position in time polynomial in $\log N$ and exponential in number of unknown blocks n. So we can formally state the formally the following result.

Corollary 1. *Let e be full bit size and $d \leq N^\delta$ with $\delta < 0.75$. Then knowledge of*

$$\left(\delta + \frac{\sqrt{1 + 4\delta}}{2} - 1\right) \log N$$

many bits of d is sufficient to factor N in time polynomial in $\log N$ and exponential in number of unknown blocks of d.

Figure 2, represents the proportion of bits of d need to be known irrespective of their position such that one can factor N in time polynomial in $\log N$ and exponential in number of unknown blocks of d.

In our experiments, Assumption 1 always holds and we have successfully collected the desired root. We present the experimental results in Table 2. LD

Table 2. Experimental results for $n = 2$ and $n = 3$

n	δ	$\sum_{i=1}^{n} \gamma_i$	(m,t)	LD	Time (Sec.)
2	0.30	0.200	(2,1)	55	99.69
2	0.35	0.145	(2,1)	55	107.94
2	0.40	0.095	(2,1)	55	114.12
2	0.45	0.060	(2,1)	55	122.82
2	0.50	0.045	(2,1)	55	114.23
2	0.55	0.010	(2,1)	55	99.68
3	0.30	0.195	(2,1)	91	911.31
3	0.35	0.140	(2,1)	91	901.11
3	0.40	0.090	(2,1)	91	1002.15
3	0.45	0.040	(2,1)	91	914.22

denotes the lattice dimension. First we take $n = 2, m = 2, t = 1$. In this case lattice dimension will be 55. When $\delta = 0.30$ i.e., bit size of d is 308, factoring N requires knowledge of $(0.30 - 0.20) \times 1024$ i.e., 103 many bits of d. When $\delta = 0.40$ and $n = 2$, we need the knowledge of $(0.40 - 0.095) \times 1024 = 313$ many bits of d. For larger values of δ or n, we need more number of bits to be known for efficient factorization.

3 Attacks Using the Partial Knowledge of k

Now consider the situation when few MSBs of d are known. Figure 3 provides a pictorial view of this case. In this situation one can also find an approximation of k.

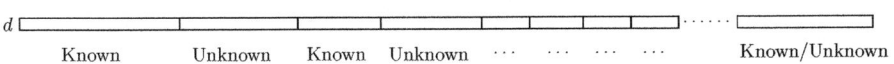

d

Known   Unknown   Known Unknown $\cdots$ $\cdots$ $\cdots$ $\cdots$ Known/Unknown

Fig. 3. Exposed and unexposed blocks of d

Theorem 3. *Let e be full bit size and $d \leq N^\delta$. Suppose one knows an approximation d_0 of d such that $|d - d_0| < N^\lambda$. Also assume that attacker knows bits of d except n many blocks and size of each such block is $\gamma_i \log N$ for $1 \leq \gamma_i \leq n$. Then he/she can factor N in time polynomial in $\log N$ and exponential in n if*

$$\sum_{i=1}^{n} \gamma_i < \frac{(1 + \beta n + \frac{n}{2}) - \sqrt{(2n^2 + 2n)\beta^2 + (n^2 + 3n + 2)\beta}}{n + 2}$$

where $\beta = \max\{\lambda, \delta - \frac{1}{2}\}$.

Proof. Since attacker knows an approximation d_0 of d, he/she can also find an approximation $k_0 = \lfloor \frac{ed_0-1}{N} \rfloor$ of k. In [1], it is proved that $|k - k_0| < N^\beta$ where $\beta = \max\{\lambda, \delta - \frac{1}{2}\}$ when e is of full bit size. Let, $k_1 = k - k_0$. Since, d is unknown for n many blocks, so one can write $d = a_0 + a_1 y_1 + \cdots + a_n y_n$, where $y_1, y_2, \ldots, y_n$ are unknown. Now from $ed = 1 + k(N + 1 - p - q)$, we have

$$ea_0 + ea_1 y_1 + \cdots + ea_n y_n - 1 - (k_0 + k_1)(N + 1 - p - q) = 0.$$

Hence, we are interested to find the root of the polynomial

$$f(x_1, \ldots, x_{n+2}) = (ea_0 - k_0 N - 1) + ea_1 x_1 + \cdots + ea_n x_n + N x_{n+1}$$
$$- k_0 x_{n+2} + x_{n+1} x_{n+2}.$$

Clearly, $f(y_1, \ldots, y_n, -k_1, 1 - p - q) = 0$.

Let $X_i = N^{\gamma_i}$, for $1 \le i \le n$. Further, suppose $X_{n+1} = N^\beta$ and $X_{n+2} = N^{0.5}$. Clearly, $X_1, \ldots, X_{n+2}$ are the upper bounds of absolute values of $y_1, \ldots, y_n$, $-k_1, 1 - p - q$, neglecting small constants. Using the extended strategy of [12, Section 2.2], we define

$$S = \bigcup_{0 \le j \le t} \{x_1^{i_1} x_2^{i_2} \cdots x_{n+1}^{i_{n+1}+j} x_{n+2}^{i_{n+2}} : x_1^{i_1} x_2^{i_2} \cdots x_{n+2}^{i_{n+2}} \text{ is a monomial of } f^m\}$$

$$M = \{\text{monomials of } x_1^{i_1} x_2^{i_2} \cdots x_{n+2}^{i_{n+2}} f : x_1^{i_1} x_2^{i_2} \cdots x_{n+2}^{i_{n+2}} \in S\}$$

where t is a non-negative integer.

Apart from f, we need to find at least $n + 1$ more polynomials $f_1, f_2, \ldots f_{n+1}$ that share the same root $(y_1, \ldots, y_n, -k_1, 1 - p - q)$ over the integers.

Considering a fixed n, we know that these polynomials can be found by LLL [14] algorithm if

$$X_1^{s_1} X_2^{s_2} \cdots X_{n+2}^{s_{n+2}} < W^s$$

for $s_j = \sum_{x_1^{i_1} \cdots x_{n+2}^{i_{n+2}} \in M \setminus S} i_j$ with $j = 1, \ldots, n + 2$, $s = |S|$, and

$$W = \|f(x_1 X_1, \ldots x_{n+2} X_{n+2})\|_\infty \ge N X_{n+1} = N^{1+\beta}.$$

From the structure of the polynomial f, it is clear that $s_1, \ldots, s_n$ are equal.

Using similar kind computation as in Theorem 2, We get

$$s \approx \frac{2m^{n+2}}{(n+2)!} + \frac{tm^{n+1}}{(n+1)!},$$

$$s_1 = \cdots = s_n \approx \frac{2m^{n+2}}{(n+2)!} + \frac{tm^{n+1}}{(n+1)!},$$

$$s_{n+1} \approx \frac{3m^{n+2}}{(n+2)!} + t\frac{2m^{n+1}}{(n+1)!} + t^2 \frac{m^n}{2n!},$$

$$s_{n+2} \approx \frac{3m^{n+2}}{(n+2)!} + \frac{tm^{n+1}}{(n+1)!}$$

Table 3. Numerical upper bound of unknown bits of d for different n using the partial knowledge of k

δ	β	γ_1	$\sum_{i=1}^{2} \gamma_i$	$\sum_{i=1}^{3} \gamma_i$
0.30	0.25	0.1424	0.1408	0.1400
0.40	0.25	0.1424	0.1408	0.1400
0.60	0.25	0.1424	0.1408	0.1400
0.75	0.25	0.1424	0.1408	0.1400
0.80	0.30	0.1101	0.1092	0.1087
0.85	0.35	0.802	0.0797	0.0794
0.90	0.40	0.0521	0.0519	0.0518
0.95	0.45	0.0255	0.0254	0.0254

Now consider $t = \tau m$, where $\tau \geq 0$ is a real number. Putting the values of X_1, $X_2, \ldots, X_{n+2}$, $s_1, \ldots, s_{n+2}$, s, and the lower bound of W in the condition

$$X_1^{s_1} X_2^{s_2} \ldots X_{n+2}^{s_{n+2}} < W^s,$$

we get

$$\left(\frac{n^2 \beta}{2} + \frac{3n\beta}{2} + \beta\right)\tau^2 + \left(n\sum_{i=1}^{n}\gamma_i + 2\sum_{i=1}^{n}\gamma_i + n\beta + 2\beta - \frac{n}{2} - 1\right)\tau$$

$$+2\sum_{i=1}^{n}\gamma_i + \beta - \frac{1}{2} < 0.$$

The optimal value of τ is $\frac{\frac{1}{2} - \sum_{i=1}^{n}\gamma_i - \beta}{n\beta + \beta}$.
Putting this optimal value of τ in Equation (2),

$$\sum_{i=1}^{n}\gamma_i < \frac{(1 + \beta n + \frac{n}{2}) - \sqrt{(2n^2 + 2n)\beta^2 + (n^2 + 3n + 2)\beta}}{n+2}.$$

$\square$

Putting $n = 1$ in Theorem 3, we get $\gamma_1 < \frac{1}{2} + \frac{\beta}{3} - \frac{1}{3}\sqrt{4\beta^2 + 6\beta}$. This bound is the same as the second result of Theorem 1. In Table 3, we present a few numerical values of the upper bound of unknown bits. In all cases we take $\lambda = 0.25$.
Now,

$$\lim_{n\to\infty} \frac{(1 + \beta n + \frac{n}{2}) - \sqrt{(2n^2 + 2n)\beta^2 + (n^2 + 3n + 2)\beta}}{n+2} = \beta + \frac{1}{2} - \sqrt{2\beta^2 + \beta}.$$

$$(2)$$

Hence, knowledge of $\left(\delta - \beta - \frac{1}{2} + \sqrt{2\beta^2 + \beta}\right) \log N$ many bits of d is sufficient to factor N in time polynomial in $\log N$ and exponential in number of unknown blocks of d. When $\lambda < \delta - \frac{1}{2}$, value of β will be $\beta = \delta - \frac{1}{2}$.

Table 4. Experimental results for $n = 2$

δ	$\sum_{i=1}^{2} \gamma_i$	(m,t)	LD	Time (Sec.)
0.38	0.116	(2,0)	50	224.28
0.38	0.120	(2,1)	70	831.55
0.40	0.093	(2,0)	50	276.32
0.40	0.098	(2,1)	70	727.19
0.45	0.058	(2,0)	50	225.38
0.45	0.065	(2,1)	70	956.81
0.50	0.040	(2,0)	50	211.85
0.50	0.044	(2,1)	70	1005.91
0.55	0.012	(2,0)	50	187.43
0.55	0.016	(2,1)	70	984.42
0.56	0.004	(2,0)	50	194.51
0.56	0.008	(2,1)	70	1155.82

So from Equation (2), we can say that in this case $\sum_{i=1}^{n} \gamma_i$ should less than $\delta - \sqrt{2(\delta - \frac{1}{2})^2 + (\delta - \frac{1}{2})}$ as $n \to \infty$.

Thus, knowing $\left(\sqrt{2(\delta - \frac{1}{2})^2 + (\delta - \frac{1}{2})}\right) \log N$ many bits of d is sufficient to factor N in time polynomial in $\log N$ and exponential in n when $\lambda < \delta - \frac{1}{2}$.

In Table 4, we present some experimental results for different sizes of d. In all cases, we assume $\lambda = 0.25$ i.e., $(\delta - 0.25) \times 1024$ many MSBs of d to be known. For larger values of δ, we need more number of bits to be known for efficient factorization.

4   Attack on ISO/IEC 9796-2

Recall that in ISO/IEC 9796-2 [11], the encoded message is of the form

$$\mu(m) = 6A_{16} \ || \ m[1] \ || \ H(m) \ || \ BC_{16},$$

where $m = m[1] \ || \ m[2]$ is split into two parts. Let the message $m = m[1] \ || \ m[2]$ be of the form $m[1] = \alpha_1 \ || \ r_1 \ || \ \cdots \ || \ \alpha_n \ || \ r_n \ || \ \alpha_{n+1}$ and $m[2]$=data, where $r_1, r_2, \cdots, r_n$ are unknown and $\alpha_1, \alpha_2, \cdots, \alpha_{n+1}$ are known to the attacker, while the attacker may or may not know $m[2]$, and does not know the hash value $H(m)$. Coron et al. [5] first studied the case when faults are injected for the same prime. Next, they considered the case when two faulty signatures occur for two different primes. In this section, we consider two faults for two different primes. We first assume that after injecting a fault, the attacker obtains a faulty signature s such that

$$s^e = \mu(m) \mod p \text{ and } s^e \neq \mu(m) \mod q.$$

Hence we have $s^e = a_0 + a_1 r_1 + \cdots + a_n r_n + a_{n+1} H(m) \bmod p$. So in this case the attacker needs to find the root $(r_1, \cdots, r_n, H(m))$ of a polynomial of the form

$$b_0 + b_1 x_1 + \cdots + b_n x_n + b_{n+1} x_{n+1}$$

in $\mathbb{Z}_p$.

Now suppose that another faulty signature, which is incorrect modulo p but correct modulo q, and n different blocks in the first part of the message are unknown. So here we have

$$s_1^e = c_0 + c_1 r_1' + \cdots + c_n r_n' + c_{n+1} H(m') \bmod q.$$

Thus one needs to find the root $(r_1', \cdots, r_n', H(m'))$ of a polynomial of the form

$$d_0 + d_1 x_1 + \cdots + d_n x_n + d_{n+1} x_{n+1}$$

in $\mathbb{Z}_q$. Hence when two faults occur, one in modulo p and the other in modulo q, we have

$$(b_0 + b_1 r_1 + \cdots + b_n r_n + b_{n+1} H(m)) \times (d_0 + d_1 r_1' + \cdots + d_n r_n' + d_{n+1} H(m'))$$
$$= 0 \bmod N.$$

So in this case the attacker needs to calculate the root of

$$f_N(z_1, \ldots, z_{n+1}, z_{n+2}, \ldots, z_{2n+2}) = (b_0 + b_1 z_1 + \cdots + b_n z_n + b_{n+1} z_{n+1}) \times$$
$$(d_0 + d_1 z_{n+2} \cdots + d_{2n+2} z_{2n+2})$$

in $\mathbb{Z}_N$.

Let $X_i = X_{n+1+i} = N^{\delta_i}$ for $1 \le i \le n$ be the upper bounds of the absolute values of $r_1, r_1', r_2, r_2' \cdots, r_n, r_n'$ respectively. Also let $X_{n+1} = X_{2n+2} = N^\gamma$ be the upper bound of $H(m)$ and $H(m')$.

Using the strategy [12, Section 2.1], we define

$$M_k = \{ z_1^{i_1} z_2^{i_2} \cdots z_{2n+2}^{i_{2n+2}} \mid z_1^{i_1} z_2^{i_2} \cdots z_{2n+2}^{i_{2n+2}} \text{ is a monomial of } f_N^m$$

$$\& \frac{z_1^{i_1} \cdots z_{2n+2}^{i_{2n+2}}}{l^k} \text{ is a monomial of } f_N^{m-k} \},$$

where m is a non-negative integer and $l = z_1 z_{n+2}$ for $0 \le k \le m$. It follows that

$$z_1^{i_1} z_2^{i_2} \cdots z_{2n+2}^{i_{2n+2}} \in M_k \Leftrightarrow \begin{cases} 0 \le i_1 + \cdots + i_{n+1} \le m, \text{ and } k \le i_1 \le m \\ 0 \le i_{n+2} + \cdots + i_{2n+2} \le m \text{ and } k \le i_{n+2} \le m \end{cases}$$

For fixed n and using the idea of [12, Section 2.1], the root of f_N can be found in time polynomial in $\log N$ but exponential in n if

$$X_1^{s_1} \cdots X_{2n+2}^{s_{2n+2}} < N^s$$

Table 5. Experimental results when two faults occur with p and q

$\log N$	n	$\delta \log N$	$\gamma \log N$	m	LD	Time (Sec)
1024	1	74	160	2	36	21.71
2048	1	278	160	2	36	98.18
2048	1	180	256	2	36	95.05

for $s_k = \sum_{z_1^{i_1} \cdots z_{2n+2}^{i_{2n+2}} \in M_0} i_k$ with $k = 1, \cdots, 2n + 2$, $s = \sum_{t=1}^{m} |M_t|$. One can check easily that

$$|M_t| = \frac{(m - t)^{2n+2}}{((n + 1)!)^2} + o((m - t)^{2n+2}).$$

Hence $s \approx \sum_{t=1}^{m} \frac{(m-t)^{2n+2}}{(((n+1)!)^2} = \frac{m^{2n+3}}{((n+1)!)^2(2n+3))}$. Also it can be easily checked that

$$s_1 = \cdots = s_{2n+2} \approx \frac{m^{2n+3}}{(n + 1)!(n + 2)!}.$$

Putting the values of $X_1, X_2, \ldots, X_{2n+2}, s_1, \ldots, s_{2n+2}, s$ in the condition

$$X_1^{s_1} X_2^{s_2} \cdots X_{2n+2}^{s_{2n+2}} < N^s,$$

and neglecting the terms of $o(m^{2n+3})$ we get the condition $\sum_{i=1}^{n} \delta_i + \gamma < \frac{n+2}{2(2n+3)}$.

So, when $n \to \infty$, one gets the upper bound 0.25. In the presence of a single fault in this situation the upper bound was 0.153 [5, Section 2.2]. Now, when $n = 1$, we get the upper bound 0.30. Using a lattice of dimension 9, Coron et al. [5, Section 2.3] obtained the upper bound 0.167. Hence we achieve a better upper bound than the work of [5]. We use Coppersmith's idea to obtain better bound than [5].

In Table 5, we present some experimental results. We first consider the case when the hash function is SHA-1. Hence $\gamma \log N = 160$. Coron et al. [5, Table 2] presented the value of $\delta_1 \log N$ to be 13 and 158 for 1024-bit and 2048-bit RSA respectively for one faulty signature. Also using the approach of Coron et al. [5, Section 2.3], for faults of two different factors, we obtain an upper bound of δ_1 as $0.167 - 0.156 = 0.011$ for 1024-bit-RSA and $0.167 - 0.078 = 0.089$ for 2048-bit-RSA. Hence upper bounds of $\delta_1 \log N$ will be 12 and 182 for 1024-bit and 2048-bit RSA respectively. In our experiments, $\delta_1 \log N$ is 71 and 278 for 1024-bit and 2048-bit RSA respectively. We also consider the case when SHA 256 is used for 2048-bit-RSA. In this case N can be factored given two faulty signatures of two different factors containing 180 random bits in less than 2 minutes.

5   Conclusion

So far, all the existing partial key exposure attacks on RSA assume a single contiguous block of unknown bits of the secret exponent when encryption exponent e is large. However, in practice, it is more likely that the attacker obtains certain bits of the secret exponent in fragmented blocks. In such a case, none of the existing methods work towards any significant cryptanalysis of the ciphers. We address this issue in our current work and generalize the above attacks, when the number of unexposed blocks can be more than one. We also study an ISO/IEC 9796-2 standard signature scheme with two faulty signatures, one modulo p and another modulo q. More than one faulty signature with same modulo was studied by Coron et al. [5]. So, it would be interesting to study the factorization of N with more than 2 signatures, some faulty modulo p and others faulty modulo q.

Acknowledgment. The authors would like to thank Prof. Subhamoy Maitra and the anonymous Indocrypt reviewers for their detailed comments on the technical issues of this paper. The authors are also grateful to Dr. Goutam Kumar Paul and Mr. Sourav Sen Gupta for their editorial contribution towards the presentation of this paper.

References

1. Blömer, J., May, A.: New Partial Key Exposure Attacks on RSA. In: Boneh, D. (ed.) CRYPTO 2003. LNCS, vol. 2729, pp. 27–43. Springer, Heidelberg (2003)
2. Boneh, D., Durfee, G., Frankel, Y.: An Attack on RSA Given a Small Fraction of the Private Key Bits. In: Ohta, K., Pei, D. (eds.) ASIACRYPT 1998. LNCS, vol. 1514, pp. 25–34. Springer, Heidelberg (1998)
3. Boneh, D., Durfee, G.: Cryptanalysis of RSA with Private Key d Less Than $N^{0.292}$. IEEE Transactions on Information Theory 46(4), 1339–1349 (2000)
4. Boneh, D., DeMillo, R.A., Lipton, R.J.: On the importance of checking cryptographic protocols for faults. Journal of Cryptology 14(2), 101–119 (2001)
5. Coron, J.-S., Joux, A., Kizhvatov, I., Naccache, D., Paillier, P.: Fault Attacks on RSA Signatures with Partially Unknown Messages. In: Clavier, C., Gaj, K. (eds.) CHES 2009. LNCS, vol. 5747, pp. 444–456. Springer, Heidelberg (2009)
6. Coron, J.-S., Naccache, D., Tibouchi, M.: Fault Attacks Against EMV Signatures. In: Pieprzyk, J. (ed.) CT-RSA 2010. LNCS, vol. 5985, pp. 208–220. Springer, Heidelberg (2010)
7. Ernst, M., Jochemsz, E., May, A., de Weger, B.: Partial Key Exposure Attacks on RSA up to Full Size Exponents. In: Cramer, R. (ed.) EUROCRYPT 2005. LNCS, vol. 3494, pp. 371–386. Springer, Heidelberg (2005)
8. Herrmann, M., May, A.: Solving Linear Equations Modulo Divisors: On Factoring Given Any Bits. In: Pieprzyk, J. (ed.) ASIACRYPT 2008. LNCS, vol. 5350, pp. 406–424. Springer, Heidelberg (2008)
9. Heninger, N., Shacham, H.: Reconstructing RSA Private Keys from Random Key Bits. In: Halevi, S. (ed.) CRYPTO 2009. LNCS, vol. 5677, pp. 1–17. Springer, Heidelberg (2009)
10. Howgrave-Graham, N.: Finding small roots of univariate modular equations revisited. In: Darnell, M.J. (ed.) Cryptography and Coding 1997. LNCS, vol. 1355, pp. 131–142. Springer, Heidelberg (1997)
11. ISO/IEC 9796-2, Information technology - Security techniques - Digital signature scheme giving message recovery, Part 2: Mechanisms using a hash-function (1997)

12. Jochemsz, E., May, A.: A Strategy for Finding Roots of Multivariate Polynomials with New Applications in Attacking RSA Variants. In: Lai, X., Chen, K. (eds.) ASIACRYPT 2006. LNCS, vol. 4284, pp. 267–282. Springer, Heidelberg (2006)
13. Kocher, P.: Timing Attacks on Implementations of Diffie-Hellman, RSA, DSS, and Other Systems. In: Koblitz, N. (ed.) CRYPTO 1996. LNCS, vol. 1109, pp. 104–113. Springer, Heidelberg (1996)
14. Lenstra, A.K., Lenstra Jr., H.W., Lovász, L.: Factoring polynomials with rational coefficients. Mathematische Annalen 261(4), 513–534 (1982)
15. Rivest, R.L., Shamir, A., Adleman, L.: A method for obtaining digital signatures and public key cryptosystems. Communications of ACM 21(2), 158–164 (1978)
16. Sarkar, S., Maitra, S.: Cryptanalysis of RSA with more than one Decryption Exponent. Information Processing Letters 110(8-9), 336–340 (2010)
17. Wiener, M.: Cryptanalysis of Short RSA Secret Exponents. IEEE Transactions on Information Theory 36(3), 553–558 (1990)

Appendix A: Detailed Calculations Related to Theorem 2

Here we present the detailed calculations for s, s_1, s_2, s_{n+2}.

Calculation of s

One may note that s is the number of solutions of $0 \leq i_1 + \cdots + i_{n+1} \leq m$, $0 \leq i_{n+2} \leq i_{n+1} + t$. Thus,

$$
\begin{aligned}
s &= \sum_{r=0}^{m} \binom{r+n-1}{r} \left(\frac{(m-r)^2}{2} + t(m-r) \right) \\
&\approx \sum_{r=0}^{m} \frac{r^{n-1}}{(n-1)!} \left(\frac{(m-r)^2}{2} + t(m-r) \right) \\
&\quad \text{(neglecting the lower order terms)} \\
&\approx \frac{m^{n+2}}{(n+2)!} + \frac{tm^{n+1}}{(n+1)!}
\end{aligned}
$$

(using Proposition 1 and neglecting the lower order terms).

Calculation of s_1

$$
\begin{aligned}
s_1 = &\sum_{i_1=0}^{m+1} \sum_{r=0}^{m-i_1+1} \sum_{i_{n+1}=0}^{m+1-i_1-r} \sum_{i_{n+2}=0}^{i_{n+1}+t} \binom{r+n-2}{r} i_1 \\
&- \sum_{i_1=0}^{m} \sum_{r=0}^{m-i_1} \sum_{i_{n+1}=0}^{m-i_1-r} \sum_{i_{n+2}=0}^{i_{n+1}+t} \binom{r+n-2}{r} i_1.
\end{aligned}
$$

Now,

$$\sum_{i_1=0}^{m}\sum_{r=0}^{m-i_1}\sum_{i_{n+1}=0}^{m-i_1-r}\sum_{i_{n+2}=0}^{i_{n+1}+t}\binom{r+n-2}{r}i_1$$

$$\approx \sum_{i_1=0}^{m}\sum_{r=0}^{m-i_1}\sum_{i_{n+1}=0}^{m-i_1-r}\frac{r^{n-2}}{(n-2)!}(i_{n+1}+t)i_1$$

(neglecting the lower order terms)

$$\approx \sum_{i_1=0}^{m}\sum_{r=0}^{m-i_1}\sum_{i_{n+1}=0}^{m-i_1-r}\frac{r^{n-2}}{(n-2)!}i_1\left(\frac{(m-i_1-r)^2}{2}+t(m-i_1-r)\right)$$

(using Proposition 1 and neglecting the lower order terms).

Let $r_1 = m - i_1$. Then

$$\sum_{i_1=0}^{m}\sum_{r=0}^{m-i_1}\sum_{i_{n+1}=0}^{m-i_1-r}\sum_{i_{n+2}=0}^{i_{n+1}+t}\binom{r+n-2}{r}i_1$$

$$\approx \sum_{i_1=0}^{m}\sum_{r=0}^{r_1}\frac{r^{n-2}}{(n-2)!}i_1\left(\frac{(r_1-r)^2}{2}+t(r_1-r)\right)$$

$$\approx \frac{1}{(n+1)!}\sum_{i_1=0}^{m}i_1(m-i_1)^{n+1}+\frac{t}{n!}\sum_{i_1=0}^{m}i_1(m-i_1)^n$$

$$. \approx \frac{m^{n+3}}{(n+3)!}+\frac{t}{(n+2)!}m^{n+2}$$

$$\text{So } s_1 \approx \frac{(m+1)^{n+3}}{(n+3)!}+\frac{t}{(n+2)!}(m+1)^{n+2}$$

$$-\frac{m^{n+3}}{(n+3)!}+\frac{t}{(n+2)!}m^{n+2}$$

$$\approx \frac{m^{n+2}}{(n+2)!}+\frac{tm^{n+1}}{(n+1)!}.$$

From the structure of the polynomial f, we have $s_1 = s_2 = \cdots = s_n$.

Calculation of s_{n+1}

Let us now consider s_{n+1}. We have,

$$s_{n+1} = \sum_{i_{n+1}=0}^{m+1}\sum_{i_{n+2}=0}^{i_{n+1}+t}\sum_{r=0}^{m+1-i_{n+1}}i_{n+1}\binom{n-1+r}{r}$$

$$-\sum_{i_{n+1}=0}^{m}\sum_{i_{n+2}=0}^{i_{n+1}+t}\sum_{r=0}^{m-i_{n+1}}i_{n+1}\binom{n-1+r}{r}.$$

Now, $\sum_{i_{n+1}=0}^{m}\sum_{i_{n+2}=0}^{i_{n+1}+t}\sum_{r=0}^{m-i_{n+1}}i_{n+1}\binom{n-1+r}{r}$

$$\approx \sum_{i_{n+1}=0}^{m}\sum_{i_{n+2}=0}^{i_{n+1}+t}\sum_{r=0}^{m-i_{n+1}}\frac{r^{n-1}}{(n-1)!}i_{n+1}$$

$$\approx \sum_{i_{n+1}=0}^{m} \frac{(m-i_{n+1})^n}{n!} i_{n+1}(i_{n+1}+t)$$

(using Proposition 1 and neglecting lower order terms)

$$\approx \frac{2m^{n+3}}{(n+3)!} + \frac{tm^{n+2}}{(n+2)!}.$$

Hence, $s_2 \approx \dfrac{2m^{n+2}}{(n+2)!} + \dfrac{tm^{n+1}}{(n+1)!}.$

Calculation of s_{n+2}

$$s_{n+2} = \sum_{i_{n+1}=0}^{m+1} \sum_{i_{n+2}=0}^{i_{n+1}+t} \sum_{r=0}^{m+1-i_{n+1}} i_{n+1}\binom{n-1+r}{r}$$

$$- \sum_{i_{n+1}=0}^{m} \sum_{i_{n+2}=0}^{i_{n+1}+t} \sum_{r=0}^{m-i_{n+1}} i_{n+1}\binom{n-1+r}{r}$$

Now, $\sum_{i_{n+1}=0}^{m} \sum_{i_{n+2}=0}^{i_{n+1}+t} \sum_{r=0}^{m-i_{n+1}} i_{n+1}\binom{n-1+r}{r}$

$$\approx \sum_{i_{n+1}=0}^{m} \frac{(m-i_{n+1})^n}{n!} \frac{(i_{n+1}+t)^2}{2}$$

$$\approx \frac{m^{n+3}}{(n+3)!} + t\frac{m^{n+2}}{(n+2)!} + t^2 \frac{m^{n+1}}{(n+1)!2}.$$

Hence, $s_{n+2} \approx \dfrac{m^{n+2}}{(n+2)!} + t\dfrac{m^{n+1}}{(n+1)!} + t^2\dfrac{m^n}{n!2}.$

The Yin and Yang Sides of Embedded Security

Christof Paar

Ruhr Universität Bochum, Germany
Christof.Paar@rub.de

Abstract. Through the prevalence of interconnected embedded systems, the vision of pervasive computing has become reality over the last few years. As part of this development, embedded security has become an increasingly important issue in a multitude of applications. Examples include the Stuxnet virus, which has allegedly delayed the Iranian nuclear program, killer applications in the consumer area like iTunes or Amazon's Kindle, the business models of which rely heavily on IP protection, and even medical implants like pace makers and insulin pumps that allow remote configuration. These examples show the destructive and constructive aspects of modern embedded security. For us embedded security researchers, the following definition of yin and yang can be useful for resolving this seemingly conflict: "The concept of yin yang is used to describe how polar opposites or seemingly contrary forces are interconnected and interdependent in the natural world, and how they give rise to each other in turn." (OK, the "natural world" part is not a 100% fit here.) In this presentation I will talk about some of our research projects over the last few years which dealt with both the yin and yang aspect of embedded security.

In 1–2 generations of automobiles, car2car and car2infrastructure communication will be available for driver-assistance and comfort applications. The emerging car2x standards call for strong security features. The large number of data of up to several 1000 incoming messages per second, the strict cost constraints, and the embedded environment makes this a challenging task. We show how an extremely high-performance digital signature engine was realized using low-cost FPGAs. Our signature engine is currently widely used in field trials in the USA. The next case study addresses the other end of the performance spectrum, namely lightweight cryptography. PRESENT, one of the smallest known ciphers which can be realized with as few as 1000 gates. The cipher was designed for extremely cost and power constrained applications such as RFID tags which can be used, e.g., as a tool for anti-counterfeiting of spare parts, or for other low-power applications. PRESENT is currently being standardized by ISO.

As "yang examples" of our research we will show how two devices with very large deployment in the real world can be broken using physical attacks. First, we show a recent attack against a modern contactless smart card equipped with 3DES. The card is widely used in authentication and payment systems. The second attack breaks the bit stream encryption of current FPGAs. These are reconfigurable hardware devices which are popular in many digital systems. We were able to extract AES and 3DES key from a single power-up of the reconfiguration process. Once the key has been recovered, an attacker can clone, reverse engineer and alter a presumingly secure hardware design.

D.J. Bernstein and S. Chatterjee (Eds.): INDOCRYPT 2011, LNCS 7107, p. 93, 2011.
© Springer-Verlag Berlin Heidelberg 2011

Mars Attacks! Revisited:

Differential Attack on 12 Rounds of the MARS Core and Defeating the Complex MARS Key-Schedule

Michael Gorski, Thomas Knapke, Eik List, Stefan Lucks, and Jakob Wenzel

Bauhaus-University Weimar, Germany
{Michael.Gorski,Thomas.Knapke,Eik.List,
Stefan.Lucks,Jakob.Wenzel}@uni-weimar.de

Abstract. The block cipher MARS has been designed by a team from IBM and became one of the five finalists for the AES. A unique feature is the usage of two entirely different round function types. The "wrapper rounds" are unkeyed, while the key schedule for the "core rounds" is a slow and complex one, much more demanding then, e.g., the key schedule for the AES. Each core round employs a 62-bit round key. The best attack published so far [KKS00] was applicable to 11 core rounds, and succeeded in recovering some 163 round key bits. But neither did it deal with inverting the key schedule, nor did it provide any other means to recover the remaining 519 round key bits in usage.

Our attack applies to 12 core rounds, needs 2^{252} operations, 2^{65} chosen plaintexts and 2^{69} memory cells. After recovering a limited number of cipher key bits, we deal with the inverse key-schedule to recover the original encryption key. This allows the attacker to easily generate all the round keys in the full.

Keywords: Differential cryptanalysis, MARS, block cipher, key recovery.

1 Introduction

Recent research provides some significant advances in the cryptanalysis of the AES, mostly for the 192-bit key and the 256-bit key variants, such as the boomerang attacks on the full-round AES-192 (12 rounds) and AES-256 (14 rounds) [BK09] and the "practical complexity attack" on AES-256 with up to 13 rounds [BK10]. Since many of these attacks are based on exploiting the AES key-schedule, which is fairly simple, we are interested in having a fresh look at other block ciphers, namely at those with a more complex key-schedule. This makes the AES finalist MARS [BCD+99] a highly relevant subject to study.

In this paper we present a differential attack on 12 core rounds of MARS. It is an improvement of the attack on 11 core rounds of MARS published in [KKS00] and the best attack known so far.

D.J. Bernstein and S. Chatterjee (Eds.): INDOCRYPT 2011, LNCS 7107, pp. 94–113, 2011.
© Springer-Verlag Berlin Heidelberg 2011

1.1   Related Work

Table 1 summarizes the results of attacks on reduced-round variants of MARS. In [KS00], Kelsey and Schneier describe several attacks on reduced-round variants of MARS. First, the authors consider two meet-in-the-middle attacks on unsymmetrically reduced versions of MARS that cover the pre-whitening, all forward mixing rounds, five core rounds, all backwards mixing rounds and the post-whitening. These attacks were applied to a version of MARS with very reduced strength, as they are limited to a small number of core rounds. An adversary guesses the pre- and post-whitening keys and can then unwrap the unkeyed mixing rounds. These attacks require time equivalent to more than 2^{232} half encryptions, more than 2^{197} bytes of memory and 8 or 2^{50} known plaintexts, respectively. In the same paper, the authors also present an attack on a more interesting version of MARS that are symmetrically reduced to three forward mixing rounds, three forward core rounds, three backward core rounds and three backward mixing rounds. This attack requires 2^{197} partial decryptions, 2^{73} bytes and 2^{69} chosen plaintexts.

In [BF00], the authors present impossible differentials on eight core rounds of MARS. The authors use a 3-round differential with probability one in combination with intrinsic properties of the E-function. The ideas seem likely that one can extend the differentials to few more rounds, but they seem unlikely to be extended to the full core of MARS.

In 2009, [Pes09] published an attack that covers the pre-whitening, eight core rounds, eight backward mixing rounds and the post-whitening. The attack uses 128-bit differences with only one non-zero bit as the input and differences with only one non-zero bit as the output of the attack. The author says, that the difference remains unchanged after the pre-whitening with probability $p = 2^{-1}$, and the same probability would apply to the output difference after the post-whitening. In the core rounds, the author uses a 3-round differential twice, for the rounds 1 - 3 and the rounds 6 - 8. The idea of the proposed differential attack says, that, with a probability of 2^{-96}, the 96 unknown bits of the difference after Round 5 can be zero and the non-zero part of the difference is then only rotated during the rounds 6 - 8.

Most of the attacks on MARS were developed during the decision progress of the AES competition. All attacks are completely unpractical and work only in the context of reduced round variants of MARS.

1.2   Our Contribution

In [KKS00], Kelsey, Kohno and Schneier published an 11-round attack against the MARS core, five forward and six backwards. In contrast to this attack, our attack covers 12 rounds of the MARS core, eight forward and four backwards. The difference results from adding one forward round (Round 3 of our distinguisher) to our attack. Furthermore, we place the two last rounds of the attack in [KKS00] at the beginning of our distinguisher. So we can get more information about the subkeys generated in the first iteration of the key expansion, which

Table 1. Attacks on reduced rounds of MARS

Type of Attack	Rounds	Texts	Memory (Bytes)	Operations	Reference
Amplified Boomerang	11C	2^{65}	2^{70}	2^{229}	[KKS00]
Amplified Boomerang	6M, 6C	2^{69}	2^{73}	2^{197}	[KS00]
MITM	16M, 5C	8	2^{236}	2^{232}	[KS00]
Differential MITM	16M, 5C	2^{50}	2^{197}	2^{247}	[KS00]
Differential	8M, 8C	2^{105}	2^{109}	2^{231}	[Pes09]
Differential	12C	2^{219}	2^{69}	2^{252}	Sec. 3.2

C: core rounds, M: mixing rounds

are closely linked with the encryption key. Moreover, it allows us to start a meet-in-the-middle attack on the MARS key schedule (see Section 4). In [KKS00] the authors can recover a total of 163 subkey bits, but they do not consider a way to attack the key schedule with this information. So they cannot recover bits from the encryption key.

1.3 Notation

The MARS cipher has a block length of 128 bit. Through the cipher, a block is split in four 32-bit words that are transformed in a Feistel structure. We denote a 128-bit difference Δ of the internal state of the cipher as a 4-tupel of four 32-bit differences $\Delta = (\alpha, \beta, \gamma, \delta)$. We denote a 32-bit difference $\delta = 0$, if all 32 bits of the difference δ are zero. We denote a 32-bit difference with a lower case letter, e. g. $\delta = a$ to indicate that a difference δ is arbitrary for our work. We specify certain bits of a 32-bit difference as $\delta = (0^i, 1, 0^j)$, to indicate, that the i most significant bits of δ are zero, followed by one 1 bit and that the sequence of the j least significant bits of δ are zero. It must hold, that $i + 1 + j = 32$. We denote a 32-bit difference $\delta = (?^i, b, 0^j)$, where we indicate a sequence of i unknown bits by $?^i$, b is a bit that can be 0 or 1 and the least significant j bits are zero.

1.4 Outline

The paper is organized as follows: In Section 2 we give a brief description of the MARS block cipher. Section 3 presents our differential attack on 12 core rounds of MARS. The attack on the MARS key scheduler is shown in Section 4 and the analysis of both attacks is given in Section 5. Section 6 concludes our paper.

2 Description of MARS

MARS is a block cipher with a block length of 128 bits and a variable key length from 128 to 448 bits, in increments of 32 bits. The design of MARS is unusual

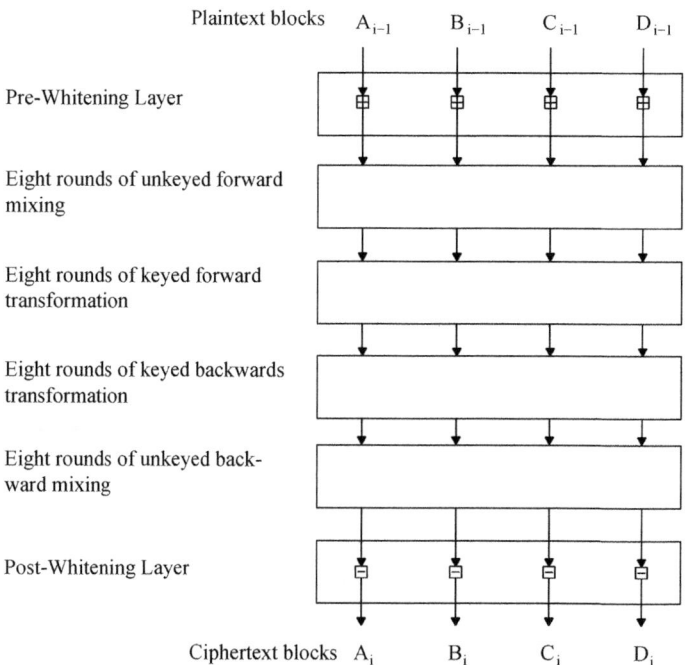

Fig. 1. The MARS structure

as it consists of two entirely different round functions. The cryptographic core consists of 16 rounds of keyed transformation, that are wrapped with a key whitening and 8 rounds of unkeyed forward mixing before, and with 8 rounds of unkeyed backward mixing and a key whitening after the core. Figure 1 visualises the structure.

The cipher is optimized for 32-bit operations. At the beginning, a 128 bit plaintext block is split into the four words A, B, C, D that are transformed during the encryption as follows:

1. **Pre-Whitening Layer**: To each of the four words A, B, C, D, a different subkey of 32 bits length is added modulo 2^{32}.
2. **Forward Mixing Layer**: Eight rounds of unkeyed mixing, using the S-box, addition and XOR operations.
3. **Forward Core Layer**: Eight rounds of keyed Feistel cipher. The core layer combines a variety of operations, including S-box lookups, multiplications, additions, XORs, fixed-value rotations and data-dependent rotation.
4. **Backward Core Layer**: Eight rounds of keyed Feistel cipher. The core layer combines a variety of operations, including S-box lookups, multiplications, additions, XORs, fixed-value rotations and data-dependent rotation.
5. **Backward Mixing Layer**: Eight rounds of unkeyed mixing, using the S-box, subtraction and xor operations.

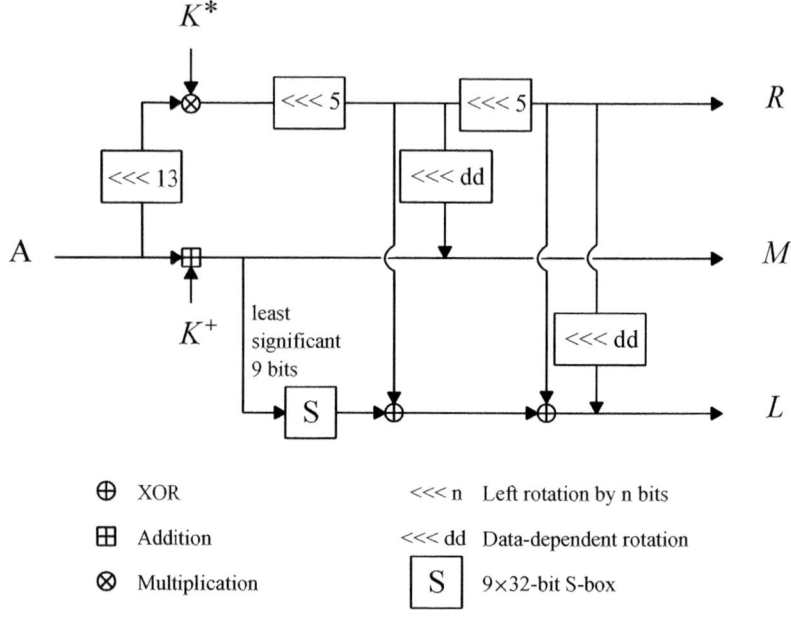

K^*

R

A

M

K^+

least significant 9 bits

L

⊕ XOR

⊞ Addition

⊗ Multiplication

<<< n   Left rotation by n bits

<<< dd   Data-dependent rotation

S   9×32-bit S-box

Fig. 2. The MARS core E-function

6. **Post-Whitening Layer**: From each of the four words A, B, C, D, a different subkey of 32 bits length is subtracted modulo 2^{32}.

2.1 The Cryptographic Core

The MARS core consists of eight forward and eight backward core rounds. In each round the cipher uses a keyed expansion function E that combines multiplication, data-dependent rotations and an S-box lookup. The cipher uses a 9×32-bit S-box that maps a 9-bit input to one of 512 32-bit outputs. In the following, we will focus on the core rounds. The Feistel network is shown in Section 3 to visualise our distinguisher. The E-function is shown in Figure 2.

In each core round of MARS one data word serves as input to the E-function. The output of the E-function is then used to modify the other words. Additionally, the source word is rotated by 13 positions to the left. One core round uses one key for addition K_i^+ and one key for multiplication K_i^*. K_i^+ has 32 bits of entropy. K_i^* has at most 30 bits of entropy, because its two least significant bits are set to 1. Thus a core round contains about 62 key bits of entropy.

3   Differential Attack on 12 Core Rounds of MARS

In this section we present the first 12-round differential attack on the reduced MARS core. Our attack itself is split into two single attacks, one considering a distinguisher over 12 rounds of the MARS core and one to recover the secret key by mounting a meet-in-the-middle attack to the first iteration of the key schedule.The first part of this section describes the distinguisher. The second part of this section describes the attack to recover the subkey candidates for some particular rounds.

3.1 The Distinguisher

The distinguisher covers 12 rounds and is partitioned in four parts with three rounds each. This section describes the differential trail used for our distinguisher. Figure 3 shows how this trail is transformed during the rounds.

$$\Delta_0 = (\alpha, \beta, \gamma, \delta) \xrightarrow{\text{Rounds 1-3}} \Delta_1 = (0, a, b, 0)$$

$$\Delta_1 = (0, a, b, 0) \xrightarrow{\text{Rounds 4-6}} \Delta_2 = (0, 0, 0, (?^7, 0^{15}, ?^{10}))$$

$$\Delta_2 = (0, 0, 0, (?^6, a, 0^6, 0^9, ?^{10})) \xrightarrow{\text{Rounds 7-9}} \Delta_3 = ((?^6, a, 0^6, ?^{19}), 0, 0, 0)$$

$$\Delta_3 = ((?^6, a, 0^6, ?^{19}), 0, 0, 0) \xrightarrow{\text{Rounds 10-11}} \Delta_4 = ((?^{15}, a, 0^6, ?^{10}), ?, ?, ?)$$

$$\Delta_4 = ((?^{15}, a, 0^6, ?^{10}), ?, ?, ?) \xrightarrow{\text{Rounds 11-12}} \Delta_5 = (?, ?, ?, (?^2, a, 0^6, ?^{23}))$$

Fig. 3. Differential trail for our 12-round distinguisher

Figure 4 shows the first part. In all figures that show the attack parts we highlight specific differences with thick lines and use normal lines to describe an unknown or a null difference.

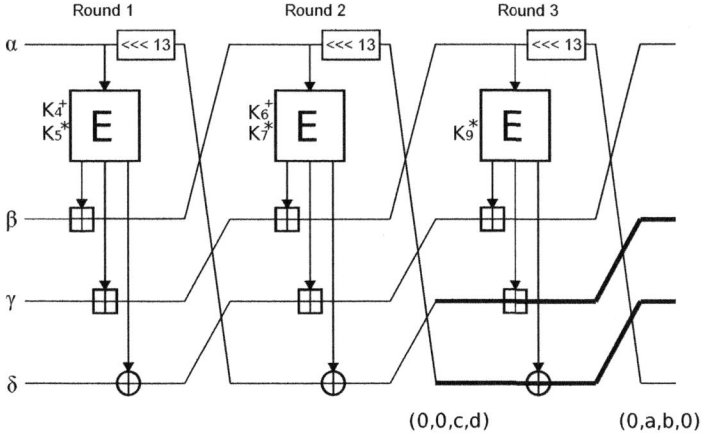

Fig. 4. Rounds 1-3 of the distinguisher

First Part. The first part consists of the rounds 1 to 3, where we have to guess both keys of the first round, both keys of the second round and the multiplication key of Round 3. Our goal is to reach a difference Δ_1 after Round 3 with the following structure,

$$\Delta_1 = (0, a, b, 0)$$

where a and b are arbitrary 32-bit differences, 0 denotes a zero 32-bit difference. To reach this difference after Round 3 we have to look at the behaviour of the first and second round. If you look at the structure of one MARS core round we can reach this difference with an input of $(0, 0, c, d)$ one round before. We have to guess K_9^* because we consider additive differences and with the right subkey K_9^* we can determine the difference d. A detailed description of getting the input difference of Round 1 can be seen in Section 3.2.

Second Part. The second part lasts from Round 4 to 6 (see Figure 5). Our goal is to reach a difference Δ_2 with the following structure after Round 6

$$\Delta_2 = (0, 0, 0, (?^7, 0^{15}, ?^{10}))$$

where a 0 stands for a zero bit or a zero 32-bit word, respectively, and a $'?'$ stands for an unknown bit difference. The structure and the chosen bits of the difference Δ_2 is based on the fact, that in Round 10 the considered bits (namely the bits $(a, 0^6)$, which occur the first time in the difference Δ_3) are placed on the least significant bit positions. This is necessary because of the multiplication in the E-function of Round 10, otherwise some unwanted difference would possibly destroy the seven considered bits during the multiplication. All following difference are chosen under this premise. In the second part, we use a three-round differential where we choose the output difference as follows: We can choose every possible

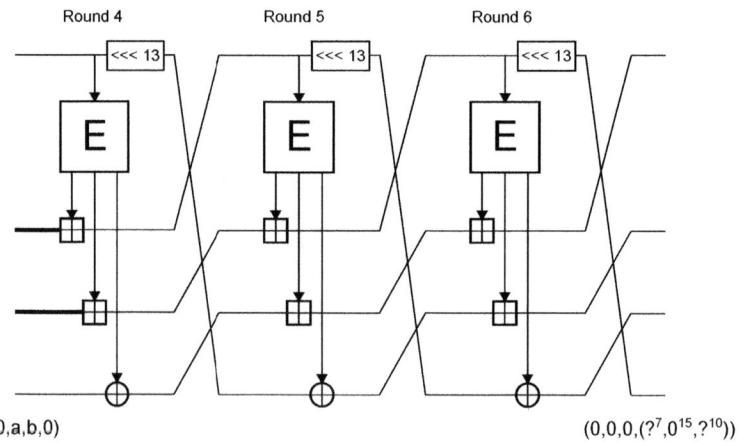

Round 4   Round 5   Round 6

(0,a,b,0)   (0,0,0,(?⁷,0¹⁵,?¹⁰))

Fig. 5. Rounds 4-6 of the distinguisher

difference after these three rounds by not specifying the input difference. This assumption is based on the following idea.

Be $(0, t, u, i)$ the input of the three-round differential, where t and u are random values and i is a counter for all possible values between 1 and $2^{32} - 1$. We denote the output after three rounds by $(v + i, w, x, y)$. All values in the output can be seen as a function of the input values t and u (resulting from the structure of a core round).

If we consider two plaintexts X and X' as $X = (0, t, u, i)$ and $X' = (0, t', u', i)$ we can reach an arbitrary difference at the output, which we can freely chose. The advantage of the counter is, that we can generate 2^{32} possible input texts with the same difference in the output, because we use the same counter in both texts, respectively. This technique allows us to choose the difference Δ_2 after Round 6.

Third Part. The third part of our distinguisher lasts from Round 7 to 9. Our goal is to reach a difference Δ_3 with the following structure after Round 9:

$$\Delta_3 = ((?^6, a, 0^6, ?^{19}), 0, 0, 0)$$

Here we use a three-round differential with a probability nearly one induced by the binomial distribution. The input difference $\Delta_2 = (0, 0, 0, (?^7, 0^{15}, ?^{10}))$ of Round 7 is only rotated by one word every round. So the non-null part of the difference will never be seen as the input of the E-function within the Rounds 7 to 9. Based on the idea of Schneider et al.(see [KKS00] for further details), we

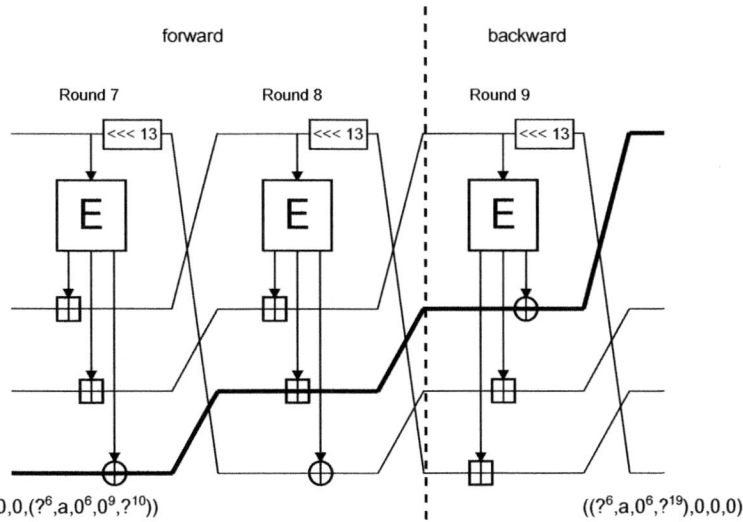

Fig. 6. Rounds 7-9 of the distinguisher

consider seven bits in the difference denoted by a followed by six zeroes, where the a-bit is not the same as in the first mentioned difference $(0, a, b, 0)$ of Round 3. So we write the new input difference Δ_2 of Round 7 as follows:

$$\Delta_2 = (0, 0, 0, (?^6, a, 0^6, 0^9, ?^{10}))$$

In Round 8 we have to assure, that the bit a and the following six zero bits remain unchanged after the addition of the third word. Before explaining details for this step of the distinguisher we have to explain what a batch of texts is. A batch B_i is considered as a set of texts more specific a set of chosen plaintexts in our case. Because of the binomial distribution (see below) we use 302 texts per batch with the following structure:

$$B_i = \{M_0^i, M_1^i, \ldots, M_{301}^i\}$$

with

$$M_j^i = (m_{j,0}^i, m_{j,1}^i, m_{j,2}^i, m_{j,3}^i)$$

and

$$M_1^i = M_0^i + 1, \ldots, M_{301}^i = M_{300}^i + 1$$

where every $m_{j,k}^i$ is the k-th 32-bit word of the j-th text in the i-th batch. The counter is added in the last word $m_{j,3}^i$ to generate a new text. So in one batch all texts are equal except for a difference in the last word. For every batch we use another arbitrary value for the text M_0^i and add the counter to produce 302 texts per batch.

For this step and the transition from Δ_2 to Δ_3 we use the binomial distribution to calculate the probability that for each corresponding text pair in the right batch pair, where our desired differential holds, the rightmost zero bit is not influenced by an occurring carry bit. The probability of getting exactly $k = 0$ successes of a randomly distributed event K in n trials with some given probability p is given by the probability mass function:

$$Pr[K = k] = \binom{n}{k} p^k (1 - p)^{n-k}$$

with

$$\binom{n}{k} = \frac{n!}{k!(n-k)!}$$

With $n = 302$ (number of texts in a batch B_i) and $p = 2^{-9}$ (which results from the amount of nine protecting zeroes in the fourth word of Δ_2) we get a probability of 55.4% (for $k = 0$) that the a bit and the following six zeroes remain unchanged. This leads in an output difference Δ_3 after Round 9. The rounds 7 to 9 can be seen in Figure 6. The number of 302 texts per batch results from the binomial distribution, where we wanted to reach a reasonable probability higher than 50%. $(a, 0^6)$ are "protected" by nine following zeroes and the probability 2^{-9} describes the event, that a carry bit will go through the nine "protecting" bits and influences the a-bit and/or the following six zeroes in the addition in Round 8.

Fourth Part. The last part of the distinguisher covers the rounds 10 to 12 and can be seen in Figure 7. Our goal is to recover our bit a and the following six zeroes after Round 11. We will have a difference after Round 11 with the structure

$$\Delta_5 = (?, ?, ?, (?^2, a, 0^6, ?^{23})).$$

We will then guess subkeys in Round 12 that allow us to recover our bits of interest. In Round 10 the non-zero part of the difference is the input of the E-

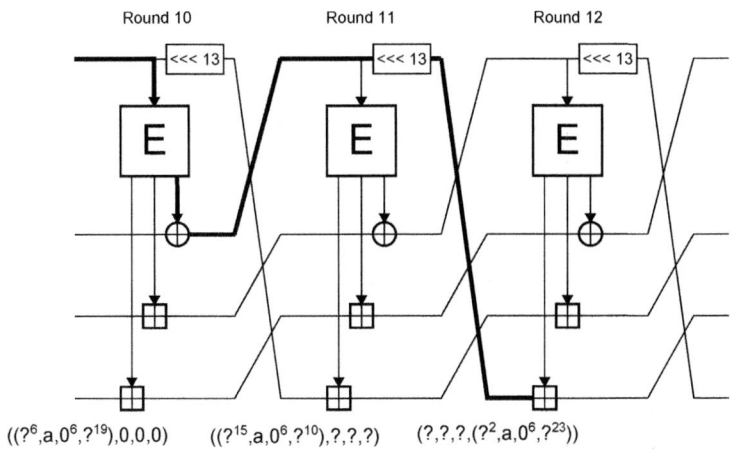

$$((?^6, a, 0^6, ?^{19}), 0, 0, 0) \qquad ((?^{15}, a, 0^6, ?^{10}), ?, ?, ?) \qquad (?, ?, ?, (?^2, a, 0^6, ?^{23}))$$

Fig. 7. Rounds 10-12 of the distinguisher

function. During the E-function the a bit and the following zero bits are rotated by 23 in the right part. We choose the a bit and the six zero bits so that they are positioned in the least significant bits before the multiplication. The output R is XORed to a zero difference from the output of Round 9(i.e., the second word of Δ_3). Because we only consider these seven specific bits we can denote the output of Round 10 as follows:

$$\Delta_4 = ((?^{15}, a, 0^6, ?^{10}), ?, ?, ?).$$

In Round 11 the considered difference is only rotated by 13 to the left and rotated in the fourth word. So we can denote the output difference of Round 11 as above

$$\Delta_5 = (?, ?, ?, (?^2, a, 0^6, ?^{23})).$$

In Round 12 we want to recover the a bit and the six following zero bits of the fourth word of Δ_5. Normally we would have to guess 39 bits within the E-function (the full 30 bits of the multiplication key and the low nine bits of the addition key, that serve as input for the S-box). These seven bits allow us to use a technique to reduce the amount of the required bits to guess. For the last

Round of the distinguisher we use the following technique to recover 2^{32} possible subkey candidates for the 39 subkey bits of K_{26}^+ and K_{27}^*, where we get 9 bits of the addition key and 30 bits of the multiplication key. We denote a ciphertext pair by (A, B, C, D) and (A', B', C', D'), where a pair (A, A') describes an additive difference of the two first words of each ciphertext.

To recover the possible subkey values we create a table for all possible input pairs (C, C') and (D, D') and all possible 2^{39} subkey values. The table holds a counter for each subkey candidate. In the table we store the resulting pair (L, L'), which is the left output of the E-function. After that we do the following computations $X = L + C$ and $X' = L + C'$ and check, whether the resulting additive difference between X and X' contains our desired values $((?^{15}, a, 0^6, ?^{10}), ?, ?, ?)$, where all bits denoted by ? are arbitrary. If so, we increase the counter for the used subkey candidate.

As we consider seven bits with fixed values $(a, 0^6)$ in this particular difference, we expect only 2^{32} instead of the full 2^{39} subkey candidates with a significantly high counter in this table. We include these subkey candidates in our further attack. We use 2^{63} pairs (C, C'), 2^{63} pairs (D, D') and 2^{39} values for the subkey candidates to create this table. This leads us to 2^{32} possible subkey values for the last round of our distinguisher, so we have to guess a total of 186 subkey bits in our distinguisher.

This table has two advantages. Firstly we reduce the possible subkey candidates by a factor of 2^{-7} from 2^{39} to 2^{32}. Secondly the creation of the table can be done in the leisure of the attacker, so the effort is additive and negligible in relation to the total effort for the attack.

3.2   Subkey Recovery

As said at the beginning of Section 3 our attack is split into two parts, the distinguisher, which leads us to the right subkey candidates, and the secret key recovery, which allows us to gain the secret key by mounting a meet-in-the-middle attack on the key schedule using the right subkey candidates. In this section we describe the way to get from the chosen plaintexts to the right subkey candidates. The attack covers 12 rounds of the MARS core, eight forward and four backwards. For this attack we use 2^{56} batches with 302 texts each. As we check 2^{154} subkey candidates, this results in a total of $2^{56} \cdot 302 \cdot 2^{154} = 2^{219}$ chosen plaintexts. We are aware that an implementation of the attack would use the entire codebook. But as the complexity is far less than the exhaustive search this does make the attack relevant for academic purpose.

Our technique for the last Round of the distinguisher (see Section 3.1) allows us to check 2^{32} subkey candidates instead of the full 2^{39} possible values. Thus we get a total of 2^{186} subkey candidates. For each of the 2^{154} subkey candidates of the first three rounds we have to do the following steps:

1. Choose 2^{56} arbitrary differences as the output of Round 3 (which is the difference described as $(0, a, b, 0)$ from Section 3.1).

2. Partially decrypt the difference $(0, a, b, 0)$ from the output of Round 3 to determine the input difference $(\alpha, \beta, \gamma, \delta)$ of the distinguisher
3. Create 2^{56} batches with 302 texts each where the difference between each of two batches is $(\alpha, \beta, \gamma, \delta)$.
4. Encrypt all plaintexts and store the resulting 2^{65} ciphertexts.
5. Partially decrypt all ciphertexts with each of the 2^{32} subkey candidates for Round 12 and extract bit a for each ciphertext.
6. Build 2^{56} 302-bit strings of the bits a for each batch.
7. Sort the resulting bit strings in order of the chosen plaintexts.
8. Compare the bit strings pairwise to identify the correct subkey candidates.

4   Attacking the MARS Key Scheduler

The key expansion of MARS expands the secret key of n 32-bit words to a total of 40 subkeys of 32-bit words. One core round uses two subkeys, one for addition and one for multiplication. The key expansion uses a temporary array T to hold the internal state of the transformed subkeys. The array T is initialized by

$$T[0 \ldots n-1] = k[0 \ldots n-1], \quad T[n] = n, \quad T[n+1 \ldots 14] = 0$$

where $n = 8$ (for 256-bit keys) and k is an array with the secret key. The key expansion repeats the following steps four times, where each iteration produces 10 subkeys:

1. **Linear transformation:** The array T is transformed by

   for $i = 0, \ldots, 14$, $T[i] = T[i] \oplus ((T[i-7 \bmod 15] \oplus T[i-2 \bmod 15]) \lll 3)$
   $\oplus (4i + j)$

2. **Four stirring rounds:** The array T is stirred using four rounds of a Feistel network [HR10]

   for $i = 0, \ldots, 14$, $T[i] = (T[i] + S[\text{low 9 bits of } T[i-1 \bmod 15]]) \lll 9$

3. **Storing keys:** The next 10 keys are stored in the array of subkeys K

   for $i = 0, \ldots, 9$, $K[10j + i] = T[4i \bmod 15]$

   where $j \in \{0, \ldots, 3\}$.

4. **Modification of multiplication keys:** To avoid weak keys, the subkeys that are used for multiplications are modified in an additional step where sequences of 10 or more equal bits are modified. The details of this operation can be found in Appendix A.

Basic Approach. The key expansion uses operations on 32-bit words, what makes it obvious to guess some words of the secret key and to guess all 32 bits of those words. As a word $T[i]$ depends not only on the word in the secret key $K[i]$ but also on the words $T[i-2 \bmod 15]$ and $T[i-7 \bmod 15]$, this approach requires the attacker to guess many words to equalize the effort of guessed bits with the effort of brute force.

The stirring rounds of the key expansion make extensive use of the S-box. If the attacker can not retrieve knowledge about the array word $T[i-1]$, one stirring round requires the attacker to guess nine bits per round to recover the value of $T[i]$. This effort of required guessing bits in the stirring rounds makes attacks on the key expansion ineffective compared to brute force.

Instead of guessing all 32 bits of some of the 15 words in the array T, our approach guesses some bits of all words in T. This approach is effective, as the words $T[8], \ldots, T[14]$ largely depend on the eight words $T[0], \ldots, T[7]$ of the secret key. We can profit from the operations in the linear transformation, as the transformation consists of XOR operations where carry bits do not occur and so, allow us to retrieve knowledge about many bits afterwards.

In the following, we are going to mount a meet-in-the-middle attack, where we guess 27 bits of almost all eight words of the secret key. We aim to guess as many bits as possible that serve as input in the S-box in the stirring rounds so we can perform two stirring rounds in the forward part of our MITM-attack.

Forward Step. We could simply guess the most significant 27 bits of $T[0], \ldots, T[7]$, if the designers would not have included a rotation by 3 in the linear transformation. Due to the rotation, the positions of the guessed bits differ after every two words by three bits because of $T[i]$ is XORed with $T[i-2 \bmod 15]$. The detailed bits we guess are visualised in Figure 8.

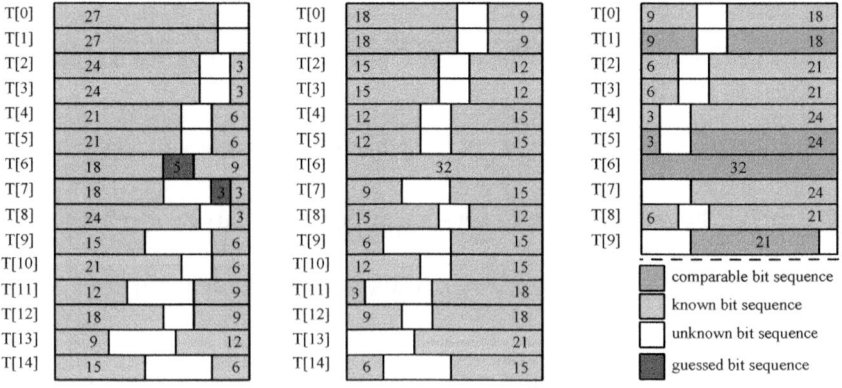

Fig. 8. The detailed bits we guess from the array T in the key schedule of MARS

In this step we will guess 210 bits of the secret key, and additionally five bits of $T[6]$ and three bits of $T[7]$. These bits do not give us information about the secret key. We guess these eight bits of values in T after the linear transformation. They help us because they influence the words $T[9], T[11], T[13]$ and $T[14]$ and allow us to determine more bits in these words. This leads us to the knowledge of all bits of $T[6]$.

After the linear transformation we can perform two stirring rounds and know any bits that go in the S-box in these two rounds. In contrast to the linear transformation, the stirring rounds use additions to transform the words $T[i]$. In each word $T[i]$ that is transformed in these rounds, there is a continuous sequence of some unknown bits. These bit sequences can lead to unpredictable carry bits after the additions of the stirring rounds. Thus, we have to consider two cases for each addition, the case in which the addition of the highest bit of such a sequence creates a carry bit and the case in which it does not.

For each addition, we need to execute the forward step twice. We perform 25 additions in the forward step, 15 additions in the first stirring round for the words $T[0], \ldots, T[14]$ and 10 additions in the second stirring round as we concentrate to recover $T[0], \ldots, T[9]$ in the forward step of our meet-in-the-middle attack. We need to consider carry bits of the additions in only 23 of the 25 times, as we know about all bits from $T[6]$ and know about occurring carry bits for $T[6]$ in the stirring rounds.

Backward Step. We can recover five subkeys from the result of our distinguisher: K_4^+, K_5^* from Round 1, K_6^+, K_7^* from Round 2 and K_9^* from Round 3, where K_i^+ denotes a subkey that is used for addition and K_i^* denotes a subkey that is used for multiplication. We will use four of these subkeys, K_4^+, K_5^*, K_6^+ and K_9^*, to revert the modification of multiplication keys and two stirring rounds for the backward part of our MITM-attack. Our recovered subkeys map to values $K[10j + i] = T[4i \bmod 15]$, so K_4^+ maps to $T[1]$, K_5^* maps to $T[5]$, K_6^+ maps to $T[9]$ and K_9^* maps to $T[6]$.

The modification of multiplication keys can not be simply inverted because it contains a feed forward where the original subkey is XORed with the modified one. We can create a lookup table that stores over all 2^{32} possible values $K[i]$ and all possible 2^5 rotation values of the patterns in B (see Appendix A). Such a table can map a subkey and retrieve the values $T[i]$ that could have created the subkey. We implemented such a table, where we counted the maximal value of possible $T[i]$s for one occurring $K[i]$. We retrieved a maximal value of 102 possible combinations of rotation values and values $T[i]$ that can result in the same value of $K[i]$. So, we estimate the effort for the inversion of one multiplication subkey with 2^7.

In each stirring round we have to guess the low nine bits from the word $T[i-1]$ that are used as input to the S-box and modify one word $T[i]$. From the word $T[5]$, we know about its low nine bits that are used to transform $T[6]$, so we need to guess only $3 \cdot 9$ bits per stirring round. Fig 9 visualises our approach.

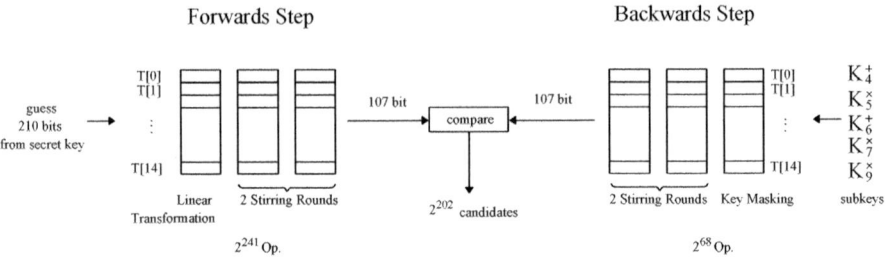

Fig. 9. Forward and backward steps of the key recovery attack on the key schedule of MARS

In the middle of the MITM attack on the key expansion, we can compare 27 bits of $T[1]$, 27 bits of $T[5]$, 32 bits of $T[6]$ and 21 bits of $T[9]$. So we can compare a total of 107 bits.

5   Analysis of the Attack

Step 1: The Distinguisher. For the distinguisher we use 2^{56} batches of 302 texts each and for the difference after Round 4 ($\Delta 2$, see Section 3.1 for further details) we use nine zeroes. This leads to a probability of 55.4% for the addition operation in Round 8 which is given through the binomial distribution with the values $p = 2^{-9}$, $n = 302$ and $k = 0$.

With 2^{56} batches with 302 texts each we are able to contruct 2^{111} different pairs of 302-bit strings. The bit strings are constructed from the extracted a bits of the corresponding texts of each batch. While looking at pairs of batches we can reduce the subkey candidates that we have to consider. The probability of finding a right pair of bit strings is 2^{-302} because every single bit (the a bit) must match in the difference between two bit strings. If we consider 2^{111} pairs of bit strings the probability of finding a right pair of 302-bit strings is:

$$2^{111} \cdot 2^{-302} = 2^{-191}$$

The amount of subkey candidates for our distinguisher is 2^{186}. The probability for a right subkey candidate (i.e., a right bit string) is 2^{-191}, which means we expect one right subkey, and expect only few false positive candidates, and we expect that testing them in an additional step requires negligible effort compared to the total complexity of the attack. The probability for a false positive subkey candidate is

$$2^{186} \cdot 2^{-191} = 2^{-5}.$$

The effort of creating the table for the last Round is 2^{165} because we have to consider 2^{63} input pairs (C, C'), 2^{63} input pairs (D, D') and 2^{39} subkey values. To obtain the 2^{32} subkey values with the highest counter we have to sort the table with an effort of $2^{165} \cdot \log(2^{165})$ which is $165 \cdot 2^{165}$. Combining these two additive effort leads us to an total effort of

$$2^{165} + 165 \cdot 2^{165} \approx 2^{173}$$

for the creation and sorting of the table. The total effort of the distinguisher is

$$2^{56} \cdot 302 \cdot 2^{186} \cdot 1.8 = 2^{251.55}.$$

If we consider the additive effort for the creation of the table for the last Round, the effort for the whole distinguisher is

$$2^{251.55} + 2^{173} \approx 2^{252}.$$

Step 2: The Key Schedule. The detailed steps of the key schedule can be found in the Appendix A. The following subkeys are given by the subkey guesses within the distinguisher.

$$[K_4, K_6, K_{26}(\text{least nine bits})]^+$$

$$[K_5, K_7, K_9, K_{27}]^*$$

The complexity to invert a subkey K^* that is used for multiplication is 2^{30}. This is a bruteforce approach with two known bits (resulting from the building structure of K^*, see Appendix A for further details).

This effort can be reduced with the help of a table with all possible input values T_i and the resulting output values at K_j^*. One needs 2^{37} operations to generate such a table to cover all 2^{32} possible values for the inputs T_i and 2^5 for the data dependent rotation. The table requires 2^{33} bits of memory because we have to save every input value of T_i and its corresponding output.

This table has two advantages. Firstly the generation can be done in the leisure of the attacker so the effort is additive and can be neglected. Secondly with the knowledge of a multiplication key we can gain knowledge about the value before Step 21 in the key expansion (see Appendix A) with a simple table lookup.

We found out, that some keys K_j result from the same value T_i. So we tested all possible values of T_i to find the maximum of amount combinations of rotation values and values T_i that can generate the same subkey K_j. We found a subkey that could be generated by 102 such combinations. A table lookup may result in 102 outputs, what increases the resulting effort by 2^7.

The key recovery step is divided in two parts, the forward step and the backward step. In the forward step we guess 210 bits of $T[0], T[1], \cdots, T[7]$, which are the values initialized by the secret key in the first iteration of the key schedule. Additionally we have to guess eight bits after the linear transformation. The addition operations in the first and second stirring round lead us to an increase of the effort by a factor of 2^{23} because we want to assure, that no carry bits occur on the specific bit positions (see Section 4 for details).

The effort for the forward round is

$$2^{210} \cdot 2^8 \cdot 2^{23} = 2^{241}.$$

In the backward step of this meet-in-the-middle attack we guess nine bits for each stirring round for each value of $T[0], T[4], T[8]$. In the two stirring rounds in the backward step, this leads us to an effort of 2^{54}. The maximum number of values $T[i]$, which can generate a multiplication key is approximately 2^7. We use two multiplication subkeys from our distinguisher, K_5^* and K_9^*, so the effort is increased by 2^{14}. Additionally we have to consider the effort to create the table, which provides the values $T[i]$ for every subkey guess. The effort is 2^{37} and the way to get to this table is described above in this section. The total effort for the backward step is

$$2^{14} \cdot 2^{27} \cdot 2^{27} + 2^{37} \approx 2^{68}.$$

In this attack we compare 107 bits in the middle of the four stirring rounds. Thus, the probability of finding two matching bit strings is 2^{-107}. If we combine the efforts of the forward step and the backward step we expect to find

$$2^{241} \cdot 2^{68} \cdot 2^{-107} = 2^{202}$$

possible candidates for the 210 bits of the secret key with an total effort of 2^{202}.

Step 3: Guessing the Remaining Secret Key. After Step 2 we got 210 bit of the encryption key with an effort of 2^{202}. This allow us to obtain the rest of the secret key via brute force, which means the total effort for Step 3 is

$$2^{202} \cdot 2^{46} = 2^{248}.$$

This is much faster than obtaining the whole secret key via brute force.

6   Conclusion

In this paper we describe a differential attack on 12 core rounds of the AES candidate cipher MARS based on an attack on 11 rounds of the MARS core that was proposed in [KKS00]. We show that the original attack did not allow the attacker to retrieve information about the encryption key from the gathered subkey material. Our attack improves the original attack, as it covers one additional core round and retrieves subkey information that can be used to recover the encryption key. For this purpose we propose a meet-in-the-middle attack that allows us to invert the key expansion of MARS more efficiently than exhaustive search.

References

[BCD+99] Burwick, C., Coppersmith, D., D'Avignon, E., Gennaro, R., Halevi, S., Jutla, C., Matyas Jr., S.M., O'Connor, L., Peyravian, M., Stafford, D., Zunic, N.: MARS - A Candidate Cipher for AES. NIST AES Proposal (1999)
[BF00]   Biham, E., Furman, V.: Impossible differential on 8-round mars' core. In: 3rd AES Candidate Conference, pp. 186–194 (2000)

[BK09]   Biryukov, A., Khovratovich, D.: Related-Key Cryptanalysis of the Full
           AES-192 and AES-256. In: Matsui, M. (ed.) ASIACRYPT 2009. LNCS,
           vol. 5912, pp. 1–18. Springer, Heidelberg (2009)
[BK10]   Biryukov, A., Khovratovich, D.: Feasible Attack on the 13-round AES-256.
           Cryptology ePrint Archive, Report 2010/257 (2010),
           http://eprint.iacr.org/
[BS90]   Biham, E., Shamir, A.: Differential Cryptanalysis of DES-like Cryptosys-
           tems. In: Menezes and Vanstone [MV91], pp. 2–21
[BS93]   Biham, E., Shamir, A.: Differential Cryptanalysis of the Data Encryption
           Standard. Springer, Heidelberg (1993)
[DR02]   Daemen, J., Rijmen, V.: The Design of Rijndael: AES - The Advanced
           Encryption Standard. Springer, Heidelberg (2002)
[HR10]   Hoang, V.T., Rogaway, P.: On generalized feistel networks (2010),
           http://eprint.iacr.org/2010/301
[KKS00]   Kelsey, J., Kohno, T., Schneier, B.: Amplified Boomerang Attacks Against
           Reduced-Round MARS and Serpent. In: Schneier, B. (ed.) FSE 2000.
           LNCS, vol. 1978, pp. 75–93. Springer, Heidelberg (2001)
[KS00]   Kelsey, J., Schneier, B.: MARS Attacks! Preliminary Cryptanalysis of
           Reduced-Round MARS Variants. In: 3rd AES Candidate Conference, pp.
           169–185 (2000)
[MV91]   Menezes, A., Vanstone, S.A. (eds.): CRYPTO 1990. LNCS, vol. 537.
           Springer, Heidelberg (1991)
[NIS00]   NIST. A Request for Candidate Algorithm Nominations for the AES
           (2000), http://www.nist.gov/aes/
[Pes09]   Pestunov, A.: Differential Cryptanalysis of the MARS Block Cipher. Prik-
           ladnaya Diskretnaya Matematika, pp. 56–63 (2009),
           http://mi.mathnet.ru/pdm157

A   MARS Key Schedule

Algorithm 1. Key Schedule of MARS

1: // n is the number of words in the key buffer $k[]$, $(4 \leq n \leq 14)i$
2: // $K[]$ is the expanded key array, consisting of 40 words
3: // T is a temporary array, consisting of 15 words
4: // B is a fixed table of four words
5: // Initialize B
6: $B[] = \{0xa4a8d57b; 0x5b5d193b; 0xc8a8309b; 0x73f9a978\}$
7: // Initialize T
8: $T[0 \ldots n-1] = k[0 \ldots n-1], T[n] = n, T[n+1 \ldots 14] = 0$
9: // Four iterations, computing 10 words of $K[]$ in each
10: **for** $j = 0$ to 3 **do**
11:     **for** $i = 0$ to 14 **do**
12:         $T[i] = T[i] \oplus ((T[i-7 \mod 15] \oplus T[i-2 \mod 15]) \lll 3) \oplus (4i+j)$
13:     **end for**
14:     **repeat four times** // four stirring rounds
15:     **for** $i = 0$ to 14 **do**
16:         $(T[i] = T[i] + S[\text{low 9 bits of } T[i-1 \mod 15]]) \lll 9$
17:     **end for**
18:     **end-repeat**
19:     **for** $i = 0$ to 9 **do**
20:         $K[10j + i] = T[4i \mod 15]$
21:     **end for**
22: **end for**
23: // Modify multiplication keys
24: **for** $i = 5, 7, \ldots, 35$ **do**
25:     $j = $ least two bits of $K[i]$
26:     $\omega = K[i]$ with both of the least two bits set to 1
27:     // Generate a bit-mask M
28:     M_l if ω_l belongs to a sequence of ten consecutive 0's or 1's in ω
29:     and also $2 \leq l \leq 30$ and $\omega_{l-1} = \omega_l = \omega_{l+1}$
30:     // Select a pattern from the fixed table and rotate it
31:     $r = $ least five bits of $K[i-1]$ // Rotation amount
32:     $p = B[j] \lll r$
33:     // Modify $K[i]$ with p under the control of the mask M
34:     $K[i] = \omega \oplus (p \wedge M)$
35: **end for**

B   Key Table

Table 2. Distribution of guessed, known and unknown bits from words in the array T in the key schedule of MARS in the linear transformation
$const$ = known constant 32-bit word, x^i = i known bits. $?^i$ = i unknown bits. x^{32} indicates that all 32 bits of a word are known. G^i = known bits that are guessed after the linear transformation and do not provide information about the secret key.

	$const \ll 3$	(x^{32})		$const \ll 3$		(x^{32})
$\oplus$	$const \ll 3$	(x^{32})	$\oplus$ $T[0/1] \ll 3$		$(x^{24}, ?^5, x^3)$	
$\oplus$	$K[0/1]$	$(x^{27}, ?^5)$	$\oplus$	$K[2/3]$		$(x^{24}, ?^5, x^3)$
$=$	$T[0/1]$	$(x^{27}, ?^5)$	$=$	$T[2/3]$		$(x^{24}, ?^5, x^3)$
	$const \ll 3$	(x^{32})		$const \ll 3$		(x^{32})
$\oplus T[2/3] \ll 3$		$(x^{21}, ?^5, x^6)$	$\oplus$	$T[4] \ll 3$		$(x^{18}, ?^5, x^9)$
$\oplus$	$K[4/5]$	$(x^{21}, ?^5, x^6)$	$\oplus$	$K[6]$		$(x^{18}, ?^5, x^9)$
$=$	$T[4/5]$	$(x^{21}, ?^5, x^6)$	$=$	$T[6]$		(x^{18}, G^5, x^9)
	$T[0] \ll 3$	$(x^{24}, ?^5, x^3)$		$T[1] \ll 3$		$(x^{24}, ?^5, x^3)$
$\oplus$	$T[5] \ll 3$	$(x^{18}, ?^5, x^9)$	$\oplus$	$T[6] \ll 3$		(x^{32})
$\oplus$	$K[7]$	$(x^{18}, ?^{11}, x^3)$	$\oplus$	$const$		(x^{32})
$=$	$T[7]$	$(x^{18}, ?^8, G^3, x^3)$	$=$	$T[8]$		$(x^{24}, ?^5, x^3)$
	$T[2] \ll 3$	$(x^{21}, ?^5, x^6)$		$T[3] \ll 3$		$(x^{21}, ?^5, x^6)$
$\oplus$	$T[7] \ll 3$	$(x^{15}, ?^8, x^9)$	$\oplus$	$T[8] \ll 3$		$(x^{21}, ?^5, x^6)$
$\oplus$	$const$	(x^{32})	$\oplus$	$const$		(x^{32})
$=$	$T[9]$	$(x^{15}, ?^{11}, x^6)$	$=$	$T[10]$		$(x^{21}, ?^5, x^6)$
	$T[4] \ll 3$	$(x^{18}, ?^5, x^9)$		$T[5] \ll 3$		$(x^{18}, ?^5, x^9)$
$\oplus$	$T[9] \ll 3$	$(x^{12}, ?^{11}, x^9)$	$\oplus$	$T[10] \ll 3$		$(x^{18}, ?^5, x^9)$
$\oplus$	$const$	(x^{32})	$\oplus$	$const$		(x^{32})
$=$	$T[11]$	$(x^{12}, ?^{11}, x^9)$	$=$	$T[12]$		$(x^{18}, ?^5, x^9)$
	$T[6] \ll 3$	(x^{32})		$T[7] \ll 3$		$(x^{15}, ?^8, x^9)$
$\oplus$	$T[11] \ll 3$	$(x^9, ?^{11}, x^{12})$	$\oplus$	$T[12] \ll 3$		$(x^{15}, ?^5, x^{12})$
$\oplus$	$const$	(x^{32})	$\oplus$	$const$		(x^{32})
$=$	$T[13]$	$(x^9, ?^{11}, x^{12})$	$=$	$T[14]$		$(x^{15}, ?^8, x^9)$

Linear Cryptanalysis of PRINTCIPHER – Trails and Samples Everywhere

Martin Ågren and Thomas Johansson

Dept. of Electrical and Information Technology, Lund University,
P.O. Box 118, 221 00 Lund, Sweden
martin.agren@eit.lth.se

Abstract. PRINTCIPHER is a recent lightweight block cipher designed by Knudsen et al. Some noteworthy characteristics are a burnt-in key, a key-dependent permutation layer and identical round keys. Independent work on PRINTCIPHER has identified weak key classes that allow for a key recovery — the obvious countermeasure is to avoid these weak keys at the cost of a small loss of key entropy. This paper identifies several larger classes of weak keys. We show how to distinguish classes of keys and give a 28-round linear attack applicable to half the keys. We show that there are several similar attacks, each focusing on a specific class of keys. We also observe how some specific properties of PRINTCIPHER allow us to collect several samples from each plaintext–ciphertext pair. We use this property to construct an attack on 29-round PRINTCIPHER applicable to a fraction 2^{-5} of the keys.

Keywords: cryptanalysis, block cipher, linear cryptanalysis, finding samples, key bit distinguisher.

1   Introduction

Over the last few years, a number of hardware-efficient block ciphers have been proposed. Some noteworthy examples are HIGHT [6], PRESENT [3], and KATAN and KTANTAN [5]. One of the most recent designs to appear is PRINT-CIPHER [8]. It is designed by Knudsen et al. and is quite similar to the well-studied PRESENT. All rounds use the same key and differ only by a round counter. The linear layer is partly key-dependent and as a result, 48-bit PRINTCIPHER uses keys of 80 bits, while 96-bit PRINTCIPHER uses 160-bit keys. We will focus exclusively on PRINTCIPHER-48 in this paper, noting that very similar results can be derived for PRINTCIPHER-96.

Our first observation relates to the key-dependent permutation: we show how there exist several linear trails in PRINTCIPHER that are biased for some keys but unbiased for most keys, allowing us to distinguish between classes of keys. In order to attack several rounds of PRINTCIPHER, we need to find many samples. Our second observation uses the identical round-structure, including identical keys, to obtain several samples per plaintext–ciphertext pair. By guessing key bits to do partial encrypting and decrypting, we eventually reach 29 rounds of 48.

D.J. Bernstein and S. Chatterjee (Eds.): INDOCRYPT 2011, LNCS 7107, pp. 114–133, 2011.
© Springer-Verlag Berlin Heidelberg 2011

Two recent attacks are similar to our work in that they identify classes of weak keys. As a fundamental idea behind PRINTCIPHER is that the key is burnt into the device, it is straightforward to protect against these attacks by avoiding the weak keys. Avoiding the 2^{52} keys attacked in [9] the size of the key space shrinks from 2^{80} to $2^{80} - 2^{52} \approx 2^{80}$ so the entropy is still 80 bits in a practical sense. Similarly, to protect against the attack in [4] the number of keys needs to be lowered to approximately $2^{79.8}$ so there is a loss of one fifth of a bit. In this independent paper, we find several classes that are very probable (e.g., probability one half), and even avoiding only the largest classes leads to a key space of size approximately 2^{78}, meaning two bits of the key entropy are effectively lost. This makes our observations very interesting compared to the previously published results.

This paper is organized as follows: Section 2 describes PRINTCIPHER. Section 3 introduces linear cryptanalysis and discusses the importance of finding many samples. Some initial, basic observations are given in Section 4, before Section 5 gives our fundamental observation: a key bit distinguisher on 23 rounds of PRINTCIPHER. Section 6 then derives attacks on 27 and 28 rounds of PRINT-CIPHER. In Section 7, we extend our observation to show that several classes of weak keys exist, making the attack very general, and to show how one can find many samples. In Section 8, we use our ability to find many samples to provide an attack on 29-round PRINTCIPHER. Section 9 concludes the paper.

2   A Description of PRINTCIPHER

We focus entirely on PRINTCIPHER-48, which uses blocks of 48 bits and 80-bit keys.

The 48-bit plaintext is loaded into the state, where we denote the 48 bit positions as (b, c), $0 \leq b < 16, 0 \leq c < 3$. The leftmost bit, also referred to as the most significant bit (msb), is $(15, 2)$ while $(0, 0)$ is the least significant bit (lsb).

There are 48 rounds where each round uses a round constant RC_i, $i = 0, \ldots, 47$ (see Table 1), a 48-bit xor-key $K^\oplus$ (the same in all rounds) and a 32-bit permutation key K^π (the same in all rounds). Each round consists of key addition, standard permutation, round constant addition, key-dependent permutation and an S-box, see Fig. 1. The S-box is given in Table 2 and takes input (x_2, x_1, x_0) to produce output (y_2, y_1, y_0).

Table 1. The round constants RC_i

i	0	1	2	3	4	5	6	7	8	9	10	11	12	13	14	15	16	17	18	19	20	21	22	23
RC_i	01	03	07	0F	1F	3E	3D	3B	37	2F	1E	3C	39	33	27	0E	1D	3A	35	2B	16	2C	18	30
i	24	25	26	27	28	29	30	31	32	33	34	35	36	37	38	39	40	41	42	43	44	45	46	47
RC_i	21	02	05	0B	17	2E	1C	38	31	23	06	0D	1B	36	2D	1A	34	29	12	24	08	11	22	04

We denote the plaintext $P = p_{47}, \ldots, p_0$ and the ciphertext (state) after r rounds of encryption $(0 < r \leq 48)$ by $C^r = c^r_{47}, \ldots, c^r_0$.

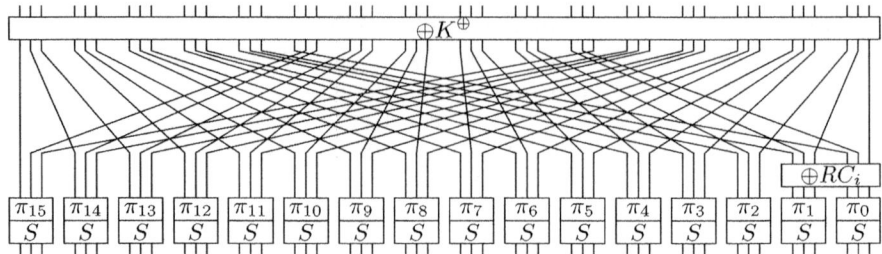

Fig. 1. One round of PRINTCIPHER.

Table 2. The S-box as values $S(x) = y = (y_2, y_1, y_0)$ corresponding to $x = (x_2, x_1, x_0)$. As an example, $S(1, 0, 0) = S(4) = 7 = (1, 1, 1)$.

x	0 1 2 3 4 5 6 7
$S(x)$	0 1 3 6 7 4 5 2

2.1 The Key

We split the key $K = K^{\oplus} || K^{\pi} = (k_{47}^{\oplus}, \ldots, k_0^{\oplus}) || (k_{31}^{\pi}, \ldots, k_0^{\pi})$ into an xor-key $K^{\oplus}$ (48 bits) and a permutation key K^{π} (32 bits).

2.2 The Standard Permutation Π

A "large" permutation Π is applied to the state of PRINTCIPHER. Each bit, at position (b, c), is moved to $(t \bmod 16, \lfloor t/16 \rfloor)$ where $t = 3b + c$. The permutation is given in Appendix A and can also be seen in Fig. 1.

2.3 The Key-Dependent Permutation π

A "small" permutation π_b is applied to each (disjoint) triplet of bits in the state of PRINTCIPHER. The new positions of bits $(b, 2), (b, 1), (b, 0)$ are $(b, c_2^b), (b, c_1^b), (b, c_0^b)$, respectively, where (c_2^b, c_1^b, c_0^b) are determined by the two key bits $(k_{2b+1}^{\pi}, k_{2b}^{\pi})$. This mapping is given in Table 3. Note in particular how one of the permutations is trivial while the others fix one bit while switching the two remaining bits. Thus, the two permutations that shift the three-bit word cyclically have been excluded from PRINTCIPHER and can not be selected by the key.

2.4 Other Notation

Let $I_r = r \bmod 2$ indicate whether r is even or odd. Partial encryption in rounds $r_1, \ldots, r_2 - 1$ is denoted by ϕ_{r_1, r_2}. Similarly, partial decryption in rounds $r_2 - 1, \ldots, r_1$ is denoted ϕ_{r_1, r_2}^{-1}.

Table 3. The key-dependent permutation. The bits at positions $(b, 2), (b, 1), (b, 0)$ are moved to positions $(b, c_2^b), (b, c_1^b), (b, c_0^b)$, respectively where (c_2^b, c_1^b, c_0^b) are determined by $(k_{2b+1}^\pi, k_{2b}^\pi)$.

$(k_{2b+1}^\pi, k_{2b}^\pi)$	(c_2^b, c_1^b, c_0^b)
$(0, 0)$	$(2, 1, 0)$
$(0, 1)$	$(1, 2, 0)$
$(1, 0)$	$(2, 0, 1)$
$(1, 1)$	$(0, 1, 2)$

2.5   Existing Work on PRINTCIPHER

Abdelraheem et al. have given a differential attack on 22-round PRINTCIPHER[1]. Using the entire code book, they study the single-bit differentials in order to learn how the bits are permuted through the entire cipher, i.e., r rounds. Finding the rth root of this permutation then gives them the single-round permutation $\pi \circ \Pi$ and thus K^π.

We note that it is straightforward to invert the last S-box upon retrieving a ciphertext (it is present only to make hardware implementations smaller as it does not require any special logic for the last round as in e.g., AES). Thus, an attacker can extend the 22-round attack to 23 rounds at very low cost: the S-box only has to be inverted if the three bits in its output are the only bits that have a difference.

Two very recent publications reach further than 22 rounds. At CRYPTO, Leander et al. [9] showed how an "invariant subspace attack" allowed for a class of 2^{52} keys to be distinguished regardless of the number of rounds, so in particular for the full PRINTCIPHER. At SAC, Karakoç et al. [4] combined differential and linear cryptanalysis to reach 29 rounds on 4.54% and 31 rounds on .036% of the keys.

3   Linear Cryptanalysis

Originally introduced by Matsui [11], linear cryptanalysis has since been applied to a large number of cryptographic primitives in many different fashions. In the original form, the goal is to find some biased linear relation on bits in the progressing state of the cipher. If key bits involved in partially encrypting and/or decrypting are guessed correctly, the bias should be observable, while for wrong guesses, the bias should not appear. As a result, the partial key guess can be verified and the rest of the key found through an exhaustive search. The end result is an attack faster than exhaustive search, but the cost is that one needs to access many plaintext–ciphertext pairs (see Section 3.1).

While it is common to study trails on linear combinations of bits in plaintext and ciphertext,

$$P\left(\boldsymbol{\alpha} \cdot P = \boldsymbol{\beta} \cdot C + \boldsymbol{\gamma} \cdot K\right) = \frac{1}{2} \pm \epsilon,$$

where $\boldsymbol{\alpha}$, $\boldsymbol{\beta}$, $\boldsymbol{\gamma}$ are bitmasks, the most simple case is to study single-bit trails such as

$$P\left(p_{47} = c_{47}\right) = \frac{1}{2} + \epsilon.$$

This paper will exclusively deal with single-bit trails, possibly involving the xor of one bit of key, although it is no doubt possible to find many more trails by using multiple-bit trails. The reason we do this is that the single-bit trails appear very naturally in PRINTCIPHER.

We refer to ϵ as the *bias* of the trail, and an attacker will naturally try to find relations with as large bias as possible. In the PRINTCIPHER specification, the designers show that the optimal linear trails over r rounds of PRINTCIPHER have probability $\frac{1}{2} + 2^{-r-1}$, i.e., bias 2^{-r-1}. In this paper, we will exclusively look at such optimal trails.

The piling-up lemma [11] tells us how the bias diminishes over more rounds. In our context it means that piling two optimal trails on r_1 and r_2 rounds into one trail on $r = r_1 + r_2$ rounds, results in a bias of $2^{-r-1} = 2^{-r_1-r_2-1}$. This is not surprising: piling two optimal trails results in an optimal trail.

Every time we look at (e.g.,) $p_{47} \oplus c_{47}$, we actually look at a *sample*, a bit that is picked from some distribution. By looking at sufficiently many samples, we can make a sufficiently good guess on which distribution we are dealing with.

3.1   On the Importance of Finding Many Samples

In order to distinguish between two distributions on $\{0, 1\}$, one with $\mathrm{Prob}(1) = \frac{1}{2} + \epsilon$ and one with $\mathrm{Prob}(1) = \frac{1}{2}$, it is commonly accepted [11,2] that one needs ϵ^{-2} samples. One actually needs $\alpha\epsilon^{-2}$ samples, but the constant α is small enough to be ignored: this allows for easier analysis and comparisons of cryptanalytic results. In this paper, we will always need to obtain 2^{2r+2} samples.

An attacker can only access 2^{48} different plaintext–ciphertext pairs on PRINT-CIPHER, which seems to indicate that only 2^{48} samples can be found and that only 23-round trails can be used, i.e., less than half the number of rounds. If we want to use a trail on $(23 + s)$ rounds, we need to obtain $2^{2(23+s)+2} = 2^{48+2s}$ samples, i.e., 2^{2s} samples per plaintext–ciphertext pair.

In this paper, we will note how some particular features of PRINTCIPHER allow us to find trails where we can access several samples per plaintext–ciphertext pair. We also see how these samples are independent (enough) to make them usable in a cryptanalytic setting.

We will only consider iterated trails, i.e., trails beginning and ending at a common bit position. This is for simplicity: iterated trails can be used to trivially create trails on larger numbers of rounds. One can also see that by using iterated trails, the number of distinct π_b involved in the trail is kept to a minimum, which keeps the involved number of key bits decently small.

Recall that a sample s_j is a bit obtained by comparing a plaintext bit to a ciphertext bit (more generally, linear combinations). Crucial in linear cryptanalysis is counting how many samples are 1 resp. 0, i.e., deriving the sum $S = \sum_j s_j$. Kaliski and Robshaw [7] note that if one can find several linear approximations that involve the exact same key bits, i.e., the same bitmask γ, so that one can get several counts $S^i = \sum_j s_j^i$, one can use a weighted sum of these counts S^i — this measurement has the same expected value but a smaller variance. In particular, when the bias is the same for all linear approximations, the weighted sum is simply the average, which up to a multiplicative constant is the same as the overall number of samples that are 1, i.e., $\sum_{i,j} s_j^i$. It is then natural to think of the different s_j^i (with varying i and j) as different samples from the same underlying distribution.

4 Some Initial Observations

4.1 The S-Box

Some single-bit trails are available on the S-box and through the remainder of this paper, we will focus on these three:

$$\mathrm{Prob}(y_2 = x_2) = \mathrm{Prob}(y_1 = x_1) = \mathrm{Prob}(y_0 = x_0 \oplus 1) = \frac{1}{2} + 2^{-2}.$$

They are conveniently all from x_i to y_i, which is not strictly necessary but simplifies the presentation of the subsequent observations and attacks.

4.2 The Permutation π_b and the S-box

One can quite easily see that with $(y_2, y_1, y_0) = S(x_2, x_1, x_0)$ and $(y_2', y_1', y_0') = S(x_2, x_0, x_1)$, we always have $y_2 = y_2'$, see Table 4. This means that if we are only interested in tap 2 out of the S-box, it does not matter if x_1, x_0 are swapped or not before entering the S-box.

As a consequence, if we

- know three bits that enter $S \circ \pi_b$,
- want to know y_2 out of the S-box, and

Table 4. The S-box evaluated for all possible permutations on the input

(x_2, x_1, x_0)	$S(x_2, x_1, x_0)$	$S(x_1, x_2, x_0)$	$S(x_2, x_0, x_1)$	$S(x_0, x_1, x_2)$
$(0,0,0)$	$(0,0,0)$	$(0,0,0)$	$(0,0,0)$	$(0,0,0)$
$(0,0,1)$	$(0,0,1)$	$(0,0,1)$	$(0,1,1)$	$(1,1,1)$
$(0,1,0)$	$(0,1,1)$	$(1,1,1)$	$(0,0,1)$	$(0,1,1)$
$(0,1,1)$	$(1,1,0)$	$(1,0,0)$	$(1,1,0)$	$(1,0,1)$
$(1,0,0)$	$(1,1,1)$	$(0,1,1)$	$(1,1,1)$	$(0,0,1)$
$(1,0,1)$	$(1,0,0)$	$(1,1,0)$	$(1,0,1)$	$(1,0,0)$
$(1,1,0)$	$(1,0,1)$	$(1,0,1)$	$(1,0,0)$	$(1,1,0)$
$(1,1,1)$	$(0,1,1)$	$(0,1,1)$	$(0,1,1)$	$(0,1,1)$

– need to guess the permutation π_b, i.e., $(k_{2b+1}^\pi, k_{2b}^\pi)$,

then we only need to make three guesses on π_b.

The same property shows up on y_0 also, but not on y_1, see Table 4. We will use this observation to reduce the amount of guesswork we need to perform during partial encryption. We will use the notation π_b^3 to mark that we only guess a ternary digit, a trit, for π_b due to these properties.

Similarly, when we guess for a partial decryption, we often do not need to guess the whole permutation π_b, i.e., two bits, but only how it permutes one particular bit. We will (e.g.,) use the notation $\pi_b(2)$ to indicate that we only guess how the bit 2 is permuted by π_b.

5   A Key Bit Distinguisher

5.1   General Attack Idea

We will use a variant of linear cryptanalysis: we study single-bit trails that are biased for certain classes of keys and non-biased for other keys. As a very non-detailed example, consider a trail from the left-most bit to the left-most bit. It is readily apparent from Fig. 1 that such a trail exists and that it is iterated (although it is of course not obvious from the figure that it has a bias). We claim that we can distinguish individual bits of K^π using this trail: it is biased for half the keys and non-biased for the other half. Thus, if we can distinguish between these two distributions (i.e., if the bias is large enough and we have sufficiently many samples) we can determine the value of this key bit.

The sample trail considered here is "simple" as it is apparent to the naked eye, but it is possible to find several such trails over considerable numbers of rounds. As a consequence, there exist many classes of weak keys in PRINTCIPHER.

5.2   A Detailed Example

We now describe how to distinguish between two distributions: one where k_{30}^π is zero, and one where it is one. This allows for a partial-key recovery, i.e., learning one bit of the key, faster than brute force.

Note that $\Pi(15,2) = (15,2)$, and that for two of four keys, $\pi_{15}(2) = 2$. This happens precisely when $k_{30}^\pi = 0$ (see Table 5).

Table 5. How the individual bits $(2,1,0)$ are moved by the key-dependent permutation π_b, and for which keys $(k_{2b+1}^\pi, k_{2b}^\pi)$ it happens

Bit Move	Possible Keys	Bit Move	Possible Keys	Bit Move	Possible Keys
$0 \to 0$	$(0,0),(0,1)$	$1 \to 0$	$(1,0)$	$2 \to 0$	$(1,1)$
$0 \to 1$	$(1,0)$	$1 \to 1$	$(0,0),(1,1)$	$2 \to 1$	$(0,1)$
$0 \to 2$	$(1,1)$	$1 \to 2$	$(0,1)$	$2 \to 2$	$(0,0),(1,0)$

Thus, with $k_{30}^\pi = 0$, $(\pi \circ \Pi)(15, 2) = (15, 2)$. The probability that this bit then passes the S-box unaltered is $\frac{3}{4}$, so after a single round of encryption, we have

$$\text{Prob}(c_{47}^1 = p_{47} \oplus k_{47}^\oplus) = \frac{1}{2} + 2^{-2}.$$

For two rounds, we have

$$\text{Prob}(c_{47}^2 = p_{47}) = \frac{1}{2} + 2^{-3}$$

as the key xors cancel and with the use of the piling-up lemma. Generalizing to any even number of rounds, we have

$$\text{Prob}(c_{47}^r = p_{47}) = \frac{1}{2} + 2^{-r-1}.$$

For PRINTCIPHER on 22 rounds, we would need almost the entire code book, 2^{46} plaintext–ciphertext pairs.

We can also use the full code book, of size 2^{48}, to attack 23 rounds. We then have an odd number of rounds, and the key bit $k_{47}^\oplus$ shows up, so we utilize the relation

$$\text{Prob}(c_{47}^r = p_{47} \oplus k_{47}^\oplus) = \frac{1}{2} + 2^{-r-1}, \tag{1}$$

with $r = 23$. Things then get slightly more tricky, as we learn more about the key but need to distinguish between three distributions:

1. $c_{47}^r = p_{47}$ with probability $\frac{1}{2}$, implying $k_{30}^\pi = 1$.
2. $c_{47}^r = p_{47}$ with "high" probability, implying $k_{30}^\pi = 0$ and $k_{47}^\oplus = 0$.
3. $c_{47}^r = p_{47}$ with "low" probability, implying $k_{30}^\pi = 0$ and $k_{47}^\oplus = 1$.

5.3 More Linear Trails on One Round of PRINTCIPHER

There are in total four iterated single-round trails, and we list them in Table 6. Some constants arise as the S-box flips bit 0 with probability $\frac{3}{4}$ rather than preserves it, and as bits of RC_i enter.

Table 6. The iterated single-round trails on PRINTCIPHER, extended to several rounds. All trails have bias 2^{-r-1}.

Trail	Requirement
$c_0^r = p_0 \oplus k_0^\oplus I_r \oplus d_r$	$k_1^\pi = 0$
$c_{23}^r = p_{23} \oplus k_{23}^\oplus I_r$	$(k_{15}^\pi, k_{14}^\pi) = (0, 1)$
$c_{24}^r = p_{24} \oplus k_{24}^\oplus I_r \oplus I_r$	$(k_{17}^\pi, k_{16}^\pi) = (1, 0)$
$c_{47}^r = p_{47} \oplus k_{47}^\oplus I_r$	$k_{30}^\pi = 0$
$d_r = \left(r + 1 + \sum_{0 \le i < r} RC_i \right) \bmod 2$	

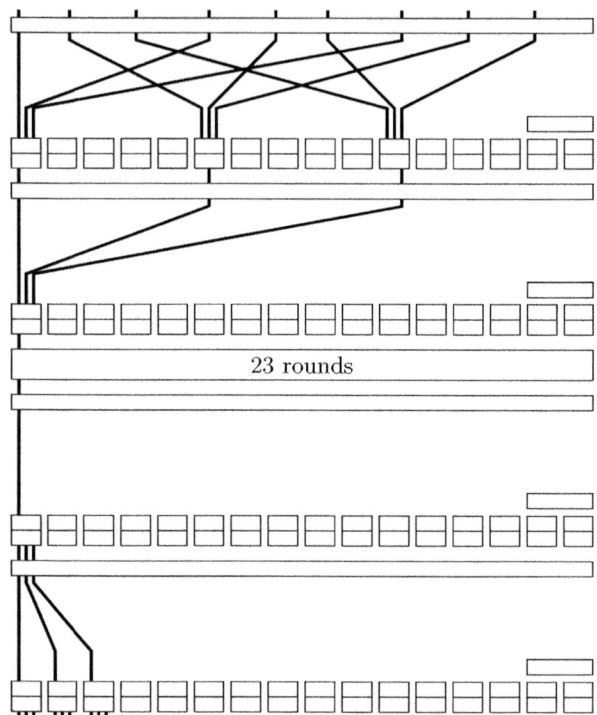

Fig. 2. Performing two rounds of partial encryption and decryption to access the bits at position $(15, 2)$

6   Guessing Keybits for Partial Encryption and Decryption

The above observation can be used as-is to mount an attack on 23-round PRINT-CIPHER, recovering up to three bits of the key, but it is straightforward to derive an even more powerful attack on 27 rounds of PRINTCIPHER: if a guessed partial key is correct, we should observe the bias, while if the guess is bad, the behaviour should be (more) random.

First, we assume that $k_{30}^\pi = 0$, meaning our attack only works for a fraction 2^{-1} of the keys. Then, we aim to decrypt two rounds at the end and encrypt two rounds at the top of PRINTCIPHER. Thus, we need to guess the bits and trits listed in Table 7. There are in total $N = 2^{13} \cdot 3^3 \approx 2^{17.8}$ guesses. See Fig. 2 for an overview of the partial calculations.

Due to the property observed in Section 4.2, we do not need to guess k_{31}^π. We have assumed $k_{30}^\pi = 0$ to fix $\pi_{15}(2) = 2$ and this is enough to predict tap 2 out of the S-box. It does not matter whether π_{15} is trivial or swaps bits 0 and 1.

We call the plaintext (resp. ciphertext) bits that affect the partial encryption (resp. decryption) to the bits we are interested in *active*. There are nine active

Table 7. The bits and trits required for encryption, decryption, and both, when encrypting/decrypting two rounds to access the bits at position $(15, 2)$

Encryption	$k_{42}^\oplus$, $k_{37}^\oplus$, $k_{31}^\oplus$, $k_{26}^\oplus$, $k_{21}^\oplus$, $k_{15}^\oplus$, $k_{10}^\oplus$, $k_{5}^\oplus$, π_{10}, π_{5}^3
Decryption	$k_{46}^\oplus$, $k_{45}^\oplus$, $\pi_{14}(2)$, $\pi_{13}(2)$
Both	$k_{47}^\oplus$

bits in the plaintext and nine in the ciphertext. For a plaintext–ciphertext pair (P, C) we can collect these bits into an eighteen-bit word $w = (p_{47}, p_{31}, \ldots, c_{39})$.

We describe the attack: Acquire all 2^{48} plaintext–ciphertext pairs (P^j, C^j). Categorize them according to the active bits, i.e., for each possible word w, count how often it appears. Denote these counters R_w. This is the data collection part of the attack.

We then begin analyzing the data. For each plaintext–ciphertext pair and for each guess of key material, denoted by G_i, $0 \le i < N$, we will calculate two rounds of encryption and decryption, $\hat{P}^j = \phi_{0,2}(G_i, P^j)$, $\hat{C}^j = \phi_{25,27}^{-1}(G_i, C^j)$, and count how often $\hat{c}_{47}^j = \hat{p}_{47}^j \oplus k_{47}^{\oplus i}$. This is done using N counters S_i. An efficient way of doing this [10] is to use the counters R_w. For each word w and each keyguess G_i, we do the partial calculations $\hat{P}^{i,w} = \phi_{0,2}(G_i, P_w)$, $\hat{C}^{i,w} = \phi_{25,27}^{-1}(G_i, C_w)$. P_w (C_w) is some plaintext (ciphertext) which has the correct active bits as determined by w. If $\hat{c}_{47}^{i,w} = \hat{p}_{47}^{i,w} \oplus k_{47}^{\oplus i}$, we add R_w to S_i.

By sorting all S_i, we can get a ranking of the different guesses. We pick the most likely guess, brute force all non-guessed bits and hopefully recover the key. Otherwise, we pick the second most likely guess, etc. The exact number of bits that need to be brute forced will be different for different guesses: where we guessed a trit (e.g.,) $\pi_{14}(2)$, we will have recovered one or two bits of (k_{29}^π, k_{28}^π). As long as the correct guess is ranked on the upper half of the sorted list of counters, the entire key will be found faster than what can be expected from a brute force (2^{78}).

The counters R_w are used for saving time [10]. Several other improvements can be made, also from [10]: We should not make $2^{18} \cdot N$ partial encryptions and decryptions. First, the plaintext and ciphertext operations can be separated completely, so that we only need to make $2^9 \cdot N + 2^9 \cdot N$ encryptions/decryptions. Second, since the overlap in encryption and decryption with respect to the guessed bits is very small, we only need to perform $2^9 \cdot N_e + 2^9 \cdot N_d$ encryptions/decryptions where N_e (N_d) is the number of key guesses that actually affect the encryption (decryption). Third, doing two complete PRINTCIPHER rounds in both directions is unneccessary as we only need to perform "partial rounds", i.e., use some small number of S-boxes.

6.1   Experimental Results

We have implemented this attack on $7 + 4 = 11$ rather than $23 + 4 = 27$ rounds of PRINTCIPHER. This means that we guess the same bits and perform the same partial encryptions, but that the bias is larger so that it is feasible for us to perform many attacks in order to gather statistics.

It turns out that over 2^{13} different weak keys, the attack works with probability 0.78. That is, almost four times out of five, the correct key ranks on the upper half of the sorted list of S_i.

6.2   Analyzing the Attack Complexity

The attack consists of data collection and data analysis. The latter in turn consists of 1) deriving two sets of counters, N_e for encryption and N_d for decryption, and 2) combining these to find N counters. If the number of active bits in the plaintext (ciphertext) is denoted a_e (a_d) and the number of active S-boxes in encryption (decryption) is denoted A_e (A_d), the time complexities are given by

$$T^{collect} = \epsilon^{-2},$$
$$T^{count} = \frac{2^{a_e} N_e A_e + 2^{a_d} N_d A_d}{16 \cdot r},$$
$$T^{combine} = 2^{a_e + a_d} N.$$

The first two measurements are normalized to r-round PRINTCIPHER evaluations, while the last describes the number of "simple" bit and integer operations needed to calculate the counters S_i.

For the specific attack detailed above, we have $N_e = 2^{11} \cdot 3$ and $N_d = 2^3 \cdot 3^2$. Since $(a_e, a_d, A_e, A_d) = (9, 9, 4, 4)$, the complexities turn out at

$$T^{collect} = 2^{2 \cdot 24} = 2^{48},$$
$$T^{count} = \frac{2^9 \cdot 2^{11} \cdot 3 \cdot 4 + 2^9 \cdot 2^3 \cdot 3^2 \cdot 4}{16 \cdot 27} \approx 2^{15},$$
$$T^{combine} = 2^{9+9} \cdot N = 2^{18} \cdot 2^{13} \cdot 3^3 \approx 2^{36}.$$

This suggests that the most time consuming part is the data collection where we need to generate and look at 2^{48} plaintext–ciphertext pairs.

6.3   Reaching the Limit: 28 Rounds

We note that in the attacks on 27 rounds, guessing and encrypting is more expensive than guessing and decrypting: during decryption, we first invert S, and then only need to control one bit in some π_b. On the other hand, during encryption, we need to fully control the permutation, so that we can calculate all three bits that go into the S-box. This leads to more expensive guesswork on π_b and especially on $K^\oplus$. Thus, the natural approach for extending the attack by one round is to add another round in the partial decryption.

In Table 8, we list the bits and trits involved in partially decrypting and encrypting from 28 rounds to 23. The attack requires $N = 2^{18} \cdot 3^8 \approx 2^{30.7}$ guesses, partitioned as $N_e = 2^{11} \cdot 3$ and $N_d = 2^9 \cdot 3^8$. With $(a_e, a_d, A_e, A_d) = (9, 27, 4, 13)$,

Table 8. The bits and trits required for encryption, decryption, and both, when encrypting/decrypting two/three rounds to access the bits at position $(15, 2)$. Note the overlap in π_{10}.

Encryption	$k_{37}^{\oplus}$, $k_{31}^{\oplus}$, $k_{26}^{\oplus}$, $k_{21}^{\oplus}$, $k_{15}^{\oplus}$, $k_{10}^{\oplus}$, $k_5^{\oplus}$, π_{10}, π_5^3
Decryption	$k_{46}^{\oplus}$, $k_{45}^{\oplus}$, $k_{44}^{\oplus}$, $k_{43}^{\oplus}$, $k_{41}^{\oplus}$, $k_{40}^{\oplus}$, $k_{39}^{\oplus}$, $\pi_{14}(2)$, $\pi_{13}(2)$, $\pi_{12}(2)$, $\pi_{11}(2)$, $\pi_{10}(2)$, $\pi_9(2)$, $\pi_8(2)$, $\pi_7(2)$
Both	$k_{42}^{\oplus}$, $k_{47}^{\oplus}$

we have

$$T^{collect} = 2^{2 \cdot 24} = 2^{48},$$

$$T^{count} = \frac{2^9 \cdot 2^{11} \cdot 3 \cdot 4 + 2^{27} \cdot 2^9 \cdot 3^8 \cdot 13}{16 \cdot 28} \approx 2^{44},$$

$$T^{combine} = 2^{9+27} \cdot N = 2^{36} \cdot 2^{18} \cdot 3^8 \approx 2^{67}.$$

28 rounds seems to be the best we can do: using a single trail, we have not been able to go beyond 28 rounds while keeping the attack costs below exhaustive search.

7   On More Rounds of PRINTCIPHER: Complementary Trails

We generalize our observation slightly and give an example two-round trail: with $(k_{23}^{\pi}, k_{22}^{\pi}, k_6^{\pi}) = (1, 1, 0)$,

$$\text{Prob}(c_{11}^2 = p_{11} \oplus k_{11}^{\oplus} \oplus k_{35}^{\oplus}) = \frac{1}{2} + 2^{-3}.$$

Note in particular how there is a complementary trail,

$$\text{Prob}(c_{35}^2 = p_{35} \oplus k_{11}^{\oplus} \oplus k_{35}^{\oplus}) = \frac{1}{2} + 2^{-3},$$

see Fig. 3. The complementary trail depends on the exact same key configuration and allows us to collect *two* samples with every plaintext–ciphertext pair. We show in Section 8.1 that this works, i.e., the samples can be considered as independent.

We do not give all the two-round trails on PRINTCIPHER, as we will not use them in the remainder of the paper. We only note that due to the structure of PRINTCIPHER, every S-box is used precisely once so far in the paper: either in one trail on one round (S-boxes 0, 7, 8, 15), or in two complementary trails on two rounds.

As a particular four-round trail that we will use later, we give

$$\text{Prob}(c_{37}^4 = p_{37} \oplus k_{37}^{\oplus} \oplus k_{17}^{\oplus} \oplus k_4^{\oplus} \oplus k_{12}^{\oplus} \oplus 1) = \frac{1}{2} + 2^{-5}, \tag{2}$$

which is activated by $(k_{25}^{\pi}, k_{24}^{\pi}, k_{10}^{\pi}, k_9^{\pi}, k_3^{\pi}) = (1, 0, 0, 0, k_2^{\pi})$.

The total number of iterated trails over various number of rounds are given in Table 9.

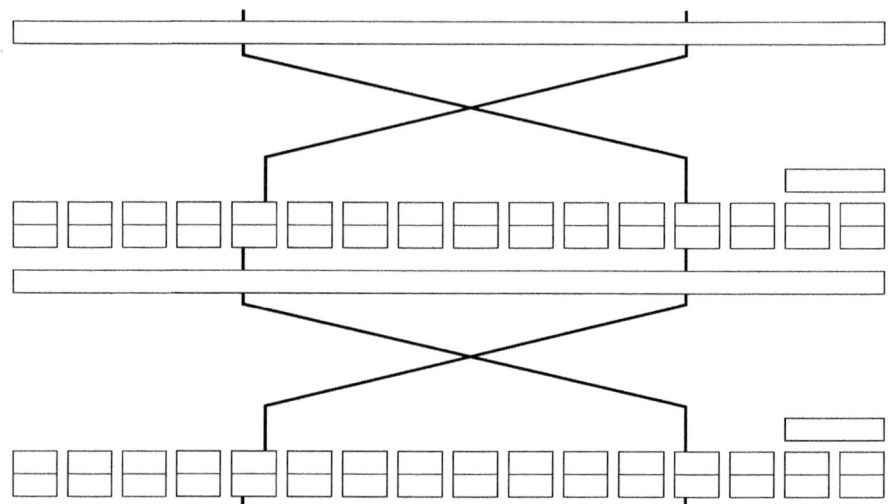

Fig. 3. Two complementary trails on two rounds of PRINTCIPHER. Both trails are activated by $(k_{23}^\pi, k_{22}^\pi, k_6^\pi) = (1, 1, 0)$.

Table 9. The number of iterated trails over various number of rounds

#Rounds	#Trails	#Rounds	#Trails	#Rounds	#Trails	#Rounds	#Trails
1	4	5	154	9	5806	13	138662
2	16	6	424	10	13366	14	283810
3	28	7	1040	11	30430	15	560608
4	96	8	2584	12	65808	16	1075000

7.1   More Attacks on 27/28 Rounds

We can use basically any trail on 23 rounds to create attacks on 27/28 rounds. Do note that the trail on bit $(15, 2)$ is very nice as the partial encryptions and decryptions involve few bits of $K^\oplus$ and K^π, due to S-box reuse. Most other trails involve more guesswork. As an example, using (2) 5.75 times yields

$$\text{Prob}(c_{12}^{25} = c_{37}^2 \oplus k_{12}^\oplus \oplus 1) = \frac{1}{2} + 2^{-24},$$

from which we can build an attack on 28 rounds. We have $N = 2^{27} \cdot 3^6 \approx 2^{36.5}$, $N_e = 2^{14} \cdot 3$, $N_d = 2^{13} \cdot 3^7$ and $(a_e, a_d, A_e, A_d) = (9, 27, 4, 13)$. The complexities are

$$T^{collect} = 2^{2 \cdot 24} = 2^{48},$$

$$T^{count} = \frac{2^9 \cdot 2^{14} \cdot 3 \cdot 4 + 2^{27} \cdot 2^{13} \cdot 3^7 \cdot 13}{16 \cdot 28} \approx 2^{46},$$

$$T^{combine} = 2^{9+27} \cdot N = 2^{36} \cdot 2^{27} \cdot 3^6 \approx 2^{73}.$$

The bits and trits guessed are listed in Appendix B.

7.2   On False Positives

By piling the single-round trail on the left-most bit, we see that e.g., $\mathrm{Prob}(c_{47}^{10} = p_{47}) = \frac{1}{2} + 2^{-11}$ when $k_{30}^{\pi} = 0$. However, there are several other ways of obtaining this distribution.

All in all, there are 102 different trails from $(15, 2)$ to $(15, 2)$ over ten rounds, each corresponding to a different class of keys. This means that a biased distribution can be explained by any of these trails, and thus any of these classes. Due to this, an attacker will prefer to use short, iterated trails involving few bits of K^{π}.

8   Using Complementary Trails to Distinguish on 24-Round Trails

We will now construct 24-round trails with bias 2^{-25}. By using trails that allow four samples per plaintext–ciphertext pair, we can get in total 2^{50} samples, allowing us to distinguish the distribution.

The best iterated trails on 24 rounds are given in Table 10. They are "best" in the sense that they use a small number of key bits (5), yet allow four complementary trails each, so that we can get the required number of samples. In fact they are constructed from iterated four-round trails, that we have piled in order to cancel the bits that appear from $K^{\oplus}$ (cf. Section 5.2).

8.1   Samples Are Independent (Enough)

The connection between the bias ϵ and the required number of samples ϵ^{-2} relies on the independence of the samples, and it is not obvious that the samples we pick are independent. Most cryptanalysis simply assumes that the samples are independent, or at least independent enough for the attacks to still be possible. Verifying the independence through simulation is common, at least on a smaller number of rounds or reduced-size versions of the algorithm ("PRINTCIPHER-12"), where it is practically possible.

We need to be a little bit more wary than usual as we pick several samples from the same plaintext–ciphertext pair — it is not hard to realize that the calculations behind the four samples have affected each other, and it is not impossible that samples obtained from the same plaintext–ciphertext pair are so dependent that they do not contribute (much) more than one sample from an information-theoretic point of view. If this is the case, we would not be able to exploit any bias smaller than (about) 2^{-23}.

Thus, we have done the following on eight-round PRINTCIPHER: We use 2^{18} plaintext–ciphertext pairs to derive equally many samples on bit $(1, 1)$, and from this we guess whether the key is in the upper-left class from Table 10 by comparing the number of samples that are 1 to some pre-defined threshold derived to yield a 50% success rate. This gives false positives/negatives with probabilities 0.03/0.50, respectively. Similar probabilities are observed for the three complementary trails, when used one on one.

Table 10. The iterated trails on eight rounds ($r = 8$) composed from four-round iterated trails, depending only on five bits of K^π. All trails have bias 2^{-r-1}, and the constants e_j^r arise from the round constants RC_i. The trails are easily extended to e.g., 24 rounds ($r = 24$), in which case only the constants need to be rechecked. (The symmetrically inclined reader have ample reasons to admire this table.)

Trail	S-boxes	Trail	S-boxes
$c_4^r = p_4 \oplus e_4^r$	4,12,5,1,4,12,5,1,...	$c_{10}^r = p_{10}$	10,14,11,3,10,14,11,3,...
$c_{12}^r = p_{12} \oplus e_{12}^r$	12,5,1,4,12,5,1,4,...	$c_{30}^r = p_{30}$	14,11,3,10,14,11,3,10,...
$c_{17}^r = p_{17} \oplus e_{17}^r$	1,4,12,5,1,4,12,5,...	$c_{35}^r = p_{35}$	3,10,14,11,3,10,14,11,...
$c_{37}^r = p_{37} \oplus e_{37}^r$	5,1,4,12,5,1,4,12,...	$c_{43}^r = p_{43}$	11,3,10,14,11,3,10,14,...
Key class		Key class	
$(k_{25}^\pi, k_{24}^\pi, k_{10}^\pi, k_9^\pi, k_3^\pi) = (1,0,0,0,k_2^\pi)$		$(k_{29}^\pi, k_{22}^\pi, k_{21}^\pi, k_7^\pi, k_6^\pi) = (k_{28}^\pi, 0,0,0,1)$	
Trail	S-boxes	Trail	S-boxes
$c_7^r = p_7$	7,7,6,2,7,7,6,2,...	$c_{24}^r = p_{24}$	8,9,13,8,8,9,13,8,...
$c_{18}^r = p_{18}$	2,7,7,6,2,7,7,6,...	$c_{25}^r = p_{25}$	9,13,8,8,9,13,8,8,...
$c_{22}^r = p_{22}$	6,2,7,7,6,2,7,7,...	$c_{29}^r = p_{29}$	13,8,8,9,13,8,8,9,...
$c_{23}^r = p_{23}$	7,6,2,7,7,6,2,7,...	$c_{40}^r = p_{40}$	8,8,9,13,8,8,9,13,...
Key class		Key class	
$(k_{15}^\pi, k_{14}^\pi, k_{13}^\pi, k_{12}^\pi, k_5^\pi) = (1,1,1,0,k_4^\pi)$		$(k_{27}^\pi, k_{19}^\pi, k_{18}^\pi, k_{17}^\pi, k_{16}^\pi) = (k_{26}^\pi, 0,1,1,1)$	
Constants ($r = 8$)		Constants ($r = 24$)	
$(e_4^8, e_{12}^8, e_{17}^8, e_{37}^8) = (1,1,1,1)$		$(e_4^{24}, e_{12}^{24}, e_{17}^{24}, e_{37}^{24}) = (1,0,1,1)$	

If we instead use only 2^{16} plaintext–ciphertext pairs, but pick 2^2 samples from each pair, we are able to carry out the attack with seemingly unchanged success: The probabilities of false positives/negatives are 0.02/0.50. These results have been obtained by attacking 2^{14} keys from each class and are listed in Table 11.

8.2 Partial Encryption and Decryption for 29 Rounds

Similar to in Section 6, we aim to guess key bits for partial encryptions and decryptions. Previously, we were able to add five rounds in this way to construct a 28-round attack using a 23-round trail. Now, using the 24-round trails, we reach 29 rounds. Again, we use the upper-left key class in Table 10.

The key observation is that we can divide all of the work, so that we deal with the four trails completely independently. If we number the trails $j = 1, 2, 3, 4$, we will have time complexities

$$T_j^{collect} = \epsilon^{-2},$$

$$T_j^{count} = \frac{2^{a_e^j} N_e^j A_e^j + 2^{a_d^j} N_d^j A_d^j}{16 \cdot r},$$

$$T_j^{combine} = 2^{a_e^j + a_d^j} N.$$

for producing the four different lists of counters S_i^j. In order to combine all counters S_i^j into N-many counters S_i we need to do N rather simple operations.

Table 11. The attack in Section 8.1 was carried out on 2^{14} different keys using either one sample per plaintext–ciphertext pair or four samples per pair but fewer pairs. "True Pos. Ratio" shows how frequently a key belonging to the keyclass was identified as such. Similarly, "True Neg. Ratio" shows how often a key not belonging to the keyclass was correctly excluded.

Trail	Pairs	Samples/Pair	True Pos. Ratio	True Neg. Ratio
$c_4^8 = p_4 \oplus 1$	2^{18}	1	0.50	0.97
$c_{12}^8 = p_{12} \oplus 1$	2^{18}	1	0.50	0.97
$c_{17}^8 = p_{17} \oplus 1$	2^{18}	1	0.51	0.97
$c_{37}^8 = p_{37} \oplus 1$	2^{18}	1	0.51	0.97
all four	2^{16}	2^2	0.51	0.98

Note that we have made related, but not identical, guesses on the permutations, e.g., by guessing $\pi_{14}(2)$ when using trail 1 and $\pi_{14}(0)$ when using trail 2. Some care must be taken here, but it does not affect the cost of this step, which remains at $T^{finalize} = N$ quite simple operations.

Specific guesses are listed in Appendix B. All $(a_e^j, a_d,^j, A_e^j, A_d^j) = (9, 27, 4, 13)$. The total attack complexities are

$$T^{collect} = 2^{2\cdot24} = 2^{48},$$

$$T^{count} = \sum_j T_j^{count} \approx 2^{50},$$

$$T^{combine} = \sum_j T_j^{combine} \approx 2^{76},$$

$$T^{finalize} = N = 2^{62} \cdot 3^3 \approx 2^{67}.$$

Although brute force costs 2^{75}, as we assume five bits of the key, we claim that 2^{76} "simple" operations compare favorably to 2^{75} evaluations of 29-round PRINTCIPHER.

Let us briefly comment on the possibility of using 25-round trails with bias 2^{-26}: if we can get 16 samples per plaintext–ciphertext pair, we have the necessary 2^{52} samples. As we need to involve all π_b, we would put restrictions on at least 16 bits of the key. This puts the brute force cost at 2^{64} or lower, which seems to be too low for the attack to be meaningful. Another obstacle to this attack is that the complementary trails are not completely identical, as different bits of $K^{\oplus}$ will appear.

9   Conclusion

Table 12 summarizes the attacks on 27–28 rounds of PRINTCIPHER outlined in this paper. Several more attacks are available for several more key classes.

We note some particular observations that all arise from the structure of PRINTCIPHER and the use of the exact same round key throughout the cipher:

Table 12. A summary of the explicit attacks on 27-, 28- and 29-round PRINTCIPHER presented in this paper. r denotes the length of the trail(s) used, and R denotes that R-round PRINTCIPHER is attacked. 'Enc' ('dec') tells how many rounds are partially encrypted (decrypted). T^{count} and $T^{combine}$ are rounded to the nearest integer power of two.

Trail	r	enc	dec	R	Key fraction	$T^{collect}$	T^{count}	$T^{combine}$
$c_{47}^{25} = c_{47}^2 \oplus k_{47}^{\oplus}$	23	2	2	27	2^{-1}	2^{48}	2^{15}	2^{36}
$c_{47}^{25} = c_{47}^2 \oplus k_{47}^{\oplus}$	23	2	3	28	2^{-1}	2^{48}	2^{44}	2^{67}
$c_{12}^{25} = c_{37}^2 \oplus k_{12}^{\oplus} \oplus 1$	23	2	3	28	2^{-5}	2^{48}	2^{46}	2^{73}
$c_4^{26} = c_4^2$ and more	24	2	3	29	2^{-5}	2^{48}	2^{50}	2^{76}

- When there is a non-decomposable, iterated r-round trail there are in fact r complementary trails, allowing r samples per plaintext–ciphertext pair.
- When we guess for a partial encryption/decryption, there is overlap between the bits that activate the trail and those we need for encryption/decryption.

With this work, linear cryptanalysis has reached 28 rounds of PRINTCIPHER. We have used weak key classes which means that we need to carry out several attacks in parallel in order to have a high probability of success. However, we have seen that there are many large key classes and in particular several of them only depend on one or two bits of the key. This means our results "invalidate" several more keys than previous results on PRINTCIPHER. The exception is the differential attack which worked on all keys but only reached 23 rounds.

We have exclusively studied PRINTCIPHER-48, but our observations are no doubt applicable to PRINTCIPHER-96 as well, where it seems reasonable that our techniques could be used to reach around 52–55 rounds. Another area of future research could be to look at the linear hull effect. The work in [9] and [4] suggest that the linear hull of PRINTCIPHER can behave in unexpected ways. It might be possible to cause peculiar effects to arise, e.g., by fixing more bits of the key, in order to reach further into PRINTCIPHER with linear cryptanalysis.

As a further research direction, we note that by inverting the $(S \circ \pi_b)$ where π_b is (partly) assumed, the number of active bits could be reduced. The technique would apply to all attacks in this paper, but the full gain of this remains to be determined.

The complementary trails that arise in PRINTCIPHER are very interesting, and allowed us to add one round to the attacks, albeit for a smaller class of keys. It would be very interesting to see if this complementary property could lead to more observations on PRINTCIPHER.

Acknowledgment. This work was supported by the Swedish Foundation for Strategic Research (SSF) through its Strategic Center for High Speed Wireless Communication at Lund. The authors wish to thank the anonymous reviewers whose comments helped improve the paper.

References

1. Abdelraheem, M.A., Leander, G., Zenner, E.: Differential Cryptanalysis of Round-Reduced PRINTCIPHER: Computing Roots of Permutation. In: Joux, A. (ed.) FSE 2011. LNCS, vol. 6733, pp. 1–17. Springer, Heidelberg (2011)
2. Baignères, T., Junod, P., Vaudenay, S.: How Far Can We Go Beyond Linear Cryptanalysis? In: Lee, P.J. (ed.) ASIACRYPT 2004. LNCS, vol. 3329, pp. 432–450. Springer, Heidelberg (2004)
3. Bogdanov, A., Knudsen, L.R., Leander, G., Paar, C., Poschmann, A., Robshaw, M.J.B., Seurin, Y., Vikkelsoe, C.: PRESENT: An Ultra-Lightweight Block Cipher. In: Paillier, P., Verbauwhede, I. (eds.) CHES 2007. LNCS, vol. 4727, pp. 450–466. Springer, Heidelberg (2007)
4. Karako, F.: Combined Differential and Linear Cryptanalysis of Reduced-Round PRINTCIPHER. In: Selected Areas in Cryptography—SAC 2011. LNCS. Springer, Heidelberg (to appear, 2011)
5. De Cannière, C., Dunkelman, O., Knežević, M.: KATAN and KTANTAN — A Family of Small and Efficient Hardware-Oriented Block Ciphers. In: Clavier, C., Gaj, K. (eds.) CHES 2009. LNCS, vol. 5747, pp. 272–288. Springer, Heidelberg (2009)
6. Hong, D., Sung, J., Hong, S.H., Lim, J.-I., Lee, S.-J., Koo, B.-S., Lee, C.-H., Chang, D., Lee, J., Jeong, K., Kim, H., Kim, J.-S., Chee, S.: HIGHT: A New Block Cipher Suitable for Low-Resource Device. In: Goubin, L., Matsui, M. (eds.) CHES 2006. LNCS, vol. 4249, pp. 46–59. Springer, Heidelberg (2006)
7. Kaliski Jr., B.S., Robshaw, M.J.B.: Linear Cryptanalysis Using Multiple Approximations. In: Desmedt, Y.G. (ed.) CRYPTO 1994. LNCS, vol. 839, pp. 26–39. Springer, Heidelberg (1994)
8. Knudsen, L., Leander, G., Poschmann, A., Robshaw, M.J.B.: PRINTCIPHER: A Block Cipher for IC-Printing. In: Mangard, S., Standaert, F.-X. (eds.) CHES 2010. LNCS, vol. 6225, pp. 16–32. Springer, Heidelberg (2010)
9. Leander, G., Abdelraheem, M.A., AlKhzaimi, H., Zenner, E.: A Cryptanalysis of PRINTCIPHER: The Invariant Subspace Attack. In: Rogaway, P. (ed.) CRYPTO 2011. LNCS, vol. 6841, pp. 206–221. Springer, Heidelberg (2011)
10. Matsui, M.: The First Experimental Cryptanalysis of the Data Encryption Standard. In: Desmedt, Y.G. (ed.) CRYPTO 1994. LNCS, vol. 839, pp. 1–11. Springer, Heidelberg (1994)
11. Matsui, M.: Linear Cryptanalysis Method for DES Cipher. In: Helleseth, T. (ed.) EUROCRYPT 1993. LNCS, vol. 765, pp. 386–397. Springer, Heidelberg (1994)

A   The Standard Permutation

Table 13 lists the key-independent permutation in PRINTCIPHER.

Table 13. The standard permutation Π in PRINTCIPHER. Bits at positions 'In' are moved to positions 'Out'.

In	Out	In	Out	In	Out	In	Out
$(0,0)$	$(0,0)$	$(4,0)$	$(12,0)$	$(8,0)$	$(8,1)$	$(12,0)$	$(4,2)$
$(0,1)$	$(1,0)$	$(4,1)$	$(13,0)$	$(8,1)$	$(9,1)$	$(12,1)$	$(5,2)$
$(0,2)$	$(2,0)$	$(4,2)$	$(14,0)$	$(8,2)$	$(10,1)$	$(12,2)$	$(6,2)$
$(1,0)$	$(3,0)$	$(5,0)$	$(15,0)$	$(9,0)$	$(11,1)$	$(13,0)$	$(7,2)$
$(1,1)$	$(4,0)$	$(5,1)$	$(0,1)$	$(9,1)$	$(12,1)$	$(13,1)$	$(8,2)$
$(1,2)$	$(5,0)$	$(5,2)$	$(1,1)$	$(9,2)$	$(13,1)$	$(13,2)$	$(9,2)$
$(2,0)$	$(6,0)$	$(6,0)$	$(2,1)$	$(10,0)$	$(14,1)$	$(14,0)$	$(10,2)$
$(2,1)$	$(7,0)$	$(6,1)$	$(3,1)$	$(10,1)$	$(15,1)$	$(14,1)$	$(11,2)$
$(2,2)$	$(8,0)$	$(6,2)$	$(4,1)$	$(10,2)$	$(0,2)$	$(14,2)$	$(12,2)$
$(3,0)$	$(9,0)$	$(7,0)$	$(5,1)$	$(11,0)$	$(1,2)$	$(15,0)$	$(13,2)$
$(3,1)$	$(10,0)$	$(7,1)$	$(6,1)$	$(11,1)$	$(2,2)$	$(15,1)$	$(14,2)$
$(3,2)$	$(11,0)$	$(7,2)$	$(7,1)$	$(11,2)$	$(3,2)$	$(15,2)$	$(15,2)$

B   Bits Involved in Given Attacks

Tables 14 and 15 lists what key material is guessed for the attacks on 28 and 29 rounds using the four-round trail 5.75 and 6 times.

Table 14. The bits and trits required for encryption, decryption, and both, when encrypting/decrypting two/three rounds to access the bits at position $(12,1)/(4,0)$

Encryption	$k_{46}^{\oplus}$, $k_{44}^{\oplus}$, $k_{41}^{\oplus}$, $k_{30}^{\oplus}$, $k_{28}^{\oplus}$, $k_{25}^{\oplus}$, $k_{9}^{\oplus}$, $k_{4}^{\oplus}$, π_{14}^{3}, π_{9},
Decryption	$k_{38}^{\oplus}$, $k_{37}^{\oplus}$, $k_{19}^{\oplus}$, $k_{18}^{\oplus}$, $k_{17}^{\oplus}$, $k_{16}^{\oplus}$, $k_{15}^{\oplus}$, $k_{13}^{\oplus}$, k_{8}^{π}, $\pi_{15}(0)$, $\pi_{14}(0)$, $\pi_{13}(0)$, $\pi_{12}(0)$, $\pi_{6}(1)$, $\pi_{3}(1)$, $\pi_{2}(1)$, $\pi_{0}(1)$
Both	$k_{36}^{\oplus}$, $k_{20}^{\oplus}$, $k_{14}^{\oplus}$, $k_{12}^{\oplus}$

Table 15. The bits and trits guessed in the attack on 29-round PRINTCIPHER by encrypting/decrypting two/three rounds

j	Pos.	N^j		Bits
1	$(12,1)$	$2^{31} \cdot 3^6$	Enc.	$k_9^\oplus$, $k_{12}^\oplus$, $k_{14}^\oplus$, $k_{20}^\oplus$, $k_{25}^\oplus$, $k_{28}^\oplus$, $k_{30}^\oplus$, $k_{36}^\oplus$, π_9, π_{14}^3
			Dec.	$k_0^\oplus$, $k_1^\oplus$, $k_2^\oplus$, $k_3^\oplus$, $k_5^\oplus$, $k_{15}^\oplus$, $k_{16}^\oplus$, $k_{17}^\oplus$, $k_{37}^\oplus$, $k_{45}^\oplus$, $k_{46}^\oplus$, $k_{47}^\oplus$, π_0, $\pi_2(0)$, $\pi_3(0)$, π_{11}^3, $\pi_{13}(2)$, $\pi_{14}(2)$, π_{15}
			Both	$k_4^\oplus$, $k_{41}^\oplus$, $k_{44}^\oplus$, $k_{46}^\oplus$
2	$(5,2)$	$2^{23} \cdot 3^9$	Enc.	$k_1^\oplus$, $k_7^\oplus$, $k_{21}^\oplus$, $k_{23}^\oplus$, $k_{28}^\oplus$, $k_{33}^\oplus$, $k_{37}^\oplus$, $k_{39}^\oplus$, $k_{44}^\oplus$, k_2^π, π_7^3
			Dec.	$k_3^\oplus$, $k_4^\oplus$, $k_9^\oplus$, $k_{10}^\oplus$, $k_{11}^\oplus$, $k_{13}^\oplus$, $k_{14}^\oplus$, $k_{15}^\oplus$, $k_{16}^\oplus$, $\pi_0(1)$, $\pi_3(0)$, k_{11}^π, $\pi_9(0)$, $\pi_{10}(0)$, $\pi_{11}(0)$, $\pi_{13}(0)$, $\pi_{14}(0)$, $\pi_{15}(0)$
			Both	$k_5^\oplus$, $k_{12}^\oplus$, $k_{17}^\oplus$
3	$(4,0)$	$2^{21} \cdot 3^6$	Enc.	$k_1^\oplus$, $k_4^\oplus$, $k_6^\oplus$, $k_{22}^\oplus$, $k_{28}^\oplus$, $k_{33}^\oplus$, $k_{44}^\oplus$, k_2^π
			Dec.	$k_{13}^\oplus$, $k_{14}^\oplus$, $k_{15}^\oplus$, $k_{16}^\oplus$, $k_{18}^\oplus$, $k_{19}^\oplus$, $k_{37}^\oplus$, $\pi_0(1)$, $\pi_2(1)$, $\pi_3(1)$, k_8^π, $\pi_{13}(0)$, $\pi_{14}(0)$, $\pi_{15}(0)$
			Both	$k_{12}^\oplus$, $k_{17}^\oplus$, $k_{20}^\oplus$, $k_{36}^\oplus$, $k_{38}^\oplus$, $\pi_6^3 = \pi_6(2)$
4	$(1,1)$	$2^{29} \cdot 3^7$	Enc.	$k_0^\oplus$, $k_1^\oplus$, $k_5^\oplus$, $k_{11}^\oplus$, $k_{16}^\oplus$, $k_{17}^\oplus$, $k_{21}^\oplus$, $k_{27}^\oplus$, $k_{32}^\oplus$, $k_{33}^\oplus$, π_0, k_2^π, π_{11}^3
			Dec.	$k_4^\oplus$, $k_{12}^\oplus$, $k_{13}^\oplus$, $k_{14}^\oplus$, $k_{36}^\oplus$, $k_{38}^\oplus$, $k_{39}^\oplus$, $k_{40}^\oplus$, $k_{41}^\oplus$, $k_{42}^\oplus$, $k_{44}^\oplus$, k_8^π, $\pi_6(2)$, $\pi_7(2)$, $\pi_8(2)$, $\pi_9(2)$, $\pi_{10}(2)$, $\pi_{11}(2)$, $\pi_{13}(2)$, $\pi_{14}(2)$
			Both	$k_{37}^\oplus$, $k_{43}^\oplus$, π_{11}
	Overall	$2^{62} \cdot 3^3$		$k_0^\oplus$, ..., $k_9^\oplus$, ..., $k_{23}^\oplus$, $k_{25}^\oplus$, $k_{27}^\oplus$, $k_{28}^\oplus$, $k_{30}^\oplus$, $k_{32}^\oplus$, $k_{33}^\oplus$, $k_{36}^\oplus$, ..., $k_{47}^\oplus$, π_0, k_2^π, π_2, π_3, k_8^π, k_{11}^π, $\pi_6^3 = \pi_6(2)$, $\pi_7^3 = \pi_7(2)$, $\pi_8(2)$, π_9, π_{10}, π_{11}, π_{13}, π_{14}, π_{15}

Practical Attack on 8 Rounds of the Lightweight Block Cipher KLEIN

Jean-Philippe Aumasson[1], María Naya-Plasencia[2,*],
and Markku-Juhani O. Saarinen[3]

[1] NAGRA, Switzerland
[2] FHNW, Windisch, Switzerland and University of Versailles, France
[3] Revere Security, USA

Abstract. KLEIN is a family of lightweight block ciphers presented at RFIDSec 2011 that combines a 4-bit Sbox with Rijndael's byte-oriented MixColumn. This approach allows compact implementations of KLEIN in both low-end software and hardware. This paper shows that interactions between those two components lead to the existence of differentials of unexpectedly high probability: using an iterative collection of differential characteristics and neutral bits in plaintexts, we find conforming pairs for four rounds with amortized cost below 2^{12} encryptions, whereas at least 2^{30} was expected by the preliminary analysis of KLEIN. We exploit this observation by constructing practical ($\approx 2^{35}$-encryption), experimentally verified, chosen-plaintext key-recovery attacks on up to 8 rounds of KLEIN-64—the instance of KLEIN with 64-bit keys and 12 rounds.

Keywords: block ciphers, cryptanalysis, lightweight cryptography.

1 Introduction

Lightweight cryptography is concerned with the design, analysis, and implementation of cryptographic schemes—such as stream ciphers or authentication protocols—that minimize the consumption of resources, mainly ROM and RAM in software, and power, energy, and area in hardware. Research in lightweight cryptography is motivated by the growing number of low-resource computing devices such as RFID tags, network sensors, or low-end embedded software systems (for example, smartphones, digital cameras, portable GPS devices). The field has gained interest these last years with a multitude of new lightweight primitives, including the block ciphers PRESENT (CHES'08) [1], KATAN and KTANTAN (CHES'09) [2], PRINTcipher (CHES'10) [3], Hummingbird-2 (RFIDsec'11) [4], LED [5], Piccolo (CHES'11) [6] and the hash functions QUARK (CHES'10) [7], PHOTON (CRYPTO'11) [8], and SPONGENT (CHES'11) [9]. Most of those

---

* Supported by the National Competence Center in Research on Mobile Information and Communication Systems (NCCR-MICS), a center of the Swiss National Science Foundation under grant number 5005-67322 and by the French Agence Nationale de la Recherche through the SAPHIR2 project under Contract ANR-08-VERS-014.

D.J. Bernstein and S. Chatterjee (Eds.): INDOCRYPT 2011, LNCS 7107, pp. 134–145, 2011.

designs are hardware-oriented, and minimize area either using a combination of a small Sbox and a simplistic linear layer, or using a shift-register-based construction.

KLEIN is a new family of lightweight block ciphers, presented at RFID-Sec 2011 by Gong, Nikova, and Law [10]. The instances KLEIN-64, KLEIN-80, and KLEIN-96 process 64-bit data blocks and respectively make 12, 16, and 20 rounds and accept keys of 64, 80, and 96 bits. Thanks to an involutive 4-bit Sbox and Rijndael's MixColumn, the KLEIN ciphers allow compact and low-memory implementations in low-end software and hardware. For example, on an Iris sensor node based on an 8-bit AVR microcontroller (ATmega128L), any of the KLEIN instances can be implemented with 97 bytes of RAM and approximately 4 KB or ROM. In 180 nm ASIC, approximately 2000 GE are needed for an implementation with 64-bit datapath.

The preliminary security analysis of KLEIN includes lower bounds on the number of active Sboxes in a differential characteristic. Namely, it is shown that any 4-round characteristic has at least 15 active Sboxes, which implies a probability below 2^{-90} for any characteristic of KLEIN-64. To the best of our knowledge, the best attack reported on KLEIN-64 is a key-recovery integral attack on five rounds with complexity 2^{48} [11, §4.2.1].

Contribution. We propose a refined differential analysis of KLEIN, showing that collections of iterative differential characteristics can be used to bypass the bound proven in the preliminary analysis. We exploit this observation by presenting practical chosen-plaintext key-recovery attacks on up to 8 rounds of KLEIN-64. Our results have been confirmed—and even refined—experimentally.

§2 starts with a brief description of KLEIN-64. Then §3 presents a high-probability differential, which is exploited in §4, first by constructing distinguishers, and then by building key-recovery attacks on top of the distinguisers. We conclude in §5.

2   Brief Description of KLEIN

Our description differs in representation from that in [10], but is functionally equivalent. This is done to make the operation of some of our attacks more apparent.

Table 1. The Sbox used by KLEIN. This Sbox is an involution.

x	0	1	2	3	4	5	6	7	8	9	a	b	c	d	e	f
$S[x]$	7	4	a	9	1	f	b	0	c	3	2	6	8	e	d	5

KLEIN is built from an involutive 4-bit Sbox (given in Table 1) and operations in $GF(2^8)$. The field representation is defined by the irreducible polynomial $x^8 + x^4 + x^3 + x + 1$ (as in Rijndael). During encryption, one can decompose the field operations to XOR and multiplication by 2. We write this multiplication operation algorithmically as a left shift with a conditional XOR:

$$L(x) = \begin{cases} (x \ll 1) & \text{if } x \wedge 80 = 00, \\ (x \ll 1) \oplus \text{1b} & \text{if } x \wedge 80 = 80. \end{cases}$$

A KLEIN round is composed of the following steps:

1. AddRoundKey, which XORs a round key to the 64-bit state.
2. SubNibbles, which applies the 4-bit Sbox to each nibble.
3. RotateNibbles, which left-rotates the state of 16 bits.
4. MixNibbles, which applies two MixColumn's in parallel.

The last round has an additional AddRoundKey operation after MixNibbles. Note that in the last round, MixNibbles is not omitted, unlike MixColumns in Rijndael.

Algorithm 1. KLEIN-64 encryption given 64-bit plaintext $P_0, \dots, P_7$

```
1: for i = 0 to 7 do
2:     V_i = P_i                                      Copy the plaintext to the state vector
3: end for
4: for r = 0 to 11 do
5:     for i = 0 to 7 do
6:         V_i = V_i ⊕ K_i^(r)                        AddRoundKey
7:     end for
8:     for i = 0 to 7 do
9:         V_{i,0} = S[V_{i,0}]                       SubNibbles (lower nibbles)
10:        V_{i,1} = S[V_{i,1}]                       SubNibbles (higher nibbles)
11:    end for
12:    for i = 0 to 7 do
13:        T_{(i+6) mod 8} = V_i                      RotateNibbles
14:    end for
15:    V_0 = L(T_0 ⊕ T_1) ⊕ T_1 ⊕ T_2 ⊕ T_3          MixNibbles lower half
16:    V_1 = L(T_1 ⊕ T_2) ⊕ T_0 ⊕ T_2 ⊕ T_3
17:    V_2 = L(T_2 ⊕ T_3) ⊕ T_0 ⊕ T_1 ⊕ T_3
18:    V_3 = L(T_3 ⊕ T_0) ⊕ T_0 ⊕ T_1 ⊕ T_2
19:    V_4 = L(T_4 ⊕ T_5) ⊕ T_5 ⊕ T_6 ⊕ T_7          MixNibbles higher half
20:    V_5 = L(T_5 ⊕ T_6) ⊕ T_4 ⊕ T_6 ⊕ T_7
21:    V_6 = L(T_6 ⊕ T_7) ⊕ T_4 ⊕ T_5 ⊕ T_7
22:    V_7 = L(T_7 ⊕ T_4) ⊕ T_4 ⊕ T_5 ⊕ T_6
23: end for
24: for i = 0 to 7 do
25:    C_i = V_i ⊕ K_i^(n)                            Final half-round and copy to ciphertext
26: end for
```

Algorithm 1 details the KLEIN-64 encryption. We index vectors as V_i where $V_{i,0}$ and $V_{i,1}$ are the low and high nibbles of byte i. Algorithm 2 describes the KLEIN key setup for the three possible key lengths: 64, 80, and 96 bits. Observe that the key setup has the following properties

- The higher and lower nibbles are not mixed at all.
- The round counter has no effect on the higher nibbles with a 64-bit key and only during last round with a 80-bit key.
- If the higher nibbles are all 0 or 7, the higher nibbles will stay as 0 or 7 throughout the key setup.
- A higher-nibble fixed point for the 64-bit key setup is

$$7000007070700000 \mapsto 7000007070700000 .$$

Algorithm 2. KLEIN-64 key setup given 64-bit key K.

1: $K^{(0)} = K$ *The first round key*
2: **for** $r = 1$ to 12 **do**
3:    **for** $i = 0$ to 4 **do**
4:        $K_i^{(r)} = K_{((i+1) \bmod 4)+4}^{(r-1)}$
5:        $K_{i+m}^{(r)} = K_{((i+1) \bmod 4)+4}^{(r-1)} \oplus K_{(i+1) \bmod 4}^{(r-1)}$
6:    **end for**
7:    $K_2^{(r)} = K_2^{(r)} \oplus r$ *Round constant*
8:    $K_{5,0}^{(r)} = S[K_{5,0}^{(r)}]$ *Nonlinear mixing*
9:    $K_{5,1}^{(r)} = S[K_{5,1}^{(r)}]$
10:   $K_{6,0}^{(r)} = S[K_{6,0}^{(r)}]$
11:   $K_{6,1}^{(r)} = S[K_{6,1}^{(r)}]$
12: **end for**

---

3   A Collection of Differential Characteristics

Our attack exploits a collection of iterative differential characteristics that have a same input difference and output differences in a specific 32-bit subspace (i.e. it is iterative). Below we first analyze the probability to follow one of those characteristics, which we successfully verified experimentally.

3.1   Observations

We first report four important observations that will allow us to identify high-probability differentials for KLEIN. We refer to MixColumn as the function executed twice within MixNibbles.

Observation 1. If the difference entering MixColumn is of the form 0000000X where X represents a non-zero difference in $\{1, \ldots, 7\}$—i.e. a nibble with null MSB—then the output difference is of the form 0Y0Y0Y0Y, where the wildcard Y represents a non-zero difference. That is, higher nibbles remain free of difference.

Observation 2. If the difference entering MixColumn is of the form 0X0X0X0X where the wildcard X represents a difference in $\{0, \ldots, 7\}$, then the output difference is of the form 0Y0Y0Y0Y, where Y represents a possibly null difference. Furthermore, the average number of non-zero Y's is 3.75, as one can experimentally verify. For example, the input difference 04020405 leads to the output difference 0f090100.

Observation 3. If the difference entering MixColumn is of the form 0X0X0X0X where the wildcard X represents a difference in $\{8, \ldots, f\}$, then the output difference is of the form 0Y0Y0Y0Y, where Y represents a (possibly zero) difference. Furthermore, the average number of non-zero Y's is 3.75. Note that, contrary to Observation 2, an X cannot be zero. For example, the input difference 0c0a080f leads to the output difference 010f0708.

Observation 4. Given a random difference, KLEIN's Sbox returns a difference in $\{1, \ldots, 7\}$ with probability $7/15 \approx 2^{-1.1}$, for a random input. If the difference is b or e, the probability is $3/4 \approx 2^{-0.42}$. These values can be verified either experimentally or using the difference distribution table in [11].

3.2   The Collection of Characteristics

Our attack exploits a truncated differential defined as a collection of (iterative) characteristics. That is, we not only set conditions on the intput/output differences, but also on the path followed to reach them.

Definition. To best exploit the first two observations, our collection of characteristics is such that higher nibbles remain inactive. A sufficient condition is that after SubNibbles the first four lower have differences either all in $\{0, \ldots, 7\}$, or all in $\{8, \ldots, f\}$. A similar condition is imposed on the last four lower nibbles. Fig. 1 gives a representation of the collection of characteristics. Note that the collection of characteristics is iterative, as it has the same conditions on the input as on the output at any round.

To maximize the probability at the first round, we choose an input difference b. At the first round, we thus have one active Sbox, and one active MixColumn. At the second round we always have four active Sboxes, and two active MixColumn's. Then we enter a state where all lower nibbles are active with high probability.

Probability analysis. We estimate the probability of our truncated differential as a collection of characteristics.

At the first round, it is sufficient that the input difference B gives a difference in $\{0, \ldots, 7\}$ after the Sbox. This occurs with probability $3/4$, thus $p_1 \approx 2^{-0.42}$.

At the second round, there are four active nibbles entering SubNibbles. RotateNibbles propagates the four active nibbles to MixNibbles, wherein two lower nibbles are inactive in each half. Thus, the differences after SubNibbles in the second round must be in $\{0, \ldots, 7\}$. Since such a difference is reached with probability $7/15$, we have $p_2 = (7/15)^4 \approx 2^{-4.40}$.

At the third round, there are on average 3.75 active nibbles coming from each MixColumn of the second round, and all four lower nibbles are active with probability $15/16$. This is necessary to obtain four active nibbles with differences in $\{8, \ldots, f\}$ after SubNibbles, but not to obtain four active nibbles with differences in $\{0, \ldots, 7\}$. The probability to obtain one the desired sets of differences in one half of the state is

$$\left(\frac{7}{15}\right)^{3.75} + \left(\frac{15}{16}\right) \times \left(\frac{8}{15}\right)^4 .$$

Recall that a difference in $\{0, \ldots, 7\}$ is reached with probability $7/15$, and one in $\{8, \ldots, \mathtt{f}\}$ with probability $8/15$. As the halves of MixNibbles behave similarly,

$$p_3 = \left(\left(\frac{7}{15} \right)^{3.75} + \left(\frac{15}{16} \right) \left(\frac{8}{15} \right)^4 \right)^2 \approx 2^{-5.82} .$$

The differential is iterated through subsequent rounds, hence we have

$$p_4 = p_5 = p_6 = p_3 .$$

Note that, for the sake of simplicity, we do not consider the case when two inactive boxes occur in the same MixColumn, as this has a negligible impact on the probability obtained. Fig. 1 shows the cumulative probabilities for a sequence of 1 to 7 rounds of KLEIN.

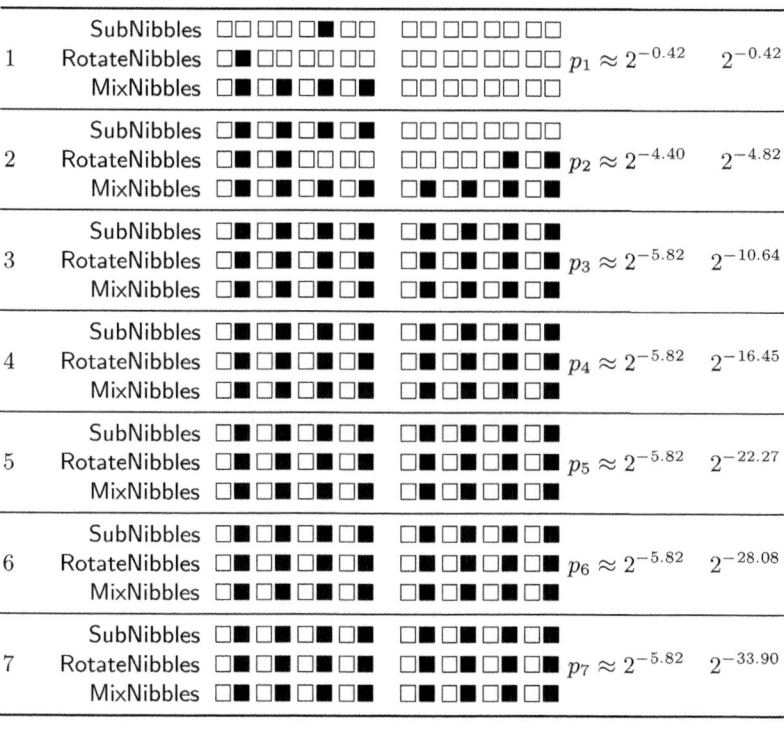

Fig. 1. Representation of the collection of differential characteristics, where a square represents a nibble (white means inactive, black means possibly active). The two right-most columns respectively show the round's probability and the cumulative probability of obtaining the differential.

3.3 Comparison with the Lower Bounds

The KLEIN paper [10] proves that any 4-round characteristic activates at least 15 Sboxes and has probability at most 2^{-30}. For comparison, our differential on 4 rounds has probability $2^{-16.45}$, yet it activates 21 Sboxes. Our result thus does not contradict the 2^{-30} bound, for it considers a collection of characteristics rather than a single one. Indeed, as argued in [10, §§3.1], *"the strength of a cipher against differential attacks is reflected by the maximum probability of a differential, i.e. a collection of characteristics"*. However, the assumption that [10, §§3.1] *"one characteristic has a much larger probability than the other characteristics of the differential"* is proven wrong by our observations. That is, the maximal probability of a characteristic cannot be taken as an accurate estimate of the maximal probability of a differential, as the above assumption would imply.

4 Attacking KLEIN

This section shows how to exploit the differential described in §3 to attack reduced version of KLEIN-64. We start with the observation that *neutral bits* can be used to reduce the cost of finding values conforming to the differential.

4.1 Finding and Exploiting Neutral Bits

The term *neutral bit* was introduced by Biham and Chen [12] in the context of SHA0 cryptanalysis. In the input of some cryptographic function, a bit is said to be *neutral* with respect to a given differential (characteristic) when flipping this bit in an input conforming to the differential (characteristic) leads to a new input also conforming to that differential. Biham and Chen actually used sets of neutral bits (called k-neutral sets). A similar technique has been used in the context of block cipher cryptanalysis, for example in Biham et al.'s analysis of Skipjack [13].

In KLEIN, one can observe that the first two and last two input bytes in a plaintext block are neutral with respect to the first two rounds' collection of characteristics. Indeed,

- in the first round, those bytes first pass through the Sbox, then after RotateNibbles they form the 4-byte half of the state that is inactive in MixNibbles
- in the second round, our neutral bytes first pass through the Sbox—still independently of the other bytes and of the difference—then they are mixed with the other bytes within MixNibbles. However, since the conformance of the output of MixNibbles only depends on the active nibbles, our bytes remain neutral up to this stage.
- in the third round, values entering the Sbox depend on the first and last two input bytes; these are thus not neutral for the third round.

Therefore, given a pair of inputs satisfying the truncated differential, 2^{32} pairs conforming to the first two rounds can be obtained by varying the first two and last two input bytes. In other words, the conformance to the differential is independent from the values of those bytes.

It follows for example that after a 2^{28} effort to find a pair satisfying the 6-round differential, one can derive 2^{32} pairs for which the full differential is followed with probability $2^{-28.06+4.80} = 2^{23.26}$. A new conforming pair can thus be found with effort $\approx 2^{23}$, instead of 2^{28} without exploiting neutral bits. With an extra effort of 2^{27} encryptions—which leaves the total complexity below 2^{29}—one expects to find 8 other conforming pairs.

4.2 Distinguisher for 7 Rounds

Based on our 6-round differential characteristic, we construct a distinguisher for 7-round KLEIN-64. Our main observation is that for a pair conforming to the 6-round differential, the SubNibbles of round 7 has all higher nibbles inactive. Although MixNibbles may activate arbitrary nibbles at round 7, one can determine the differences before MixNibbles given only the output after 7 rounds, thanks to the linearity of MixNibbles. In other words, one can check that only lower nibbles were active after SubNibbles. A conforming pair is expected to be detected after 2^{28} observations, against 2^{32} ideally, which constitutes the distinguisher.

The distinguisher is actually more powerful: once a conforming pair is found in 2^{28}, one can produce approximately 8 other pairs with negligible extra cost, as explained in §§4.1.

4.3 Distinguisher for 8 Rounds

The distinguisher for 7 rounds consisted in finding one (and possibly many) conforming pairs at a lower cost than for an ideal cipher. For 8 rounds, the distinguisher consists in finding several pairs (rather than one) with reduced complexity.

First, one collects approximately $2^{33.90}$ pairs, and records the ones that conform to the output difference as per the collection of characteristics in §3. One expects to record approximately 4 pairs satisfying the difference by chance, and one conforming to the collection of characteristics. Observe that the conforming pair can be identified using the neutral bits, as it is the only pair for which neutral bits will lead to an additional conforming pairs in approximately $2^{33.90-4.80} = 2^{29.10}$ trials.

It follows that by testing 2^{32} derived pairs for each of the (say) 5 pairs obtained initially, one new conforming pair is expected for the pairs obtained by chance, and about 8 new pairs for the one conforming to the characteristics. Therefore, with about $2^{33.90} + 4 \times 2^{32} \approx 2^{35}$, one expects to find twice more conforming pairs than ideally (16 vs. 8).

4.4 Key-Recovery for 7 Rounds

Our key-recovery attack for 7 rounds starts by using the distinguisher of §§4.2 to detect a pair satisfying the 6-round differential. Then, we exploit the invertibility of the final MixNibbles and RotateNibbles to determine the output differences of each nibble after the last SubNibbles (i.e. that of the seventh round. These differences should be null for all higher nibbles.

Then, the attack tries values of the lower nibbles (i.e., linear combinations of key bits) and pass them through the Sbox; the difference obtained is inverted through MixColumn; if the difference obtained has only lower nibbles active, then the guess is considered as possible.

Since the inverse MixNibbles produces lower only active nibbles given lower only active nibble with probability 2^{-3}, we can reduce the search space from 2^{16} to $2^{16}/2^3 = 2^{13}$ for each of the two MixColumn instances. Overall, this reduces the cost of key-recovery to 2^{58} trials. The attack always succeeds, as all candidate keys are tried within the 2^{58} trials.

The attack can be improved by using several conforming pairs. Using neutral bits, one can generate 8 more pairs in 2^{27}. It is expected that 6 are sufficient to identify the correct combination of key bits, by taking the intersection of the 2^{13}-element sets determined for each conforming pair. One can thus recover the 32 bits corresponding to the XOR between the lower nibbles of the ciphertexts, and those after the last SubNibbles. Since these bits are a linear combination of the key bits, it is equivalent to recovering 32 key bits (due to the linear independency of the 32 equations). The 32 key bits left can then be bruteforced in 2^{32}.

The attack thus recovers the complete 64-bit key with fewer than 2^{33} encryptions.

4.5   Key-Recovery for 8 Rounds

One can extend the strategy of the 7-round attack to 8 rounds, using the trick mentioned in §§4.3 to detect the pair conforming to the collection of characteristics when "false alarms" (i.e. values conforming to the input/output differential but not necessarily to the collection of characteristics). Within fewer than 2^{34} encryptions, one thus identifies a conforming pair with high probability.

Using neutral bits, one expects to produce approximately 8 other conforming pairs after 2^{32} trials. This is more than enough to identify with certainty 32 bits of the last subkey, as done in the attack on 7 rounds. Overall, the 64 bits of the last subkey (and thus of the original key) can be found with complexity below 2^{35} encryptions.

4.6   Experimental Verification

We experimentally verified the correctness of the probablities reported in §3 as well as the correctness of the distinguishers and key recovery attacks claimed. Namely, we implemented the chosen-plaintext attack that aims to recover the lower nibbles of the last subkey. As exact complexities cannot be fully confirmed experimentally, we just checked that the order of magnitude was consistent with the expected complexities.

Since no reference code of KLEIN is published by its designers, we wrote our own reference C implementation of KLEIN-64 and made sure it matched the test vectors provided in [11]. A reference implementation of KLEIN is very easily written: it took us less than one hour to implement KLEIN-64, by reusing available code of MixColumn to implement MixNibbles, and implementing Sub-Nibbles with a look-up table.

Based on our reference KLEIN-64 code, we implemented the method that recovers the combinations of subkey bits that are XORed with the higher nibbles of the state after the last SubNibbles: a first conforming pair is found by bruteforce, then neutral bits are used to find five more conforming pairs. Below we copy examples of outputs of our program for the attack on 6 and 7 rounds, reporting the number of trials done for finding each of the pairs, the value found—to make sure that all are distinct—as well as the total time of the attack:

```
$ ./attack 6
test vector ok
soundness ok
Pair found in 2^22.85: c093c2304ac8b7ca
Pair found in 2^18.41: ccc0c2304ac855a6
Pair found in 2^15.81: b4efc2304ac8fa2d
Pair found in 2^17.42: bbddc2304ac81c53
Pair found in 2^14.72: 9b53c2304ac8bdd2
Pair found in 2^19.26: 40c9c2304ac86349
Subkey lower nibbles recovered:
745a
     a1ba
Actual subkey lower nibbles:
745a a1ba
10 seconds elapsed


$ ./attack 7
test vector ok
soundness ok
Pair found in 2^27.29: 0e45d5ed12117e30
Pair found in 2^23.78: dd50d5ed12114908
Pair found in 2^23.74: d78dd5ed12112a02
Pair found in 2^24.45: fdb7d5ed1211f745
Pair found in 2^22.69: a4e3d5ed121123bc
Pair found in 2^15.82: 6286d5ed12116f2c
Subkey lower nibbles recovered:
1bda
     5d7d
Actual subkey lower nibbles:
1bda 5d7d
296 seconds elapsed
```

In the list of pairs found, the first one is found by bruteforce and the subsequent ones are derived using neutral bits, thereby reducing the cost by a factor $2^{4.82}$ on average. We used an Athlon64 X2 Dual Core 4400+. Although our code is slightly optimized (e.g. with SubNibbles implemented as 8-bit look-ups)

and gcc-compiled with speed-optimization flags (-O3 -m64 -march=athlon64
-fomit-frame-pointer -funroll-loops), the attacks can probably be sped
up further.

The 8-round attack took much more time to verify than the attack on 7 rounds,
due to the possible finding of false alarms. For example, one failed experiment
returned after a few hours

```
Pair found in 2^32.27: 0beeadb61e7d4787
Pair found in 2^29.80: 0beeadb61e7d4787
Pair found in 2^31.61: 0beeadb61e7d4787
```

Such results suggest that the first pair found was not conforming to the collection of characteristics, as the tentative use of neutral bits failed to find distinct
conforming pairs. Nevertheless, we were able to find conforming pairs for the
8-round attack. We were particularly lucky with the following experiment:

```
$ ./attack 8
test vector ok
soundness ok
Pair found in 2^28.21: fb5248c1a424ca3e
Pair found in 2^26.43: 00b848c1a424882f
Pair found in 2^28.54: 180b48c1a4245a09
Pair found in 2^26.78: 1ee948c1a4246b1d
Pair found in 2^25.81: 226848c1a424362e
Pair found in 2^27.56: 2e3548c1a424f161
Subkey lower nibbles recovered:
d42c
     d515
Actual subkey lower nibbles:
d42c d515
1344 seconds elapsed
```

The experiments reported above demonstrate that the attacks described in §4
do work and succeed in recovering the key with a complexity that seems to be
in line with our analytical estimates.

5   Conclusion

We presented practical, experimentally verified attacks on the lightweight cipher KLEIN-64 reduced to up to 8 rounds, out of 12 in total. Our attack is
made possible by a high-probability differential described as a large collection
of differential characteristics. Our results suggest that combining a 4-bit Sbox
(as used in Serpent) with the byte-oriented MixColumn linear layer (as used in
Rijndael/AES) is not an optimal strategy, as far as security is concerned. This
work is the first third-party analysis of KLEIN published, to our best knowledge.
Future works may seek to extend our attacks to more rounds of KLEIN.

Acknowledgments. We would like to thank the reviewers of INDOCRYPT 2011 for their insightful comments.

References

1. Bogdanov, A., Knudsen, L.R., Leander, G., Paar, C., Poschmann, A., Robshaw, M.J.B., Seurin, Y., Vikkelsoe, C.: PRESENT: An Ultra-Lightweight Block Cipher. In: Paillier, P., Verbauwhede, I. (eds.) CHES 2007. LNCS, vol. 4727, pp. 450–466. Springer, Heidelberg (2007)
2. De Cannière, C., Dunkelman, O., Knežević, M.: KATAN and KTANTAN — A Family of Small and Efficient Hardware-Oriented block Ciphers. In: Clavier, C., Gaj, K. (eds.) CHES 2009. LNCS, vol. 5747, pp. 272–288. Springer, Heidelberg (2009)
3. Knudsen, L., Leander, G., Poschmann, A., Robshaw, M.J.B.: PRINTCIPHER: A Block Cipher for IC-Printing. In: Mangard, S., Standaert, F.-X. (eds.) CHES 2010. LNCS, vol. 6225, pp. 16–32. Springer, Heidelberg (2010)
4. Engels, D., Saarinen, M.J.O., Smith, E.M.: The Hummingbird-2 lightweight authenticated encryption algorithm. In: RFIDsec (2011)
5. Guo, J., Peyrin, T., Poschmann, A., Robshaw, M.: The LED Block Cipher. In: Preneel, B., Takagi, T. (eds.) CHES 2011. LNCS, vol. 6917, pp. 326–341. Springer, Heidelberg (2011)
6. Shibutani, K., Isobe, T., Hiwatari, H., Mitsuda, A., Akishita, T., Shirai, T.: *piccolo*: An ultra-lightweight blockcipher. In: Preneel, B., Takagi, T. (eds.) CHES 2011. LNCS, vol. 6917, pp. 342–357. Springer, Heidelberg (2011)
7. Aumasson, J.-P., Henzen, L., Meier, W., Naya-Plasencia, M.: QUARK: A lightweight Hash. In: Mangard, S., Standaert, F.-X. (eds.) CHES 2010. LNCS, vol. 6225, pp. 1–15. Springer, Heidelberg (2010)
8. Lindell, Y., Oxman, E., Pinkas, B.: The IPS Compiler: Optimizations, Variants and Concrete Efficiency. In: Rogaway, P. (ed.) CRYPTO 2011. LNCS, vol. 6841, pp. 259–276. Springer, Heidelberg (2011)
9. Bogdanov, A., Knežević, M., Leander, G., Toz, D., Varıcı, K., Verbauwhede, I.: SPONGENT: A Lightweight Hash Function. In: Preneel, B., Takagi, T. (eds.) CHES 2011. LNCS, vol. 6917, pp. 312–325. Springer, Heidelberg (2011)
10. Gong, Z., Nikova, S., Law, Y.W.: KLEIN: a new family of lightweight block ciphers. In: RFIDSec (2011)
11. Gong, Z., Nikova, S., Law, Y.W.: KLEIN: a new family of lightweight block ciphers (2011), http://doc.utwente.nl/73129/
12. Biham, E., Chen, R.: Near-Collisions of SHA-0. In: Franklin, M. (ed.) CRYPTO 2004. LNCS, vol. 3152, pp. 290–305. Springer, Heidelberg (2004)
13. Biham, E., Biryukov, A., Dunkelman, O., Richardson, E., Shamir, A.: Initial Observations on Skipjack: Cryptanalysis of Skipjack-3XOR. In: Tavares, S., Meijer, H. (eds.) SAC 1998. LNCS, vol. 1556, p. 362. Springer, Heidelberg (1999)

On Related-Key Attacks and KASUMI: The Case of A5/3

Phuong Ha Nguyen[1], Matthew J.B. Robshaw[2], and Huaxiong Wang[1]

[1] Nanyang Technological University, Singapore
[2] Applied Cryptography Group, Orange Labs, France
NG0007HA@e.ntu.edu.sg, hxwang@ntu.edu.sg
matt.robshaw@orange-ftgroup.com

Abstract. Due to its widespread deployment in mobile telephony, the block cipher KASUMI is a prominent target for cryptanalysts. While the cipher offers excellent resistance to differential and linear cryptanalysis, in the related-key model there have been several impressive cryptanalytic results. In this paper we revisit these related-key attacks and highlight a small, but important, detail in the specification of KASUMI for the algorithm A5/3; namely that a 64- and not a 128-bit session key is used. We show that existing related-key attacks on KASUMI in the literature are (negatively) impacted by this feature and we provide evidence that repairing these attacks will be difficult.

Keywords: Block-cipher, KASUMI, A5/3, related-key attack, 64-bit key version of KASUMI.

1 Introduction

Underpinning the success of mobile phone telephony are a range of authentication and confidentiality services. For confidentiality, several encryption algorithms have been specified and, in GSM networks, the encryption algorithms are referred to as A5/x with x taking values from one to four.

A5/2 offered very little security while the limited security offered by A5/1 has been confirmed in a range of papers [6,1]. Both algorithms were shift register based stream ciphers. In third generation networks, encryption is provided by a different set of algorithms that includes the block cipher KASUMI [9]. Given the existing and widespread deployment of KASUMI in 3G networks it is natural to consider that it be reused in existing 2G networks in the guise of two algorithms called A5/3 and A5/4.

KASUMI is an important block cipher and it has, appropriately enough, been the subject of cryptanalysis over recent years. The elegant design of the algorithm, closely based on MISTY [19], means that the ciphers' resistance to classical block cipher attacks such as differential [5] and linear cryptanalysis [18] is well-understood. However the key schedule of KASUMI offers a very clean framework for analysis and several related-key attacks have been proposed. These began as attacks on reduced-round KASUMI [16,17], advancing to sophisticated

D.J. Bernstein and S. Chatterjee (Eds.): INDOCRYPT 2011, LNCS 7107, pp. 146–159, 2011.

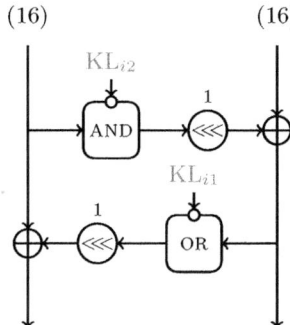

Fig. 1. Details of the FL function in the KASUMI cipher. Numbers in brackets indicate the size of the inputs in bits.

related-key attacks on the full eight rounds of KASUMI [2], and finally culminating in related-key attacks on the full cipher with a time complexity that permitted experimental verification [8].

The purpose of this paper is to further explore the applicability of related-key attacks to KASUMI. It is well-known, but not always made explicit in cryptanalysis papers, that while related-key attacks on KASUMI are of considerable academic interest they cannot be applied directly within mobile telephony networks. However, there is an additional factor that can limit the applicability of KASUMI related-key attacks and this appears to have been overlooked so far; namely that when KASUMI is used as A5/3 then backward compatibility for session key generation requires that KASUMI be used with 64- and not 128-bit keys. The purpose of this paper, therefore, is to see whether the related-key attacks in the literature can be applied to KASUMI when used as A5/3.

2   MISTY and KASUMI

The original basis for KASUMI is the block cipher MISTY [19]. This was one of the first block ciphers to provide an element of provable resistance to both linear and differential cryptanalysis. While the cipher might at first sight appear to be rather complicated, its innovative recursive design—with a round function that calls upon two mini-Feistel networks FI and FO—is easy to work with and yields useful bounds on the probability and bias of differential and linear characteristics. KASUMI inherits all the significant features of the MISTY design and the overall form of encryption with KASUMI is shown in Figs. 2 and 3.

The key schedule for KASUMI is very simple and is a slightly modified version of the schedule proposed for MISTY. The 128-bit user-provided key K is partitioned into 16-bit words K_0, K_1, ..., K_7 and subkeys for round i are given by $KL_i = KL_{i1} \| KL_{i2}$, $KO_i = KO_{i1} \| KO_{i2} \| KO_{i3}$, and $KI_i = KI_{i1} \| KI_{i2} \| KI_{i3}$. This derivation as illustrated in Fig. 4.

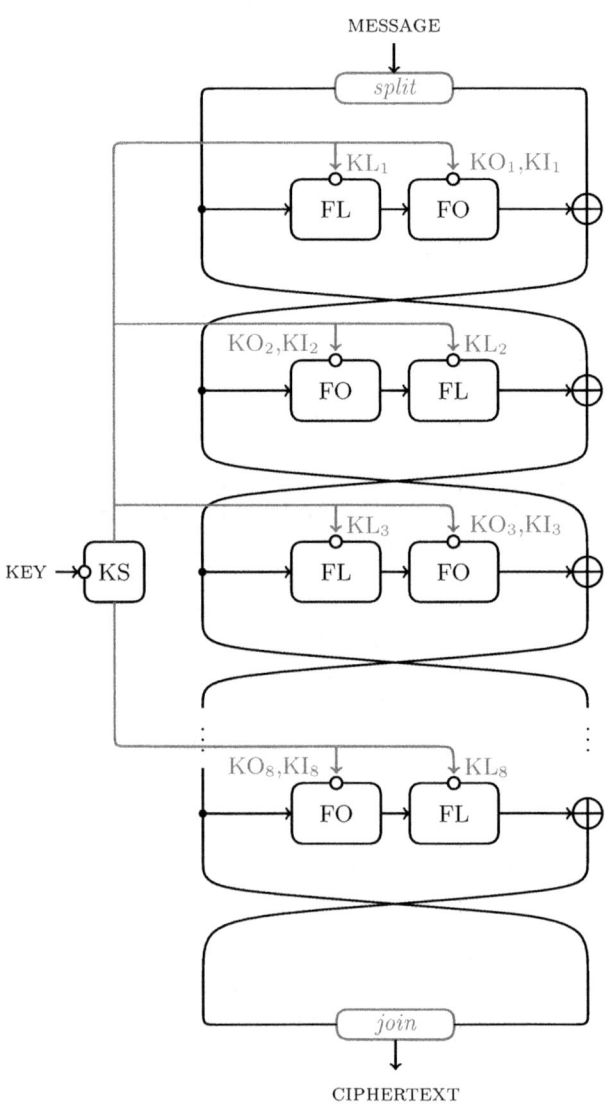

Fig. 2. Computational graph for the encryption process of the KASUMI cipher. Details of the function FO is given in Fig. 3.

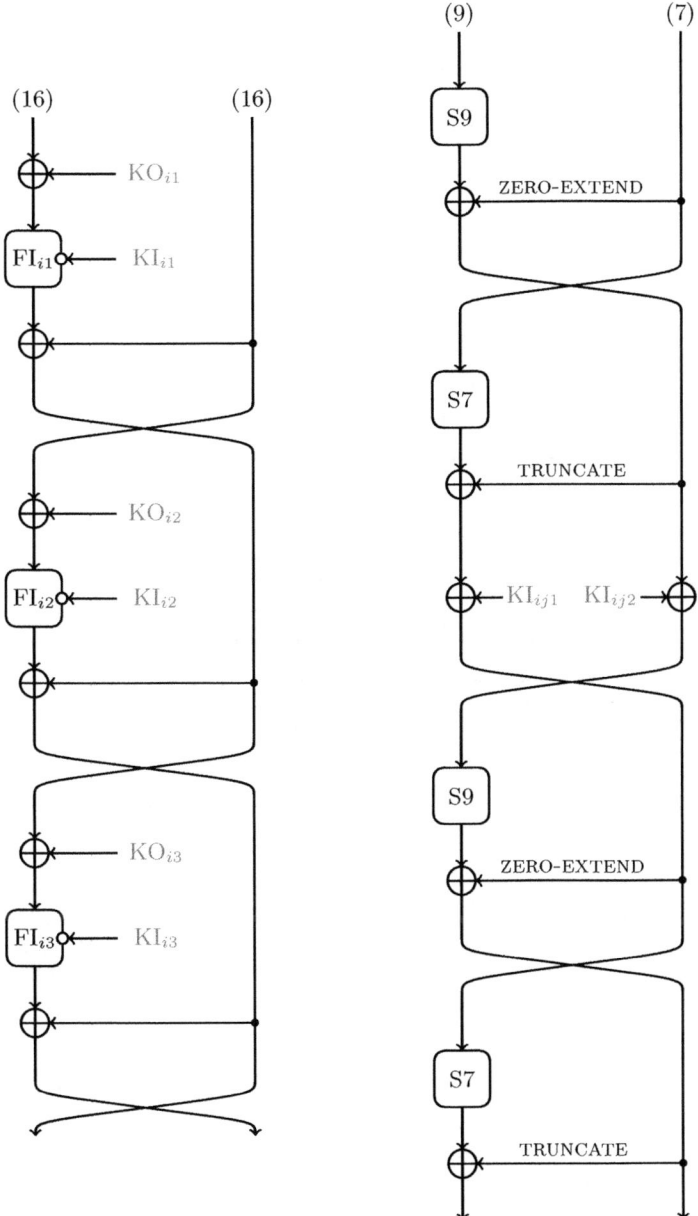

Fig. 3. The FO_i function (left) and FI_{ij} function (right) used in round i of the KASUMI cipher. Numbers in brackets indicate the size of the inputs in bits. The 48-bit round key KO_i is interpreted as $KO_{i1}\|KO_{i2}\|KO_{i3}$ while the 48-bit round key KI_i is interpreted as $KI_{i1}\|KI_{i2}\|KI_{i3}$. Within FI_{ij} the 16-bit KI_{ij} is interpreted as $KI_{ij1}\|KI_{ij1}$.

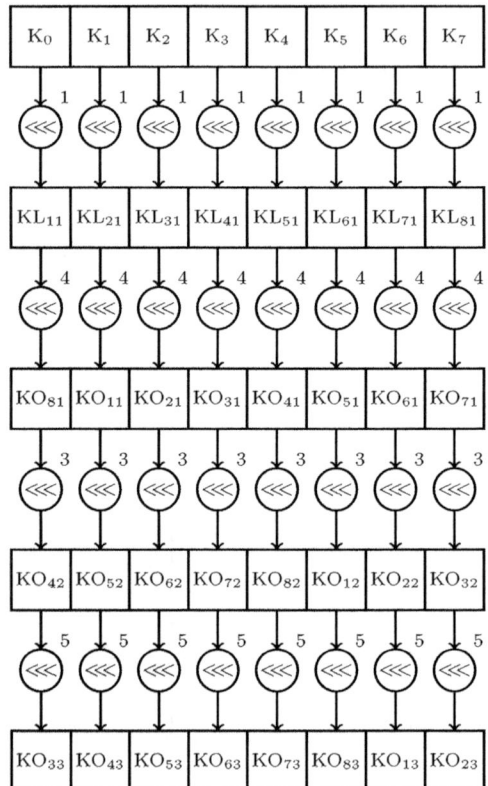

round	1	2	3	4
KL_{i2}	$K_2 \oplus$ 89ab	$K_3 \oplus$ cdef	$K_4 \oplus$ fedc	$K_5 \oplus$ ba98
KI_{i1}	$K_4 \oplus$ fedc	$K_5 \oplus$ ba98	$K_6 \oplus$ 7654	$K_7 \oplus$ 3210
KI_{i2}	$K_3 \oplus$ cdef	$K_4 \oplus$ fedc	$K_5 \oplus$ ba98	$K_6 \oplus$ 7654
KI_{i3}	$K_7 \oplus$ 3210	$K_0 \oplus$ 0123	$K_1 \oplus$ 4567	$K_2 \oplus$ 89ab

round	5	6	7	8
KL_{i2}	$K_6 \oplus$ 7654	$K_7 \oplus$ 3210	$K_0 \oplus$ 0123	$K_1 \oplus$ 4567
KI_{i1}	$K_0 \oplus$ 0123	$K_1 \oplus$ 4567	$K_2 \oplus$ 89ab	$K_3 \oplus$ cdef
KI_{i2}	$K_7 \oplus$ 3210	$K_0 \oplus$ 0123	$K_1 \oplus$ 4567	$K_2 \oplus$ 89ab
KI_{i3}	$K_3 \oplus$ cdef	$K_4 \oplus$ fedc	$K_5 \oplus$ ba98	$K_6 \oplus$ 7654

Fig. 4. Details of the key schedule for the KASUMI cipher. The 128-bit user-supplied key is divided into eight 16-bit words, with all the resultant subkeys being generated as a result of bitwise rotations or the addition of a constant. The use of each subkey is illustrated in Fig. 2.

3   Related-key Attacks and KASUMI

To begin, we follow [15] and establish the basic features of a related-key attack.

Definition 1. *We say that a related-key differential $\alpha \to \beta$ with the key differ- ence ΔK holds for cipher E with probability p, if*

$$\Pr_{P,K}[E_K(P) \oplus E_{K \oplus \Delta K}(P \oplus \alpha) = \beta] = p,$$

where $E_K(\cdot)$ denotes encryption through E with the key K. For the ease of ex- position, we denote this event by $Pr(\alpha \xrightarrow[\Delta K]{E} \beta)$.

The ΔK that appears in related-key attacks offers the attacker considerable power in his cryptanalysis. To see how things change, consider the following two lemmas. The first lemma illustrates the bound on a differential characteris- tic to the FI function under conventional cryptanalysis; it is stated as a result in [11]. When we allow related key attacks, as in Lemma 2, the bound is severely degraded. The bounds for both lemmas are tight, *i.e.* there exist differential characteristics with the stated probability when averaged over all inputs to the function FI.

Lemma 1. *For any active input difference to the KASUMI function FI with key difference $\Delta(\mathrm{KI}) = 0$, the probability of a differential characteristic is $\leq 2^{-14}$.*

Lemma 2. *For any (active or inactive) input difference to the KASUMI func- tion FI with key difference $\Delta(\mathrm{KI}) \neq 0$, the probability of a differential character- istic is $\leq 2^{-6}$.*

Proof. Denote the action of the FI function on text input $\alpha \| \beta$, where α and β are 16-bit values, as

$$\mathrm{FI}\,(\,\alpha\|\beta\,,\,\mathrm{KI}_{i,j,1}\,,\,\mathrm{KI}_{i,j,2}\,) = \gamma\|\delta.$$

Inspection, see Fig. 3, reveals that there are only two instances where a single S-box S7 is differentially active. The first is when

$$\mathrm{FI}\,(\,\Delta\alpha\|\Delta\beta\,,\,\Delta\mathrm{KI}_{i,j,1}\,,\,\Delta\mathrm{KI}_{i,j,2}\,) = 0\|0$$

with $\Delta\alpha = 0$ and $\Delta\mathrm{KI}_{i,j,2} = \textsc{zero-extend}(\Delta\beta)$. The second is when

$$\mathrm{FI}\,(\,0\|0\,,\,\Delta\mathrm{KI}_{i,j,1}\,,\,\Delta\mathrm{KI}_{i,j,2}\,) = \Delta\gamma\|\Delta\delta$$

with $\Delta\mathrm{KI}_{i,j,2} = 0$ and $\Delta\delta = \textsc{zero-extend}(\Delta\mathrm{KI}_{i,j,1})$. The result, and the fact that the bound is tight, follows. $\square$.

Already we see the resistance of FI to differential-attack is greatly reduced under related keys.

But there are other reasons to wonder about the potential for related-key cryptanalysis against KASUMI. In particular, a glance at the key schedule (see

Fig. 4) reveals that the process of subkey generation preserves byte boundaries. This means that a difference confined to a single byte of the 128-bit user-supplied key will always lead to a single byte difference in any subsequent subkey in which it appears.

Finally, there is an interesting interaction between related keys and the structure of KASUMI and the FL function. In one way or another, all KASUMI related-key attacks in the literature use the following observation which was first used in [7]. Consider a difference in word K_2 of the user-supplied key $K_0 \| K_1 \| \cdots \| K_7$. This word of key is used within the FL function in the first round and it is possible that identical round inputs give identical round outputs even when these two different round keys are used. These identical outputs from the FL function are generated with a probability that depends on the Hamming weight of the key difference[1]. See Fig. 1 to verify this observation. In the next round, a bitwise rotation of the same key difference is used as the first subkey to the FO function. But with a pair of judiciously-chosen plaintexts, and exploiting the Feistel-like structure of KASUMI, the difference introduced by the subkey can be canceled by an identical difference in the text inputs to FO. In the third round, another bitwise rotation of the difference in subkey K_2 is used as input to the FL function. This is illustrated in Fig. 5 and the net result, in typical related-key attacks on KASUMI, is that when Δ is chosen as a single-bit difference the three-round characteristic holds, in general, with probability $\frac{1}{4}$.

Clearly any three-round differential holding with a probability of at least $\frac{1}{4}$ is going to be very useful in an attack against an eight-round cipher. In a regular related-key attack, as explored by Blunden and Escott [7], this good start can be extended at some cost to the probability to give an attack on up to five and six rounds. However we can go further if we turn to some more sophisticated attacks and take further advantage of the structure in KASUMI.

Related-key boomerang and rectangle attacks. As explained in [15] a *boomerang attack* [22] exploits two short differentials rather than a single longer one. Such an attack is achieved using a combination of chosen plaintext and adaptively chosen ciphertext. Alternatives to this approach, *i.e.* variants that rely solely on chosen plaintext, have been named the *amplified boomerang attack* [14] and *rectangle attack* [3]. Extending these attacks to a related-key setting is, at least in principle, straightforward.

In KASUMI the three-round related key differential running from rounds one to three, see Fig. 5, has a close relative in other rounds. A very similar differential holds across rounds four to six when using the same key difference in K_6. This can be exploited in the reverse direction and, with hindsight, it is now clear that several factors are aligning in the cryptanalysts' favour. In particular we have two high-probability differentials (in the related-key model) that cover, together, six out of eight rounds. If we can find a way to define a distinguisher that incorporates one additional round then, in principle, key counting techniques could be used to recover user-supplied key from the entire cipher. The very

---

[1] By fixing the value of bits of the input this probability can be 1 in the first round.

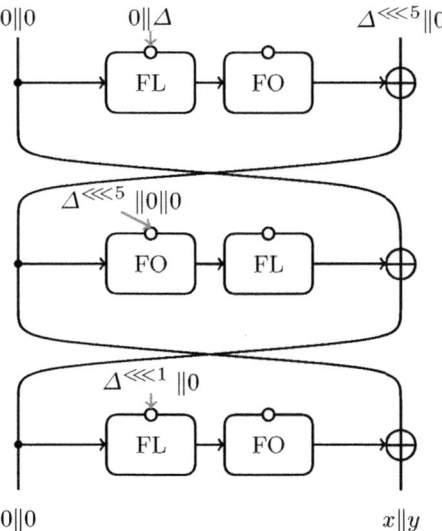

Fig. 5. The basic three-round related-key differential for KASUMI with each identified value being 16 bits long. Input keys KO_i to FO are illustrated as inputs from the left while the input keys KI_i come from the right. The probability of the differential for given $x\|y$ is $\geq 2^{-2\mathrm{Hwt}(\Delta)}$.

elegant work of Dunkelman *et al.* [8] provides one way to do this. The "filling" in the so-called *sandwich attack* [8] provides the missing round and allows two three-round related-key differentials to be connected, while key-counting on the resultant seven-round distinguisher allows the encryption key to be recovered with very practical resource requirements.

Related-key attacks in practice. At this point it should be observed that the ETSI SAGE designers of KASUMI have always been open in stating that related-key attacks are not a particular concern. This is because session keys are derived in a pseudo-random way and it is not possible for an adversary to choose how session keys might be related. Further, KASUMI is specified for use in a mode that generates keystream and this doesn't allow an adversary to specify inputs to the encryption process itself. But even if we were to set aside these operational objections to related-key attacks, there is a more fundamental reason for the failure of this attack when using A5/3.

4   Related-key Attacks and A5/3

When KASUMI is used as A5/3 it is stipulated that 64-bit keys be used; the associated 128-bit key is obtained by concatenating two copies of the 64-bit key [10]. This restriction on the form of the key is relaxed within the KASUMI-based specifications A5/4, but we are nevertheless left with an interesting question: Given

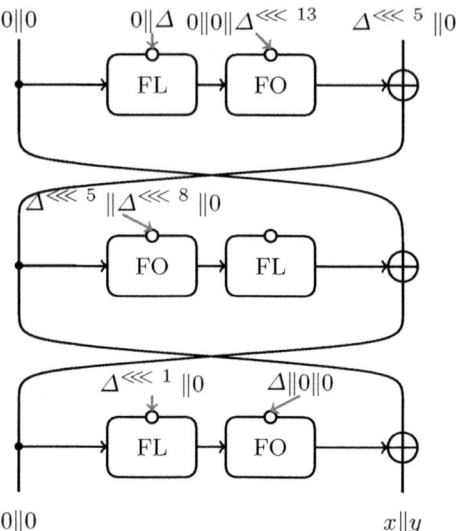

Fig. 6. Active key differences for a repeated 64-bit key and the basic three-round related-key differential of KASUMI. Keys KO_i are illustrated as inputs from the left while keys KI_i come from the right.

that A5/3 is likely to be an encryption algorithm in widespread use, how do related-key attacks on KASUMI translate to this particular choice of algorithm?

4.1 Existing Related-key Attacks

The 64-bit key k that is supplied to the KASUMI encryption unit is treated as a 128-bit key of the form $k\|k$ [10]. The immediate consequence is that any related key difference gives an active difference in two, four, six, or eight words. These are listed in detail in Table 1. It is easy to see the effect of this change on the basic three-round related-key differential and this is illustrated in Fig. 6. To bound the probability of the differential characteristic now, we can appeal to Lemma 2 from earlier and a more involved lemma given here:

Lemma 3. *Write the key inputs to FO as (KO_1, KO_2, KO_3) and (KI_1, KI_2, KI_3). For any (active or inactive) text input to FO, and for any active key difference in at least one of $(KO_1, KO_2, or KO_3)$ there must be at least one FI function that is differentially active except in the following three cases:*

1. $\Delta(KO_1) \neq 0$, $\Delta(KO_2) = 0$, and $\Delta(KO_3) = 0$.
2. $\Delta(KO_1) = 0$, $\Delta(KO_2) \neq 0$, and $\Delta(KO_3) \neq 0$.
3. $\Delta(KO_1) \neq 0$, $\Delta(KO_2) \neq 0$, and $\Delta(KO_3) \neq 0$.

Proof. Cases can be enumerated and evaluated by hand using Fig. 3.   □

Lemma 3 concentrated on how inputs to FO might cause the inner function FI to become differentially active. Of course, independently of whether KO_1, KO_2, or KO_3 is active, an FI function can become active directly via differences in the keys (KI_1, KI_2, KI_3). So an immediate consequence of Lemmas 2 and 3 is the following. The three-round related-key differential characteristic that held with probability $\frac{1}{4}$ in KASUMI with 128-bit keys holds with probability $\leq 2^{-18}$ in A5/3. It is unlikely that this bound is tight.

On its own, this result seriously degrades the effectiveness of related-key attacks when applied against KASUMI as used in A5/3. The existence of a high probability three-round differential was key to the attack in [8] where, in effect, the differential (and close relatives) are used four times. With the drop in the probability for this three-round differential when moving to 64-bit keys, and since $(2^{-18})^4 = 2^{-72} < 2^{-64}$, we conclude (under typical and reasonable assumptions of independence of differentials) that the specific related-key attacks of [8] cannot be mounted against A5/3 without solving some challenging problems. Seeing that existing attacks need a radical rethink is clear; our goal in the next section is to give evidence that the task might be difficult.

4.2   Revised Related-key Attacks

In this section we derive an upper bound on the probability of *all* three-round related-key differential in A5/3. Since three-round differentials (or longer) feature prominently in all related-key attacks on KASUMI, we show that their effectiveness is severely reduced in the context of A5/3. For the particularly efficient attacks described in [8], this new bound on any three-round related-key differential is sufficient to demonstrate that the attack, as described in [8], cannot be currently applied to A5/3.

Lemma 4. *In a round of KASUMI, if FO has one active ΔKI then the maximum probability of a differential characteristic is 2^{-6}. If there are at least two active ΔKI then the maximum probability of a differential characteristic is 2^{-12}.*

Proof. In round i, i odd, we have for $\Delta K \neq 0$, that

$$\mathrm{Pr.}(\alpha \xrightarrow{i} \beta) = \sum_{X=0}^{2^{32}} \mathrm{Pr.}(\alpha \xrightarrow{FL} X) \cdot \mathrm{Pr.}(X \xrightarrow{FO} \beta)$$

$$\leq \mathop{\mathrm{Pr.}}_{max}(X \xrightarrow{FO} \beta) \sum_{X=0}^{2^{32}} Pr(\alpha \xrightarrow{FL} X) \leq \mathop{\mathrm{Pr.}}_{max}(X \xrightarrow{FO} \beta)$$

where α and β are the input and output difference to the i-th round and X is any given output difference of FL. Similar reasoning holds for i even with X being the output difference from FO. The result now follows since, by Lemma 2, $\mathop{\mathrm{Pr.}}_{max}(X \xrightarrow{FO} \beta) \leq 2^{-6}$ when one FI is differentially active and $\leq 2^{-12}$ when two FI functions are differentially active. $\square$.

Lemma 5. *For any three-round differential of KASUMI across rounds i, $i+1$, and $i+2$, the probability of the differential (in a related-key setting) is upper-bounded by*

$$\min\{\Pr_{max}.(\Delta^i) \times \Pr_{max}.(\Delta^{i+1}), \Pr_{max}.(\Delta^{i+1}) \times \Pr_{max}.(\Delta^{i+2}), \Pr_{max}.(\Delta^i) \times \Pr_{max}.(\Delta^{i+2})\}$$

where $\Pr_{max}.(\Delta^i)$ denotes the maximum probability of any non-trivial differential characteristic across round i in the related-key setting.

Proof. Consider three consecutive rounds of KASUMI and, to give additional precision, denote the input to the three-round differential by $\alpha_L \| \alpha_R$ and the output after three rounds, without a swap in the third round, as $\beta_L \| \beta_R$. Each of these indicated quantities is 32 bits in length. Then we wish to establish an upper-bound on the probability of the three round differential

$$\Pr.(\alpha_L \| \alpha_R \xrightarrow{i,\ i+1,\ i+2} \beta_L \| \beta_R)$$

in the related-key setting. For a given $\Delta K \neq 0$, we have

$$\Pr.(\alpha_L \| \alpha_R \xrightarrow{i,\ i+1,\ i+2} \beta_L \| \beta_R) =$$

$$\sum_{X=0}^{2^{32}-1} \Pr.(\alpha_L \xrightarrow{i} X) \cdot \Pr.(\alpha_R \oplus X \xrightarrow{i+1} \alpha_L \oplus \beta_L) \cdot \Pr.(\beta_L \xrightarrow{i+2} \alpha_R \oplus X \oplus \beta_R)$$

where X is the output difference of FO in round i for i odd and the output difference of FL in round i for i even. Without loss of generality $\Pr.(\alpha_L \xrightarrow{i} X) < \Pr_{max}.(\Delta^i)$ and $\Pr.(\alpha_R \oplus X \xrightarrow{i+1} \alpha_L \oplus \beta_L) < \Pr_{max}.(\Delta^{i+1})$ for any X. So we have that

$$\sum_{X=0}^{2^{32}-1} \Pr.(\alpha_L \xrightarrow{i} X) \cdot \Pr.(\alpha_R \oplus X \xrightarrow{i+1} \alpha_L \oplus \beta_L) \cdot \Pr.(\beta_L \xrightarrow{i+2} \alpha_R \oplus X \oplus \beta_R)$$

$$\leq \Pr_{max}.(\Delta^i) \times \Pr_{max}.(\Delta^{i+1}) \times \sum_{X=0}^{2^{32}-1} \Pr.(\alpha_L \xrightarrow{i} X)$$

$$\leq \Pr_{max}.(\Delta^i) \times \Pr_{max}.(\Delta^{i+1})$$

The other two cases are treated similarly and the result follows.   □

Theorem 1. *The probability of any three-round related-key differential over KASUMI, when used as A5/3, is $\leq 2^{-18}$.*

Proof. We appeal to Lemmas 3, 2, 4, and 5. First we construct Table 1 where we note that, due to rotational symmetries in the way subkeys are used, it suffices to consider the first three rounds only. There are 15 cases to consider, depending on which pairs $\{k_0, k_4\}$, $\{k_1, k_5\}$, $\{k_2, k_6\}$, or $\{k_3, k_7\}$ are active. However these are easily broken down into a few cases and enumerated.

Table 1. Key differences in the 64-bit user-supplied key lead to *at least* the above-noted subkeys being differentially active in the specified round

round	$\{k_0, k_4\}$	$\{k_1, k_5\}$	$\{k_2, k_6\}$	$\{k_3, k_7\}$
1	KI_1	KO_1, KO_2	KO_3	KI_2, KI_3
2	KI_2, KI_3	KI_1	KO_1, KO_2	KO_3
3	KO_3	KI_2, KI_3	KI_1	KO_1, KO_2
4	KO_1, KO_2	KO_3	KI_2, KI_3	KI_1
5	KI_1	KO_1, KO_2	KO_3	KI_2, KI_3
6	KI_2, KI_3	KI_1	KO_1, KO_2	KO_3
7	KO_3	KI_2, KI_3	KI_1	KO_1, KO_2
8	KO_1, KO_2	KO_3	KI_2, KI_3	KI_1

- If either of the pairs $\{k_0, k_4\}$ or $\{k_1, k_5\}$ are active, then the result follows from Lemmas 2, 4, and 5.
- If the pair $\{k_2, k_6\}$ or $\{k_1, k_5\}$ are active then the result follows from Lemmas 2, 3, 4, and 5.   □.

The results in this section show that any three-round differential for A5/3, even in the related-key model, must have a probability that is upper-bounded by 2^{-18}. Since success in the work of [8] relies on the probability of a three-round differential being raised to the fourth power, and since this probability will be $\leq 2^{-72} < 2^{-64}$, it seems that existing related-key techniques cannot be used against A5/3 without potentially significant adjustment.

Note that we have not gone to great lengths to reduce the upper bound beyond what we need. Instead, for clarity, we have restrained ourselves to what we can achieve in a straightforward manner. The bound therefore is unlikely to be tight and there may be considerable scope for improvement.

4.3   Implications and Observations

One conclusion from the work in this paper is that the redundancy in creating a 128-bit key from a duplicated 64-bit key can be very useful, particularly in the related-key setting where any differences in the user-supplied key are immediately amplified. Certainly it seems that it would be very difficult to patch existing related-key attacks on KASUMI to give attacks that are anyway near as effective on A5/3 as those that exploit full 128-bit keys, particularly since the work effort in attacking A5/3 would be limited to 2^{64} operations.

The design of key schedules for block ciphers is notoriously *ad hoc* and while there are some guiding principles, such as to use as much of the user-supplied key as possible in different parts of the encryption process (and to do so often), there remain a very broad range of proposals in the literature from the ultra-lightweight [20] to the computationally demanding [21]. Given the interaction between hash function and block cipher design and analysis, and observing a range of new proposals that incorporate some measure of resistance to related-key attacks, *e.g.* [13], it seems that an improved understanding of the role of the key schedule would be very useful to designers and analysts alike.

5   Conclusion

While KASUMI is proposed for use as A5/3, an important operational restriction lends, perhaps ironically, some additional cryptographic protection. The restriction is that the user-supplied key in A5/3 is 64 and not 128 bits long and this forces redundancy into the typical KASUMI key schedule in the form of repeated key material. For related-key attacks this can be a severe drawback since the cryptanalyst would like to avoid the proliferation of differences during the encryption process, be they from the text or the key.

In this paper we have considered recent related-key attacks on KASUMI and demonstrated that for A5/3 they do not apply. It is an open question whether they can be patched and re-applied to A5/3. It is also unclear whether this duplication of key material might open other opportunities to the attacker. Nevertheless, we have used the elegant structure of MISTY/KASUMI to derive upper-bounds on all three-round differentials for A5/3 in the related-key setting, and these bounds indicate that developing a patch might not be straightforward.

Acknowledgments. This work was supported in part by the Singapore National Research Foundation under Research grant NRF-CRP2-2007-03. The first author is supported by the Singapore International Graduate (SINGA) Scholarship.

References

1. Barkan, E., Biham, E., Keller, N.: Instant Ciphertext-Only Cryptanalysis of GSM Encrypted Communication. In: Boneh, D. (ed.) CRYPTO 2003. LNCS, vol. 2729, pp. 600–616. Springer, Heidelberg (2003)
2. Biham, E., Dunkelman, O., Keller, N.: A Related-Key Rectangle Attack on the Full KASUMI. In: Roy, B. (ed.) ASIACRYPT 2005. LNCS, vol. 3788, pp. 443–461. Springer, Heidelberg (2005)
3. Biham, E., Dunkelman, O., Keller, N.: The Rectangle Attack - Rectangling the Serpent. In: Pfitzmann, B. (ed.) EUROCRYPT 2001. LNCS, vol. 2045, pp. 340–357. Springer, Heidelberg (2001)
4. Biham, E., Dunkelman, O., Keller, N.: New Results on Boomerang and Rectangle Attacks. In: Daemen, J., Rijmen, V. (eds.) FSE 2002. LNCS, vol. 2365, pp. 1–16. Springer, Heidelberg (2002)
5. Biham, E., Shamir, A.: Differential Cryptanalysis of the Data Encryption Standard. Springer, Heidelberg (1993)
6. Biryukov, A., Shamir, A., Wagner, D.: Real Time Cryptanalysis of A5/1 on a PC. In: Schneier, B. (ed.) FSE 2000. LNCS, vol. 1978, pp. 37–44. Springer, Heidelberg (2001)
7. Blunden, M., Escott, A.: Related Key Attacks on Reduced Round KASUMI. In: Matsui, M. (ed.) FSE 2001. LNCS, vol. 2355, pp. 277–285. Springer, Heidelberg (2002)
8. Dunkelman, O., Keller, N., Shamir, A.: A Practical-Time Related-Key Attack on the KASUMI Cryptosystem Used in GSM and 3G Telephony. In: Rabin, T. (ed.) CRYPTO 2010. LNCS, vol. 6223, pp. 393–410. Springer, Heidelberg (2010)

9. ETSI. TS 135 202 V7.0.0: Universal Mobile Telecommunications System (UMTS); Specification of the 3GPP confidentiality and integrity algorithms; Document 2: KASUMI specification (3GPP TS 35.202 version 7.0.0 Release 7), http://www.etsi.org
10. ETSI. TS 55.216 V6.2.0: 3rd Generation Partnership Project; Technical Specification Group Services and System Aspects; 3G Security; Specification of the A5/3 Encryption Algorithms for GSM and ECSD, and the GEA3 Encryption Algorithm for GPRS; Document 1: A5/3 and GEA3 Specifications (Release 6), http://www.etsi.org
11. ETSI. TS 55.919 V6.1.0: 3rd Generation Partnership Project; Technical Specification Group Services and System Aspects; 3G Security; Specification of the A5/3 Encryption Algorithms for GSM and ECSD, and the GEA3 Encryption Algorithm for GPRS; Document 4: Design and evaluation report (Release 6), http://www.etsi.org
12. ETSI. 3rd Generation Partnership Project; Technical Specification Group Services and System Aspects; 3G Security; Specification of the A5/4 Encryption Algorithms for GSM and ECSD, and the GEA4 Encryption Algorithm for GPRS (Release 9), http://www.etsi.org
13. Guo, J., Peyrin, T., Poschmann, A., Robshaw, M.: The LED Block Cipher. In: Preneel, B., Takagi, T. (eds.) CHES 2011. LNCS, vol. 6917, pp. 326–341. Springer, Heidelberg (2011)
14. Kelsey, J., Kohno, T., Schneier, B.: Amplified Boomerang Attacks Against Reduced-Round MARS and Serpent. In: Schneier, B. (ed.) FSE 2000. LNCS, vol. 1978, pp. 75–93. Springer, Heidelberg (2001)
15. Kim, J., Hong, S., Preneel, B., Biham, E., Dunkelman, O., Keller, N.: Related-key boomerang and rectangle attacks, http://eprint.iacr.org/2010/019.pdf
16. Kühn, U.: Cryptanalysis of Reduced-Round MISTY. In: Pfitzmann, B. (ed.) EUROCRYPT 2001. LNCS, vol. 2045, pp. 325–339. Springer, Heidelberg (2001)
17. Kühn, U.: Improved Cryptanalysis of MISTY1. In: Daemen, J., Rijmen, V. (eds.) FSE 2002. LNCS, vol. 2365, pp. 61–75. Springer, Heidelberg (2002)
18. Matsui, M.: Linear Cryptanalysis Method for DES Cipher. In: Helleseth, T. (ed.) EUROCRYPT 1993. LNCS, vol. 765, pp. 386–397. Springer, Heidelberg (1994)
19. Matsui, M.: New Block Encryption Algorithm MISTY. In: Biham, E. (ed.) FSE 1997. LNCS, vol. 1267, pp. 54–68. Springer, Heidelberg (1997)
20. National Security Agency (NSA). SKIPJACK and KEA algorithm specifications (May 1998), http://csrc.ncsl.nist.gov/encryption/skipjack-1.pdf
21. Rivest, R.L.: The RC5 encryption algorithm. In: Preneel, B. (ed.) FSE 1994. LNCS, vol. 1008, pp. 86–96. Springer, Heidelberg (1995)
22. Wagner, D.: The Boomerang Attack. In: Knudsen, L.R. (ed.) FSE 1999. LNCS, vol. 1636, pp. 156–170. Springer, Heidelberg (1999)

Cryptology: Where Is the New Frontier?

Ross Anderson

University of Cambridge, UK
Ross.Anderson@cl.cam.ac.uk

Abstract. Twenty years ago, the crypto community was relatively ho-
mogeneous, with the people who went to Crypto and Eurocrypt spanning
everything from theory to applications. Now it's much more diverse, with
several underlying bodies of theory (from complexity to protocol analysis) and a great variety of applications. Where should a young researcher
focus?

Doing good cryptographic engineering to support complex socio-
technical systems is hard, and I will discuss three examples. First, payment protocols such as EMV (which is just being adopted in India) and
the more recent work in mobile wallets, have a major problem in managing complexity. Second, infrastructure protection such as DNSSEC and
BGPSEC is a good thing but often runs up against a lack of deployment
incentives. Finally, the UEFI proposal for authenticated boot revives
many of the questions of trust that were previously discussed during the
crypto wars, during the debate over "Trusted Computing", and in the
context of SSL CAs. The lesson is that the security and cryptology research communities in India should engage with the policy and economic
implications of our field. Although India's situation may be different from
America's or Europe's, many of the same issues of trust, control, innova-
tion and privacy will surely come round again and again. What's more,
good research tends to come from real problems; researchers who engage
with the real world can spot these more quickly.

D.J. Bernstein and S. Chatterjee (Eds.): INDOCRYPT 2011, LNCS 7107, p. 160, 2011.
© Springer-Verlag Berlin Heidelberg 2011

Analysis of the Parallel
Distinguished Point Tradeoff

Jin Hong[1,*], Ga Won Lee[1], and Daegun Ma[2]

[1] Department of Mathematical Sciences and ISaC,
Seoul National University, Seoul 151-747, Korea
{jinhong,gwlee87}@snu.ac.kr
[2] Department of Mathematics,
Konkuk University, Seoul 143-701, Korea
ma.daegun@gmail.com

Abstract. Cryptanalytic time memory tradeoff algorithms are tools for quickly inverting one-way functions and many consider the rainbow table method to be the most efficient tradeoff algorithm. However, it was recently announced, mostly based on experiments, that the parallelization of the perfect distinguished point tradeoff algorithm brings about an algorithm that is 50% more efficient than the perfect rainbow table method. Motivated by this claim, we provide an accurate theoretic analysis of the parallel version of the non-perfect distinguished point tradeoff algorithm.

Performance differences between different tradeoff algorithms are usually not very large, but even these small differences can be crucial in practice. So we take care not to ignore the side effects of false alarms while analyzing the online time complexity of the parallel distinguished point tradeoff algorithm. Our complexity results are used to compare the parallel non-perfect distinguished point tradeoff against the non-perfect rainbow table method. The two algorithms are compared under identical success rate requirements and the pre-computation efforts are taken into account. Contrary to our anticipation, we find that the rainbow table method is superior in typical situations, even though the parallelization did have a positive effect on the efficiency of the distinguished point tradeoff algorithm.

Keywords: time memory tradeoff, parallel distinguished point, distinguished point, rainbow table.

1 Introduction

Cryptanalytic time memory tradeoff algorithms are tools for quickly inverting one-way functions with the help of pre-computed tables. By changing the algorithm parameters, one is able to balance the size of the stored pre-computed table against the average time required for each inversion. The tradeoff techniques are typically used to recover passwords from the information of password

---

* Corresponding author.

D.J. Bernstein and S. Chatterjee (Eds.): INDOCRYPT 2011, LNCS 7107, pp. 161–180, 2011.

hashes and there are commercial implementations which allow one to defeat the access control mechanisms that are embedded in widely used document file formats. The tradeoff technique is not only used among hackers, but is also an important tool for the law enforcement agencies.

It has long been known that the tradeoff attacks can be prevented by designing the security system to incorporate what are referred to as random *salts*, but there still are many systems in use today that do not incorporate this countermeasure, and the only obstacle in applying the tradeoff attacks to these systems is the large pre-computation requirement. However, as was shown in [13], even this barrier is lowered when the security ultimately relies on human generated passwords. Hence, many current security systems are susceptible to the tradeoff attacks, and finding the exact capabilities and limitations of the tradeoff algorithms remains an interesting subject of study.

The first explicit time memory tradeoff algorithm was invented by Hellman [8]. Shortly thereafter, the distinguished point (DP) technique was introduced. This idea, attributed to Rivest in [7], greatly reduces the table lookup requirements of Hellman's original algorithm. The tradeoff algorithm most widely known to the public today is the rainbow table method [14]. When the search space size is N, a typical tradeoff algorithm allows storage requirement M and inversion time T to be balanced, subject to the equation $TM^2 \approx N^2$, through the choice of associated parameters. Rough analyses of the tradeoff algorithms show that the *tradeoff coefficient* $\mathsf{X}_{tc} = TM^2/N^2$ is close to 1 for reasonable choices of parameter sets. However, it is different for each algorithm and does change with the choice of the parameter set.

The tradeoff coefficient X_{tc} is a measure of how efficiently an algorithm balances storage against online time, with a smaller number corresponding to better tradeoff efficiency. In a recent work [12], that builds on the works [1, 10], the value of X_{tc} was accurately computed for the Hellman, DP, and rainbow tradeoffs, and then the tradeoff coefficients of the three algorithms were compared against each other. Unlike previous attempts, the comparison of [12] took the inversion success rate, pre-computation cost, and the number of bits required to store each table entry fully into account. Their conclusion, in oversimplified terms, was that the classical Hellman and the DP tradeoffs perform comparable to each other and that the rainbow table method outperforms these two. Even though this was in agreement with what many had believed for some time, the performances of the algorithms were shown to be quite close to each other for moderate success rates, justifying the need for such careful analyses before any comparison.

In this work, we give a treatment analogous to [12] of the DP tradeoff variant that processes its multiple pre-computation tables in parallel. We present an accurate theoretic online time complexity analysis of the parallel DP tradeoff and compare its performance with that of the rainbow tradeoff. Clearly, the wall-clock running time of the parallel DP tradeoff will be much shorter than the original serial DP tradeoff, but our focus will be on the combined running time of all processors, which is the measure of algorithm execution complexity that is widely used in cryptology.

The conclusions are that parallelization has a positive effect, not only on the wall-clock running time, but also on the combined processor time of the DP tradeoff. However, for typical range of parameters, the gain in performance obtained is shown to be insufficient to overcome the superiority of the rainbow table method over the DP tradeoff. If the wall-clock running time is more important than the total processor time, depending on the degree of parallelism available, there can be situations where the parallel DP tradeoff is desirable over the rainbow tradeoff. However, if the combined processor time is more important, one should use the rainbow method rather than the parallel DP method.

Our analysis may be seen as theoretic or practically limited in two respects. First, the above mentioned combined processing time may not directly correspond to the real-world cost of running an algorithm, but this issue is clearly outside the scope of this paper. Second, we do not discuss whether implementing the parallel DP tradeoff is practical on various specific platforms. In fact, the demand for online memory by the parallel DP tradeoff is higher than other tradeoff algorithms and this could be critical in certain environments. However, aside from issues that are specific to implementation environments, we are as practical as possible in treating the two compared algorithms, in that the success probabilities, pre-computation costs, and bits required per table entry are taken into account during algorithm comparison.

We acknowledge that parallel processing of tradeoff tables is not a new idea. Distributed key search with a central pre-computation table repository is mentioned in [2, 3] as an application for DP tradeoffs. It was noted in [14] that processing multiple rainbow tables in parallel will reduce the total combined processor time. In fact, the rainbow tradeoff is usually taken to process its tables in parallel, even though the number of its tables is much smaller than that of the DP tradeoff and allows lower degree of straightforward parallelization. In [10] one can find a very rough heuristic argument as to why the classical Hellman tradeoff will not benefit from parallelization. Finally, the work [9, 15] claims that the parallel version of the *perfect*[1] DP tradeoff is twice as efficient as the perfect rainbow table method, mainly based on experiment results.

The current work was motivated by the above mentioned [15], which announced the perfect table parallel DP tradeoff as the "New World Champion" of tradeoff algorithms. However, all the algorithms considered in this paper are the non-perfect table versions. Let us explain our choice to analyze the non-perfect table version of the parallel DP tradeoff rather than the perfect table version.

The perfect DP tradeoff was first studied in [2, 3]. They left some unresolved problems and many view [16] as completing the analysis. However, the analysis of [16] does not provide figures that are accurate enough for use in this paper as data representing the performance of a competing algorithm. For example,

---

[1] The use of perfect tables is not stated explicitly in [9, 15], but was clarified to us during private communication. We also learned that short chain length bounds were used in their experiments.

an entry in Table 2 of the paper gives 21.6425 and 21.1357 as experimental and corresponding theoretic figures concerning a certain storage count. On the surface, the two values seem to be in good agreement, but one must note that they are presented in logarithmic scale. We cannot explain the details here, but when the two storage figures representing experiment and theory are translated to tradeoff efficiency figures, they become the ratio $\left(\frac{2^{21.6425}}{2^{21.1357}}\right)^2 \approx 2.02$. Theoretic result of such inaccuracy may be acceptable for many applications, but is not appropriate for the purpose of comparing different algorithms, since algorithm performances are likely to differ by a comparable ratio.

A perfect DP table is obtained by removing redundancies present in a non-perfect DP table and this implies that each perfect table requires more pre-computation to construct than the non-perfect version, if the two are to bring about the same success rate. Even though the perfect table DP tradeoff is expected to be more efficient than the non-perfect table version, with the current state of knowledge explained above, it is hard to judge whether the degree of enhancement in efficiency justifies the additional pre-computation involved.

Since the state of current knowledge is far from ready for the treatment of perfect table parallel DP tradeoffs, and since it is unclear if the efficiency advantage of the perfect version will be worth the higher pre-computation cost, the non-perfect table parallel DP tradeoff is treated in this paper. The perfect table parallel DP tradeoff is certainly of interest and is left as a subject of future study, which can be approached only after a more accurate analysis of the perfect table (serial) DP tradeoff has been developed. The initial view of the effects of parallelism we obtain and the method of approaching we develop in dealing with the non-perfect case will be of guidance in studying the perfect case.

The rest of this paper is organized as follows. We fix the basic terminology and recall some previous results in Section 2. The parallel DP tradeoff algorithm is made explicit in Section 3 and its theoretic analysis is given in Section 4. After verifying our theoretic developments with experiments in Section 5, the efficiency figures obtained through our analysis are compared against those of the original DP and rainbow table methods in Section 6. The final section contains a summary of our results.

2   Preliminaries

Throughout this paper $F : \mathcal{N} \to \mathcal{N}$ will be a function acting on a set $\mathcal{N}$ of size N. As is done by any theoretic analysis of tradeoff algorithms, we take F to be the random function during our theoretic arguments.

Inversion Problem. Let us first clarify the inversion problem we are considering, as there are two versions that need to be distinguished in any accurate analysis of the tradeoff algorithms. Given the inversion target $\mathbf{y} = F(\mathbf{x})$, the first version asks the algorithm to return any single $x \in \mathcal{N}$ satisfying $F(x) = \mathbf{y}$. In the second version, the tradeoff algorithm is allowed to return multiple $x \in \mathcal{N}$ satisfying $F(x) = \mathbf{y}$ and is declared successful as soon as the specific $\mathbf{x}$ that was used to create $\mathbf{y}$ is returned. The two inversion objectives clearly require different amounts of resources to achieve. If applications of the tradeoff technique to password recovery from sufficiently long password hashes are to be considered, the second version is the correct inversion problem to study [12]. In the interest of practical applicability, the current paper deals with the second inversion problem.

Terminology. We assume that the reader has basic knowledge of the DP and rainbow tradeoff algorithms. Below, we quickly review the terminology and fix notation, mainly focusing on the DP tradeoff, but analogous terminology and notation will be used with the rainbow tradeoff. All tradeoff algorithms considered in this paper are the non-perfect table versions.

The correct answer to be recovered and the inversion target will always be denoted by $\mathbf{x}$ and $\mathbf{y} = F(\mathbf{x})$, respectively. Each *DP table*, consisting of starting and ending point pairs of pre-computed *DP chains*, will contain approximately m entries. What is meant by the term approximately will be explained below. The probability of DP occurrence is set to $\frac{1}{t}$, so that the average chain length is roughly t. We distinguish between a DP table and a *DP matrix*, which is the complete set of m pre-computed chains. An online chain *merging* into a pre-computation chain will bring about a *false alarm*. We omit any mentioning of reduction functions, as our arguments will mostly focus on a single table. Rather than using exactly t tables, we allow more flexibility and use ℓ tables, where $\ell \approx t$. Flexibility is also given to the matrix stopping rule, so that $mt^2 = \mathsf{D}_{msc}N$, with a *matrix stopping constant* $\mathsf{D}_{msc} \approx 1$. The condition $\mathsf{D}_{msc} \approx 1$ can be justified as was originally done by Hellman [8] or through a birthday paradox argument as given in [5, 6]. In all cases, we assume that the parameters are reasonable in the sense that $m, t \gg 0$, which, in particular, implies $t \ll \sqrt{N}$ through the matrix stopping rule.

Any implementation of the DP tradeoff will place a *chain length bound* $\hat{t}$ to detect chains falling into loops that do not contain DPs. If one generates m_0 chains with the chain length bound $\hat{t}$, one can expect to collect $m_0\{1 - (1 - \frac{1}{t})^{\hat{t}}\}$ DP chains. Rather than requiring a DP table to contain the information of exactly m DP chains, we take the approach that the chains are always generated from $m_0 = m/\{1 - (1 - \frac{1}{t})^{\hat{t}}\}$ starting points and that the resulting approximately m DP chains are accounted for by each DP table.

Note that the expected number of chains that do not reach a DP within the $\hat{t}$ iterations is $m_0(1 - \frac{1}{t})^{\hat{t}}$ so that the ratio of wasted iterations $m_0\hat{t}(1 - \frac{1}{t})^{\hat{t}}$ over the approximate total pre-computation effort m_0t is $\frac{\hat{t}}{t}(1 - \frac{1}{t})^{\hat{t}}$. This quickly approaches zero as $\frac{\hat{t}}{t}$ is increased. In the interest of practical applications of the tradeoff technique, we will focus on the situation where $\frac{\hat{t}}{t}$ is sufficiently large, so that most of the pre-computation effort is put to use. However, since the mentioned approach to zero is extremely fast, we treat $\hat{t}$ and t as being of somewhat similar order, even when assuming $\frac{\hat{t}}{t}$ to be sufficiently large. This allows us to assume $\hat{t} \ll \sqrt{N}$ and freely use the approximation $(1 - \frac{1}{t})^{\hat{t}} \approx e^{-\hat{t}/t}$. In practice, setting $\hat{t} = 10\,t$ should be large enough for most purposes.

Extensions to the DP Tradeoff. There are a few tricks that can be used with the DP tradeoff to increase its efficiency.

1. Starting points that require less storage than random ones are used [2, 6]. In particular, sequential starting points [1] allow each starting point to be recorded in $\log m_0 \approx \log m$ bits rather than $\log N$ bits.
2. Information concerning the ending points that can be recovered from the DP definition is not recorded [6]. This reduces $\log t$ bits of storage per ending point.
3. The index file (or hash table) technique [6] is used to remove almost $\log m$ bits per ending point without any loss of information.
4. The ending points are simply truncated before storage [4, 6]. Information is lost through this process and since a partial match between the end of an online chain and a DP table entry will falsely be interpreted as a chain collision, a new type of false alarm appears. However, these can be resolved in exactly the same way as with the exiting false alarms. Using the analysis given in [12], one can maintain the side effects of additional false alarms to a manageable level by controlling the degree of truncation.
5. Suppose that the online chain has become a DP chain of length i and has produced an alarm. When regenerating the associated pre-computation chain to resolve the alarm, one need not continue any further than $\hat{t} - i$ iterations [12].

From now on, the term *(original) DP tradeoff* will refer to the algorithm variant that utilizes all five techniques described above.

When the first four tricks described above, which reduce storage, are applied, each DP table entry can be stored in slightly more than $\log m$ bits. The 1-st, 3-rd, and 4-th items can also be applied to the rainbow tradeoff, after which each rainbow table entry consumes slightly over $\log m$ bits. However, the m for the rainbow tradeoff should be taken to be of order similar to mt of the DP tradeoff, so the rainbow table entries take up larger space than the DP table entries. If the DP tradeoff parameters m and t are roughly of the same order and corresponding rainbow tradeoff parameters are used, each rainbow table entry will occupy twice the number of bits required of a DP table entry. This discussion concerning the difference in bits required per table entry was first made in [4] and was theoretically confirmed in [12].

Previous Results. The original DP tradeoff, as defined above, was analyzed in [12]. In the remainder of this section, we review results from [12] that are required in this paper and introduce some more notation.

Given a DP matrix, its *coverage rate* is defined to be the number of distinct nodes that appear among the DP chains as inputs to the one-way function F, divided by mt. Note that the DPs ending each pre-computation chain are not counted in this definition. The expected coverage rate can be computed through the formula

$$\mathsf{D}_{cr} = \frac{2}{\sqrt{1 + 2\mathsf{D}_{msc}} + 1}, \tag{1}$$

when $\frac{\hat{t}}{t}$ is sufficiently large. In particular, the coverage rate can be seen as a function of the single variable $D_{msc} = \frac{mt^2}{N}$, rather than the separate parameters m, t, and N.

Let $D_{pc} = \frac{mt\ell}{N}$ be the pre-computation coefficient so that $D_{pc}N$ is the pre-computation cost. It is not difficult to show that the probability of success for the DP tradeoff can be expressed as $D_{ps} = 1 - e^{-D_{pc}D_{cr}}$. If we rewrite this equation in the form

$$D_{pc} = -\frac{\ln(1 - D_{ps})}{D_{cr}} = -\frac{\ln(1 - D_{ps})}{2} \left(\sqrt{1 + 2D_{msc}} + 1 \right), \qquad (2)$$

we can see that, under a fixed requirement on the success rate of the DP tradeoff, the pre-computation coefficient D_{pc} is a function of the matrix stopping constant D_{msc}.

Finally, when $\frac{\hat{t}}{t}$ is sufficiently large, the time memory tradeoff curve for the original DP tradeoff is given by $TM^2 = D_{tc}N^2$, where the tradeoff coefficient is

$$D_{tc} = \left(2 + \frac{1}{D_{msc}} \right) \frac{1}{D_{cr}^3} D_{ps} \left\{ \ln(1 - D_{ps}) \right\}^2. \qquad (3)$$

When placed under a fixed success rate requirement D_{ps}, this can also be seen as a function of D_{msc} through a substitution of (1).

3   Parallel DP

The details of the DP tradeoff algorithm that processes its tables in parallel are made explicit in this section. The following two further extensions to the DP tradeoff appear in [9, 15].

6. A full record of the online chain is maintained during the online phase. When required to resolve an alarm, one compares the current end of the regenerated pre-computation against the complete online chain, rather than against just y. This allows one to stop at the exact position of chain merge, rather than at the end of the pre-computation chain.
7. The ℓ DP tables are processed in parallel, rather than serially. This causes relatively more time to be spent in dealing with short online chains and brings about a reduction in the number of alarms.

The DP tradeoff that incorporates all seven DP extension techniques discussed so far will be referred to as the *parallel DP* tradeoff, or pD in short. The purpose of this paper is to analyze the pD tradeoff. Analogous to the D_{msc} notation introduced for the DP tradeoff, we use pD_{msc} to denote the matrix stopping constant associated with the pD tradeoff.

Since the pD and the original DP tradeoffs share the same pre-computation algorithm, their respective coverage rates and pre-computation costs are equal, when identical parameters m, t, and ℓ are used. The online phase algorithms of the two tradeoffs are different, but the differences have no effect on the success probability. Thus, equations (1) and (2) are also valid for the pD tradeoff, when the variable D_{msc} is replaced by pD_{msc}. Hence, the notation D_{cr}, D_{ps}, and D_{pc} will also be used to denote coverage rate, probability of success, and pre-computation coefficient for the pD tradeoff. However, equation (3) is not applicable to pD and obtaining its analogue for the pD tradeoff is the main objective of the next section.

The 6-th item will surely increase the efficiency of the tradeoff algorithm, but the effect of the 7-th item is not as clear. Working with shorter chains will reduce collisions, but each collision is likely to require a longer chain regeneration before it can be ruled out as a false alarm. Predicting its positive effect with certainty does not seem possible without a rigorous analysis, as will be done in this paper.

One can easily argue that a straightforward application of the final two additional DP extensions would require online memory that is of $t\ell = \Theta(T)$ order size, which cannot be acceptable. The number of expected accesses to the online memory is also of the same order and can become a problem. The work [9] explains that an application of a secondary DP definition for selective recording of the online chain can overcome these problems. However, even the resulting vastly reduced requirements for online memory and its accesses will still be somewhat larger than those of the original DP tradeoff. Discussions of how practical or impractical such modified approaches will be on specific implementation environments are outside the scope of this paper. In this work, we simply treat the online memory issue as accesses to acceptably sized fast memory.

The 7-th item requires more explanation. The number of DP tables is roughly of $O(N^{\frac{1}{3}})$ order, which is likely to be larger than the number of available processors for N of interest, implying that each processor will be assigned to multiple tables. In such a situation, we require each processor to work with its share of assigned tables in a round-robin fashion. A processor should process a single iteration for a table and then move onto the next table it was assigned, rather than take the approach of fully processing one table and then fully processing its next assigned table.

We have partially clarified how DP should be parallelized, but there still is an issue concerning the resolving of false alarms. Consider, for the moment, a fully parallel system, where all the DP tables are distributed to different processors. When a processor encounters an alarm, it will regenerate a pre-computation chain, during which time period other processors will continue with their respective online chain iterations. By the time the alarm is resolved, many of the other processors would have reached the end of the online chain creation. This shows that the approach of the 7-th trick in trying to have more time spent on short online chains fails in the fully parallel environment.

Fortunately, each processor is likely to be assigned multiple tables in practice. We assume this situation and, in implementing the 7-th DP extension, each processor is made to resolve any alarm that it encounters, before processing any more online chain iterations. Then, since each processor will be struggling to resolve its share of alarms, further iterations of the online chains are effectively postponed until many of the alarms are resolved. If a set of tables allocated to a certain processor rarely produces alarms, online chain iterations for this set of tables will proceed faster than those of other sets of tables, but the overall behavior will be as if the online chain iterations were delayed until current alarms are resolved.

During our analysis, when counting the total function iterations, we shall take the simplified view that the i-th online chain iterations for all tables are executed simultaneously and that the $(i+1)$-th simultaneous iterations are executed only after all alarms encountered at the i-th iterations are resolved. This view correctly reflects the parallelization details discussed so far.

4   Complexity of the pD Tradeoff

To analyze the tradeoff efficiency of the pD tradeoff, we need to compute the expected time complexity of its online phase. This will be computed as a sum of two parts. The first part is the time taken for the online chain creation. An extremely rough approximation for this would be t times the number of tables ℓ, but we want to be much more precise. The second part is the extra cost of resolving alarms, which has been ignored in many existing analyses of tradeoff algorithms. Three lemmas will be prepared before we state the tradeoff coefficient of pD.

Since the pD tradeoff processes all the tables in parallel, the online phase is likely to terminate with a correct answer before any of the tables are processed in full. Hence, to compute the online time complexity, we need to understand the success probability associated with the processing of each column of the DP matrix rather than with the complete processing of a DP table.

Lemma 1. *Visualize a DP matrix as having been aligned at the ending points. The number of distinct points found in a column of distance i from the ending points is expected to be*

$$\overline{m}_i = \mathsf{D}_{cr}\, m \left(1 - \frac{1}{t}\right)^{i-1},$$

when $\frac{\hat{t}}{t}$ is sufficiently large.

To roughly verify the correctness of this lemma, first notice that, by the definition of D_{cr}, we can expect there to be $\mathsf{D}_{cr}mt$ distinct points in a single DP matrix. Among these, a ratio of $\frac{1}{t}$ points are expected to reach DPs at their next iterations. Hence, there are $\mathsf{D}_{cr}m$ points that are 1-iteration away from the ending point DPs. This count is what is claimed by this lemma as $\overline{m}_1$. We can generalize this approach to obtain the number of points that lie further iterations away from the DPs. A full proof of this lemma can be found in [11].

The sum $\sum_{j=1}^{i} \overline{m}_j$, which may be computed from this lemma, allows us to express the probability for the answer $\mathbf{x}$ not to be found in any of the DP tables within the first i iterations. By suitably combining this with the probability $\left(1 - \frac{1}{t}\right)^{i-1}\frac{1}{t}$ for an online chain to reach a DP at the i-th iteration, it should be possible to obtain the probability for the online chain creation for a specific table to terminate at the i-th iteration. This leads to the expected online chain creation time of pD and is summarized below. We remind the readers that, as was discussed in the previous section, the online chain creation is to be stopped as soon as the alarm for the correct answer is encountered (and resolved).

Lemma 2. *The online chain creation of the pD tradeoff is expected to require*

$$t^2 \frac{D_{ps}}{pD_{msc}D_{cr}}$$

invocations of F, when $\frac{\hat{t}}{t}$ is sufficiently large.

We refer the interested readers to [11] for a detailed proof.

Our first goal, i.e., expressing the online chain creation effort, has been reached. The second part, which is the cost of resolving alarms, is given next.

Lemma 3. *The number of iterations required by the pD tradeoff in dealing with alarms is expected to be*

$$t^2 \frac{\ln(1 - D_{ps})}{D_{cr}} \int_0^1 (1 - D_{ps})^{1-u} \ln u \, du,$$

when $\frac{\hat{t}}{t}$ is sufficiently large.

Let us only briefly explain how one might prove this lemma. The full technical proof is given in [11]. As was seen while obtaining the previous lemma, we already have access to the probability for the i-th iteration of a table to be processed. This i-th iteration generates work related to an alarm if and only if the online chain becomes a DP chain at precisely the i-th iteration and it merges with a pre-computation chain. The probability to encounter an online DP chain of length i is easily written as $(1 - \frac{1}{t})^{i-1}\frac{1}{t}$. For the merging part, we turn things around and view the pre-computation chain as colliding into the online chain of length i. This allows us to keep track of the length of the pre-computation chain up to the point of collision while computing the collision probability. Note that the length up to collision is equal to the work factor when the 6-th DP extension is used. To arrive at the final statement, some computation is required after combining the three parts that we have explained.

We have gathered enough material to compute the efficiency of pD in balancing storage against online time.

Theorem 1. *The time memory tradeoff curve for the pD tradeoff is $TM^2 = pD_{tc}N^2$, where the tradeoff coefficient is given by*

$$pD_{tc} = \left(\frac{\ln(1 - D_{ps})}{D_{ps}} \int_0^1 (1 - D_{ps})^{1-u} \ln u \, du + \frac{1}{pD_{msc}} \right) \frac{1}{D_{cr}^3} D_{ps} \left\{ \ln(1 - D_{ps}) \right\}^2,$$

when $\frac{\hat{t}}{t}$ is sufficiently large.

Proof. The expected online time complexity of the pD tradeoff is the sum of its online chain creation and alarm treatment costs, which are given by Lemma 2 and Lemma 3, respectively. Thus, the time complexity may explicitly be written as

$$T = t^2 \Big(\frac{\mathsf{D}_{ps}}{\mathsf{pD}_{msc}\mathsf{D}_{cr}} + \frac{\ln(1 - \mathsf{D}_{ps})}{\mathsf{D}_{cr}} \int_0^1 (1 - \mathsf{D}_{ps})^{1-u} \ln u \, du \Big).$$

Since the storage size is $M = m\ell$, we find

$$TM^2 = (mt\ell)^2 \Big(\frac{\mathsf{D}_{ps}}{\mathsf{pD}_{msc}\mathsf{D}_{cr}} + \frac{\ln(1 - \mathsf{D}_{ps})}{\mathsf{D}_{cr}} \int_0^1 (1 - \mathsf{D}_{ps})^{1-u} \ln u \, du \Big).$$

To arrive at the claimed statement, it suffices to substitute $mt\ell = \mathsf{D}_{pc}N$ and suitably combine the result with $\mathsf{D}_{ps} = 1 - e^{-\mathsf{D}_{pc}\mathsf{D}_{cr}}$, which was mentioned above (2) and stated as also being correct for pD in Section 3.   □

Note that even though the definite integral appearing in this theorem cannot be simplified any further, it can be treated as an explicit constant, as soon as the requirement for inversion success rate D_{ps} is fixed.

5   Experiment Results

This section presents two sets of experiments that were conducted to test some of our theoretic arguments made in the previous section. In all tests, the one-way function was taken to be the key to ciphertext mapping computed with AES-128. Different fixed plaintexts were used to create multiple one-way functions. Zero padding of keys and truncation of ciphertexts were used to control the size of the space the one-way function acted on.

During our theoretic developments we dealt with two DP extension techniques that were not treated in existing analyses of tradeoff algorithms. The first concerns the 6-th DP extension that shortens the regeneration of pre-computation chains, and we had to compute the reduced expected cost of dealing with alarms. The second hurdle concerns the 7-th DP extension that changed the order of online chain iteration executions, and we had to work out the probability of inversion success associated with each column of a DP matrix, as opposed to that associated with a whole DP matrix. We tested the correctness of our arguments concerning these two issues with experiments.

Our theory surrounding the online chain record is hidden from view behind Lemma 3. Using arguments made during its proof we can explicitly write out the expected cost of resolving alarms associated with a *single* table as

$$\mathsf{D}_{msc} t \int_0^{\hat{t}/t} x\, e^{-x} - \frac{\hat{t}/t}{e^{\hat{t}/t} - 1} \big(1 - e^{-x} \big)\, dx = \mathsf{D}_{msc} t \Big\{ 1 - \frac{1}{e^{\hat{t}/t}} - \frac{(\hat{t}/t)^2}{e^{\hat{t}/t} - 1} \Big\}. \quad (4)$$

More precisely, this ignores whether or not the correct answer was found among other tables, and is the expected cost of resolving false alarms during the complete processing of a single table, when all DP extensions up to the 6-th trick are

Table 1. Cost of resolving alarms when fully processing a single DP table with an online chain record

N	m	t	$\hat{t}/t$	D_{msc}	theory	test	N	m	t	$\hat{t}/t$	D_{msc}	theory	test
2^{28}	512	512	1	0.5	12.84	13.02	2^{28}	768	512	1	0.75	19.26	19.47
2^{28}	512	512	5	0.5	210.86	210.19	2^{28}	768	512	5	0.75	316.29	315.67
2^{28}	512	512	10	0.5	254.83	254.70	2^{28}	768	512	10	0.75	382.24	382.12
2^{29}	1536	512	1	0.75	19.26	19.47	2^{28}	1536	512	1	1.5	38.51	38.88
2^{29}	1536	512	5	0.75	316.29	316.28	2^{28}	1536	512	5	1.5	632.58	631.08
2^{29}	1536	512	10	0.75	382.24	381.68	2^{28}	1536	512	10	1.5	764.48	763.41

applied. Rather than testing Lemma 3 directly, which could hide small details through the averaging effect over multiple tables, we tested our theoretic treatment of the 6-th DP extension through (4) that predicts behavior seen while processing a single table. Testing this equation is also more appropriate in that it still retains the dependence on the chain length bound parameter $\hat{t}$.

Experimental verification of (4) is summarized in Table 1. Each table entry corresponds to 2,000 randomly generated DP tables and 5,000 online chain creations per table. All iterations spent in dealing with alarms that were generated during this whole process were counted and divided by $2,000 \times 5,000$. The tests were conducted with different values of $\frac{\hat{t}}{t}$ and D_{msc}, so as to verify (4) at various inputs. We also tried different m and t pairs that give the same D_{msc} value to verify that the formula is indeed a function of D_{msc}. All the experiment results are very close to our theory.

The second experiment validates our treatment of the parallel processing of tables. At the core of our associated argument is an equation obtained during the proof of Lemma 2, that expresses the probability for the processing of a table to stop at its i-th iteration. The theoretically obtained probability was compared with values obtained through experiments. Working with the i-th termination probability, rather than Lemma 2, which gives the number of one-way function applications summed over all i, allows us more direct verification of finer details.

After fixing various parameters, we first executed a complete pre-computation phase, thus preparing a set of ℓ tables. A simple XOR with a random fixed constant was used as the reduction function for each table. Then, we ran the pD online phase algorithm with a randomly generated inversion target and recorded the iteration count at which the processing of each table was terminated. The online phase was repeated with a certain number of random inversion targets. The process described so far, starting from the pre-computation phase to the multiple target inversions, was repeated a small number of times. Let us denote the number of inversion targets tried per pre-computation set by *keys* and the number of complete pre-computation phases that were executed as *rpt*. The number of times each specific i was recorded, summed over all tables and test trials, was divided by $\ell \times keys \times rpt$, and was taken as the probability obtained through the experiment.

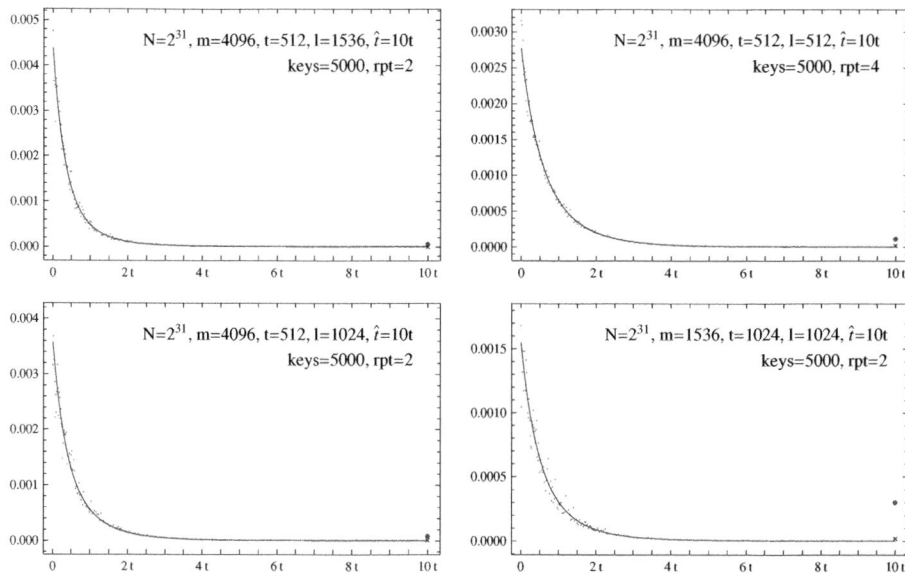

Fig. 1. Probability for pD to stop processing of a specific table at the i-th iteration; Only approximately 240 out of the $\hat{t}$ test values are plotted as dots in each graph (x-axis: i; y-axis: probability)

Test results are depicted as graphs in Figure 1. In each framed graph box, the tiny irregular (red) dots represent the experiment results and the smooth (blue) curve represents the theory. Because plotting all test results made the dots too densely packed and hard to see, we only plotted approximately 240 points among the $\hat{t}$ points of each graph, with the intervals between plotted points shorter at smaller i values. The theoretic value for the $i = \hat{t}$ position, which is expressed as a separate equation, is marked with an ×-sign and the corresponding experiment result is marked with a dot that is slightly larger than others. It is clear that the test results and theory are mostly in good agreement. However, the experiment data and theoretic value are visibly different at $i = \hat{t}$ and this requires explanation.

The disagreement at $i = \hat{t}$ is the largest for the right bottom graph. In this case the chain length bound $\hat{t} = 10t = 10240 \approx 2^{13.3}$ is rather close to $\sqrt{N} = 2^{15.5}$, and since such a choice does not satisfy the assumptions made in Section 2, this does not indicate a problem with our analysis. We were forced to use such small parameters by lack of computational resources, but this should not be an issue in practical applications of the tradeoff technique, where N would be much larger.

Let us discuss this matter slightly further. Recall that we had repeatedly used $\left(1 - \frac{1}{t}\right)^{i}$ as the probability for a chain not to reach a DP until the i-th iteration. This value disregards the possibility of the chain looping back onto itself and a more exact expression is

$$\prod_{j=1}^{i}\left(1-\frac{1}{t}-\frac{j}{N}\right)+\sum_{k=1}^{i}\frac{k}{N}\prod_{j=1}^{k-1}\left(1-\frac{1}{t}-\frac{j}{N}\right).$$

The value $\left(1-\frac{1}{t}\right)^{i}$ is a good approximation of this, as long as $i \ll \sqrt{N}$, but one knows from the birthday paradox that $\left(1-\frac{1}{t}\right)^{i}$ will slowly deviate from the value given by the more complicated expression as i approaches $\sqrt{N}$. We conducted a supplementary test, which we do not explain here due to page limits, to verify that our use of $\left(1-\frac{1}{t}\right)^{i}$ as the ratio of non-DP chains remaining after the i-th iteration was indeed the cause of the discrepancy between theory and experiment at $i=\hat{t}$, when $\hat{t}$ is close to $\sqrt{N}$.

6   Comparison of Tradeoff Algorithms

The analysis given in Section 4 contains enough information for us to make comparisons between algorithms. We will first present a direct comparison between the pD and the original DP tradeoffs. Then, we will present the range of design options made available by the two DP variants and the rainbow tradeoffs compactly as graphs. These graphs will be of more practical value than the initial direct comparisons.

6.1   pD versus DP

As discussed in Section 3, the pD and the original DP tradeoffs will display identical coverage rate, require identical pre-computation cost, and succeed in recovering $\mathbf{x}$ with identical probability, when they are executed under the same set of parameters m, t, and ℓ. The two algorithms also require the same amount of physical storage for DP tables and they only differ in the online execution time. Since the tradeoff coefficients are given as their respective $\frac{TM^{2}}{N^{2}}$ values, the online time complexities may directly be compared through pD_{tc} and D_{tc}. Here, a smaller coefficient implies a more efficient tradeoff algorithm.

After reviewing the tradeoff coefficients as given by (3) and Theorem 1, one can see that they are expressed in a very similar form. In fact, the only difference is in the first constant that sits inside the first set of parentheses. To compare the pD tradeoff against the original DP tradeoff, it suffices to evaluate $\frac{\ln(1-\mathrm{D}_{ps})}{\mathrm{D}_{ps}}\int_{0}^{1}(1-\mathrm{D}_{ps})^{1-u}\ln u\,du$ at various success rates. Some explicit values are 0.9293, 0.8345, 0.6879, 0.5283, 0.4332, and 0.2830 at respective success rates 25%, 50%, 75%, 90%, 95%, and 99%. Since all of these are strictly less than 2, the corresponding constant of the original DP tradeoff, we have a sure indication that the pD tradeoff will outperform the original DP tradeoff.

We have clearly ranked pD over DP, but there is one issue that could affect this conclusion. The online memory access bottleneck of the pD tradeoff could remain a non-negligible overhead, even when the secondary DP technique mentioned in Section 3 is applied, and one iteration of the pD tradeoff could take longer, on average, than that of the DP tradeoff. Should the circumstances be such that a marked difference in this iteration time is inevitable, the difference must be taken into account when interpreting a tradeoff coefficient ratio as a performance ratio. For example, if the pD tradeoff requires α time units per iteration and the DP tradeoff requires β time units per iteration, then one must compare the values $\alpha \cdot \mathsf{pD}_{tc}$ and $\beta \cdot \mathsf{D}_{tc}$ against each other rather than compare pD_{tc} against D_{tc}.

After seeing that pD outperforms DP, one naturally questions whether the improvement comes from the 6-th or the 7-th DP extension, i.e., whether keeping a record of the online chain or the parallel processing of tables was the main cause for the improvement. Let us refer to the DP variant that uses the first six of the seven DP extensions as the DP-OCR tradeoff. Details are not provided in this paper due to the page limit, but we have also analyzed DP-OCR in full. Results are that the tradeoff coefficient for DP-OCR falls between those of pD and DP, if all three are subject to the same parameters m, t, and ℓ. When parameters are such that only low inversion success rates (50%) are expected, DP-OCR performs quite close to pD, implying that the online chain record accounts for most of the improvement. However, when parameter achieving high success rates (99%) are chosen, DP-OCR stands approximately halfway between original DP and pD, implying that the parallelization does play a significant role in increasing efficiency.

6.2   pD versus Rainbow

Our next goal is to include the rainbow tradeoff in the comparison. This is not as straightforward as the comparisons between the two DP variants, mainly because of the large structural differences between the DP and rainbow tradeoffs. Below, we quickly review the approach of [12] before following it to present a fair comparison. The information required to draw the graphs for the rainbow tradeoff was also taken from [12]. Notation R_{pc} and R_{tc} will be used to denote the pre-computation coefficient and the tradeoff coefficient of the rainbow tradeoff, and the notation X_{pc} and X_{tc} will be used when referencing the pre-computation and tradeoff coefficients that are not specific to a tradeoff algorithm.

The tradeoff coefficient $\mathsf{X}_{tc} = \frac{TM^2}{N^2}$ is a measure of how efficiently the algorithm balances online time against storage requirements. A smaller X_{tc} implies a more efficient tradeoff between online time and storage. However, better tradeoff efficiency usually requires a higher pre-computation cost and is not always desirable in practice. The optimal balance point between the tradeoff efficiency X_{tc} and pre-computation cost X_{pc} is a subjective matter that cannot be arbitrarily set in this paper. What can be done objectively is to present the range of

"tradeoff efficiency X_{tc} can be utilized, if pre-computation effort X_{pc} is invested"

options that are made available by each algorithm, as a graph. Then, the implementer can make subjective decisions based on this compact display of possible choices and the physical resources that are available to him.

The above discussion would be sufficient for anyone interested in choosing parameters for one fixed tradeoff algorithm. However, a little more care is required when comparing different algorithms. The range of possible options for each algorithm must be presented in a consistent manner. A common requirement on the success rate must be taken and the unit used to express the tradeoff coefficients must be unified. We have already seen that a direct comparison between pD_{tc} and D_{tc} is justifiable. As for comparisons of these two against the rainbow tradeoff, let us simply state that pD_{tc} and D_{tc} should be compared against $4R_{tc}$ (as opposed to R_{tc}) to be fair. The main reason is that, as discussed in Section 2, the number of bits per table entry required by a rainbow tradeoff is roughly twice that of a DP tradeoff. The decision to compare pD_{tc} against $4R_{tc}$ involves certain assumptions, such as that m and t are of similar order and that the bookkeeping of online chain records causes negligible overhead, but they are assumptions that are typically made during theoretic analysis of tradeoff algorithms and the details should be clear to anyone with a full understanding of [12]. This concludes our short review of the approach to fair tradeoff comparison given by [12].

The range of (X_{pc}, X_{tc}) options each tradeoff algorithm is capable of providing is given in Figure 2. To draw the graph, for example, corresponding to the DP tradeoff, one refers to (2) for the x-coordinate (pre-computation cost), substitutes (1) into (3) for the y-coordinate (tradeoff coefficient), and then plots the curve parameterized by D_{msc}.

In each framed graph box, the two curves and the sequence of dots should be seen as extending infinitely upwards. However, the right ends of the three graphs are either clearly marked or clearly visible. The curves start to go back up beyond the marked right ends, so that these marks correspond to the minimum tradeoff coefficient achievable by each algorithm. As going beyond this minimum implies using larger pre-computation while obtaining worse tradeoff efficiency, parameters corresponding to the parts that are not drawn should not be used.

We are now ready to discuss the implications of the graphs given in Figure 2. The graphs for D_{tc} and pD_{tc} are given by the dashed and solid lines, respectively. The possible $(R_{pc}, 4R_{tc})$ choices are given by the discrete sequence of dots. Each dot corresponds to the use of a certain number of rainbow tables and since these table counts tend to be small, especially at low success rate requirements, the possible options appear spaced apart from each other. If one is required to fill in the space between the dots, one may extend horizontal lines to the right of each dot, until the line reaches over the dot to its right. Each box corresponds to a certain requirement on the probability of successful inversions.

In all the graph boxes, the graphs for the pD tradeoff sit further away from the bottom left corner than the dots for the rainbow tradeoff. Being closer to the bottom left corner implies that the same tradeoff efficiency can be obtained at a smaller pre-computation cost and that a better tradeoff efficiency can be obtained at equal pre-computation cost. Hence a very rough conclusion would be that the rainbow tradeoff is the best among the three algorithms.

Let us discuss this in more detail, starting with the case of success rate set to 25%. The optimal tradeoff coefficient reachable by the pD tradeoff is $pD_{tc} = 0.10$.

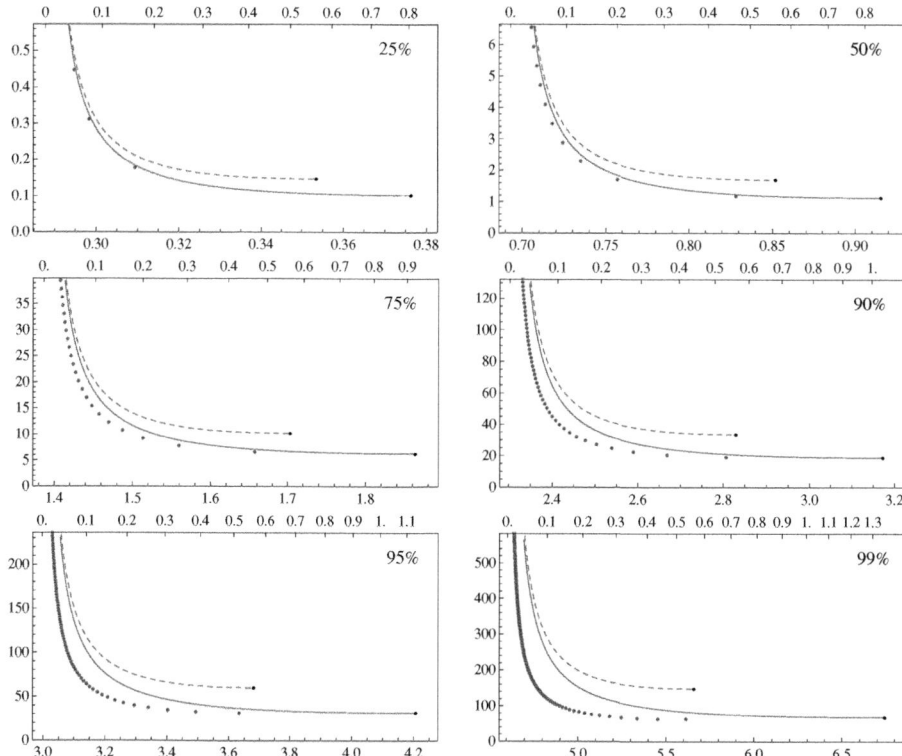

Fig. 2. Tradeoff coefficients for the DP (dashed), pD (solid), and rainbow (large dots) tradeoffs in relation to pre-computation cost, at various success rates; Numeric values on each frame represent pre-computation iterations in units of N (bottom), tradeoff coefficient values D_{tc}, pD_{tc}, $4R_{tc}$ (left), and the matrix stopping constants D_{msc}, pD_{msc} that served as parameters for drawing the curves (top)

This tradeoff efficiency can be used if the available resources permit $0.376N$ iterations of pre-computation. In comparison, the rainbow tradeoff achieves optimal tradeoff coefficient $4R_{tc} = 0.18$ at $0.309N$ pre-computation iterations. Even though $pD_{tc} = 0.10$ and $4R_{tc} = 0.18$ represent online time ratio of 1.8 at equal physical storage size, depending on the resources available to the tradeoff implementer, the advantage of pD in tradeoff efficiency may or may not be worth its disadvantage in the pre-computation cost one must accept. So far, neither of the two tradeoffs can be said to be clearly superior over the other.

However, another issue that is evident in the 25% case is that the rainbow tradeoff provides much less flexibility in options than the pD tradeoff. For example, the option of using $pD_{tc} = 0.11$ at $D_{pc} = 0.336$ is available with the pD algorithm. Compared to the optimal efficiency of $pD_{tc} = 0.10$ at $D_{pc} = 0.376$, this gives a valuable reduction in pre-computation cost at a small degradation of tradeoff efficiency. Unless the cost of pre-computation is extremely cheap, most implementers of the tradeoff algorithm will prefer to use $pD_{tc} = 0.11$ over the optimal $pD_{tc} = 0.10$. The rainbow tradeoff does not allow such a freedom of choice at the 25% probability of success.

Since the dots for the rainbow tradeoff are very close to the curve for the pD tradeoff, one can say that every option provided by the rainbow tradeoff can (nearly) be provided by the pD tradeoff. Since the pD tradeoff provides higher flexibility and even the possibility of a lower tradeoff coefficient, it seems safe to conclude that the pD tradeoff is preferable over the rainbow tradeoff at 25% success rate.

Even though we have explained at length that the pD tradeoff could be preferable over the rainbow tradeoff, the observations made for the 25% success rate are not very applicable to any of the other graph boxes. In the 50% success rate case, the optimal pD option of $pD_{tc} = 1.12$ at $D_{pc} = 0.915$ is not a very attractive choice over the rainbow option of $4R_{tc} = 1.17$ at $R_{pc} = 0.828$, which achieves similar tradeoff efficiency at a visibly lower pre-computation cost. Similarly, we have $pD_{tc} = 6.19$ at $D_{pc} = 1.86$ versus $4R_{tc} = 6.48$ at $R_{pc} = 1.66$ for the 75% success rate, and $pD_{tc} = 18.5$ at $D_{pc} = 3.17$ versus $4R_{tc} = 18.7$ at $R_{pc} = 2.81$ for the 90% success rate. At these moderate success rates, the minimum pD_{tc} is slightly better than the minimum $4R_{tc}$, but its use cannot be justified when pre-computation cost is taken into account. In fact, as discussed in the 25% success rate, implementers are likely to choose parameters somewhat away from the optimal tradeoff efficiency points, where the rainbow tradeoff is clearly advantageous over the pD tradeoff. As for higher success rates 95% and 99%, even the minimum $4R_{tc}$ is smaller than the minimum pD_{tc}, so that there is no reason to prefer the pD tradeoff over the rainbow tradeoff.

7   Conclusion

The parallel DP tradeoff studied in this work is a cryptanalytic time memory tradeoff algorithm that adds two extra techniques to the more widely known DP tradeoff. The first is to keep a full record of the online chain so that alarms can be resolved earlier during the pre-computation chain regeneration. The second idea is to process the multiple DP tables in parallel. This allows for more time to be spent in dealing with relatively shorter chains so that false alarms are hopefully reduced. We have confirmed that both of these ideas have positive effects on the efficiency of the DP tradeoff.

Our analysis of the pD tradeoff did not ignore the time taken to resolve false alarms and is accurate enough to provide multiple significant digits. Results of the analysis were used to provide a comparison between the pD and rainbow tradeoffs. The comparison of tradeoff efficiency was done in a fair manner in the sense that factors such as the success probability of inversion, the storage size in number of bits, and pre-computation cost were all taken into account. Hence, the comparison results have practical implications on the choice of which tradeoff algorithm to use.

Comparisons show that, even with the extra enhancements, the pD tradeoff is not likely to be preferable over the rainbow tradeoff under most situations.

The only exception is when the success rate requirement is very low. For example, when dealing with multi-target time memory tradeoffs [5], where the rainbow tradeoff is known to be much less efficient than both the original Hellman and DP tradeoffs, our analysis is an indication that the use of the pD tradeoff could be advantageous over the original DP tradeoff. At moderate success rate requirements, the pD tradeoff can be slightly more efficient than the rainbow tradeoff, but the choice to use the pD tradeoff cannot be justified when the pre-computation cost is taken into account.

In short, when reduction in wall-clock running time is very important and one is willing to parallelize the online phase to a very high degree, depending on the degree of parallelization available, variants of the DP tradeoff could be a reasonable choice. However, if total CPU time is more important than wall-clock time, one should work with the rainbow tradeoff. Still, the pD tradeoff is more efficient than the usual DP tradeoff, in that it requires a smaller total number of function iterations, when the two are provided with the same pre-computation table.

The theoretic analysis and the resulting concrete graphs of this paper can easily be adjusted to cope with various specific situations and allow for educated decisions. For example, when run on resource constrained environments such as GPUs, iterations of pD may take longer than those of DP due to pD's higher demands for online memory. In this situation, it suffices to scale the tradeoff coefficients of pD and DP according to their respective average iteration timings before comparing their graphs to conclude whether the online memory requirement undermines the small advantage of pD over DP.

Acknowledgements. JH and GWL were supported by Basic Science Research Program through the National Research Foundation of Korea(NRF) funded by the Ministry of Education, Science and Technology(2011-0005764). GWL was also supported by the National Institute for Mathematical Sciences (NIMS) grant funded by the Korean government (No. A21101).

References

1. Avoine, G., Junod, P., Oechslin, P.: Characterization and improvement of time-memory trade-off based on perfect tables. ACM Trans. Inform. Syst. Secur., 11(4), 17:1–17:22 (2008); Preliminary version presented at INDOCRYPT 2005
2. Borst, J.: Block Ciphers: Design, Analysis, and Side-Channel Analysis. Ph.D. Thesis, Katholieke Universiteit Leuven (September 2001)
3. Borst, J., Preneel, B., Vandewalle, J.: On the time-memory tradeoff betweeen exhaustive key search and table precomputation. In: Proceedings of the 19th Symposium on Information Theory in the Benelux, WIC (1998)
4. Barkan, E., Biham, E., Shamir, A.: Rigorous Bounds on Cryptanalytic Time/Memory Tradeoffs. In: Dwork, C. (ed.) CRYPTO 2006. LNCS, vol. 4117, pp. 1–21. Springer, Heidelberg (2006)
5. Biryukov, A., Shamir, A.: Cryptanalytic Time/Memory/Data Tradeoffs for Stream Ciphers. In: Okamoto, T. (ed.) ASIACRYPT 2000. LNCS, vol. 1976, pp. 1–13. Springer, Heidelberg (2000)

6. Biryukov, A., Shamir, A., Wagner, D.: Real Time Cryptanalysis of A5/1 on a PC. In: Schneier, B. (ed.) FSE 2000. LNCS, vol. 1978, pp. 1–18. Springer, Heidelberg (2001)

7. Denning, D.E.: Cryptography and Data Security. Addison-Wesley (1982)

8. Hellman, M.E.: A cryptanalytic time-memory trade-off. IEEE Trans. on Infor. Theory 26, pp. 401–406 (1980)

9. Hoch, Y.Z.: Security analysis of generic iterated hash functions. Ph.D. Thesis, Weizmann Institute of Science (August 2009)

10. Hong, J.: The cost of false alarms in Hellman and rainbow tradeoffs. Des. Codes Cryptogr. 57(3), 293–327 (2010)

11. Hong, J., Lee, G.W., Ma, D.: Analysis of the Parallel Distinguished Point Tradeoff. Cryptology ePrint Archive. Report 2011/387

12. Hong, J., Moon, S.: A comparison of cryptanalytic tradeoff algorithms. Cryptology ePrint Archive. Report 2010/176

13. Narayanan, A., Shmatikov, V.: Fast dictionary attacks on passwords using time-space tradeoff. In: Proceedings of the 12th ACM CCS, pp. 364–372. ACM (2005)

14. Oechslin, P.: Making a Faster Cryptanalytic Time-Memory Trade-Off. In: Boneh, D. (ed.) CRYPTO 2003. LNCS, vol. 2729, pp. 617–630. Springer, Heidelberg (2003)

15. Shamir, A.: Random Graphs in Security and Privacy. Invited talk at ICITS 2009 (2009)

16. Standaert, F.-X., Rouvroy, G., Quisquater, J.-J., Legat, J.-D.: A Time-Memory Tradeoff Using Distinguished Points: New Analysis & FPGA Results. In: Kaliski Jr., B.S., Koç, Ç.K., Paar, C. (eds.) CHES 2002. LNCS, vol. 2523, pp. 593–609. Springer, Heidelberg (2003)

On the Evolution of GGHN Cipher

Subhadeep Banik, Subhamoy Maitra, and Santanu Sarkar

Applied Statistics Unit, Indian Statistical Institute, 203 B T Road, Kolkata 700108
{s.banik_r,subho}@isical.ac.in, sarkar.santanu.bir@gmail.com

Abstract. In this paper we study the GGHN stream cipher presented in CISC 2005. This cipher has been motivated from RC4 with the idea to obtain further speed-up by considering word-oriented keystream output instead of byte-oriented ones. We prove that there exist a large number of short cycles of length equal to the length of the state array used in the cipher. Then towards having a theoretical analysis of GGHN type evolution, we study a randomized model of this cipher. Using Markovian process, we show how this model evolves to all zero state much faster than what is expected in an ideal scenario.

Keywords: Cryptanalysis, GGHN, RC4, Short Cycles, Stream Cipher.

1 Introduction

The GGHN (name follows from the first character of the surnames of the four coauthors Gong, Gupta, Hell and Nawaz) cipher [2] was proposed from the motivation of extending RC4 from the byte-oriented model to word-oriented model. This cipher has received attention in literature as evident from the cryptanalytic results by Paul and Preneel (Asiacrypt 2006) [7], Tsunoo, Saito, Kubo and Suzaki (IEEE-IT, 2007) [9] and Kircanski and Youssef (CCDS 2010) [4].

It is needless to mention that RC4 is the most popular stream cipher in the area of cryptology from commercial as well as theoretical interest. The algorithm has two parts: the Key Scheduling Algorithm (KSA) and the Pseudo-Random Generation Algorithm (PRGA), as given below in Algorithm 1 and Algorithm 2. Since the structure of RC4 is quite elegant, it has attracted in-depth cryptanalysis in many papers (one may refer to the recent papers [6,5,8] and the references therein). Still the cipher is quite secure for 128-bit key-size if used with proper care.

One limitation of RC4 is that its output generation is in bytes (8-bits) and it is natural to extend the idea to word oriented (4 or 8 bytes in general) output that will produce the keystream much faster. The 32 or 64-bit processors are ready for such applications. For storing 8-bit permutations, one needs only 2^8 bytes, but for storing 32-bit permutations, a huge array of size 2^{32} words (1 word = 4 bytes, here) is required. As this is impractical, there are several efforts to work with small random arrays only, instead of complete random permutations. As example one may refer to the GGHN cipher [2] which is indeed as simple and elegant looking as RC4. On the other hand, as it will be shown in this paper, it is a classical example that a simple and elegant cipher can have quite a few weaknesses. Before proceeding further, let us first describe the GGHN cipher.

D.J. Bernstein and S. Chatterjee (Eds.): INDOCRYPT 2011, LNCS 7107, pp. 181–195, 2011.
© Springer-Verlag Berlin Heidelberg 2011

Input: Secret Key K.
Output: S-Box S generated by K.

Initialize $S = \{0, 1, 2, \ldots, N - 1\}$;
Initialize counter: $j = 0$;
for $i = 0, \ldots, N - 1$ **do**
    $j = (j + S[i] + K[i]) \bmod N$;
    Swap $S[i] \leftrightarrow S[j]$;
end

Input: S-Box S, from KSA output.
Output: Pseudorandom stream z.

Initialize the counters: $i = j = 0$;
while *keystream is required* **do**
    $i = (i + 1) \bmod N$;
    $j = (j + S[i]) \bmod N$;
    Swap $S[i] \leftrightarrow S[j]$;
    $z = S[(S[i] + S[j]) \bmod N]$;
end

Algorithm 1. RC4 KSA   **Algorithm 2.** RC4 PRGA

Input: S-Box S: an m-bit integer array of $N = 2^n$ locations, initially loaded
        with certain predefined value;
Input: Key array K: an m-bit integer array of l locations;
Input: An integer r for number of rounds;
Output: Pseudorandom S-Box S and a pseudorandom m-bit integer k;

$j = 0, k = 0, t = 0, M = 2^m$;
while $t < r$ **do**
    for $i = 0, \ldots, N - 1$ **do**
        $j = (j + S[i] + K[i \bmod l]) \bmod N$;
        Swap $S[i] \leftrightarrow S[j]$;
        $S[i] = (S[i] + S[j]) \bmod M$;
        $k = (k + S[i]) \bmod M$;
    end
    $t = t + 1$;
end

Algorithm 3. GGHN-KSA(n, m)

Input: S-Box S and integer k from GGHN-KSA(n, m);
Output: m-bit Keystream words z

$i = 0, j = 0, M = 2^m$;
while *Keystream is generated* **do**
    $i = (i + 1) \bmod N$;
    $j = (j + S[i]) \bmod N$;
1   $k = (k + S[j]) \bmod M$;
    $z = (S[(S[i] + S[j]) \bmod N] + k) \bmod M$;
2   $S[(S[i] + S[j]) \bmod N] = (k + S[i]) \bmod M$;
end

Algorithm 4. GGHN-PRGA(n, m)

Regarding the security of GGHN(8, 32) [2], the following results are available in literature:

- a $2^{32.89}$ distinguisher [7] based on the observation that the least significant bit of the keystream words are biased to zero;
- a $2^{30.02}$ distinguisher [9] based on the equality of the least significant bits of the first two keystream words;
- structural weakness of GGHN(8, 32) has been presented in [4], where it has been shown that out of 8240 bits ($256 \cdot 32$ bits for S, 32 bits for k and $8 \cdot 2$ bits for i, j) of the internal state if 2064 bits can be obtained (a fault attack is presented too on how to obtain those bits) then the complete state can be explored effectively.

One should also note the following comment related to certain weakness of the cipher pointed out by the authors themselves [2, Section 4.7].

"When all entries are even and k is even, then all outputs as well as all future entries will be even, resulting in a biased keystream. The state update function in RC4(n, m) is not an invertible mapping so it will always be possible to enter one of these weak states. However the probability that all state entries, as well as k are even is very low, 2^{-257}. From this we can conclude that these weak states are of no concern to the security of the cipher."

In [2], the authors called GGHN(n, m) as RC4(n, m) and while putting the probability 2^{-257}, they considered $n = 8, m = 32$. We would like to remind that $N = 2^n, M = 2^m$.

1.1   Organization of the Paper

In Section 2, we study the evolution of GGHN PRGA. We first note that once all the words become even, half of the state array will not be updated further. More importantly, we identify a lot of short cycles in this cipher. As a general result, given a GGHN(n, m) cipher, we prove by construction that there are cycles of length 2^n which starts from $i = 0, j = 0, k \equiv -1 \bmod 2^n, S[r] \equiv 1 \bmod 2^n \ \forall r \in [0, 2^n - 1], r \neq 2$ and $S[2] \equiv 0 \bmod 2^n$ with the constraint that

$$k_0 \equiv -\left(s_0 + 3 \cdot s_1 + \sum_{r=3}^{2^n-1} s_r\right) \bmod 2^m, \quad s_2 \equiv -\left(3 \cdot s_1 + \sum_{r=3}^{2^n-1} s_r\right) \bmod 2^m.$$

As we obtain several weaknesses of the GGHN(n, m) cipher, we got motivated to dig into a theoretical model of this cipher where all the indices will be chosen uniformly at random. This is presented in Section 3, where we study a randomized version of GGHN cipher. Using Markovian process, we show that the number of iterations to reach the all even state (N bits taking each LSB of S and 1 bit for the LSB of k, thus $N + 1$ bits in total and all of them need to be zero) is much less than 2^{N+1}. For $N = 256$, instead of 2^{257}, it is only $2^{142.16}$. In fact, our analysis shows, that the expected number of iterations to reach the all zero state is $32 \cdot 2^{142.16} = 2^{147.16}$ only.

We conclude the paper in Section 4.

2   Short Cycles in GGHN(n, m)

Refer to GGHN-PRGA(n, m) as in Algorithm 4. Immediately one can note the following result.

Lemma 1. *Consider the situation when all the elements of S and k have evolved to even values. Then half of the elements of S which are indexed by odd values will never be modified further.*

Proof. Since all the values of S array are even, $S[i] + S[j]$ is also even and so is $(S[i]+S[j]) \bmod N$. Thus, in the line marked 4 of Algorithm 4, the left hand side $S[(S[i] + S[j]) \bmod N]$ will be $S[l]$ where l is even. Thus only the even indexed locations will be updated and the odd indexed locations will not be modified at all.   □

This is indeed a weakness of the design. However, we now present more important results related to short cycles of the cipher. One may note that there are well known short cycles for RC4 which are famous as Finney cycles, but the initial conditions in RC4 PRGA are so chosen that such cycles cannot occur [1]. We now show that the GGHN PRGA(n, m) algorithm may fall into a cycle where a given state repeats after a certain number of iterations. To explain things clearly, we move step by step with several intermediate results and examples. We will first illustrate the case for GGHN$(2, 2)$, i.e., with the array S having 4 cells each containing 2 bit data.

Proposition 1. *For the GGHN PRGA$(2, 2)$, the initial condition*

$$S[0] = 1, \ S[1] = 1, \ S[2] = 0, \ S[3] = 1, \ i = j = 0, \ k = 3$$

forms a cycle of length 4.

Proof. The algorithm starts from the given state $i = 0$, $j = 0$, $k = 3$, $S[0] = 1$, $S[1] = 1$, $S[2] = 0$, $S[3] = 1$ and goes through the following states before returning to the initial one:
$$i = 1, \ j = 1, \ k = 0, \ S[0] = 1, \ S[1] = 1, \ S[2] = 1, \ S[3] = 1,$$
$$i = 2, \ j = 2, \ k = 1, \ S[0] = 1, \ S[1] = 1, \ S[2] = 2, \ S[3] = 1,$$
$$i = 3, \ j = 3, \ k = 2, \ S[0] = 1, \ S[1] = 1, \ S[2] = 3, \ S[3] = 1.$$
    It is interesting to see that the value $S[i] + S[j] \equiv 2 \bmod 4$ is an invariant, and thus at every stage of the cycle, only the value $S[2]$ gets altered in the array S. Also $i = j$ at all the stages of the cycle.   □

Now let us present a more generalized result.

Lemma 2. *For the GGHN PRGA$(2, m)$ with $m \geq 2$, consider an initial condition of the form $i = j = 0$, $S[0] = s_0 \equiv 1 \bmod 4$, $S[1] = s_1 \equiv 1 \bmod 4$, $S[2] = s_2 \equiv 0 \bmod 4$, $S[3] = s_3 \equiv 1 \bmod 4$, $k = k_0 \equiv 3 \bmod 4$. If*

$$k_0 \equiv -(s_0 + 3s_1 + s_3) \bmod 2^m \ \text{and}$$

$$s_2 \equiv -(3s_1 + s_3) \bmod 2^m$$

are satisfied, then a cycle of length 4 will be formed.

Proof. It is clear that the values in S array and k are in the range $[0, \ldots, 2^m - 1]$ and i, j are in $[0, \ldots, 3]$. Because of the restrictions placed on $S[r]$, $r = 0, 1, 2, 3$ and k the values of i and j will always remain same during the evolution. Furthermore, the value $S[i] + S[j] \equiv 2 \bmod 4$ is an invariant, and hence, $S[2]$ is the only location of the array S that undergoes alteration at each step of the cycle as follows. In the following state sequence '+' stands for addition modulo 2^m.

$$\left(\begin{array}{c} i = 1, \ j = 1, \ k = k_0 + s_1 \\ S[0] = s_0, \ S[1] = s_1, \ S[2] = k_0 + 2s_1, \ S[3] = s_3 \end{array} \right)$$

$$\left(\begin{array}{c} i = 2, \ j = 2, \ k = 2k_0 + 3s_1 \\ S[0] = s_0, \ S[1] = s_1, \ S[2] = 3k_0 + 5s_1, \ S[3] = s_3 \end{array} \right)$$

$$\left(\begin{array}{c} i = 3, \ j = 3, \ k = 2k_0 + 3s_1 + s_3 \\ S[0] = s_0, \ S[1] = s_1, \ S[2] = 2k_0 + 3s_1 + 2s_3, \ S[3] = s_3 \end{array} \right)$$

$$\left(\begin{array}{c} i = 0, \ j = 0, \ k = 2k_0 + 3s_1 + s_3 + s_0 \\ S[0] = s_0, \ S[1] = s_1, \ S[2] = 2k_0 + 3s_1 + s_3 + 2s_0, \ S[3] = s_3 \end{array} \right)$$

For this to represent a cycle the conditions $2k_0 + 3s_1 + s_3 + s_0 \equiv k_0 \bmod 2^m$ and $2k_0 + 3s_1 + s_3 + 2s_0 \equiv s_2 \bmod 2^m$ must hold. That is to say k_0 needs to satisfy the modular equation $k_0 \equiv -(s_0 + 3s_1 + s_3) \bmod 2^m$ and s_2 needs to satisfy the equation $s_2 \equiv -(3s_1 + s_3) \bmod 2^m$.

Note that as $m \geq 2$, $k_0 \equiv -(s_0 + 3s_1 + s_3) \bmod 4 \equiv 3 \bmod 4$ and $s_2 \equiv -(3s_1 + s_3) \bmod 4 \equiv 0 \bmod 4$ that are indeed consistent with the condition presented in the statement of this lemma.   □

To explain further, if we choose any arbitrary $s_0, s_1, s_3 \in [0, 2^m - 1]$ satisfying $s_0, s_1, s_3 \equiv 1 \bmod 4$, we can easily calculate s_2 and k from the above equations such that the state $i = 0$, $j = 0$, $k = k_0$, $S[0] = s_0$, $S[1] = s_1$, $S[2] = s_2$, $S[3] = s_3$ is the initial state of a cycle of length 4. For example, in the system GGHN(2, 8) if we take $s_0 = 69$, $s_1 = 141$, $s_3 = 9$, using the above equations we get $k_0 = 11$ and $s_2 = 80$ and so $i = 0$, $j = 0$, $k = 11$, $S[0] = 69$, $S[1] = 141$, $S[2] = 80$, $S[3] = 9$ forms a cycle of length 4.

In fact, for GGHN(2, m), there are other examples not covered by Lemma 2 for cycles of length 4. One can start with any of the following cases and then proceed similar to Lemma 2 to get different conditions:

(i) $i = 0, j = 2, k \equiv 3 \bmod 4, S[0] \equiv 1 \bmod 4, S[1] \equiv 3 \bmod 4, S[2] \equiv 0 \bmod 4, S[3] \equiv 3 \bmod 4$,

(ii) $i = 0, j = 2, k \equiv 0 \bmod 4, S[0] \equiv 3 \bmod 4, S[1] \equiv 0 \bmod 4, S[2] \equiv 3 \bmod 4, S[3] \equiv 2 \bmod 4$,

(iii) $i = 0, j = 0, k \equiv 1 \bmod 4, S[0] \equiv 3 \bmod 4, S[1] \equiv 3 \bmod 4, S[2] \equiv 0 \bmod 4, S[3] \equiv 3 \bmod 4$,

(iv) $i = 0, j = 1, k \equiv 0 \bmod 4, S[0] \equiv 2 \bmod 4, S[1] \equiv 1 \bmod 4, S[2] \equiv 2 \bmod 4, S[3] \equiv 2 \bmod 4$.

Similarly the GGHN(3, m) PRGA has a cycle of length 8 of the form $i = 0$, $j = 7$, $k \equiv 0 \bmod 8$ and S array as $\{0, 3, 0, 0, 3, 4, 0, 3\}$ modulo 8; and also the initial state variables $S[i] = s_i$ and $k = k_0$ need to satisfy the following modular matrix equation.

$$
\begin{pmatrix}
0 & 0 & 3 & 0 & 0 & 0 & 0 & 0 & 1 \\
0 & 1 & 0 & 0 & 0 & 0 & 0 & 0 & 0 \\
0 & 0 & 1 & 0 & 0 & 0 & 0 & 0 & 0 \\
0 & 0 & 18 & 0 & 45 & 0 & 0 & 7 \\
0 & 0 & 0 & 0 & 1 & 0 & 0 & 0 & 0 \\
0 & 0 & 0 & 0 & 0 & 1 & 0 & 0 & 0 \\
0 & 0 & 15 & 0 & 45 & 7 & 0 & 6 \\
0 & 0 & 5 & 0 & 1 & 2 & 0 & 0 & 2 \\
0 & 0 & 15 & 0 & 45 & 0 & 0 & 6
\end{pmatrix}
\cdot
\begin{pmatrix}
s_0 \\ s_1 \\ s_2 \\ s_3 \\ s_4 \\ s_5 \\ s_6 \\ s_7 \\ k_0
\end{pmatrix}
=
\begin{pmatrix}
s_0 \\ s_1 \\ s_2 \\ s_3 \\ s_4 \\ s_5 \\ s_6 \\ s_7 \\ k_0
\end{pmatrix}
\bmod 2^m
$$

In these equations s_1, s_2, s_4, s_5 can be chosen from $[0, 2^m - 1]$ such that $s_1 \equiv 3 \bmod 8, s_2 \equiv 0 \bmod 8, s_4 \equiv 3 \bmod 8, s_5 \equiv 4 \bmod 8$. The remaining state variables are then calculated by solving the above equations giving the complete state which forms a cycle of length 2^3. For example, in GGHN(3, 8), letting $s_1 = 3, s_2 = 24, s_4 = 83, s_5 = 20$ we can use the above modular matrix equation to get $k_0 = s_6 = 200, s_3 = 216, s_0 = 16, s_7 = 131$.

Now we present a result for GGHN(n, n).

Lemma 3. *The state $i = 0, \ j = 0, \ k = 2^n - 1, S[r] = 1 \ \forall r \in [0, 2^n - 1], r \neq 2, \ and \ S[2] = 0$ forms a cycle of length 2^n in GGHN(n, n) PRGA algorithm.*

Proof. Given this particular initial state, the algorithm runs through the states as given below

$$i = 1, j = 1, k = 0, S[r] = 1 \forall r \in [0, 2^n - 1], r \neq 2, S[2] = 1,$$

$$i = 2, j = 2, k = 1, S[r] = 1 \forall r \in [0, 2^n - 1], r \neq 2, S[2] = 2,$$

$$\vdots$$

$$i = 2^n - 1, j = 2^n - 1, k = 2^n - 2, S[r] = 1 \forall r \in [0, 2^n - 1], r \neq 2, S[2] = 2^n - 1.$$

before returning to the initial state. As before the quantity $S[i] + S[j] \equiv 2 \bmod 2^n$ is an invariant throughout the cycle, and hence $S[2]$ is the only location of the array S to undergo changes.   □

Now we construct a cycle in the most general case for GGHN(n, m).

Theorem 1. *In the GGHN(n, m) PRGA algorithm, one can obtain a cycle of length 2^n starting with the initial state $i = 0, \ j = 0, \ k = k_0 \equiv -1 \bmod 2^n, S[r] = s_r \equiv 1 \bmod 2^n \ \forall r \in [0, 2^n - 1], r \neq 2, \ and \ S[2] = s_2 \equiv 0 \bmod 2^n \ under \ the conditions*

$$k_0 \equiv - \left(s_0 + 3 \cdot s_1 + \sum_{r=3}^{2^n - 1} s_r \right) \bmod 2^m \ and$$

$$s_2 \equiv - \left(3 \cdot s_1 + \sum_{r=3}^{2^n - 1} s_r \right) \bmod 2^m.$$

Proof. One can check that the value $S[i] + S[j] = 2 \bmod 2^n$ is an invariant. Now the states are evolved as follows. In the following state sequence '+' stands for addition modulo 2^m.

$$\left(\begin{array}{c} i = j = 1, \ k = k_0 + s_1 \\ S[r] = s_r \forall r \in [0, 2^n - 1], r \neq 2, \ S[2] = k_0 + 2s_1 \end{array} \right)$$

$$\left(\begin{array}{c} i = j = 2, \ k = 2k_0 + 3s_1 \\ S[r] = s_r \forall r \in [0, 2^n - 1], r \neq 2, \ S[2] = 3k_0 + 5s_1 \end{array} \right)$$

$$\left(\begin{array}{c} i = j = 3, \ k = 2k_0 + 3s_1 + s_3 \\ S[r] = s_r \forall r \in [0, 2^n - 1], r \neq 2, \ S[2] = 2k_0 + 3s_1 + 2s_3 \end{array} \right)$$

$$\vdots$$

$$\left(\begin{array}{c} i = j = i_0, \ k = 2k_0 + 3s_1 + \sum_{r=3}^{i_0} s_r \\ S[r] = s_r \forall r \in [0, 2^n - 1], r \neq 2, \ S[2] = 2k_0 + 3s_1 + \sum_{r=3}^{i_0} s_r + s_{i_0} \end{array} \right)$$

$$\vdots$$

$$\left(\begin{array}{c} i = j = 0, \ k = 2k_0 + s_0 + 3s_1 + \sum_{r=3}^{2^n - 1} s_r \\ S[r] = s_r 0 \forall r \in [0, 2^n - 1], r \neq 2, \ S[2] = 2k_0 + 2s_0 + 3s_1 + \sum_{r=3}^{2^n - 1} s_r \end{array} \right)$$

Thus to get a cycle of length 2^n we need to satisfy $k_0 \equiv 2k_0 + s_0 + 3s_1 + \sum_{r=3}^{2^n - 1} s_r \bmod 2^m$ and $s_2 \equiv 2k_0 + 2s_0 + 3s_1 + \sum_{r=3}^{2^n - 1} s_r \bmod 2^m$. From which we get

$$k_0 \equiv - \left(s_0 + 3 \cdot s_1 + \sum_{r=3}^{2^n - 1} s_r \right) \bmod 2^m \text{ and}$$

$$s_2 \equiv - \left(3 \cdot s_1 + \sum_{r=3}^{2^n - 1} s_r \right) \bmod 2^m.$$

□

It is possible to choose $S[r] \equiv 1 \bmod 2^n$ for all $r \in [0, 2^n - 1]$ except $r = 2$. Then one can obtain the values satisfying the conditions mentioned in Theorem 1. For example, in GGHN(8, 32) if $s_r = 1 \ \forall \ r \in [0, 2^n - 1]$ except $r = 2$, then we would obtain $k_0 = 2^{32} - (2^8 + 1)$ and $s_2 = 2^{32} - 2^8$.

This is the first time such short 2^n-length cycles are demonstrated for the cipher GGHN(n, m). That is, the cycles are as short as the length of the S array itself. One may easily see that for $r \neq 2$, $S[r] \equiv 1 \bmod 2^n \ \forall r \in [0, 2^n - 1]$. Thus, for each location of S array (except the 2nd one), we have 2^{m-n} options. Thus there are at least $(2^{m-n})^{2^n - 1} = 2^{(m-n)(2^n - 1)}$ many initial conditions for which GGHN(n, m) will land into a short cycle equal to the S-array length. This number is however only a small fraction of the total number of possible initial states (2^{m2^n}) of the GGHN PRGA cipher. One may also start with several other conditions and explore different kinds of short cycles.

3   Evolution of a Randomized Variant of GGHN Cipher

We have so far studied the exact GGHN cipher. It has been clearly shown that the cipher has weaknesses due to the short cycles. Now to have a view of the evolution of similar kinds of ciphers, we would like to present a theoretical model where the indices are chosen independently and uniformly at random from $[0, N-1]$. In this regard, we make the assumption that

$$j, (S[i] + S[j]) \bmod N \text{ and } i$$

are chosen independently and uniformly at random from $[0, \ldots, N-1]$ and we call them a, b, c respectively. In fact, it is clear from the design of GGHN PRGA [2] that they tried to simulate $j, (S[i] + S[j]) \bmod N$ as pseudorandom indices and we consider the same here in the theoretical model. However, the index i has been incremented by one in each step of GGHN [2] and making a difference, the index i is taken as a random index in our theoretical model. We here consider the evolution from the situation when the KSA is completed and we assume that the values of the array S as well as k are chosen uniformly at random from m-bit integers. Let us now describe the algorithm.

> **while** *Keystream is generated* **do**
>   Select a, b, c uniformly at random from $[0, N-1]$;
> 1 $k = (k + S[a]) \bmod M$;
> 2 $z = (S[b] + k) \bmod M$;
> 3 $S[b] = (k + S[c]) \bmod M$;
> **end**

Algorithm 5. RAND-GGHN-PRGA(n, m)

We will first show that there is a function $f(N)$ such that it is expected that in $f(N)$ many steps the S-Box S of RAND-GGHN-PRGA(n, m) will be even. For this we can work with the following algorithm considering S as a bit-array and k as a single bit. The bits of the S array as well as k are initially chosen uniformly at random from $\{0, 1\}$. Since we are working with the least significant bit only, it is enough to use EXOR (i.e., GF(2) addition) as the least significant bit is same for both EXOR and modulo 2 addition. In this case $m = 1$. We want to find the expected number of iterations required after which all the elements in S as well as k become zero. We use results related to Markovian process and one may refer [3, Chapter 11] for more details of this technique.

Theorem 2. *Following BIT-RAND-GGHN-PRGA$(n, 1)$, let S be a binary array of length N. The elements of S as well as k are filled independently and uniformly at random from $\{0, 1\}$. Then there is a function $f(N)$ such that it is expected that all elements of S as well as k will be zero in $f(N)$ many steps.*

> **while** *the loop is required to be run* **do**
>   Select a, b, c uniformly at random from $[0, N - 1]$;
> 1 $k = (k \oplus S[a])$;
> 2 $S[b] = (k \oplus S[c])$;
> **end**

Algorithm 6. BIT-RAND-GGHN-PRGA$(n, 1)$

Proof. Consider that q denotes the number of 1's in S. Hence it should be sufficient to analyze the Markov chain with state $(q, k) \in \{0, N\} \times \{0, 1\}$. At the t-th stage, denote q_t as the number of 1's in S and k_t as the value of k. The transitions of this chain are given by

$$k_{t+1} = \begin{cases} 1 \oplus k_t, & \text{with probability } \frac{q_t}{N}, \\ k_t, & \text{otherwise} \end{cases} \tag{1}$$

and

$$q_{t+1} = \begin{cases} q_t + 1, & \text{with probability } \frac{q_t}{N}\left(1 - \frac{q_t}{N}\right) \text{ if } k_{t+1} = 0, \\ q_t - 1, & \text{with probability } \frac{q_t}{N}\left(1 - \frac{q_t}{N}\right) \text{ if } k_{t+1} = 0, \\ q_t + 1, & \text{with probability } \left(1 - \frac{q_t}{N}\right)^2 \text{ if } k_{t+1} = 1, \\ q_t - 1, & \text{with probability } \left(\frac{q_t}{N}\right)^2 \text{ if } k_{t+1} = 1, \\ q_t & \text{otherwise.} \end{cases} \tag{2}$$

If $q = 0$ and $k = 0$, then the process terminates as it will remain in the same state. We obtain the expected number of steps t to hit $(q_t, k_t) = (0, 0)$ given any initial state.

Note that q follows Binomial$(N, \frac{1}{2})$ and k is uniform in $\{0, 1\}$. It is convenient to write the transition matrix M as $M_f = M_q M_k$, where M_k (respectively M_q) gives the transition probability for Step 6 (respectively Step 6) of Algorithm 6. It is clear to note that both M_k and M_q are $(2N + 1) \times (2N + 1)$ matrices.

Now we describe the matrices following (1) and (2). Let j be the column index from 1 to $2N + 1$. Let $v_j = \lceil \frac{j-1}{2} \rceil$. The diagonal entries of M_k are $1 - \frac{v_j}{N}$ for the column j. The off diagonal entries of M_k are $\frac{v_j}{N}$ with one entry for each column, just below the diagonal if j is even and just above if j is odd. Here we follow the ordering of (q_t, k_t) as $(0, 1)$, $(1, 0)$, $(1, 1)$, $(2, 0)$, $(2, 1)$, $\ldots$, $(n, 1)$.

The diagonal elements of M_q are $1 - \frac{2v_j}{N}\left(1 - \frac{v_j}{N}\right)$ for j-th column with even j and $1 - \left(\frac{v_j}{N}\right)^2 - \left(1 - \frac{v_j}{N}\right)^2$ for odd j. The off diagonal elements of M_q are exactly two rows above and two rows below the diagonal element. These values are equal to $\frac{v_j}{N}\left(1 - \frac{v_j}{N}\right)$ for even j in both upper and lower side. For odd j the values are $\left(\frac{v_j}{N}\right)^2$ and $\left(1 - \frac{v_j}{N}\right)^2$ for upper and lower sides respectively.

Here M_k is a (right) stochastic matrix, i.e., each of its column values sum to one, while M_q (and hence M_f also) is stochastic except for second column, whose sum is one minus the probability of transition to $(q = 0, k = 0)$ state.

To calculate the expected survival time of the process, we need the fundamental matrix $F = (I - M_f)^{-1}$. The element F_{ij} gives the expected number of times the process visits state i given the process started from the state j. Summing over the columns of F, we get the expected total survival time for each initial state (except for which $q = 0, k = 0$, as the survival time is 0 for that case).

Now consider a vector E of length $2N + 1$. Each element of the vector E is the column sum of the matrix F, i.e., $E = \mathbf{1}^T F$. We interpret E as a vector indexed by 1 to $2N + 1$ and also consider the initial state distribution vector v of $2N + 1$ length as $v[1] = \frac{1}{2^N+1}$ and for $q \geq 1$, $v[2q] = v[2q + 1] = \binom{N}{q}\frac{1}{2^N+1}$. Then $f(N) = E \cdot v$ gives the expected survival time which we denote by $f(N)$. □

Now we will describe the approach of Theorem 2 in detail with the following example.

Example 1. Let us consider the case $N = 4$. The entries of M_k, M_q are the transition probabilities of Steps 6 and 6 of Algorithm 6 respectively. Here the ordering of the states as $(0, 1)$, $(1, 0)$, $(1, 1)$, $(2, 0)$, $(2, 1)$, $(3, 0)$, $(3, 1)$, $(4, 0)$ and $(4, 1)$. The transition matrices will be as follows.

$$M_k = \frac{1}{4}\begin{pmatrix} 4 & 0 & 0 & 0 & 0 & 0 & 0 & 0 & 0 \\ 0 & 3 & 1 & 0 & 0 & 0 & 0 & 0 & 0 \\ 0 & 1 & 3 & 0 & 0 & 0 & 0 & 0 & 0 \\ 0 & 0 & 0 & 2 & 2 & 0 & 0 & 0 & 0 \\ 0 & 0 & 0 & 2 & 2 & 0 & 0 & 0 & 0 \\ 0 & 0 & 0 & 0 & 0 & 1 & 3 & 0 & 0 \\ 0 & 0 & 0 & 0 & 0 & 3 & 1 & 0 & 0 \\ 0 & 0 & 0 & 0 & 0 & 0 & 0 & 0 & 4 \\ 0 & 0 & 0 & 0 & 0 & 0 & 0 & 4 & 0 \end{pmatrix}, M_q = \frac{1}{16}\begin{pmatrix} 0 & 0 & 1 & 0 & 0 & 0 & 0 & 0 & 0 \\ 0 & 10 & 0 & 4 & 0 & 0 & 0 & 0 & 0 \\ 16 & 0 & 6 & 0 & 4 & 0 & 0 & 0 & 0 \\ 0 & 3 & 0 & 8 & 0 & 3 & 0 & 0 & 0 \\ 0 & 0 & 9 & 0 & 8 & 0 & 9 & 0 & 0 \\ 0 & 0 & 0 & 4 & 0 & 10 & 0 & 0 & 0 \\ 0 & 0 & 0 & 0 & 4 & 0 & 6 & 0 & 16 \\ 0 & 0 & 0 & 0 & 0 & 3 & 0 & 16 & 0 \\ 0 & 0 & 0 & 0 & 0 & 0 & 1 & 0 & 0 \end{pmatrix}.$$

Hence the final transition matrix $M_q M_k$ will be

$$M_f = \frac{1}{64}\begin{pmatrix} 0 & 1 & 3 & 0 & 0 & 0 & 0 & 0 & 0 \\ 0 & 30 & 10 & 8 & 8 & 0 & 0 & 0 & 0 \\ 64 & 6 & 18 & 8 & 8 & 0 & 0 & 0 & 0 \\ 0 & 9 & 3 & 16 & 16 & 3 & 9 & 0 & 0 \\ 0 & 9 & 27 & 16 & 16 & 27 & 9 & 0 & 0 \\ 0 & 0 & 0 & 8 & 8 & 10 & 30 & 0 & 0 \\ 0 & 0 & 0 & 8 & 8 & 18 & 6 & 64 & 0 \\ 0 & 0 & 0 & 0 & 0 & 3 & 9 & 0 & 64 \\ 0 & 0 & 0 & 0 & 0 & 3 & 1 & 0 & 0 \end{pmatrix}.$$

The fundamental matrix (elements written upto two decimal places) will be

$$F = \begin{pmatrix} 1.33 & 0.25 & 0.33 & 0.29 & 0.29 & 0.29 & 0.29 & 0.29 & 0.29 \\ 5.33 & 5.97 & 5.33 & 5.65 & 5.65 & 5.65 & 5.65 & 5.65 & 5.65 \\ 5.33 & 3.41 & 5.33 & 4.37 & 4.37 & 4.37 & 4.37 & 4.37 & 4.37 \\ 6.00 & 5.10 & 6.00 & 7.80 & 6.80 & 6.99 & 7.11 & 7.11 & 7.11 \\ 10.00 & 8.02 & 10.00 & 10.76 & 11.76 & 11.57 & 11.45 & 11.45 & 11.45 \\ 5.33 & 4.37 & 5.33 & 6.19 & 6.19 & 7.83 & 7.21 & 7.21 & 7.21 \\ 5.33 & 4.37 & 5.33 & 6.19 & 6.19 & 7.01 & 8.03 & 8.03 & 8.03 \\ 1.33 & 1.09 & 1.33 & 1.55 & 1.55 & 1.83 & 1.93 & 2.93 & 2.93 \\ 0.33 & 0.27 & 0.33 & 0.39 & 0.39 & 0.48 & 0.46 & 0.46 & 1.46 \end{pmatrix}.$$

So E, whose entries are sum of columns, will be

$$E = \mathbf{1}^T F = \begin{bmatrix} 40.33 & 32.87 & 39.33 & 43.19 & 43.19 & 46.02 & 46.52 & 47.52 & 48.52 \end{bmatrix}$$

Here the initial state distribution vector $v = \frac{1}{32}(1, 4, 4, 6, 6, 4, 4, 1, 1)$. Hence $E \cdot v = 41.05$ gives the expected number of steps where all elements of S and also k are zero. □

For BIT-RAND-GGHN-PRGA$(n, 1)$, in Table 1, we present theoretical bounds and experimental observations on the expected number of iterations required to get all entries of S as well as k to be zero for different values of array length N. The theoretical exercises related to formation of matrices are done with the help of SAGE [10]. For experiments, the programs are written in C language (the random number generator of "GNU project C compiler in Linux environment Ubuntu 11.04" is used to get a, b, c) and the average is taken over 10^5 many runs. For the cases $N \geq 32$, we could not perform the experiments because of the long execution time and hence we put '-' marks there. One may note that we get quite close results in theory and experiment.

As $62301892884735739250717421213653150644452309 \approx 2^{142.16}$, following Table 1, we can say that when length of S is 256, all the elements of S as well as

Table 1. Theoretical bounds and experimental values of $f(N)$ for different values of $N = 2^n$ in BIT-RAND-GGHN-PRGA$(n, 1)$

N	$f(N)$	Experiment
4	41.05	41.02
8	280.49	279.89
12	1463.27	1469.15
16	7118.88	7111.03
20	33836.28	33433.15
32	3423401.56	-
64	619282894484.52	-
128	149191364194358609155574.98	-
256	62301892884735739250717421213653150644452309.75	-

k become zero within expected $2^{142.16}$ many steps for $N = 256$. Note that, for a properly designed cipher one can expect that it will reach the all even state in $2^{N+1} = 2^{257}$ iterations as there are 256 LSBs of the 256 locations in the state array S and 1 LSB of k. In the randomized model, the expected time to get into such an even state requires much less iteration than 2^{257}.

Now consider Algorithm 5 and here we have the following result.

Lemma 4. *In the algorithm RAND-GGHN-PRGA(n, m), all the elements of S as well as the integer k are expected to be zero in $m \cdot f(N)$ many iterations.*

Proof. From Theorem 2, for Algorithm 6, it is expected that in $f(N)$ many iterations all entries of S as well as k will be zero. Now consider Algorithm 5. When the Least Significant Bits (LSBs) of all the elements of S and k become zero, it will remain zero for all further steps. It is expected that LSBs of the elements of S and k will be zero after $f(N)$ many steps. Then the operations will not affect the next significant bit and for those bits of S and k, one can expect to get zero in the next $f(N)$ many iterations again. Proceeding in this way, after $m \cdot f(N)$ many steps all elements S and as well as the integer k will be zero.   □

For RAND-GGHN-PRGA(n, m) (in Algorithm 5), where $n = 8$ and $m = 32$ (i.e., $N = 2^8$ and $M = 2^{32}$), one can expect after $32 \times 2^{142.16} = 2^{147.16}$ iterations, all entries of S and k will become zero. For an ideally designed cipher one can expect that it will reach the all zero state in expected $2^{(N+1)m} = 2^{8224}$ iterations as there are $256 \cdot 32$ bits of the 256 length state array S and additional 32 bits of k. In this randomized model, the expected time to get into such an all zero state requires much less iteration than 2^{8224}.

3.1   Towards Estimating the Actual GGHN PRGA

It is interesting to see how similar or dissimilar the evolution of RAND-GGHN-PRGA is when compared to the actual PRGA of GGHN. One striking dissimilarity, already noted in Lemma 1, is that after the actual PRGA of GGHN the state array S evolves to the all even state and the cells of the state array marked by odd indices do not undergo any further change. Hence the PRGA of actual GGHN does not evolve to the all zero state (unless the odd-indexed places of S had already become zero, which is extremely less probable). Hence, it is not possible to do a similar analysis for the exact cipher.

However one can analyze and compare the evolution of several randomized variants of the original PRGA. In this regard, once again we would like to remind that the variables

$$a, b, c \text{ of the RAND-PRGA-GGHN algorithm correspond to } j, S[i] + S[j], i$$

of actual GGHN PRGA. In the RAND-PRGA-GGHN algorithm we have made the assumption that the variables a, b, c are chosen uniformly at random. One can come up with two other variants of this cipher RAND-GGHN-PRGA' and

```
1  c = 0;
   while Keystream is generated do
       Select a, b uniformly at random from [0, N − 1];
2      c = (c + 1) mod N;
3        k = (k + S[a]) mod M;
4        z = (S[b] + k) mod M;
5          S[b] = (k + S[c]) mod M;
   end
```

Algorithm 7. RAND-GGHN-PRGA′(n, m)

```
1  c = a = 0;
   while Keystream is generated do
       Select b uniformly at random from [0, N − 1];
2      c = (c + 1) mod N;
3        a = (a + S[c]) mod N;
4          k = (k + S[a]) mod M;
5          z = (S[b] + k) mod M;
6            S[b] = (k + S[c]) mod M;
   end
```

Algorithm 8. RAND-GGHN-PRGA″(n, m)

RAND-GGHN-PRGA″ given below which are closer to the actual GGHN PRGA. In RAND-GGHN-PRGA′, the variable c (corresponding to i of the original GGHN PRGA) is not selected randomly but incremented by 1 after each iteration as in the original GGHN PRGA. In RAND-GGHN-PRGA″, both c and a (corresponding to i and j of the original GGHN PRGA) are not chosen randomly but incremented as in the original cipher. If the variable b too were to be not chosen randomly and incremented as in the original cipher, then the resulting PRGA would be exactly same as the original GGHN PRGA, but as already explained the analysis of a such a cipher can not be performed due to restrictions of the model. As it turns out RAND-GGHN-PRGA″ is the model which most closely resembles the original cipher.

To have a view of how the theoretical result of Lemma 4 matches with practice we consider the case of $n = 4, m = 16$ for RAND-GGHN-PRGA(n, m). From Table 2 we note that the number steps for one additional set of bits to be zero is almost equal for RAND-GGHN-PRGA. This corresponds to Lemma 4. In the experiments for RAND-GGHN-PRGA(n, m), we have observed that when the t-th LSBs of each location of the S array and k become all zero for the first time, at that point the $(t + u)$-th LSBs look random for $u > 0$.

We also consider the other two randomized variants of the cipher and note the number of iterations each variant of the cipher takes to reach the all zero state. The results are average of 10^5 many independent runs. The LSBs of all the entries of S become zero in expected 7130.88 many steps RAND-GGHN-PRGA

Table 2. Average number of steps required for t many LSBs to be zero for all the elements of S as well as the integer k. The algorithms considered are RAND-GGHN-PRGA$(4, 16)$, RAND-GGHN-PRGA$'(4, 16)$ and RAND-GGHN-PRGA$''(4, 16)$.

t	RAND-GGHN-PRGA	RAND-GGHN-PRGA$'$	RAND-GGHN-PRGA$''$
0	7130.88	10906.15	9089.36
1	14235.94	21826.94	14517.11
2	21352.26	32824.21	17271.46
3	28483.93	43716.31	18821.33
4	35613.17	54635.29	20092.96
5	42710.16	65527.85	21320.74
6	49791.33	76423.01	22544.97
7	56895.35	87361.42	23766.03
8	63981.16	98236.69	24985.67
9	71040.10	109174.21	26199.41
10	78158.61	120082.29	27418.34
11	85253.82	131058.35	28637.54
12	92328.69	141963.34	29855.91
13	99448.96	152865.53	31075.49
14	106477.86	163751.12	32296.63
15	113568.96	174659.31	33510.00

whereas it takes 10906.15 many steps in RAND-GGHN-PRGA$'$ and 9089.36 many steps in RAND-GGHN-PRGA$''$; the next significant bit of those become zero in expected 14235.94 steps in RAND-GGHN-PRGA and so on. From experiments, we note that RAND-GGHN-PRGA$'$ evolves to all zero state after more number of steps than RAND-GGHN-PRGA. However, RAND-GGHN-PRGA$''$, the closest model to actual GGHN PRGA, evolves to all zero state much faster than RAND-GGHN-PRGA.

4   Conclusion

In this paper we revisit the stream cipher GGHN [2] and present several cryptanalytic results. Our main motivation is to study the evolution of the cipher and we show that the cipher has several weaknesses that include a large family of short cycles of length equal to the length of the state array. We also concentrate on a theoretical model of the cipher (referred as RAND-GGHN-PRGA) considering the indices of the cipher are chosen uniformly at random from a certain range of values. Our analysis shows that such model evolves to all zero state much faster than what is expected in an ideal cipher. We have also presented experimental results related to two other models RAND-GGHN-PRGA$'$ and RAND-GGHN-PRGA$''$, the second one being very close to the actual GGHN PRGA and it evolves to the all-zero state much faster. However, theoretical modelling of these two requires further investigation. The analysis in this paper may be used by the future designers to consciously avoid the GGHN kind of evolution while designing a modified version of RC4.

Acknowledgment. The authors would like to thank the anonymous reviewers for their detailed comments that improved the technical as well as editorial quality of the paper.

References

1. Finney, H.: An RC4 cycle that can't happen. Posting to sci.crypt. (September 1994)
2. Gong, G., Gupta, K.C., Hell, M., Nawaz, Y.: Towards a General RC4-Like Keystream Generator. In: Feng, D., Lin, D., Yung, M. (eds.) CISC 2005. LNCS, vol. 3822, pp. 162–174. Springer, Heidelberg (2005)
3. Grinstead, C.M., Snell, J.L.: Introduction to Probability (2006), http://www.dartmouth.edu/~chance/teaching_aids/books_articles/ probability_book/book.html
4. Kircanski, A., Youssef, A.M.: On the structural weakness of the GGHN stream cipher. Cryptography and Communications (Discrete Structures, Boolean Functions and Sequences) 2(1), 1–17 (2010)
5. Maitra, S., Paul, G., Sen Gupta, S.: Attack on Broadcast RC4 Revisited. In: Joux, A. (ed.) FSE 2011. LNCS, vol. 6733, pp. 199–217. Springer, Heidelberg (2011)
6. Maximov, A., Khovratovich, D.: New State Recovery Attack on RC4. In: Wagner, D. (ed.) CRYPTO 2008. LNCS, vol. 5157, pp. 297–316. Springer, Heidelberg (2008)
7. Paul, S., Preneel, B.: On the (In)security of Stream Ciphers Based on Arrays and Modular Addition. In: Lai, X., Chen, K. (eds.) ASIACRYPT 2006. LNCS, vol. 4284, pp. 69–83. Springer, Heidelberg (2006)
8. Sepehrdad, P., Vaudenay, S., Vuagnoux, M.: Statistical Attack on RC4. In: Paterson, K.G. (ed.) EUROCRYPT 2011. LNCS, vol. 6632, pp. 343–363. Springer, Heidelberg (2011)
9. Tsunoo, Y., Saito, T., Kubo, H., Suzaki, T.: A Distinguishing Attack on a Fast Software-Implemented RC4-Like Stream Cipher. IEEE Transactions on Information Theory 53(9), 3250–3255 (2007)
10. http://www.sagemath.org

HiPAcc-LTE: An Integrated High Performance Accelerator for 3GPP LTE Stream Ciphers

Sourav Sen Gupta[1,*], Anupam Chattopadhyay[2], and Ayesha Khalid[2]

[1] Applied Statistics Unit, Indian Statistical Institute, Kolkata, India
[2] MPSoC Architectures, UMIC, RWTH Aachen University, Germany
sg.sourav@gmail.com, {anupam,ayesha.khalid}@umic.rwth-aachen.de

Abstract. Stream ciphers SNOW 3G and ZUC are the major players in the domain of next generation mobile security as both of them have been included in the security portfolio of 3GPP LTE-Advanced, the potential candidate for 4G mobile broadband communication standard. In this paper, we propose HiPAcc-LTE, a high performance integrated design that combines the two ciphers in hardware, based on their structural similarities. The integrated architecture reduces the area overhead significantly compared to two distinct cores, and also provides *almost double* throughput in terms of keystream generation. This is in comparison with the state-of-the-art implementations of the individual ciphers, both in the academic literature as well as in the commercial domain. We present detailed description of the design idea, optimization techniques and comparison results in this paper. Long term vision of this hardware integration approach for cryptographic primitives is to build a flexible core supporting multiple designs having similar algorithmic structures.

Keywords: 3GPP, LTE-Advanced, Stream Cipher, SNOW 3G, ZUC, Hardware Implementation, ASIC, High Throughput, Area Efficiency.

1 Introduction

The world is expecting a completely new mobile broadband technology in the near future. The 3rd Generation Partnership Project (3GPP) had submitted Long Term Evaluation (LTE) Advanced [2] as a candidate for the 4G system, and LTE-Advanced (Release 10) [2] has been standardized by 3GPP as a major enhancement of their LTE standard. The main contenders for the new 3GPP LTE security suite had been EEA1/EIA1 [15] (identical to earlier protocols UEA2/UIA2) based on SNOW 3G [16], and EEA2/EIA2 based on block cipher AES-128. Later, a new set of security algorithms EEA3/EIA3 [17], based on another stream cipher ZUC [18], was proposed for inclusion in LTE, and it has been considered for verification and absorption in the LTE standards. At present, the block cipher AES-128 and the two stream ciphers SNOW 3G and ZUC constitute the core of the LTE security algorithms [1].

---

* This work was done during Sourav Sen Gupta's visit to MPSoC Architectures, UMIC, RWTH Aachen University, Germany as a Summer Intern during June–August 2011.

D.J. Bernstein and S. Chatterjee (Eds.): INDOCRYPT 2011, LNCS 7107, pp. 196–215, 2011.
© Springer-Verlag Berlin Heidelberg 2011

Motivation. It can be observed that the new security architecture for 4G is going to have two different stream ciphers, which are very unlikely to be used simultaneously. However, the hardware implementations of these two ciphers have always been done independently, as distinct cores. Be it in academic literature [12,13] or in commercial products [5,6,7,8,9,10,11], even the single-platform implementations of LTE-Advanced security suite seem to use two distinct cores for the two stream ciphers. We, on the other hand, focus on merging the two towards a compact hardware solution for the LTE suite.

Our main objective, however, is to explore a more general problem in cryptographic hardware. One may want to implement two or more ciphers on the same platform, and in such a scenario, the following question arises:

> If there is a requirement to implement an array of ciphers on the same platform, how should one approach the hardware design?

A general solution towards this direction is to incorporate custom instructions for the individual ciphers into a general purpose processor and thus facilitate it to run any cipher independently. However, this kind of an implementation may not always be the best choice in terms of throughput and area, as a general purpose processor with custom instructions do not provide the implementor with full freedom to explore the design space for an optimal solution.

We approach this problem from a completely different angle. Implementing custom instructions in a processor attempts to merge the ciphers from a hardware implementation point of view. We take a step back, and try to merge the ciphers from an algorithmic point of view first. Once this is accomplished, one may design an integrated custom accelerator for the ciphers such that each of the algorithms can be accessed individually. This approach offers the flexibility of

- sharing of resources, both storage and logic,
- optimization of mutual critical path,
- throughput vs. area optimization at the base level, and
- potential mechanism for combined protection against fault attacks.

The process of integration at both algorithm and hardware levels produce the best solutions in terms of throughput and area, and provides the designer with handles on both. It is quite surprising why this kind of a hybrid approach has never been considered for integrated design of cryptographic accelerators.

Contribution. In this paper, we take up the LTE Stream Ciphers as a case study for our idea of hardware integration. There has been a few academic publications towards hardware implementation of the individual ciphers. Especially, Kitsos et al [12] provide us with a high performance ASIC implementation of SNOW 3G and recently Liu et al [13] have published an efficient FPGA based implementation of ZUC. However, the state-of-the-art hardware implementation of both the ciphers come from the commercial domain, especially from Elliptic Technologies Inc. [5,6,7,8,9,10] and IP Cores Inc. [11], the established brands in the field of hardware security solutions.

In each of the above cases, the accelerators for SNOW 3G and ZUC have been developed separately as individual cores, whereas the ciphers are going to be used on the same platform. Moreover, the two ciphers have significant structural similarities to facilitate an integrated design. This is the driving factor behind our attempt to construct a unified accelerator that would provide higher throughput compared to all existing designs.

We design an integrated high performance accelerator (HiPAcc-LTE) for SNOW 3G and ZUC (version 1.5, as in LTE Release 10 and beyond), targeted towards the 4G mobile broadband market. We merge the two ciphers within a single core by sharing resources among them, thereby reducing the area overhead compared to two independent implementations. HiPAcc-LTE provides *almost twice* the throughput for both the ciphers compared to any existing architecture for the individual algorithms. We also provide the user with the flexibility to choose the 'area vs. throughput' trade-off for a customized design.

We also provide a combined fault detection and protection mechanism in HiPAcc-LTE. In case of SNOW 3G, we provide tolerance against the known fault attack by Debraize and Corbella [3]. For ZUC, as there are no known fault attacks till date, we just leave the room for future fault protection requirements.

Long term vision of this hardware integration approach for cryptographic algorithms is to have a flexible core supporting multiple designs including intermediate design points. This strategy will provide the developer to design a unified architecture with optimal performance for a number of cryptographic primitives with similar structural and algorithmic construct. To the end user, the integrated core presents a fast platform to design, validate and benchmark upcoming cipher primitives as well.

Organization. The technical content of this paper is organized as follows:

Section 2 presents a brief overview of the ciphers SNOW 3G and ZUC. We also present some initial observations regarding the structural similarities and dissimilarities of the two that will help us later in their integration.

Section 3 contains the main theoretical idea of this paper. We discuss
- the approach for integration and restructure the hardware designs of the two ciphers to exploit the similarities to the fullest,
- construction of the optimal pipeline for the unified hardware, and
- the final top-level design and pipeline architecture for HiPAcc-LTE.

Section 4 deals with simulation, testing and synthesis of HiPAcc-LTE.
- We present a detailed account of the base implementation and further optimization of the proposed hardware.
- We also present a comparison of our design in multiple technologies with existing architectures in the academic and commercial domains.
- Furthermore, we provide a combined fault detection and correction facility in our architecture that prevents the known fault attack against SNOW 3G and allows prevention of future fault attacks against ZUC.

Section 5 concludes the paper by providing a future direction of research oriented around the idea of hardware integration proposed in this paper.

2   Preliminaries

Before diving into the main technical content of this paper, let us first recall the design and constructs of the stream ciphers SNOW 3G and ZUC. This individual overview of the ciphers will help us identify the similarities and dissimilarities in their designs, which will later lead to their high performance integration.

2.1   Brief Overview of SNOW 3G and ZUC

SNOW 3G [16] is an LFSR based stream cipher designed by ETSI-SAGE, largely based on the cipher SNOW 2.0 [4] by Ekdahl and Johansson. The cipher generates a keystream of 32-bit words using an LFSR of size 16 words, that is $16 \times 32 = 512$ bits. The FSM of this design consists of three 32-bit registers which are updated based on two different S-boxes S1, S2. The LFSR update function depends on a couple of field operations (multiplication and division by field element α) and XOR combinations.

   Alike most stream ciphers, SNOW 3G has two distinct modes of operation. During the initialization mode, the LFSR is initiated using a 128-bit key and a 128-bit initialization variable (IV), and the output of the FSM is XOR-ed with the LFSR update function in the feedback loop for the first 32 iterations. Thereafter, in the keystream generation mode, the output of the FSM is combined with the first LFSR location s_0 to produce the output keystream word. The operation of the cipher in keystream generation mode is illustrated in Fig. 1.

ZUC [18] is also an LFSR based word oriented stream cipher, designed by the Data Assurance and Communication Security Research Center of the Chinese Academy of Sciences (DACAS). This cipher produces a keystream of 32-bit words, and is executed in two stages (initialization and keystream generation). The LFSR for ZUC consists of 16 blocks, each of length 31 bits, and the update

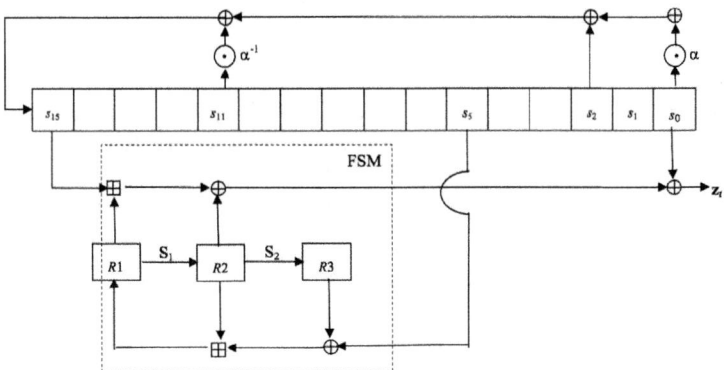

Fig. 1. SNOW 3G cipher in Keystream Generation mode [16]

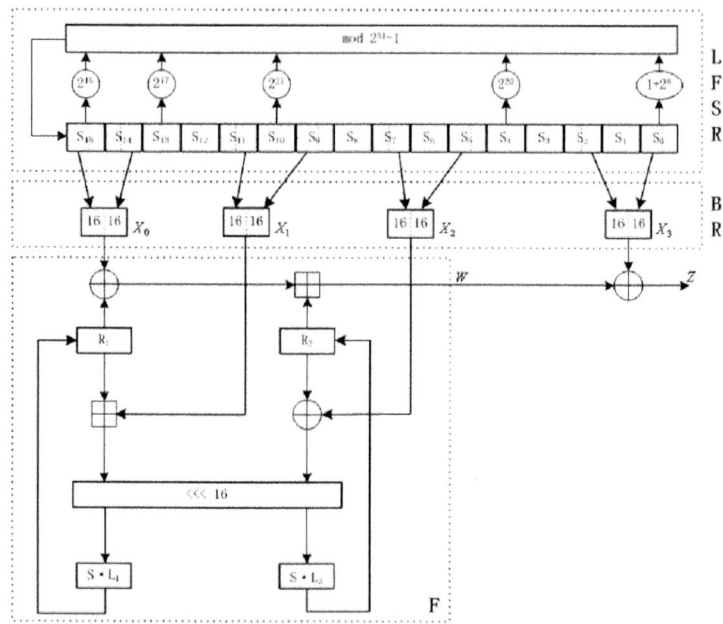

Fig. 2. ZUC cipher in Keystream Generation mode [18]

function of the LFSR is based on a series of modulo $2^{31} - 1$ (this is a prime) multiplications and additions. The FSM takes as input 32-bit words constructed from the LFSR (through a routine called Bit Reorganization or BR) and outputs a 32 bit word as well. It consists of two 32-bit registers R1 and R2 which are updated using two different linear functions L1, L2 and the same S-box S.

The initial state of the LFSR is constructed using a 128 bit key and a 128 bit IV, and during the first 32 iterations, the output of the FSM is added (modulo $2^{31} - 1$ addition after right shift by 1 place) to the feedback loop for LFSR update. In the keystream generation mode, the output of the FSM is combined with the word X_3, constructed from the LFSR places s_0 and s_2, to produce the final output. The keystream generation mode of ZUC is illustrated in Fig. 2.

2.2 SNOW 3G and ZUC: Similarities and Dissimilarities in Design

In this section, we put the two ciphers SNOW 3G and ZUC side by side for a structural comparison in the designs. Observations made in this section will help us later in building the integrated platform for the LTE stream ciphers.

Similarities. The reader may easily spot the inherent structural similarity in the designs of the two ciphers SNOW 3G and ZUC. This is mainly because both ciphers are based on the same principle of combining an LFSR with an FSM, where the LFSR feeds the next state of the FSM. In the initialization mode, the output of the FSM contributes towards the feedback cycle of the LFSR, and in the keystream generation mode, the FSM contributes towards the keystream.

A top level structure for both the ciphers can hence be represented as in Fig. 3. The figure on the left indicates the initialization mode of operation while the figure on the right demonstrates the operation during keystream generation. In Fig. 3, the combination of the LFSR update and the FSM during initialization mode is represented by C, which is either an XOR or a shift and addition modulo $2^{31} - 1$ for SNOW 3G and ZUC respectively. In the keystream generation mode, the combination of the LFSR state with the FSM output is denoted as K, which is an XOR for SNOW 3G and a bit reorganized XOR for ZUC. The operations are individually presented in the previous subsections for the two ciphers. Z represents the output keystream for both the ciphers.

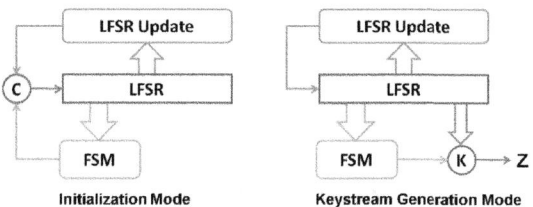

Fig. 3. Top level structure of both SNOW 3G and ZUC

The key point to observe in Fig. 3 is that we have a similar 3-layer structure for both the ciphers SNOW 3G and ZUC. Note that we have not considered Bit Reorganization of ZUC as a special stage, but have taken it as a part of the FSM, thus exhibiting better structural similarity with SNOW 3G.

Dissimilarities. As we probe deeper into the individual components of the design, the dissimilarities start appearing one by one. Let us categorize the dissimilarities in the two designs according to the main stages of the ciphers.

1. **LFSR update routine** is fundamentally different for the two ciphers. While SNOW 3G relies on field multiplication/division along with XOR for the LFSR feedback, ZUC employs addition modulo the prime $p = 2^{31} - 1$. Another point to note is that the new updated value s_{15} is required for the next feedback in case of ZUC, whereas SNOW 3G does not have this dependency. This creates a major difference in designing the combined architecture.
2. **The main LFSR** is slightly different for the ciphers as well, although both SNOW 3G and ZUC output 32-bit words. SNOW 3G uses an LFSR of 16 words, each of size 32 bits, whereas ZUC uses an LFSR of 16 words, each of size 31 bits. However, the bit organization stage of ZUC builds 32 bit words from the LFSR towards FSM update and output generation.
3. **FSM operations** of SNOW 3G and ZUC are quite different as well, though they use similar resources. SNOW 3G has three registers R1, R2 and R3 where the updation dependency $R1 \rightarrow R2 \rightarrow R3 \rightarrow R1$ is cyclic with the last edge depending on the LFSR as well. In case of ZUC, there are only two registers R1 and R2. The updation of each depends on its previous state as

well as that of the other register. And of course, the LFSR also feeds the state updation process, as in the case of SNOW 3G.

In the next section, we will try to merge the designs of SNOW 3G and ZUC in such a fashion that the similarities are exploited to the maximum extent, and the common resources are shared. The dissimilarities that we have discussed above will be treated specially for each of the ciphers.

3   Integration of SNOW 3G and ZUC

In this section, we present our main idea behind the architectural integration of SNOW 3G and ZUC. We will attempt this merger in three parts, each corresponding to the major structural blocks of the two designs; namely, the main LFSR, the LFSR update function and the FSM.

3.1   Integrating the Main LFSR

Recall that the LFSR of SNOW 3G has 16 words of 32 bits each, while that of ZUC has 16 words of 31 bits each. Our first goal is to share this resource among the two ciphers. If we do a naive sharing by putting the 31 bit words of ZUC in the same containers as those for the 32 bit words of SNOW 3G, 1 bit per word is left unused in ZUC. Hence, our first target was to utilize this bit in such a way that reduces the critical path in the overall implementation.

Motivation. In Section 4, while discussing the pipeline structure, we will note that the critical path flows through the output channel, that is, through the bit reorganization for s_{15}, s_{14} and s_2, s_0, and the FSM output of W. In fact, bit reorganization is also required for the FSM register update process. Keeping this in mind, we tried to remove the bit reorganization process from the FSM.

Restructuring the LFSR. In this direction, we construct the LFSR as 32 registers of 16 bits each. The 32 bit words for SNOW 3G would be split in halves and stored in the LFSR registers naturally. For ZUC, we split the 31 bit words in 'top 16 bit' and 'bottom 16 bit' pieces, and store them individually in the 16 bit LFSR registers. The organization of bits is shown in the middle column of Fig. 4, where the two blocks share the center-most bit of the 31 bit original word. Notice that we do not require the bit reorganization any more in the FSM operation, as it reduces to simple *read* from two separate registers in our construction. The modified bit reorganization model is illustrated in Fig. 4.

However, note that the LFSR update function of ZUC uses the 31 bit words for the modulo $2^{31}-1$ addition. Thus, we have actually moved the bit reorganization stage to the LFSR update stage instead of keeping it in the FSM. The effects of our design choices will be discussed later in Remark 1.

3.2   Integrating the FSM

Although the FSM of the two ciphers do not operate the same way, they share similar physical resources. Thus, our main goal for the integrated design is to

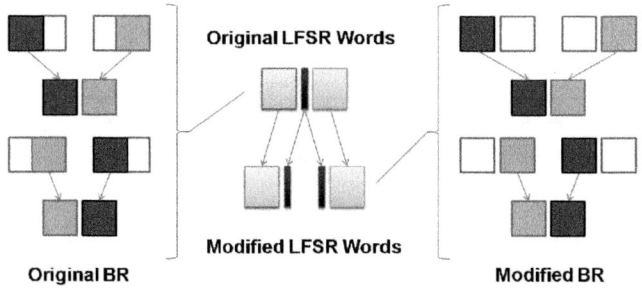

Fig. 4. Modified bit reorganization for ZUC after LFSR integration

share all possible resources between them. Note that the bit reorganization stage is not present in the ZUC FSM any more, due to our LFSR reconstruction.

Register Sharing. One can straight away spot the registers R1, R2 and R3 for potential sharing. We share R1 and R2 between SNOW 3G and ZUC, while R3 is needed only for the former. If required, R3 can be utilized in ZUC for providing additional buffer towards fault protection, discussed in Section 4.

Sharing the Memory. During the FSM register update process, both SNOW 3G and ZUC use S-box lookup. In the software version of the ciphers, SNOW 3G [16] uses S_R, S_Q and ZUC [18] uses S_0, S_1. However, for efficient hardware implementation of SNOW 3G with memory access, we choose to use the tables S1_T0, S1_T1, ..., S2_T3, as prescribed in the specifications [16]. This saves a lot of computations after the memory read, and hence reduces the critical path to a considerable extent. We store the 8 tables in a data memory of size 8 KByte.

For ZUC, however, we can not bypass the lookup to S_0 and S_1. But one may note that these tables are accessed 4 times each during the FSM update. So, to parallelize the memory access, we store 4 copies of each table (thus 8 in total) in the same 8 KByte of data memory that we have allocated for SNOW 3G. Note that we are not using the full capacity of the memory in ZUC, as we store 1 byte in each location (as in S_0 and S_1) whereas it is capable of accommodating 4 bytes in each (as in S1_T0, S1_T1, ..., S2_T3).

By duplicating the ZUC tables in the 8 distinct memory locations, we have restricted the memory read requests to 1 call per table in each cycle of FSM. This makes possible the sharing of memory access between SNOW 3G and ZUC as well. We use only a single port to read from each of the tables, and that too is shared between the ciphers for efficient use of resources. This in turn reduces the multiplexer logic and area of the overall architecture.

Pipeline based on Memory Access. Now that we have memory lookup during the FSM update, we partition the pipeline according to it. We simulate the memory by a synchronous SRAM with single-cycle read latency. To optimize the efficiency with an allowance for the latency in memory read, we split the pipeline in two stages, keeping the memory read request and read operations in the middle. The structure of our initial pipeline idea is shown in Fig. 5.

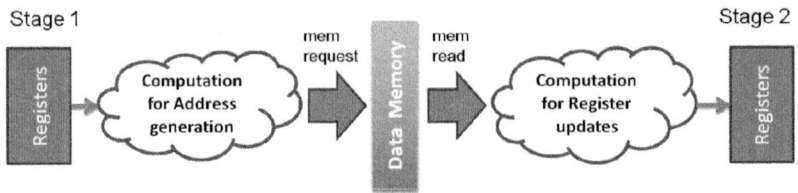

Fig. 5. Pipeline structure based on Memory Access

This pipeline is organized around the memory access, where we perform

- the memory read request and LFSR update in Stage 1, and
- the memory read and output computation in Stage 2.

For SNOW 3G, the computation for memory address generation is a simple partitioning of R1 and R2 values in bytes. The computation for register update however, requires an XOR after the memory read. In case of ZUC, the computation for address generation is complicated, and depends on the LFSR as well as R1 and R2. However, the computation for register update is a simple concatenation of the values read from memory.

Remark 1. So far, we have made a few design choices in integrating the two ciphers. In a nutshell, the choices provide

- *reduction in the critical path by reducing the memory and LFSR read times,*
- *reduced critical path by moving the bit reorganization away from FSM, and*
- *an efficient method for combined fault protection in both the ciphers.*

The effect of these choices will be reflected in the critical path and fault tolerance mechanism, discussed later in Section 4 of this paper.

Next, we deal with the integration of the most crucial part of the two ciphers: the LFSR update and shift operations. The final structure of the pipeline will evolve during this phase as we deal with the intricate details in the design.

3.3   Integrating the LFSR Update Function

The LFSR update function is primarily different for the two ciphers. The only thing in common is the logic for LFSR update during initialization, and this poses a big problem with our earlier pipeline idea based on memory access (Fig. 5).

Pipeline restructuring for Key Initialization. In the initialization mode of the two ciphers, the FSM output W is fed back to the LFSR update logic. The update of s_{15} takes place based on this feedback, and in turn, this controls the next output of the FSM (note that W depends on R1, R2 and s_{15} in both ciphers). This is not a problem in the keystream mode as the LFSR update path is independent of the output of FSM. However, during initialization, it creates a combinational loop from Stage 2 to Stage 1 in our earlier pipeline organization (Fig. 5). This combinational loop in memory access due to dependencies prohibits us from keeping the memory access and memory read in two different stages of the pipeline. Thus, we design a new structure as follows:

- Stage 1: Initial computation for memory access and LFSR shift.
- Stage 2: Memory read, LFSR update and subsequent memory read request.

This new pipeline structure allows us to resolve the memory access dependencies within a single stage and the independent shift of the LFSR occurs in the other. Now, the main goal is to orient the LFSR update logic around this pipeline structure, or to redesign the pipeline according to the LFSR update function.

Pipeline organization for LFSR update. The LFSR update logic of SNOW 3G is easier to deal with. The update depends upon the LFSR positions s_0, s_2 and s_{11}, and also on the FSM output W during key initialization. A part of s_0 and s_{11} each undergoes a field operation (MUL_α and DIV_α respectively), and the other part gets XOR-ed thereafter. To reduce the combinational logic of realizing the field operations, two lookup tables are prescribed in the specifications [16]. For an efficient implementation in hardware, we follow this idea and store the two tables MUL_{alpha} and DIV_{alpha} in two 1 KByte memory locations. These are also read-only memories with single-cycle read latency. Now, we can fit the update routine for SNOW 3G within the two stage pipeline proposed earlier.

- Stage 1: Precompute the simple XOR involving s_0, s_2 and s_{11}, and generate the addresses for memory read requests to tables MUL_{alpha} and DIV_{alpha}.
- Stage 2: Perform memory read and XOR with the previous XOR-ed values to complete the LFSR feedback path, run the FSM and complete the LFSR update of s_{15} depending on W.

Note that this pipeline structure works both for initialization as well as keystream generation, as it takes into account all possible values required for the LFSR update. Thus, in terms of SNOW 3G, we stick to our 2-stage pipeline.

In case of ZUC however, the LFSR update logic is quite complicated. This is mostly because of the additions modulo the prime $p = 2^{31} - 1$. Liu et al [13] had proposed a single adder implementation of this addition modulo prime, and this logic has also been included in the specifications [18]. We use the same for our hardware, at least at this initial phase. In the same line, we first try a 5-stage pipeline, similar to the one proposed in [13] for LFSR update of ZUC.

The initial idea for 5-stage pipeline is shown as Pipeline 1 in Fig. 6. All the adders are modulo prime, similar to the ones in [13], and the variables a, b, c, d, e, f represent $s_0, 2^8 s_0, 2^{20} s_4, 2^{21} s_{10}, 2^{17} s_{13}, 2^{15} s_{15}$ (modulo $p = 2^{31} - 1$) respectively. Variable g denotes the FSM output W, which is added with the cumulative LFSR feedback, and is then fed back to s_{15} in the LFSR itself.

However, Pipeline 1 creates a combinational loop between Stage 5 and Stage 4 in the key initialization phase. The final output in Stage 5 of the addition pipeline has to be fed back to s_{15} that controls the input f in Stage 4. This loop is shown by the curvy solid line in Fig. 6, and it occurs due to mutual dependency of FSM and LFSR update during initialization. The authors of [13] also observed this dependency, and they proposed the 32 rounds of key initialization to be run in software in order to achieve one-word-per-cycle using their structure.

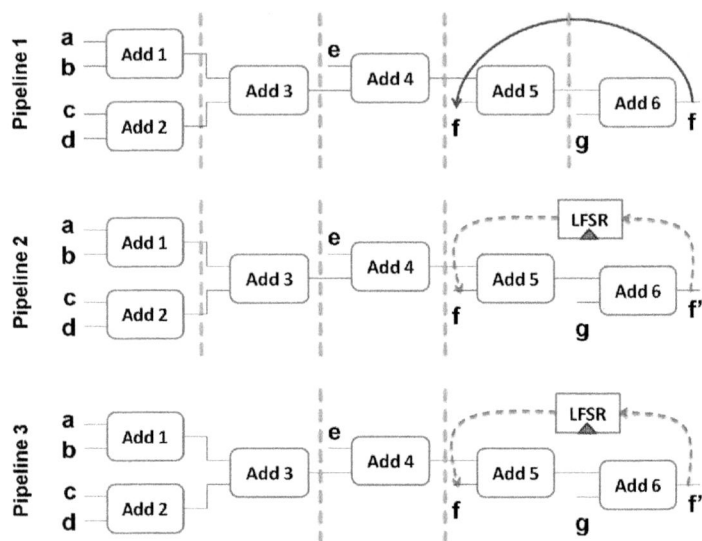

Fig. 6. Pipeline structure reorganization for LFSR update of ZUC

Our challenge was to integrate this phase into the hardware without losing the throughput. The main motivation is to restrain the use of an external aide for the initialization mode. There are two direct ways of resolving this issue:

1. Allow a bypass logic for the f value across the stages
2. Restructure the pipeline to merge the last two stages

We choose the second one and reorganize the pipeline. As the dependency discussed so far occurs in between the last two stages of the pipeline, we merge those to resolve the inter-stage combinational loop. In this case, the output f' of this stage is written into the s_{15} location of the LFSR, and read back as f at the next iteration. This is shown as Pipeline 2 in Fig. 6.

The reader may note that we have two adders (modulo prime p) in series at the last stage of Pipeline 2 (Fig. 6). So, we can put two adders in any other stage as well, without affecting the critical path. We decide to merge Stages 1 and 2 to have two adders in parallel followed by an adder in series in the first stage. This does not increase the critical path, which still lies in the last stage due to the two adders and some associated combinational logic. The final structure of the LFSR update pipeline for ZUC is shown in Fig. 6 as Pipeline 3. In the next section, we design the integrated pipeline structure combining all components.

3.4   Final Design of the Pipeline

In this section, we present the final pipeline structure for the integrated architecture. In the previous sections, we have already partitioned the components into pipeline stages as follows.

- FSM: Two stages - initial computations for address generation in the first stage, and memory access and related computations in the second stage.
- LFSR Movement: Two stages - shift in first stage and s_{15} write in second.
- LFSR Update: Two stages for SNOW 3G and three stages for ZUC.

Here, we combine all three components of SNOW 3G and ZUC and design the final pipeline for our proposed hardware implementation, as shown in Fig. 7.

Fig. 7. Final 3-stage Pipeline Structure for the Integrated Design

The stages of SNOW 3G and ZUC are different only in case of the LFSR update routine, and we show these separately in the figure. The pipeline behavior of the LFSR shift and write operations, as well the FSM precomputation and update routines are almost same for both the ciphers, and hence we show single instances of these in Fig. 7. In the next section, we discuss the practical issues with the final ASIC implementation of our integrated hardware.

4   ASIC Implementation of the Integrated Hardware

In this work, we utilized the hardware generation environment and simulation framework from LISA, the Language for Instruction-Set Architectures, for designing the accelerator. The complete automatic generation environment is commercially available via Synopsys Processor Designer [19]. The accelerator in our case is designed as a state machine. This allowed fast exploration of design alternatives and ease of high level modeling for making pipelining and resource organization decisions. The language allows full control over minute design decisions and preserves the overall structural organization neatly in the generated hardware description. This is especially important for verifying the design costs (area, timing) and accordingly modifying the design at high level. Such a capability of strong designer interaction with the tool during high level synthesis is not common among automatic C to HDL flows [14], thereby forcing designers to go through time consuming and error prone low-level design iterations.

The gate-level synthesis was carried out using Synopsys Design Compiler Version D-2010.03-SP4, using topographical mode for a 65 nm target technology library. The area results are reported using equivalent 2-input NAND gates. The total lines of LISA code for our best implementation is 1131, while the total lines of auto-generated HDL code is 13440 for the same design. The modeling, implementation, optimization and tests were completed over a span of two weeks.

In this section, we first discuss the issues with the *critical path* in our design, and the optimizations thereof. This will be followed by a set of detailed implementation results and comparisons with the existing designs.

4.1   Critical Path

After the initial synthesis of our design using LISA modeling language, we identified the critical path to occur in the key initialization phase of ZUC. Fig. 8 depicts the critical path using the curvy dashed line. To understand the individual components in the critical path, let us first associate the pieces in Fig. 8 to the original initialization routine of ZUC, as described in its specification [18].

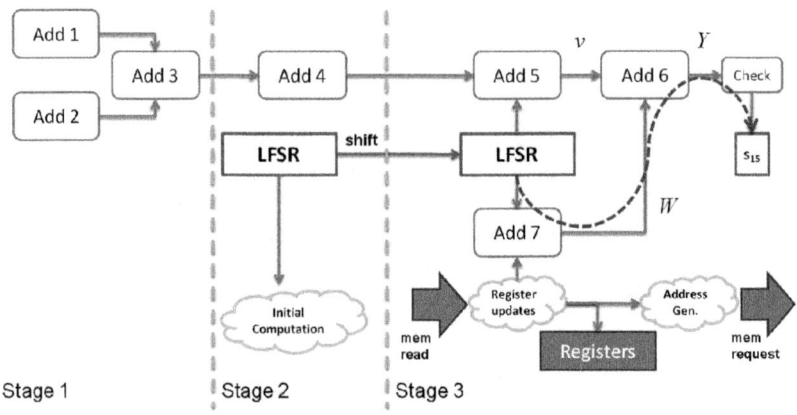

Fig. 8. Critical path in the Key Initialization of ZUC (curvy dashed line)

ZUC Key Initialization Routine. The following is the key initialization routine of ZUC, as per our notation and pipeline orientation. Note that the operation is the same as in the LFSRWithInitialisationMode() function of [18].

LFSR_Key_Initialization (W)
1. $v = 2^{15}s_{15} + 2^{17}s_{13} + 2^{21}s_{10} + 2^{20}s_4 + 2^8 s_0 + s_0 \pmod{2^{31}-1}$
2. $Y = v + (W \gg 1) \pmod{2^{31}-1}$
3. If $Y = 0$, then set $Y = 2^{31} - 1$
4. Write Y to location s_{15} of the LFSR

In Fig. 8, the first five adders Add 1 to Add 5 are part of the general LFSR feedback loop in ZUC, and they compute the value

$$v = 2^{15}s_{15} + 2^{17}s_{13} + 2^{21}s_{10} + 2^{20}s_4 + 2^8 s_0 + s_0 \pmod{2^{31}-1}.$$

The LFSR is also accessed to run the FSM and the adder Add 7 at the bottom of Stage 3 computes the FSM output $W = (X_0 \oplus R_1) + R_2$, where this addition is a normal 32-bit addition. The special operation in LFSR update of ZUC in its initialization mode is to compute $Y = v + (W \gg 1) \pmod{2^{31} + 1}$, realized by the adder Add 6 on the top layer of Stage 3. If this sum $Y = 0$, it is replaced by $Y = 2^{31} - 1$ in the 'Check' module of Fig. 8. Finally, this 31 bit value Y is written to s_{15} of the LFSR, thus completing the LFSR update loop. The critical path, as shown by the curvy dashed line in Fig. 8, is as follows:

LFSR Read $\rightarrow$ 32-bit Add $\rightarrow$ Modulo Add $\rightarrow$ Check $\rightarrow$ LFSR Write

In this section, we try all possible optimizations to reduce the critical path.

LFSR Read Optimization. At first, we implemented the LFSR as a register array. However, different locations of the LFSR are accessed at different stages of the pipeline we have designed, and the LFSR read will be faster if we allow the individual LFSR cells to be placed independently in the stages. This motivated us to implement the LFSR as 32 distinct registers of size 16 bits each. Furthermore, we shadowed the last two locations, i.e., s_{15} of the LFSR, so that it can be read instantaneously from both Stage 4 and Stage 5. This led to a reduction in the critical path. Though this optimization is targeted towards physical synthesis, the gate-level synthesis results indicated strong improvement as well.

Modulo p Adder Optimization. Initially, we designed the modulo $p = 2^{31} - 1$ adder as prescribed in [13]. This looks like the circuit on the left of Fig. 9. However, one may bypass the multiplexer (MUX) by simply incrementing the sum by the value of the carry bit. That is, if the carry bit is 1, the sum gets incremented by 1, and it remains the same otherwise. The modified design (right side of Fig. 9) slightly reduces the critical path and we replace all the modulo p adders in our design (except for Add 6) by this modified circuit.

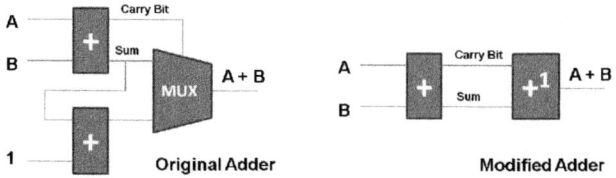

Fig. 9. Modulo p Adder optimization for ZUC

Check Optimization. The 'Check' block in the critical path actually has two checks in series; one due to Add 6 where the increment is based on the carry bit, and the second check is for the $Y = 0$ situation. We try to optimize as follows.

- Carry $= 0$: We just require to check if $Y = 0$. If so, set $Y = 2^{31} - 1$.
- Carry $= 1$: We just require to set $Y = Y + 1$ without any further checks.

Table 1. Synthesis results for HiPAcc-LTE with 10 KByte memory

Frequency (MHz)	Area (equivalent NAND Gates)		
	Total	Sequential	Combinational
200	11699	5540	6159
500	13089	5540	7549
800	14102	5541	8561
1000	15696	5541	10155
1050	16055	5554	10501
1090	16886	5568	11318

The first case is obvious, as the sum would remain unchanged if the carry is 0. In the second case, note that the inputs v and $(W \gg 1)$ to Add 6 are both less than or equal to $2^{31} - 1$. Thus, the sum Y is bounded from above by $2^{32} - 2$. Even if the carry is 1, the incremented value of sum will be bounded from above by $2^{32} - 1$, which can never have the lower 31 bits all equal to 0. Thus, we do not even require the 'Check' block in this situation. This optimization simplifies the logic and reduces the critical path considerably.

4.2   Performance Results

After performing all the optimizations discussed in the previous section, we still find the critical path flowing through the same components. We proceed for our final synthesis and performance results based on the current state of the design.

Table 1 presents all the architecture design points for HiPAcc-LTE that we have implemented using the 65 nm technology. The area-time chart for the design points of HiPAcc-LTE is shown in Fig. 10.

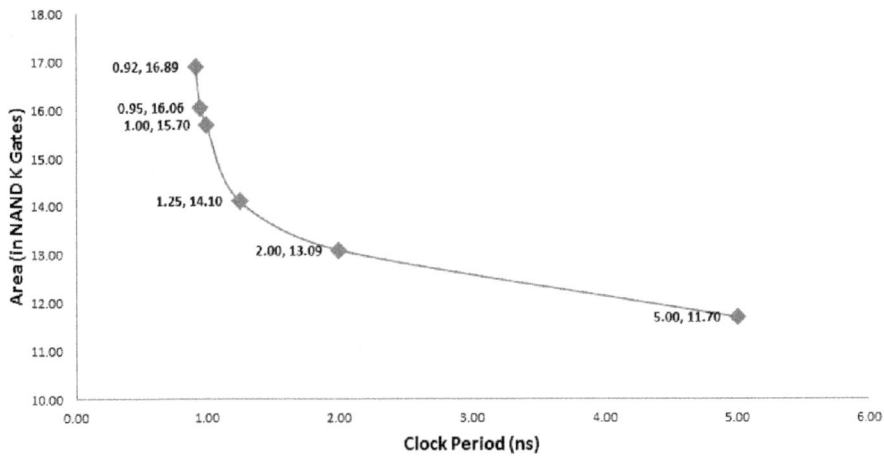

Fig. 10. Area-Time chart for HiPAcc-LTE (10 KByte memory) using 65 nm technology

Table 2. Synthesis results for alternate design of HiPAcc-LTE with 3 KByte memory

Frequency (MHz)	Area (equivalent NAND Gates)		
	Total	Sequential	Combinational
200	10519	5548	4971
500	13090	5540	7550
800	14103	5541	8562
1000	15696	5541	10155
1090	16887	5568	11319

The maximum frequency we could achieve is 1090 MHz, which corresponds to a critical path length of approximately 0.92 ns. This provides us with a net throughput of 34.88 Gbps, with 1 keystream word per cycle. The total area is about 17 KGates NAND equivalent and 10 KByte of data memory is required.

Experiments with Reduced Data Memory. In the original HiPAcc-LTE design as above, the static data for S-box and field operations have been stored in external data memory. While SNOW 3G utilizes the complete 10 KByte memory, ZUC requires only about 2 KByte of the allocated space. This motivated us to experiment with an alternate design that requires less data memory.

In the alternate design, we use S-box tables S_R, S_Q for SNOW 3G [16] instead of the tables S1_T0, S1_T1, ..., S2_T3, as in the previous case. During the sharing of memory, the ZUC tables S_0, S_1 fit exactly in the space for S_R, S_Q as they are of the same size, 256 bytes each. There are exactly 4 calls to each table per cycle, and we store two copies of each table in dual-port RAMs to get optimum throughput. This amounts to a data memory of $2 \times (256 + 256)$ bytes $= 1$ KByte. The MUL_{alpha} and DIV_{alpha} tables (size 1 KByte each) in case of SNOW 3G could not be avoided due to the complicated combinational logic involved in these field operations. The total data memory for this alternate design sums up to 3 KByte, and the details for all design points are presented in Table 2.

This alternate design retains the maximum frequency of 1090 MHz, which provides us with a net throughput of 34.88 Gbps, with 1 word per cycle. The area figure is still about 17 KGates NAND equivalent, but only 3 KByte of external data memory is required. It is interesting to note that the combinational area remained almost similar even after introducing the computations for S-boxes. This is possibly due to the availability of high-speed, area-efficient library cells in our target technology library and efficient design style.

It is expected that our design will be practically deployed in a system-on-chip setting, where the memory requirement of 3 KByte is quite reasonable. With this alternate design of HiPAcc-LTE having 3 KByte of memory, the performance of the individual ciphers SNOW 3G and ZUC are also tested in stand-alone mode. The synthesis results in this direction are presented in Table 3.

4.3   Comparison with Existing Designs

To put the performance of HiPAcc-LTE into perspective, we compare it with the state-of-the-art architectures available in academia and the commercial sector.

Table 3. Synthesis results for stand-alone mode in HiPAcc-LTE with 3 KByte memory

Cipher	Frequency (MHz)	Area (equivalent NAND Gates)		
		Total	Sequential	Combinational
SNOW 3G	500	6867	5061	1807
	1000	7033	5062	1971
ZUC	500	9555	4798	4757
	1000	11412	4811	6601

Comparison with Academic Literature. In the domain of published academic results, we could not find an ASIC implementation of ZUC, and neither could we find a 65 nm technology implementation of SNOW 3G. The only hardware realizations for ZUC have been done in FPGA [13] so far. Thus, we could not compare HiPAcc-LTE to any academic results in terms of ZUC.

In case of SNOW 3G, the best academic publication is [12] that uses 130 nm technology. To compare with this result, we synthesized our proposed design (with 10 KByte data memory) in 130 nm, and the comparison is as follows.

- SNOW 3G of [12]: 7.97 Gbps with 249 MHz max. freq. and 25 KGates area
- Our HiPAcc-LTE: 24.0 Gbps with 750 MHz max. freq. and 18 KGates area

Both designs use about 10 KByte of external data memory for look-up tables. It is clear that we achieve surprisingly better throughput from HiPAcc-LTE due to our careful pipeline design. Our integrated implementation for both the LTE stream ciphers even outperforms the single stand-alone core in terms of area.

Comparison with Commercial Designs. In the commercial arena, the best architectures available for SNOW 3G and ZUC are from IP Cores Inc. [11] and Elliptic Tech Inc. [8] respectively. Both provide stand-alone solutions for the individual stream ciphers and match our technology of 65 nm. One tricky issue in the comparison is the area required for the memory. It is not always clear from the product white-paper whether additional memories have been used.

For the sake of fairness, we first compare our designs using 3 KB memory with existing stand-alone ZUC and SNOW 3G implementations. The memory is synthesized with Faraday Memory Compiler in 65 nm technology node. Further, we replace the S-Box SRAM implementations with hard macros in the RTL design and obtained the gate-level synthesis results. From the commercial designs, the designs with best performance claims in 65 nm technology node are selected. We provide the detailed comparison and analysis in Table 4.

Area comparison: Around an operating frequency of 200-500 MHz, if one uses the two best cores separately, the combined area comes around 18-20 KGates. HiPAcc-LTE synthesizes within 16-18 KGates in this frequency zone (using hard macros), hence offering about 10% reduction in area. Even with this reduced area figure, HiPAcc-LTE offers the same throughput as CLP-410 [8] and *more than double* throughput compared to SNOW3G1 [11].

Throughput comparison: The best throughput (1 word/cycle) is provided by the CLP-410 ZUC core from Elliptic Tech. However, they just quote a figure of 6

Table 4. Comparison of HiPAcc-LTE with existing 65 nm commercial designs

Performance of Commercial Designs					
Cipher	Name of Design	Designer	Max. Freq. (MHz)	Throughput (Gbps)	Total Area (KGates)
SNOW 3G	SNOW3G1 [11]	IP Cores Inc.	943	7.5	8.9
ZUC	CLP-410 [8]	Elliptic Tech.	500	-	10-13

Performance of HiPAcc-LTE					
Cipher	Mode of Design for Static Tables	Frequency (MHz)	Throughput (Gbps)	Memory (KGates)	Total Area (KGates)
SNOW 3G	3 KByte memory	1000	32.0	43.0	50.0
ZUC	3 KByte memory	1000	32.0	26.8	38.2
Both	3 KByte memory	1090	34.9	43.0	59.9
SNOW 3G	Hard macro	1650	52.8	-	18.1
ZUC	Hard macro	920	29.4	-	20.6
Both	Hard macro	920	29.4	-	23.9

Gbps for 200 MHz [8]. A simple scaling to their maximum frequency of 500 MHz would translate this to an estimate of 15 Gbps. Even in this case, the throughput 29.4 Gbps of HiPAcc-LTE (in hard macro design) is *almost double* compared to any of the commercial stand-alone implementations of the ciphers.

For a very rough estimate, if one wants to achieve a comparable throughput (approx. 30 Gbps) using the existing stand-alone modules, then 4 parallel blocks of SNOW3G1 [11] and 2 parallel blocks of CLP-410 [8] would be required. This amounts to a total area of roughly 56-62 KGates, while HiPAcc-LTE achieves the same using only 23.9 KGates (almost 57% reduction) for the hard macro based design. For the sake of fairness, one may also note that we have a comparable area figure of 59.9 KGates for an even higher throughput (34.9 Gbps) using 3 KByte of external data memory. If the extreme throughput is not required for communication purpose, it may facilitate a scaling in frequency/voltage for reduced power consumption.

4.4   Fault Detection and Protection in HiPAcc-LTE

Till date, no significant fault attack has been mounted on ZUC, and the best fault attack against SNOW 3G has been reported in [3]. In HiPAcc-LTE, we provide detection and protection against this fault attack of SNOW 3G, and provide room for tolerance against future fault attacks on ZUC, if any.

In [3], the authors themselves propose a method to prevent their fault attack in hardware. They have shown that if one *shadows* the five LFSR locations s_0, s_1, s_2, s_3, s_4 continuously, the attack becomes impossible [3, Section X].

In the hardware implementation of HiPAcc-LTE, we have additionally implemented this shadowing mechanism as well. This is realized by keeping a buffer register of $5 \times 32 = 160$ bits which continuously shadows the five LFSR locations by shifting the array by one word in sync with the LFSR shift, and by recording the value of s_5 in the array during Stage 2 of the pipeline (note that this becomes the shadowed value of s_4 in Stage 3). A fault is detected in this locations by comparing the values in the LFSR with the shadowed values from the buffer array, and the keystream byte is not produced if a fault is detected.

The fault tolerance mechanism does not affect the critical path, and HiPAcc-LTE still achieves a maximum frequency of 1090 MHz. However, the area figures rise slightly, as expected. Compared to the original HiPAcc-LTE, the new area figures increase by approximately 1.5 KGates at 1090 MHZ in the 65 nm technology, when the design is implemented using external data memory.

The design automatically provides a mechanism for 160 bit shadowing for ZUC, if required, and this is where our earlier design choices for resource sharing (as mentioned in Remark 1) prove to be effective.

5   Conclusion

In this paper, we propose a novel idea for unified cryptographic hardware accelerator design based on the algorithmic and structural similarities between the ciphers to be implemented. As a practical case study of our proposal, we present HiPAcc-LTE, an integrated high performance hardware accelerator for 3GPP LTE stream ciphers SNOW 3G and ZUC. Through a careful design of the pipeline structure and storage organization, we achieve significantly higher throughput than state-of-the-art implementations.

A detailed physical synthesis and post-layout validation of the proposed design HiPAcc-LTE is on our roadmap. Furthermore, the design principle applied to the LTE stream ciphers can be exploited towards several similar hardware designs in the domain of cryptography. In particular, we would like to explore the application of our approach towards an integrated accelerator for other stream ciphers, block ciphers and hash functions with structural similarities.

Acknowledgment. The authors are grateful to the anonymous reviewers of Indocrypt 2011 for their critical review comments that helped in improving the technical quality of this paper. The first author would also like to express his gratitude to MPSoC Architectures, UMIC, RWTH Aachen University, Germany for hosting him during June–August 2011, when this work was completed.

References

1. 3GPP TS 33.401 v11.0.1. 3rd Generation Partnership Project, Technical Specification Group Services and Systems Aspects. 3GPP System Architecture Evolution (SAE): Security Architecture (Release June 11, 2011)
2. 3rd Generation Partnership Project. Long Term Evaluation Release 10 and beyond (LTE-Advanced). Proposed to ITU at 3GPP TSG RAN Meeting, Spain (2009)
3. Debraize, B., Corbella, I.M.: Fault Analysis of the Stream Cipher Snow 3G. In: Fault Diagnosis and Tolerance in Cryptography, FDTC 2009 (September 2009)

4. Ekdahl, P., Johansson, T.: A New Version of the Stream Cipher SNOW. In: Nyberg, K., Heys, H.M. (eds.) SAC 2002. LNCS, vol. 2595, pp. 47–61. Springer, Heidelberg (2003)
5. Elliptic Technologies Inc. CLP-41: SNOW 3G Flow Through Core., http://elliptictech.com/products-clp-41.php (retrieved on August 5, 2011)
6. Elliptic Technologies Inc. CLP-400: SNOW 3G Key Stream Generator
7. Elliptic Technologies Inc. CLP-403: SNOW 3G Look Aside Core, http://elliptictech.com/products-clp-403.php (retrieved on August 5, 2011)
8. Elliptic Technologies Inc. CLP-410: ZUC Key Stream Generator, http://elliptictech.com/products-clp-410.php (retrieved on August 5, 2011)
9. Elliptic Technologies Inc. CLP-411: ZUC Look Aside Core, http://elliptictech.com/products-clp-411.php (retrieved on August 5, 2011)
10. Elliptic Technologies Inc. CLP-412: ZUC Flow Through Core, http://elliptictech.com/products-clp-412.php (retrieved on August 5, 2011)
11. IP Cores Inc. SNOW 3G Encryption Core, http://ipcores.com/Snow3G.htm (retrieved on August 5, 2011)
12. Kitsos, P., Selimis, G., Koufopavlou, O.: High Performance ASIC Implementation of the SNOW 3G Stream Cipher. In: IFIP/IEEE VLSI-SOC 2008 - International Conference on Very Large Scale Integration, Greece (2008)
13. Liu, Z., Zhang, L., Jing, J., Pan, W.: Efficient Pipelined Stream Cipher ZUC Algorithm in FPGA. In: First Int'l Workshop on ZUC Algorithm, China (2010)
14. Schaumont, P.R., Kuo, H., Verbauwhede, I.M.: Unlocking the design secrets of a 2.29 Gb/s Rijndael processor. In: Design Automation Conf. (DAC 2002), USA (2002)
15. Specification of the 3GPP Confidentiality and Integrity Algorithms UEA2 & UIA2. Document 1: UEA2 and UIA2 Specification. ETSI/SAGE Specification, Version: 1.1 (September 6, 2006)
16. Specification of the 3GPP Confidentiality and Integrity Algorithms UEA2 & UIA2. Document 2: SNOW 3G Specification. ETSI/SAGE Specification, Version: 1.1 (September 6, 2006)
17. Specification of the 3GPP Confidentiality and Integrity Algorithms 128-EEA3 & 128-EIA3. Document 1: 128-EEA3 and 128-EIA3 Specification. ETSI/SAGE Specification, Version: 1.5 (January 4, 2011)
18. Specification of the 3GPP Confidentiality and Integrity Algorithms 128-EEA3 & 128-EIA3. Document 2: ZUC Specification. ETSI/SAGE Specification, Version: 1.5 (January 4, 2011)
19. Synopsys Processor Designer. Synopsys Inc., http://www.synopsys.com/

Addressing Flaws in RFID Authentication Protocols

Mohammad Hassan Habibi[1], Mohammad Reza Aref[1], and Di Ma[2]

[1] ISSL Lab, EE Department, Sharif University of Technology, Tehran, Iran
mohamad.h.habibi@gmail.com, aref@sharif.edu
[2] SAFE Lab, Department of Computer and Information Science, University of
Michigan-Dearborn, Michigan, USA
dmadma@umich.edu

Abstract. The development of RFID systems in sensitive applications
like e-passport, e-health, credit cards, and personal devices, makes it necessary to consider the related security and privacy issues in great detail.
Among other security characteristic of an RFID authentication protocol,
untraceability and synchronization are the most important attributes.
The former is strongly related to the privacy of tags and their hold-
ers, while the latter has a significant role in the security and availability
parameters. In this paper, we investigate three RFID authentication protocols proposed by Duc and Kim, Song and Mitchell, and Cho, Yeo and
Kim in terms of privacy and security. We analyze the protocol proposed
by Duc and Kim and present desynchronization and traceability attacks.
By initiating traceability, backward traceability and desynchronization
attacks, we show that the protocol proposed by Song and Mitchell lacks
location privacy and availability. In addition, we study the weaknesses
in Cho et al.'s protocol and address its defects by applying desynchronization, traceability and backward traceability attacks. We also propose
revisions to secure the Cho et al.'s protocol against the cited attacks.

Keywords: RFID, authentication protocol, privacy analysis, desynchronization attack.

1 Introduction

Radio frequency Identification (RFID) systems have developed rapidly in the last
decade. They are used to identify objects wirelessly in many applications such
as passports, e-health, public transportation, and supply chain management. A
typical RFID system consists of tags (transponders), readers (transceivers), and
a back-end database. The history of using RFID technology is begins during
World War II when a system called "IFF" (Identification, Friend or Foe) was
applied. The IFF system allowed the British military to determine if incoming
aircraft were friendly or hostile [6]. However, due to the cost of developing RFID
systems, their use was uncommon until low-cost RFID tags were produced in
the last decade.

D.J. Bernstein and S. Chatterjee (Eds.): INDOCRYPT 2011, LNCS 7107, pp. 216–235, 2011.

Although RFID systems are becoming more widespread, security and privacy are still challenging issues related to these systems. To provide secure communication in the application layer of an RFID system, an authentication protocol is designed and used. In recent years, a variety of authentication protocols have been proposed to assure security and privacy in RFID systems [11,25,35]. Nonetheless, a secure RFID authentication protocol should fulfil some requirements, though providing some of these requirements or providing them in sufficient degree is dependent on the application type. We study the most important requirements of these protocols as follows:

- The first requirement is *confidentiality*. Confidentiality has been defined as "ensuring that information is accessible only to those authorized to have access" [16]. In the RFID context, it means that a protocol must be proof against leaking any information to an unauthorized party. Confidentiality can be categorized according to whether it is related to data privacy or location privacy. The former is related to limiting access to tag's secret values such as identity, and cryptographic keys, while the latter aims to prevent the disclosure of sensitive location information.
  Due to the wireless communication in RFID systems, the messages exchanged between reader and tag are easily accessible by an adversary. Eavesdropping may occur on the channel and the messages may be retrieved or information may be leaked [19] to instigate different attacks on the protocol. Some attacks that violate confidentiality are key recovery [15], ID recovery [1], traceability [36], and cloning [17] attacks.
- *Integrity* is another requirement that should be satisfied by an authentication protocol. Integrity entails "safeguarding the accuracy and completeness of information and processing methods" [16]. In RFID protocols integrity indicates that the origin of data is proven and the messages transmitted between two parties are not modified in transit whether by accidental or by hostile activity. This absence of modification means that the authentication attribute is satisfied and the authenticated message exchange is provided. In these protocols, integrity can be violated by some attacks, such as spoofing [23], man-in-the-middle [13], and replay attack [37].
- Although *availability* is more relevant to the system functionality, it is also an important requirement for authentication protocols. Availability has been defined as "ensuring that authorized users have access to information and associated assets when required" [16]. In RFID authentication protocols, availability means that a legitimate tag can access the system services efficiently at will. The efficiency and complexity parameters of a protocol are defined in the availability field. In this context, the most dangerous threat is caused by desynchronization attacks, in which an adversary tries to make a legitimate tag and the server inconsistent [20]; thus, the target tag cannot be identified in subsequent sessions and will be unavailable.

To evaluate the cited requirements, in this paper we study three recent proposed RFID authentication protocols and prove that they fail to satisfy any of them.

More precisely, we first analyze an RFID authentication protocol proposed by Duc and Kim (O-FRAP$^+$) [12] and exploit the defects in its update mechanism to apply desynchronization and traceability attacks. Other RFID authentication protocols proposed by Song and Mitchell (hereafter denoted as SMP in reference to Song-Mitchell Protocol) [34] and Cho et al. (hereafter referred to as HRAP, in reference to the Hash-based RFID Authentication Protocol) [9] are analyzed in terms of synchronization and location privacy. We show that these protocols fail to fulfil integrity and availability, inasmuch as they suffer from a basic weakness in their update process that results in susceptibility to desynchronization attacks. Moreover, we show that the cited protocols lack confidentiality and location privacy. In particular, we present traceability and backward traceability attacks on them in a formal model. Then we propose our revisions to improve the HRAP protocol in terms of security and privacy and explain why the revised protocol is secure.

Organization of the paper is as follows: We review related works in Section 2 and explain the privacy model in Section 3. The O-FRAP$^+$, SMP and HRAP protocols are introduced and analyzed in Sections 4, 5 and 6 respectively. The revised HRAP protocol is proposed in Section 7, and, finally, we present our conclusions in Section 8.

2   Related Works

In this section, we briefly study the history of formal privacy analysis studies of RFID protocols as well as review some previous works related to RFID authentication protocols.

2.1   Privacy Works

The design of privacy-preserving RFID protocols requires a methodical analysis of its characteristics. One of the first works on the formalization of privacy in RFID protocols was made by Avoine [2] and was continued in his later work [3,4]. Avoine and Oechslin also presented a multilayer approach to RFID privacy and studied different layers of RFID systems in terms of privacy issues [5]. The Avoine model was extended in Juels and Weis's work [18] by their introducing the strong privacy notion and considering side channel information. This approach was followed by Lim and Kwon [21] who gave a formal definition of the forward and backward untraceability notions. In this direction, Oufi and Phan published a few papers [26,27,29] in which they introduced a modified version of [21] as well as presented privacy analysis on a variety of RFID protocols. However, Vaudenay proposed a comprehensive privacy model [39] that abstracted all previous models as well as introduced some new notions. He defined eight privacy classes for RFID protocols and proved the strongest class, strong privacy, is impossible to achieve. Later, Ng, Susilo, Mu, and Safavi-Naini [24] showed the highest achievable privacy levels for synchronized RFID authentication protocols are weak and narrow-forward privacy in the Vaudenay's model. Other works on formal privacy models can be found in [10,14,22].

2.2   O-FRAP and O-RAP Protocols

Burmester, van Le, and de Medeiros proposed an RFID authentication protocol (O-FRAP) in which a server identifies and authenticates RFID tags [7]. Each tag and the server share a key K_i and a pseudonym r_i which are updated in every session. Additionally, they use a function F to compute the authentication messages and new secrets. O-RAP is a simplified version of O-FRAP which is essentially O-FRAP but without a key updating procedure [7].

In [27], Ouafi and Phan presented a traceability attack on O-FRAP by causing a desynchronization problem on the target tag. They also showed O-FRAP fails to provide forward security by corrupting the target tag at the point after the tag outputs Accept. Another analysis on O-FRAP and O-RAP was made by Duc and Kim [12] in which they addressed a scalability issue on these protocols.

2.3   Song-Mitchell's Protocols

In Wisec 2008, Song and Mitchell presented an efficient RFID authentication protocol [32]. Their protocol made use of a hash function, a pseudorandom number generator, and bitwise operations such as XOR and circular shift. Afterwards, Song proposed an ownership transfer protocol [31] which was based completely on the authentication protocol. The authors of [32] also introduced an RFID pseudonym protocol in [33].

In [30], Rizomiliotis, Rekleitis, and Gritzalis exploited some flaws in designing the messages of [32] and presented a reader impersonation attack on this protocol that desynchronizes the tag and server. Another security analysis on the Song-Mitchel protocol [32] was made by van Deursen and Radomirović [38]. They make use of the weak security properties of bitwise XOR and rotation to present tag impersonation, reader impersonation, desynchronization, and traceability attacks on the authentication protocol.

Peris-Lopez et al. [28] completed the work in [30,38] and showed new security flaws in the presented protocols in [32,31]. They proved that the authentication and ownership protocols fail to provide location and information privacy. Furthermore, they proved the update protocol is vulnerable to desynchronization attack.

3   Privacy Model

In recent years, the privacy issue has been become a major research field in many communication and computation systems. Due to the ubiquitous presence of RFID systems, it is vital to take privacy issues into consideration. Nonetheless, we need formal models to evaluate the privacy level of a variety of RFID authentication protocols. Thus, we first explain the privacy model proposed by Oufi-Phan [26], and then present our traceability attacks based on this model. The model presented in [26] is based on [18] and [21], and is summarized below.

3.1   Ouafi-Phan Model

The protocol parties are $Tags(\mathcal{T} \in Tags)$ or $Readers(\mathcal{R} \in Readers)$ which interact in protocol sessions. In this model an adversary $\mathcal{A}$ controls the communication channel between all parties by interacting either passively or actively with them. The adversary $\mathcal{A}$ is allowed to run the following queries.

- Execute($\mathcal{R}, \mathcal{T}, i$) **query.** This query models passive attacks. The adversary $\mathcal{A}$ eavesdrops on the communication channel between $\mathcal{T}$ and $\mathcal{R}$ and gets read access to the messages exchanged between the parties in session i of a truthful protocol execution.
- Send(U, V, m, i) **query.** This query models active attacks by allowing the adversary $\mathcal{A}$ to impersonate some reader $U \in Readers$ (respectively tag $U \in Tags$) in some protocol session i and send a message m of its choice to an instance of some tag $V \in Tags$ (respectively reader $V \in Readers$). Furthermore, the adversary $\mathcal{A}$ is allowed to block or alert the message m that is sent from U to V (respectively V to U) in session i of a truthful protocol execution.
- Corrupt($\mathcal{T}, K$) **query.** This query allows the adversary $\mathcal{A}$ to learn the stored secret K of the tag $\mathcal{T} \in Tags$, and which further sets the stored secret to K'. Corrupt query means that the adversary has physical access to the tag, i.e., the adversary can read and tamper with the tag's permanent memory.
- Test($\mathcal{T}_0, \mathcal{T}_1, i$) **query.** This query does not correspond to any of $\mathcal{A}$'s abilities, but it is necessary to define the untraceability test. When this query is invoked for session i, a random bit $b \in \{0,1\}$ is generated, and then $\mathcal{A}$ is given $\mathcal{T}_b \in \{\mathcal{T}_0, \mathcal{T}_1\}$. Informally, $\mathcal{A}$ wins if she can guess the bit b.

Untraceable privacy (UPriv) is defined using the game $\mathcal{G}$ played between an adversary $\mathcal{A}$ and a collection of the reader and the tag instances. The game $\mathcal{G}$ is divided into the three following phases.

- **Learning phase :** $\mathcal{A}$ is given tags $\mathcal{T}_0$ and $\mathcal{T}_1$ randomly, and she is able to send any Execute, Send, and Corrupt queries of its choice to $\mathcal{T}_0$, $\mathcal{T}_1$, and the reader.
- **Challenge phase :** $\mathcal{A}$ chooses two fresh tags $(\mathcal{T}_0, \mathcal{T}_1)$ to be tested and sends a Test($\mathcal{T}_0, \mathcal{T}_1, i$) query. Depending on a randomly chosen bit $b \in \{0,1\}$, $\mathcal{A}$ is given a tag $\mathcal{T}_b$ from the set $\{\mathcal{T}_0, \mathcal{T}_1\}$. $\mathcal{A}$ continues to make any Execute and Send queries at will.
- **Guess phase :** Finally, $\mathcal{A}$ terminates the game $\mathcal{G}$ and outputs a bit $b' \in \{0,1\}$, which is her guess of the value of b.

The success of $\mathcal{A}$ in winning game $\mathcal{G}$ and thus breaking the notion of *UPriv* is quantified in terms of $\mathcal{A}$'s advantage in distinguishing whether she receives $\mathcal{T}_0$ or $\mathcal{T}_1$ and is denoted by $\mathsf{Adv}_{\mathcal{A}}^{\mathsf{UPriv}}(k)$ where k is the security parameter.

$$\mathsf{Adv}_{\mathcal{A}}^{\mathsf{UPriv}}(k) = \mid pr(b' = b) - pr(random\ flip\ coin) \mid = \mid pr(b' = b) - \frac{1}{2} \mid$$

where $0 \leq \mathsf{Adv}_{\mathcal{A}}^{\mathsf{UPriv}}(k) \leq \frac{1}{2}$.

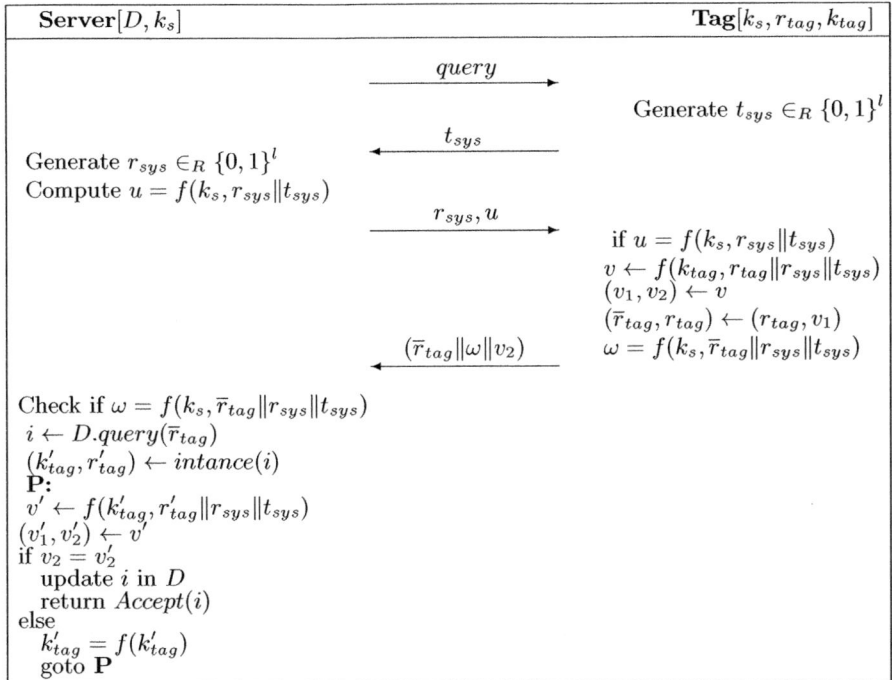

Fig. 1. The O-FRAP$^+$ protocol

The notion of backward untraceability is defined such that "even if given all the internal states of a target tag at time t, the adversary shouldn't be able to identify the target tag's interactions that occur at time t', $t' < t$." [21].

4   The O-FRAP$^+$ Protocol

Duc and Kim proposed O-FRAP$^+$ [12] to improve the O-FRAP protocol [7] in terms of availability and location privacy. In this section, we first introduce the protocol and then present our attacks on O-FRAP$^+$.

4.1   Review O-FRAP$^+$

In O-FRAP$^+$ [12], each tag keeps (k_s, k_{tag}, r_{tag}) wherein k_s is a permanent key to authenticate the server and can be common for all tags or a local group of tags. The secret key and pseudonym of each tag are k_{tag} and r_{tag}, respectively, both with bit length l. These values are updated by the tag in every session after server authentication. The server keeps a pair of (secret key, pseudonym)=(k_{tag}, r_{tag}) for each tag in its database (D). In O-FRAP$^+$, the server is authenticated first, and then a tag updates its secrets before the server.

In the authentication phase, both parties use a pseudorandom function $f(.)$ to compute the messages and secret values. In addition, the $D.query(.)$ notation stands for the database querying procedure. The authentication process is shown in Fig. 1.

4.2   Our Attacks on O-FRAP$^+$

Though the authors of [12] tried to overcome desynchronization and traceability attacks on O-FRAP by proposing O-FRAP$^+$, this protocol is vulnerable to the cited attacks. The main problem with O-FRAP$^+$ arises from its pseudonym update process. This problem results in desynchronization and traceability attacks as described below.

Desynchronization Attack

1. Let a scenario in which the server and a target tag are synchronous with values (r_{tag}^1, k_{tag}^1).
2. The server and tag start a new authentication session while the adversary eavesdrops on the channel and controls it. They exchange the messages $\{query, t_{sys}, (r_{sys}, u)\}$ and the tag updates its secret values to (r_{tag}^2, k_{tag}^2); then it sends the server the message $(\bar{r}_{tag}^1 \| \omega^1 \| v_2^1)$ to authenticate itself.
3. The adversary blocks the tag's response to prevent the server from updating; as a result, the server fails to update its secrets and still keeps the values (r_{tag}^1, k_{tag}^1) for this tag in its database
4. At the next session, the tag sends the server $(\bar{r}_{tag}^2 \| \omega^2 \| v_2^2)$ to authenticate itself. At first, the server computes $\omega^2 = f(k_s, \bar{r}_{tag}^2 \| r_{sys}^2 \| t_{sys}^2)$ and checks whether the tag is a legal one. Then, it performs $D.query(.)$ to get a $(k_{tag}, \bar{r}_{tag}^2)$ pair of the tag i stored in database. However, the server has no entry including the value $(\bar{r}_{tag}^2)$ in its database and finds no match. Hence, it rejects the tag in this session and all future sessions.

Although the authors of O-FRAP$^+$ specified a way to resynchronize k_{tag} between a tag and the server via computing $k_{tag}^{new} = f(k_{tag})$ by the server, they have no proposal to solve the case in which the pseudonym r_{tag} is desynchronized. Thus, after the cited attack, the target tag and server cannot recover their synchronization in any of the subsequent sessions.

Besides the attack just described, with respect to using of a common key k_s among all tags or a local group of tags, if a tag is compromised by an adversary, then the security of other tags will be violated. Indeed, the adversary can impersonate the server via the compromised key k_s and desynchronize every tag in that group.

Traceability Attack. Due to the wireless communication in RFID systems, location privacy is one of the most challenging problems in these systems. Generally, the location privacy notion in RFID applications is equivalent to the notion of tag untraceability. The untraceability notion itself is based on the notions of *anonymity* and *unlinkability*. The anonymity notion means an adversary cannot identify a target tag while the unlinkability notion refers to the fact that the tag interactions cannot be linked.

In this context and due to the attack explained in Subsection 4.2, we can apply the traceability attack to O-FRAP$^+$ as follows:

- In the Learning phase the adversary chooses two fresh tags $\mathcal{T}_0$ and $\mathcal{T}_1$ that are synchronous with the sever. The adversary desynchronizes the tag $\mathcal{T}_0$ with a Send query according to the attack presented in Subsection 4.2. Then, she terminates the Learning phase. In the challenge phase, when the adversary is given the tag $\mathcal{T}_b$, she simply allows $\mathcal{T}_b$ and the server to perform an authentication session and obtain the sever acceptance output with an Execute query. Now, the adversary knows she is interacting with the tag $\mathcal{T}_1$ if the server accepts $\mathcal{T}_b$ in the challenge phase, and knows she is interacting with the tag $\mathcal{T}_0$ otherwise.

With respect to the similar structure used in O-FRAP$^+$ and O-RAP$^+$, all presented attacks in this section are also applicable to the O-RAP$^+$ protocol proposed in [12].

5 The SMP Protocol

Song and Mitchell introduced an RFID pseudonym protocol and a tag ownership transfer protocol in their paper [34]. In this section, the RFID pseudonym protocol (SMP) is reviewed, and our attacks on it are presented.

5.1 Review SMP

SMP uses a precomputed look-up table which contains a number of hash-chain elements as identifiers for each tag. A keyed hash function $e_k(.) : \{0,1\}^* \times \{0,1\}^l \longrightarrow \{0,1\}^l$ is used to create each hash-chain, using a secret key k shared by the tag and server. The hash-chain length, m, determines the number of tag identifiers, x_i, that can be generated using any one key.

The server chooses values for l, l_r, l_m and functions e, f, g and h as follows:

- l is the bit-length of a tag identifier x_i, and a shared key k;
- $l_r \leq l$ is the bit-length of a random string r;
- l_m is the bit-length of an integer m, that defines the length of the hash-chain used for tag identifiers;
- $e : \{0,1\}^* \times \{0,1\}^l \longrightarrow \{0,1\}^l, f : \{0,1\}^* \times \{0,1\}^l \longrightarrow \{0,1\}^l$ and $g : \{0,1\}^* \times \{0,1\}^l \longrightarrow \{0,1\}^{2l+l_m}$ are keyed hash functions; and
- $h : \{0,1\}^* \longrightarrow \{0,1\}^l$ is a hash function.

The server chooses random l-bit strings s and x_0. Then it computes l-bit key k, $k = h(s)$, and the hash chain values $x_i, x_i = e_k(x_{i-1})$ for $1 \leq i \leq m$. The server stores s, k and the identifiers $\langle x_0, x_1, ..., x_m \rangle$ as the corresponding entry for each tag in its look-up table. Each tag stores k, x (initially set to x_0) and a counter c (set to m).

The SMP authentication phase is performed in three different cases as depicted in Fig. 2.

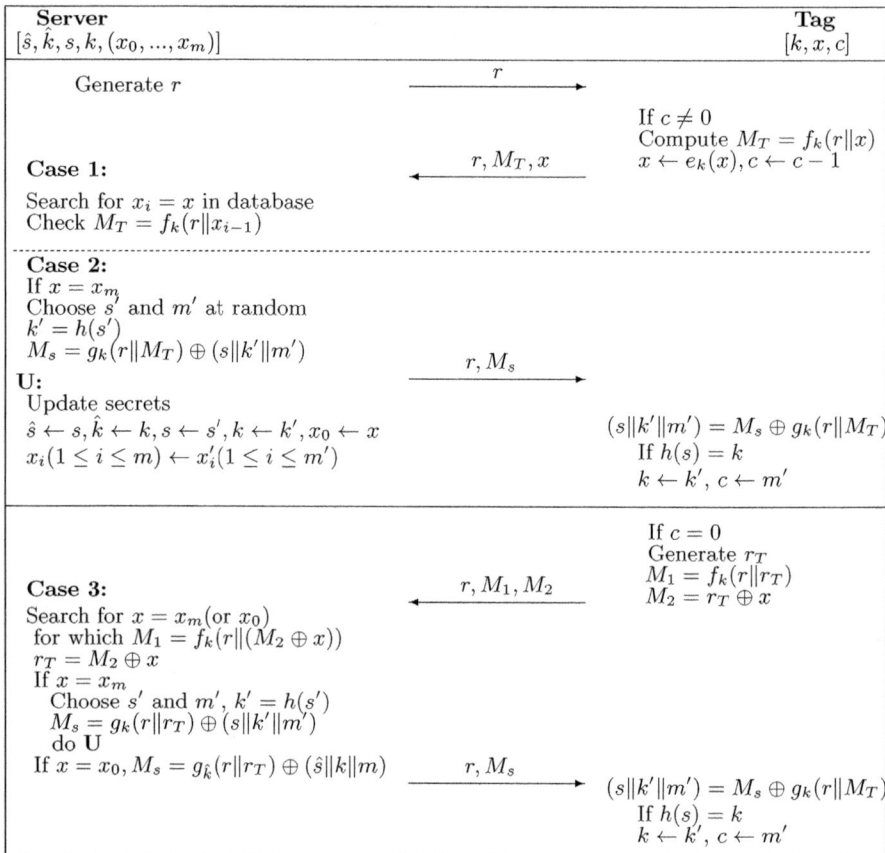

Fig. 2. The SMP protocol

5.2 Our Attacks on SMP

Privacy and synchronization are two important attributes of an RFID authentication protocol. Unfortunately, not only does SMP fail to provide privacy but it also suffers from a desynchronization problem. In this section, first we present a desynchronization attack on SMP which makes the target tag and the reader inconsistent in the future. Then, the traceability and backward traceability attacks are applied to SMP in a formal model.

Desynchronization Attack. Keeping synchronization between the server and a tag is a significant characteristic of RFID protocols. In this context, SMP has a considerable flaw. More precisely, when the first stage of SMP is completed, the server performs the update process to renew the secrets and transmit them to the tag. At this point the new secrets are transmitted to the tag in an insecure way, inasmuch as an adversary can manipulate the transmitted message without awareness of the target tag. In fact, the security flaw appears in the construction of the message M_s which is transmitted to update the secrets in cases 2 and 3. The message M_s is a $2l+l_m$ length bit string such as $M_s = g_k(r\|M_T) \oplus (s\|k'\|m')$ in case 2 and $M_s = g_{\hat{k}}(r\|r_T) \oplus (\hat{s}\|k\|m)$ in case 3. In this message, $g_k(.)$ is the output of a keyed hash function with length $2l + l_m$ as the length of M_s. We can parse the string $g_k(.)$ as $g_k(.) = L_1\|L_2\|L_3$ where $|L_1| = |L_2| = l$ and $|L_3| = l_m$. The substring L_1 is the first l bit of $g_k(.)$, L_2 is the next l bit of $g_k(.)$ and L_3 is the last l_m bit of $g_k(.)$. Thus, with regard to the concatenation property, we can construct M_s as $M_s = L_1 \oplus s\|L_2 \oplus k'\|L_3 \oplus m'$ in case 2 and $M_s = L_1 \oplus \hat{s}\|L_2 \oplus k\|L_3 \oplus m$ in case 3. Now we explain the attack on the *Case 2* of the protocol. The attack on case 3 is similar to this attack.

1. The adversary eavesdrops on the communication channel between a target tag $\mathcal{T}$ and the server until the first stage concludes. Consequently, the update procedure is performed via the sending of the message M_s by the server.
2. At this point the adversary transforms M_s to M'_s as described below:

$$M'_s = M_s \oplus ((0)^l\|N_2\|N_3) = L_1 \oplus s \oplus (0)^l\|L_2 \oplus k' \oplus N_2\|L_3 \oplus m' \oplus N_3 \Rightarrow$$

$$M'_s = L_1 \oplus s\|L_2 \oplus k' \oplus N_2\|L_3 \oplus m' \oplus N_3$$

   Where $(0)^l$ is a zero bit string with length l, N_2 is an arbitrary bit string with length l, and N_3 is an arbitrary bit string with length l_m. Then, the adversary sends (r, M'_s) to the target tag $\mathcal{T}$.
3. Receiving (r, M'_s), $\mathcal{T}$ computes

$$g_k(r\|M_T) \oplus M'_s = (L_1\|L_2\|L_3) \oplus (L_1 \oplus s\|L_2 \oplus k' \oplus N_2\|L_3 \oplus m' \oplus N_3) =$$

$$s\|k' \oplus N_2\|m' \oplus N_3$$

4. The tag obtains the secret s from the computed message. Since $h(s) = k$, $\mathcal{T}$ authenticates the server and updates its secrets to $k' \oplus N_2$ and $m' \oplus N_3$ while the server has stored (s, k, s', k') in its database for the tag $\mathcal{T}$. As a result, when $\mathcal{T}$ tries to authenticate itself in the subsequent sessions, it is rejected by the server and they have no way to resynchronize themselves.

This desynchronization attack is also applied to the secret update protocol of [34] in a similar way.

Traceability Attack. Besides the desynchronization attack described above, SMP is also vulnerable to traceability attacks. The problem occurs when the update procedure begins. In fact, case 1 of SMP involves two information flows, the flow of sending (r) by the server and the flow of replying (r, M_T, x) by the tag. However, when the protocol move on to the update procedure, case 2 or case 3, it takes a third flow of sending by the server for the purpose of updating. As a result, an adversary can exploit this point to trace a target tag as follows:

- In the Learning phase, She chooses two fresh tags $(\mathcal{T}_0, \mathcal{T}_1)$ which appear at the onset of the authentication stage (*Case 1*). She sends several consecutive Send($\mathcal{R}, \mathcal{T}_0, r, i$) queries to drive $\mathcal{T}_0$ into *Case 3* (c = 0).
- In the Challenge phase, $\mathcal{A}$ makes an Execute($\mathcal{R}, \mathcal{T}_b, i+1$) query to obtain a transcript of the executed protocol between $\mathcal{T}_b$ and the server in the $(i+1)$th round. As a result, she receives a transcript with either two or three flows.

- $\mathcal{A}$ terminates the game $\mathcal{G}$ and guesses the correct tag with overwhelming probability. In particular, $\mathcal{A}$ recognizes $\mathcal{T}_0$ provided that the transcript received involves three information flows and $\mathcal{T}_1$ otherwise. In fact, when the session performed between the server and $\mathcal{T}_b$ has a transcript with three flows, it means that $\mathcal{T}_b$ is in the case 2 or case 3. However, the adversary knows (from the Learning phase) that $\mathcal{T}_1$ is in the beginning of the *Case 1* and its related transcript involves two information flows. Thus, she can guess the correct tag with the probability of 1.   □

6   The HRAP Protocol

Cho et al. proposed a hash-based authentication protocol for RFID systems [9]. This protocol is explained and analyzed in this section.

6.1   Review HRAP

At the beginning, the server and each tag share a hash function $h(.)$. The security of the protocol relies on two values RID_i and S. RID_i is the group ID of a random number. Each group is divided into random numbers based on a certain value, after their being sorted sequentially. Each group of random numbers has a group $ID(RID_i)$ and the range of each group is determined by the secret value S, which is randomly specified by the server at the end of each session. The secret value S can be any natural number within the range $2^1 - 2^{86}$, but it is not necessarily a multiple of two. Each RID_i is made by using the MSB (48 bits) of the minimum value and the LSB (48 bits) of the maximum value of its group: $RID_i = R_{i_{min}}(0 : 47) \| R_{i_{max}}(48 : 95)$. In the jth round of the protocol, the secret values ID and S_j are stored on each tag and the server keeps $(ID, S_{j-1}, S_j, Data)$ for that tag.

After the initialization phase, the server and tag perform the authentication phase to authenticate each other as illustrated in Figure 3.

6.2   Our Attacks on HRAP

In this section, the significant security flaws in HRAP are shown. Indeed, this protocol is vulnerable to desynchronization and traceability attacks. In the following, first we present a desynchronization attack on HRAP which makes the target tag and the reader inconsistent in the next sessions. Then, traceability and backward traceability attacks are applied on HRAP.

Desynchronization Attack. In HRAP, the update mechanism suffers from a basic weakness which results in vulnerability to a desynchronization attack. More precisely, the secret value S_{j+1} is chosen by the server and is transmitted via the message $h(\beta \oplus RID_i) \| R_t \oplus S_{j+1}$ on an insecure channel to update the tag's secret values. This enables an adversary to manipulate the transmitted message and causes a desynchronization problem as follows:

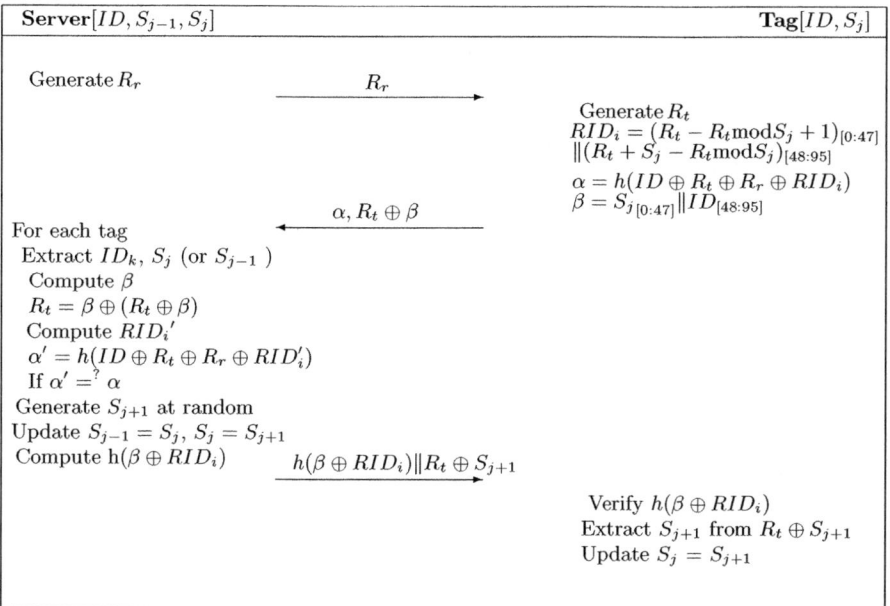

Fig. 3. The HRAP protocol

1. An adversary eavesdrops on the communication channel. When the server transmits the massage $h(\beta \oplus RID_i)\|R_t \oplus S_{j+1}$, $\mathcal{A}$ changes $R_t \oplus S_{j+1}$ to $R_t \oplus S_{j+1} \oplus R'$ where R' is an arbitrary bit string. Then she sends $h(\beta \oplus RID_i)\|R_t \oplus S_{j+1} \oplus R'$ to the target tag.

2. Receiving $h(\beta \oplus RID_i)\|R_t \oplus S_{j+1} \oplus R'$, the tag authenticates the server via $h(\beta \oplus RID_i)$. Then, it extracts $S_{j+1} \oplus R'$ from $R_t \oplus S_{j+1} \oplus R'$ and updates its S_j to $S_{j+1} \oplus R'$. Since the server keeps only the values S_j and S_{j+1}, the target tag will be rejected in all future sessions.

Traceability Attack. In this context, we address another weakness in the HRAP, namely vulnerability to traceability attack. In fact, the problem occurs in the choosing of the secret value S_j when the authors state that the value S_j need not necessarily be a multiple of two. We show that an adversary can exploit this point to trace a target tag by way of a passive attack with overwhelming probability. In the following, the formal analysis is presented.

- **Learning phase:** The adversary chooses a fresh tag $\mathcal{T}_0$ with which to interact. $\mathcal{A}$ makes an Execute($\mathcal{R}, \mathcal{T}_0, j$) query and obtains the messages $M_1 = (R_{t,j}^{\mathcal{T}_0} \oplus \beta_j^{\mathcal{T}_0})$ and $M_2 = (R_{t,j}^{\mathcal{T}_0} \oplus S_{j+1}^{\mathcal{T}_0})$ from the exchanged messages between $\mathcal{T}_0$ and the server. The notation $X_i^{\mathcal{T}_x}$ indicates to the vale X related to the tag $\mathcal{T}_x$ in round i of the protocol.

- **Challenge phase:** $\mathcal{A}$ chooses two fresh tags $(\mathcal{T}_0, \mathcal{T}_1)$ to be tested and sends a Test($\mathcal{T}_0, \mathcal{T}_1, j{+}1$) query. Depending on a randomly chosen bit $b \in \{0,1\}$, $\mathcal{A}$

is given a tag $\mathcal{T}_b$ from the set $\{\mathcal{T}_0, \mathcal{T}_1\}$. $\mathcal{A}$ makes an Execute($\mathcal{R}, \mathcal{T}_b, j+1$) query and is given the messages $M_3 = (R_{t,j+1}^{T_b} \oplus \beta_{j+1}^{T_b})$ and $M_4 = (R_{t,j+1}^{T_b} \oplus S_{j+2}^{T_b})$ from the messages exchanged between $\mathcal{T}_b$ and the server.

- **Guess phase:** finally, $\mathcal{A}$ terminates the game $\mathcal{G}$ and outputs a bit $b' \in \{0, 1\}$ as her guess of the value of b. In particular, $\mathcal{A}$ utilizes the following simple decision rule:

$$b' = \begin{cases} 0 & \text{if} \;\; [M_1 \oplus M_2]_{[95]} = [M_3 \oplus M_4]_{[95]} \\ 1 & \text{otherwise} \end{cases}$$

where $X_{[95]}$ is the *LSB* of the bit string X.

Hence,

$$\mathsf{Adv}_\mathcal{A}^{\mathsf{UPriv}}(k) = |\, pr(b' = b) - \frac{1}{2} \,| = |\, \frac{3}{4} - \frac{1}{2} \,| = \frac{1}{4} \gg \varepsilon$$

Proof: By obtaining the messages M_1, M_2, M_3, and M_4, the adversary proceeds as follows:

$$M_1 \oplus M_2 = (R_{t,j}^{T_0} \oplus \beta_j^{T_0}) \oplus (R_{t,j}^{T_0} \oplus S_{j+1}^{T_0}) = \beta_j^{T_0} \oplus S_{j+1}^{T_0}$$

$$= (S_{j[0:47]}^{T_0} \| ID_{[48:95]}^{T_0}) \oplus (S_{j+1[0:47]}^{T_0} \| (S_{j+1[48:95]}^{T_0})$$

$$\Longrightarrow [M_1 \oplus M_2]_{[95]} = ID_{[95]}^{T_0} \oplus S_{j+1[95]}^{T_0}$$

$$M_3 \oplus M_4 = (R_{t,j+1}^{T_b} \oplus \beta_{j+1}^{T_b}) \oplus (R_{t,j+1}^{T_b} \oplus S_{j+2}^{T_b}) = \beta_{j+1}^{T_b} \oplus S_{j+2}^{T_b}$$

$$= (S_{j+1[0:47]}^{T_b} \| ID_{[48:95]}^{T_b}) \oplus (S_{j+2[0:47]}^{T_b} \| (S_{j+2[48:95]}^{T_b})$$

$$\Longrightarrow [M_3 \oplus M_4]_{[95]} = ID_{[95]}^{T_b} \oplus S_{j+2[95]}^{T_b}$$

The authors of [9] have stated that the secret value S_j need not be a multiple of two; therefore, the adversary knows:

$$S_{j+1[95]}^{T_0} = S_{j+2[95]}^{T_b} = 1 \Rightarrow \begin{cases} [M_1 \oplus M_2]_{[95]} = \overline{ID_{[95]}^{T_0}} \\ [M_3 \oplus M_4]_{[95]} = \overline{ID_{[95]}^{T_b}} \end{cases}$$

Now, if we consider uniform probability distribution on bit string ID^{T_x}, we have:

$$pr(b' = b) = \frac{1}{2} \times pr(b' = 0 | b = 0) + \frac{1}{2} \times pr(b' = 1 | b = 1)$$

$$= \frac{1}{2} \times pr(ID_{[95]}^{T_b} = ID_{[95]}^{T_0} | \mathcal{T}_b = \mathcal{T}_0) + \frac{1}{2} \times pr(ID_{[95]}^{T_b} \neq ID_{[95]}^{T_0} | \mathcal{T}_b = \mathcal{T}_1)$$

$$= \frac{1}{2} \times 1 + \frac{1}{2} \times \frac{1}{2} = \frac{3}{4} \qquad \square$$

Backward Traceability Attack. In this section, we show another weakness in the protocol of Cho et al. which leads to a backward traceability attack. Since ID is constant in all rounds of the protocol, an adversary $\mathcal{A}$ can track the target tag with a high probability.

- **Learning phase:** The adversary makes a $\mathsf{Corrupt}(\mathcal{T}_0, K)$ query in the jth round of the protocol. As a result, she is given the $\mathcal{T}_0$'s security parameters $(ID^{\mathcal{T}_0}, S_j^{\mathcal{T}_0})$.

- **Challenge phase:** $\mathcal{A}$ sends a $\mathsf{Test}(\mathcal{T}_0, \mathcal{T}_1, j\text{-}1)$ query and receives a tag $\mathcal{T}_b \in \{\mathcal{T}_0, \mathcal{T}_1\}$ randomly. $\mathcal{A}$ makes an $\mathsf{Execute}(\mathcal{R}, \mathcal{T}_b, j\text{-}1)$ query to get a transcript of the executed protocol between $\mathcal{T}_b$ and the reader in the $(j-1)$th round. The adversary stores the messages $M_1'=(R_{t,j-1}^{\mathcal{T}_b} \oplus \beta_{j-1}^{\mathcal{T}_b})$ and $M_2'=(R_{t,j-1}^{\mathcal{T}_b} \oplus S_j^{\mathcal{T}_b})$ from the received transcript.

- **Guess phase:** After finishing the game $\mathcal{G}$, $\mathcal{A}$ outputs a bit $b' \in \{0,1\}$ as her guess of the value of b as following.

$$b' = \begin{cases} 0 & if \;\; [M_1' \oplus M_2']_{[48:95]} = [ID^{\mathcal{T}_0} \oplus S_j^{\mathcal{T}_0}]_{[48:95]} \\ 1 & otherwise \end{cases}$$

The $\mathcal{A}$'s success probability is:

$$\mathsf{Adv}_{\mathcal{A}}^{\mathsf{UPriv}}(k) = |\, pr(b'=b) - \frac{1}{2} \,| = |\, (1 - 2^{-49}) - \frac{1}{2} \,| = \frac{1}{2} - 2^{-49} \gg \varepsilon$$

Proof: By $Xoring$ the messages M_1' and M_2', we have

$$M_1' \oplus M_2' = (R_{t,j-1}^{\mathcal{T}_b} \oplus \beta_{j-1}^{\mathcal{T}_b}) \oplus (R_{t,j-1}^{\mathcal{T}_b} \oplus S_j^{\mathcal{T}_b}) = \beta_{j-1}^{\mathcal{T}_b} \oplus S_j^{\mathcal{T}_b}$$

$$= (S_{j-1}^{\mathcal{T}_b}{}_{[0:47]} \| ID_{[48:95]}^{\mathcal{T}_b}) \oplus (S_{j[0:47]}^{\mathcal{T}_b} \| (S_{j[48:95]}^{\mathcal{T}_b}) \Longrightarrow$$

$$[M_1' \oplus M_2']_{[48:95]} = ID_{[48:95]}^{\mathcal{T}_b} \oplus S_{j[48:95]}^{\mathcal{T}_b}$$

Now, if the uniform probability distribution on bit strings $ID^{\mathcal{T}_X}$ is considered, the following equations obtain:

$$pr(b'=b) = \tfrac{1}{2} \times pr(b'=0|b=0) + \tfrac{1}{2} \times pr(b'=1|b=1)$$

$$= \tfrac{1}{2} \times pr([ID^{\mathcal{T}_b} \oplus S_j^{\mathcal{T}_b}]_{[48:95]} = [ID^{\mathcal{T}_0} \oplus S_j^{\mathcal{T}_0})]_{[48:95]} | \mathcal{T}_b = \mathcal{T}_0)$$

$$+ \tfrac{1}{2} \times pr([ID^{\mathcal{T}_b} \oplus S_j^{\mathcal{T}_b})]_{[48:95]} \neq [ID^{\mathcal{T}_0} \oplus S_j^{\mathcal{T}_0}]_{[48:95]} | \mathcal{T}_b = \mathcal{T}_1)$$

$$= \tfrac{1}{2} \times 1 + \tfrac{1}{2} \times (1 - 2^{-48}) = 1 - 2^{-49} \qquad \square$$

7   The Improved HRAP Protocol

In this section, we first address the defects of HRAP in detail and then introduce an improved protocol to remedy the revealed flaws.

7.1   HRAP's Defects

The first obvious weakness of HRAP is in its update mechanism. In fact, the update process is based on the value S_j which is selected by the server and transmitted to the tag on insecure channel. This allows an adversary to manipulate the new secret and make the tag and server inconsistent. To solve this problem, we change the update mechanism in such a way that two parties can compute the new secret value S_j independently.

The other two weaknesses are due to the permanent ID and β's construction. In the original HRAP, the value β is computed as: $\beta = S_{j[0:47]}\|ID_{[48:95]}$ with a permanent ID. This construction together with the message $(R_t \oplus S_{j+1})$ leads to traceability attack, as we showed in the Subsection 6.2.

We also exploit the permanent ID and update mechanism to apply a backward traceability attack on HRAP as in Subsection 6.2.

To remedy the security and privacy flaws identified, we renew the value ID in each session and change the update mechanism and β's construction in the revised protocol.

7.2   Procedure of the Improved HRAP

The protocol has an initial phase in which the parameters are chosen by the server. Then, the authentication phase is performed.

Set up Phase. In the improved protocol, we use the same notations, hash function, group $ID(RID_i)$, secret value (S_j) and ID as the original HRAP (see Subsection 6.1). For each tag, the server stores $[\ ID_{old}, ID_{new}, S_{j-1}, S_j)]$ as an entry for the corresponding tag . The values (ID_{old}, S_{j-1}) represent the values that were used in the previous session. Each tag stores the values $[ID, S_j]$ in its memory.

Authentication Phase. The authentication phase is performed as follows and is depicted in Figure 4.

1. The server sends a random number R_r to the tag.
2. The tag generates a random number R_t and computes RID_i of group involving R_t as
   $$RID_i = (R_t - R_t \bmod S_j + 1)_{[0:47]}\|(R_t + S_j - R_t \bmod S_j)_{[48:95]};$$
   then it creates α and β where
   $$\alpha = h(ID \oplus R_t \oplus R_r \oplus RID_i)$$
   $$\beta = ID_{[48:95]}\|S_{j[0:47]};$$
   the tag sends $(\alpha, R_t \oplus \beta)$ to the server.
3. The server performs the following steps to authenticate the tag:
   (a) Extract ID_k and S_j of a tag in the table and generates β.
   (b) Extract R_t from the received $R_t \oplus \beta$ and computes $RID_i{}'$ using extracted S_j and R_t.
   (c) Compute α' using $R_r, R_t, RID_i{}'$ and ID_k.
   (d) Repeat steps (a) to (c) until it finds that α' matches to α.
4. If the same α' is found, the server computes $h(\beta \parallel RID_i)$ and sends it to the tag. Then, the server updates the secret values of the relevant tag as follows:
   If $ID = ID_{new}$:

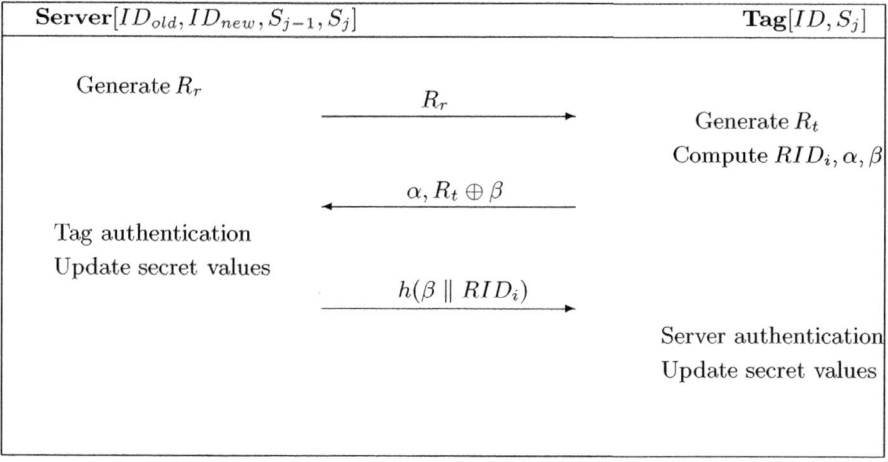

Fig. 4. The improved HRAP

$$S_{j-1} \leftarrow S_j, \quad S_j \leftarrow h(S_j \parallel RID_i), \quad ID_{old} \leftarrow ID, \quad ID_{new} \leftarrow h(ID \parallel S_j)$$
If $ID = ID_{old}$:
$$S_j \leftarrow h(S_j \parallel RID_i), \quad ID_{new} \leftarrow h(ID \parallel S_j)$$

5. If the same α' is not found, the server extracts S_{j-1} from the relevant entry and repeats steps 1-4.
6. The tag authenticates the server via $h(\beta \parallel RID_i)$; if the server is authenticated, the tag updates its ID and S_j as

$$S_j \leftarrow h(S_j \parallel RID_i), \quad ID \leftarrow h(ID \parallel S_j)$$

7.3   Security and Privacy Analysis

Our revisions to HRAP improve its security and privacy and eliminate its vulnerability to desynchronization, traceability and backward traceability attacks. In the following section, we discuss why the improved HRAP is secure.

– *Desynchronization attack.* To remove the desynchronization problem, we have designed a new update mechanism in which both the tag and sever are capable to compute the new secret. As a result, the adversary cannot cause a desynchronization problem by manipulating the exchanged messages. In addition, the update process is flexible and resistant to blocking attacks. In sum, we have designed a two-state update mechanism.
  The first state is occurred in the regular cases in which two parties are synchronous. Hence, the server stores the current values (S_j, ID) in the (S_{j-1}, ID_{old}) positions to recover synchronization in the future, and updates the current (S_j, ID) to the new ones. However, when the parties are asynchronous, the second state is performed. In this case, the tag has not updated

its secrets in the previous session, and the shared values between parties are (S_{j-1}, ID_{old}). Thus, the server maintains these values in its database until a synchronized case is encountered, and then it simply updates the current values (S_j, ID) to the new ones. This process means that even if an adversary blocks the message $h(\beta \parallel RID_i)$ several times to desynchronize the server and a target tag, the server is still able to identify the target tag.

- *Traceability attack.* Although HRAP is vulnerable to traceability attacks due to the construction of β and the message $(R_t \oplus S_{j+1})$, vulnerability to this type of attack has been removed in the revised protocol. In fact, in the improved HRAP, the construction of β has been changed to $\beta = ID_{[48:95]} \parallel S_{j[0:47]}$ and the message $(R_t \oplus S_{j+1})$ has been omitted due to the new update mechanism. Hence, even if an adversary makes many Execute and Send queries when interacting with target tags in the game $\mathcal{G}$, she has no way to recognize the correct tag in the guess phase.

  Moreover, with respect to including the random value R_t in the construction of the RID_i, when an adversary queries a target tag consecutively, she receives different responses. Thus, she cannot trace that tag by means of this method.

- *Backward traceability attack.* We imposed a backward traceability attack on the HRAP with using the permanent ID and the message $(R_t \oplus S_{j+1})$. However, the improved HRAP is resistant to backward traceability attack, since in the revised protocol the value ID is renewed in each session, as well as the omission of the message $(R_t \oplus S_{j+1})$.

  Now, consider an adversary that compromises a tag $\mathcal{T}_0$ in the jth round, and obtains $(S_j^{\mathcal{T}_0}, ID_j^{\mathcal{T}_0})$. Nevertheless, the previously exchanged messages between $\mathcal{T}_0$ and the server have been constructed using the values $(S_{j-1}^{\mathcal{T}_0}, ID_{j-1}^{\mathcal{T}_0})$ that are out of the reach of adversary. Thus, the adversary has no way to recognize the $\mathcal{T}_0$'s previous sessions using the values $(S_j^{\mathcal{T}_0}, ID_j^{\mathcal{T}_0})$.

In addition, most massage structures in the improved HRAP are similar to those in the original HRAP that make the revised protocol resistant to other attacks according to the proofs presented in [9].

8   Conclusion

The increasing use of RFID systems in many applications makes it necessary to pay more attention on the related security and privacy issues. In the last decade, many authentication protocols have been proposed to provide desirable security parameters in RFID systems. However, due to the restrictions on computational capabilities of low-cost tags, many of the proposed protocols have failed to satisfy the minimum security and privacy requirements. Nonetheless, researchers continue their efforts toward achieving security and privacy in RFID systems.

In this paper, we analyzed three recent proposed RFID authentication protocols by Duc and Kim (O-FRAP$^+$)[12], Song and Mitchell (SMP)[34] and Cho et al. (HRAP)[9] in terms of availability and location privacy. First we showed that the O-FRAP$^+$ protocol is vulnerable to desynchronization and traceability attacks related to a weakness in its pseudonym update process. Then, we addressed a basic flaw in SMP and HRAP's update mechanisms and proved that an adversary can exploit these defects to simply desynchronize a target tag and the server. Indeed, the update mechanism in SMP and HRAP is based on a secret value chosen by the server and transmitted to the tag on an insecure channel.

Generally, as a result of our experience, we do not recommend using this method to update new secrets in RFID protocols, as this update method makes it easy for an adversary to cause desynchronization problem by manipulating the transmitted secret value. Instead, it is more desirable to design an update mechanism in which two parties are capable of computing new secrets by themselves.

Moreover, we performed a formal privacy analysis on SMP and HRAP and presented traceability and backward traceability attacks on them. We exploited some flaws in the their message structures to trace a tag forward and backward. Finally, we stated the problems with HRAP in details and proposed an improved protocol to eliminate the cited attacks.

Acknowledgment. The work of Mohammad Hassan Habibi and Mohammad Reza Aref was partially supported by Iran National Science Fund (INSF)-cryptography chair. The work of Di Ma was partially supported by SAFE Lab, University of Michigan-Dearborn, Michigan, USA.

References

1. Alomair, B., Lazos, L., Poovendran, R.: Passive Attacks on a Class of Authentication Protocols for RFID. In: Nam, K.-H., Rhee, G. (eds.) ICISC 2007. LNCS, vol. 4817, pp. 102–115. Springer, Heidelberg (2007)
2. Avoine, G.: Adversarial model for radio frequency identification. Cryptology ePrint Archive, Report 2005/049 (2005), http://eprint.iacr.org/2005/049
3. Avoine, G.: Cryptography in radio frequency identification and fair ex-change protocols. Phd Thesis no. 3407, EPFL (2005), http://library.epfl.ch/theses/?nr=3407
4. Avoine, G., Dysli, E., Oechslin, P.: Reducing Time Complexity in RFID Systems. In: Preneel, B., Tavares, S. (eds.) SAC 2005. LNCS, vol. 3897, pp. 291–306. Springer, Heidelberg (2006)
5. Avoine, G., Oechslin, P.: RFID Traceability: A Multilayer Problem. In: S. Patrick, A., Yung, M. (eds.) FC 2005. LNCS, vol. 3570, pp. 125–140. Springer, Heidelberg (2005)
6. Banks, J., Pachano, M., Thompson, L., Hanny, D.: RFID Applied. John Wiley & Sons, Inc., Hoboken, New Jersey (2007)
7. Burmester, M., Van Le, T., De Medeiros, B., Tsudik, G.: Universally composable RFID identification and authentication protocols. ACM Transactions on Information and Systems Security 12(4), (Article 21) (2009)
8. Burmester, M., van Le, T., de Medeiros, B.: Universally composable and forward-secure RFID authentication and authenticated key exchange. In: Proc. of ASIACCS, pp. 242–252. ACM Press, New York (2007)

9. Cho, J.-S., Yeo, S.-S., Kim, S.K.: Securing against brute-force attack: A hash-based RFID mutual authentication protocol using a secret value. Computer Communications 34(3), 391–397 (2011)

10. Deng, R.H., Li, Y., Yung, M., Zhao, Y.: A New Framework for RFID Privacy. In: Gritzalis, D., Preneel, B., Theoharidou, M. (eds.) ESORICS 2010. LNCS, vol. 6345, pp. 1–18. Springer, Heidelberg (2010)

11. Dimitriou, T.: A lightweight RFID protocol to protect against traceability and cloning attacks. In: Proceedings of SecureComm 2005, pp. 59–66 (2005)

12. Duc, D.N., Kim, K.: Defending RFID authentication protocols against DoS attacks. Computer Communications 34(3), 384–390 (2011)

13. Gilbert, H., Robshaw, M., Sibert, H.: An active attack against $HB^{?+?}$ -A provably secure lightweight authentication protocol. In: Cryptology ePrint Archive, http://eprint. iacr.org/2005/23.pdf

14. Ha, J., Moon, S.-J., Zhou, J., Ha, J.C.: A New Formal Proof Model for RFID Location Privacy. In: Jajodia, S., Lopez, J. (eds.) ESORICS 2008. LNCS, vol. 5283, pp. 267–281. Springer, Heidelberg (2008)

15. Hernandez-Castro, J.C., Peris-Lopez, P., Phan, R.C.-W., Tapiador, J.M.E.: Cryptanalysis of the David-Prasad RFID Ultralightweight Authentication Protocol. In: Ors Yalcin, S.B. (ed.) RFIDSec 2010. LNCS, vol. 6370, pp. 22–34. Springer, Heidelberg (2010)

16. ISO/IEC 17799: Information technology-security techniques-code of practice for information security management. International Organization for Standardization (2005)

17. Juels, A.: Strengthening EPC Tags Against Cloning. In: Ngu, A.H.H., Kitsuregawa, M., Neuhold, E.J., Chung, J.-Y., Sheng, Q.Z. (eds.) WISE 2005. LNCS, vol. 3806, Springer, Heidelberg (2005)

18. Juels, A., Weis, S.: Defining strong privacy for RFID. Cryptology ePrint Archive, Report 2006/137 (2006), http://eprint.iacr.org/2006/137

19. Li, L., Deng, R.H.: Vulnerability analysis of EMAP-An efficient RFID mutual authentication protocol. In: AReS 2007: Second International Conference on Availability, Reliability and Security (2007)

20. Li, T., Wang, G., Deng, R.H.: Security analysis on a family of ultra-lightweight RFID authentication protocols. Journal of Software 3(3), 1–10 (2008)

21. Lim, C.H., Kwon, T.: Strong and Robust RFID Authentication Enabling Perfect Ownership Transfer. In: Ning, P., Qing, S., Li, N. (eds.) ICICS 2006. LNCS, vol. 4307, pp. 1–20. Springer, Heidelberg (2006)

22. Ma, C., Li, Y., Deng, R., Li, T.: RFID privacy: Relation between two notions, minimal condition, and efficient construction. In: ACM CCS (2009)

23. Mitrokotsa, A., Rieback, M.R., Tanenbaum, A.S.: Classifying RFID attacks and defenses. Information Systems Frontiers - ISF 12(5), 491–505 (2010)

24. Ng, C.Y., Susilo, W., Mu, Y., Safavi-Naini, R.: New Privacy Results on Synchronized RFID Authentication Protocols Against Tag Tracing. In: Backes, M., Ning, P. (eds.) ESORICS 2009. LNCS, vol. 5789, pp. 321–336. Springer, Heidelberg (2009)

25. Ohkubo, M., Suzuki, K., Kinoshita, S.: Efficient hash-chain based RFID privacy protection scheme. In: Davies, N., Mynatt, E.D., Siio, I. (eds.) UbiComp 2004. LNCS, vol. 3205, Springer, Heidelberg (2004)

26. Ouafi, K., Phan, R.C.-W.: Privacy of Recent RFID Authentication Protocols. In: Chen, L., Mu, Y., Susilo, W. (eds.) ISPEC 2008. LNCS, vol. 4991, pp. 263–277. Springer, Heidelberg (2008)

27. Ouafi, K., Phan, R.C.-W.: Traceable Privacy of Recent Provably-Secure RFID Protocols. In: Bellovin, S.M., Gennaro, R., Keromytis, A.D., Yung, M. (eds.) ACNS 2008. LNCS, vol. 5037, pp. 479–489. Springer, Heidelberg (2008)
28. Peris-Lopez, P., Hernandez-Castro, J.C., Estevez-Tapiador, J.M., Ribagorda, A.: Vulnerability analysis of RFID protocols for tag ownership transfer. Computer Networks 54(9), 1502–1508 (2010)
29. Phan, R.C.-W., Wu, J., Ouafi, K., Stinson, D.R.: Privacy analysis of forward and backward untraceable RFID authentication schemes. Wireless Personal Communications 54(2) (2010), doi:10.1007/s11277-010-0001-0
30. Rizomiliotis, P., Rekleitis, E., Gritzalis, S.: Security analysis of the Song-Mitchell authentication protocol for low-cost RFID tags. IEEE Communications Letters 13(4), 274–276 (2009)
31. Song, B.: RFID tag ownership transfer. In: Proceedings of Workshop on RFID Security (RFIDsec 2008), Budapest, Hungary (2008)
32. Song, B., Mitchell, C.J.: RFID authentication protocol for low-cost tags. In: Gligor, V.D., Hubaux, J., Poovendran, R. (eds.) ACM Conference on Wireless Network Security WiSec 2008, pp. 140–147. ACM Press, New York (2008)
33. Song, B., Mitchell, C.J.: Scalable RFID pseudonym protocol. In: Proceedings of the Third International Conference on Network and System Security NSS 2009, pp. 216–224. IEEE Computer Society (2009)
34. Song, B., Mitchell, C.J.: Scalable RFID security protocols supporting tag ownership transfer. Computer Communications 34(4), 556–566 (2011)
35. Tsudik, G.: YA-TRAP: Yet Another Trivial RFID Authentication Protocol. In: Proceedings of PerCom 2006, pp. 640–643 (2006)
36. van Deursen, T., Mauw, S., Radomirović, S.: Untraceability of RFID Protocols. In: Onieva, J.A., Sauveron, D., Chaumette, S., Gollmann, D., Markantonakis, K. (eds.) WISTP 2008. LNCS, vol. 5019, pp. 1–15. Springer, Heidelberg (2008)
37. van Deursen, T., Radomirović, S.: Algebraic Attacks on RFID Protocols. In: Markowitch, O., Bilas, A., Hoepman, J.-H., Mitchell, C.J., Quisquater, J.-J. (eds.) Information Security Theory and Practice. LNCS, vol. 5746, pp. 38–51. Springer, Heidelberg (2009)
38. van Deursen, T., Radomirovic, S.: Attacks on RFID protocols. Cryptology ePrint archive, Report 2008/310 (2008), http://eprint.iacr.org/2008/310
39. Vaudenay, S.: On Privacy Models for RFID. In: Kurosawa, K. (ed.) ASIACRYPT 2007. LNCS, vol. 4833, pp. 68–87. Springer, Heidelberg (2007)

Practical Analysis of Reduced-Round KECCAK*

María Naya-Plasencia[1,3,**], Andrea Röck[2,***], and Willi Meier[1,†]

[1] FHNW, Windisch, Switzerland
[2] Aalto University School of Science, Finland
[3] University of Versailles, France

Abstract. KECCAK is a finalist of the SHA-3 competition. In this paper we propose a practical distinguisher on 4 rounds of the hash function with the submission parameters. Recently, the designers of KECCAK published several challenges on reduced versions of the hash function. With regard to this, we propose a preimage attack on 2 rounds, a collision attack on 2 rounds and a near collision on 3 rounds of $\lfloor$KECCAK$\rfloor_{224}$ and $\lfloor$KECCAK$\rfloor_{256}$. These are the first practical cryptanalysis results on reduced rounds of the hash function scenario. All of our results have been implemented.

Keywords: hash function, KECCAK, practical cryptanalysis, SHA-3.

1 Introduction

Cryptographic hash functions are one of the three main branches of symmetric cryptography. They are deterministic functions, $\mathcal{H}$, that given an input or message M of an arbitrary length, return a short pseudo-random value of fixed length ℓ that must verify certain properties. This value must be easy to compute and is typically called digest or hash value, h. Hash functions have many important applications like authentication, integrity check of executables, digital signatures, etc. A hash function can normally be defined by an iterative construction and a compression function.

The digest should verify certain properties so that the hash function can be considered secure. The classical security requirements of a hash function are:

1. Collision resistance: finding two message M_1 and M_2 so that $\mathcal{H}(M_1) = \mathcal{H}(M_2)$ must be "hard". The generic collision attack, that applies to all hash functions, requires $2^{\ell/2}$ calls to the compression function.

---

* This work was partially supported by the European Commission through the ICT programme under contract ICT-2007-216676 ECRYPT II.
** Supported by the National Competence Center in Research on Mobile Information and Communication Systems (NCCR-MICS), a center of the Swiss National Science Foundation under grant number 5005-67322 and Partially supported by the French Agence Nationale de la Recherche through the SAPHIR2 project under Contract ANR-08-VERS-014.
*** Supported by the Academy of Finland under project 122736.
† Supported by GEBERT RÜF STIFTUNG, project no. GRS-069/07

D.J. Bernstein and S. Chatterjee (Eds.): INDOCRYPT 2011, LNCS 7107, pp. 236–254, 2011.
© Springer-Verlag Berlin Heidelberg 2011

2. Second preimage resistance: Given a message M_1 and its hash value $h = \mathcal{H}(M_1)$, finding another message M_2 so that $\mathcal{H}(M_2) = h$ must be "hard". The generic second preimage attack requires 2^ℓ calls to the compression function.

3. Preimage resistance: Given a hash value h, finding a message M_1 so that $\mathcal{H}(M_1) = h$ must be "hard". The generic preimage attack requires 2^ℓ calls to the compression function.

Defining what "hard" means in the previous concepts is a difficult task. In a strict way, we can ask that building a collision or a (second) preimage on the hash function must require at least as many calls to the compression function as the generic attacks. Beside these properties, a hash function must verify some other conditions, like for example generating hash values that are random-looking.

Recently, a big number of cryptanalysis results on hash functions have appeared, including the ones on the standards MD5 [13] and SHA-1 [12]. The confidence in the standard SHA-2 has then been undermined due to its resemblance with SHA-1. Because of this, the American National Institute of Standards and Technology (NIST) decided to launch in 2008 a competition to find a new hash function standard, SHA-3. From the 64 initial submissions, two rounds and three years later, only five candidates remain in the final round of this competition. One of them is KECCAK.

KECCAK is a sponge based hash function. The main cryptanalytic results published so far on KECCAK are results on building blocks, that is, on the permutation involved in KECCAK and not on the hash function. In [6], a zero-sum distinguisher on all 24 rounds of the permutation is proposed. This distinguisher has a complexity of 2^{1590}.

On the hash function setting, which is arguably a more interesting one, the only known results are marginally better than generic preimage attack for 6, 7 and 8 rounds [1] for the 512 bit version, having complexities of 2^{506}, 2^{507} and $2^{511.5}$ respectively. In [8], a practical preimage attack is proposed on three rounds of a modified KECCAK, using different parameters than the recommended ones, like for example, a hash size of 1024 bits, which weakens considerably the hash function.

We believe that due to the lack of results on the hash setting of reduced rounds of the hash function, the authors of KECCAK proposed a number of challenges for finding practical collisions and (second) preimages on reduced round versions of KECCAK. Inspired by these challenges, we decided to study in detail the hash function setting, trying to find practical results. We present in this paper a distinguisher on the recommended hash functions $\lfloor \text{KECCAK}[1088,512] \rfloor_{256}$ and $\lfloor \text{KECCAK}[1152,448] \rfloor_{224}$ when reduced to 4 rounds, a second preimage on two rounds, a collision on 2 rounds and a near collision on 3 rounds. These are the first practical results of cryptanalysis of the KECCAK hash function setting where all the parameters but the number of rounds remain unchanged. Note that the challenges proposed by the KECCAK designers have smaller hash values (80 bits for a preimage, 160 bit for a collision) and a smaller capacity, $c = 160$, and are thus easier instances of the preimage and collision problem.

Table 1. Best known cryptanalysis results on the KECCAK hash function. We omit the analysis on the building blocks and detail all the results on the hash function setting.

Rounds	Version	Time	Memory	Generic	Type	Reference
6/7/8	512	$2^{506}/2^{507}/2^{511.5}$	–	2^{512}	Second Preimage Attack	[1]
4	256/224	2^{25}	–	2^{36}	Hash Function Distinguisher	Section 3
3	256/224	2^{25}	–	2^{64}	Hash Function Near-Collision	Section 4
2	256/224	2^{33}	–	2^{128}	Hash Function Collision	Section 5
2	256/224	2^{33}	2^{29}	2^{256}	Hash Function (Second) Preimage	Section 6

In [11], practical attacks on the compression functions of other hash functions were presented, but KECCAK was not one of them. Our analysis methods are based on different techniques, and propose a deep study of reduced-round KECCAK and its resistance to attacks on the hash function scenario, which are stronger results than compression function ones. For the sake of simplicity, in this paper we present the results on the recommended hash function with 256 bits of output, to which we refer at as KECCAK during the analysis. The equivalent results for the 224 bit version are similar to them.

The paper is organized as follows: In Section 2, a description of KECCAK and the notations that we use through the paper are given. Section 3 describes a differential distinguisher on 4 rounds of the recommended hash function KECCAK-256, that is extended in Section 4 to a near-collision attack on 3 rounds. Section 5 presents a hash function collision on 2 rounds and Section 6 describes how to build a hash function (second) preimage for two rounds.

2   KECCAK Description and Notations

KECCAK is a family of sponge hash functions [4]. A sponge hash function absorbs a message block of r bits into its internal state and subsequently applies an internal permutation to the state. This step is repeated until the all blocks of the message to hash have been treated. Next, in the squeezing phase, r bits are generated from the state before each new permutation application, until the number of the wanted output bits has been generated. In the following we will describe the recommended KECCAK versions for the SHA-3 competition, which are the ones that we have considered in our analysis. All the versions use the same internal permutation: KECCAK$-f[1600]$.

The full KECCAK$-f[1600]$ state is composed of 1600 bits, organized in 64 slices of 5×5 bits. The position of a bit in a slice can be given either by its x and y value or by its bit-number. The two notations are given in Table 2. The z coordinate gives the number of the slice $0 \le z \le 63$. Most of the steps in the round function of KECCAK are invariant to a translation in z direction. The only part non-invariant is the round constant addition ι.

The bits in the state can be also numbered from 0 to 1599. The conversion from x, y, z coordinate to global state bit position is done as follows:

$$\text{global pos} = 64(5y + x) + z.$$

Table 2. Bit notation in a slice

	x = 3	x = 4	x = 0	x = 1	x = 2
y = 2	bit 1	bit 2	bit 3	bit 4	bit 5
y = 1	bit 6	bit 7	bit 8	bit 9	bit 10
y = 0	bit 11	bit 12	bit 13	bit 14	bit 15
y = 4	bit 16	bit 17	bit 18	bit 19	bit 20
y = 3	bit 21	bit 22	bit 23	bit 24	bit 25

The inner permutation of the full Keccak hash function consists of 24 iterations of the round function. The round function itself is composed of five steps:

1. θ: Xor to each bit the XOR of two *columns* (column = same x value, y from 0 to 4). The first column is in the same slice as the bit and the second column is in the slice before the bit.
2. ρ: Translate a bit in z direction.
3. π: Permute the bits within a slice.
4. χ: Apply a 5×5 S-box on one row (row = same y value, x from 0 to 4).
5. ι: Addition of round constant.

Each of the versions (224, 256, 384 and 512 output bits) has a different block message size r. The capacity in a sponge construction is the size of the internal state minus the size of a message block. Consequently, they all have a different capacity c:

- For an output of 224 bits, $r = 1152$ and $c = 448$.
- For an output of 256 bits, $r = 1088$ and $c = 512$.
- For an output of 384 bits, $r = 832$ and $c = 768$.
- For an output of 512 bits, $r = 576$ and $c = 1024$.

For more details we refer to [3]. As previously said, in this paper we analyze the 256 bit version, that we will denote simply by Keccak in the following. All of our results can be directly extended to the 224 bit-output version.

3 Differential Distinguisher

In this section we first present an efficient way of searching low weight differential paths. This method was used to find the differential paths that we apply for the distinguisher on 4 rounds, the collision on 2 rounds and the near collision on 3 rounds. For the distinguisher we use in addition the concept of free bits as was used in [7].

3.1 Searching Differential Paths

As the authors define in [2], a state-difference is a *kernel* if it is invariant to the function θ, e.g. in each column we have a difference in zero or in an even number of bits. If we have a column where we have a difference in an odd number of bits, θ will spread this difference to 10 bits. Thus, for a low weight differential path we would like the state-differences to stay a kernel as long as possible. The designers of KECCAK show in [2] that it is not possible to construct low weight differentials that are a kernel for three states in a row, however two states in a row is possible, though they are not given in the documentations. We will denote the two kernels in a row a *double kernel*.

For our search we use the special property of χ that every 1-bit difference in a row constructed before χ will produce the same 1-bit difference after χ with probability 2^{-2}. Thus such a 1-bit difference will be invariant to the only non-linear part with probability 2^{-2}. If in addition we have a kernel, *i.e.* the difference is invariant to θ, we can concentrate on the functions ρ and π to find a double kernel.

For finding a double kernel we use the following procedure. At first, we fix in how many slices we want a difference in the first state. An example for a kernel in 3 slices is given in Figure 1. We start by choosing one bit in slice $z = 0$. The following algorithm will be repeated for all bits in slice $z = 0$. From our chosen bit we compute the position after one application of ρ and π. For this new bit position we check all bits in the same column and compute its position back by applying π^{-1} and ρ^{-1}. We once again check all possible bits in the same column and compute their position after applying ρ and π. We continue this procedure until we have touched the wanted number of slices. We will find a double kernel if after the last step we are again at the original slice at the right column.

This basic method allows us to find all double kernels which have k active slices in each of the two kernels with a complexity of $25 * 4^{2k-1}$. Every solution will be found $2 * k$ times, since every point of the first kernel can be a starting point.

By this method we can find very fast all possible differential paths that are a kernel for two states in a row and have low hamming weight. We can use this method for example to find a differential path over 4 rounds which has a probability of 2^{-142} and a double-kernel on 3 slices in the first two non-linear layers, and a differential path over 5 rounds with a probability of 2^{-510} by computing one step back and two steps forwards from a double-kernel on 6 slices. However, the objective of our paper is not to find the best differential path. We used the method to find suitable differential paths for our analysis.

3.2 Conditional Differentials and Free Bits

In [7], the concept of conditional differentials applied to NLFSR-based systems is introduced. The main idea is to consider a differential path with a good probability and try to control the first rounds of the path by imposing some conditions on the values of the internal state, so that the path with these conditions is verified with probability 1. For detecting a bias after the biggest number of rounds

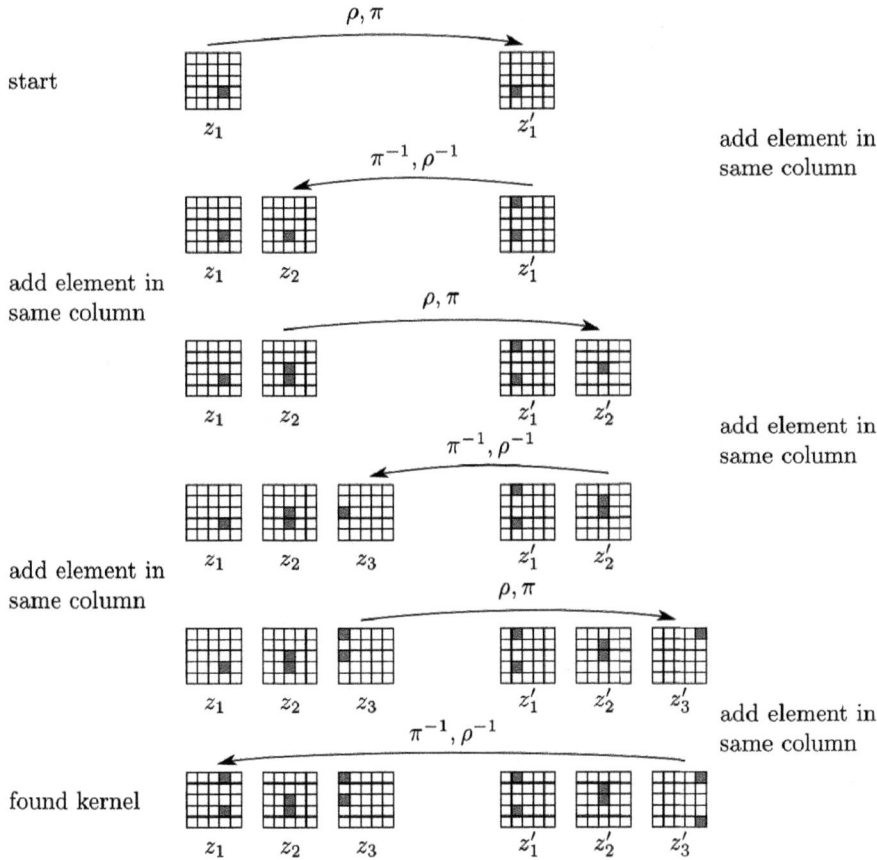

Fig. 1. Successful search of a double kernel on 3 slices

possible, the concept of free bits is introduced. In [10,5], some related work can be found.

The free bits [7] are the input bits such that, once we have found a pair of inputs that satisfies the differential path for the first rounds and so the corresponding needed conditions, can take any values without contradicting the conditions needed to satisfy the differential path. This means that once we have found a pair of inputs that follows the differential characteristic, we can change these bits and the new pairs will still follow the characteristic for these these first rounds.

3.3   Best Differential Paths

We have tested all low weight differential paths starting with a double kernel and checked which of them have an input difference that fits into the message part of the hash function. This is needed in order to perform a distinguisher in the hash function, as the difference will be introduced by the last round message block. Each path has a probability of being verified, i.e. the probability a random pair following the characteristic, that is given by the χ transformations behaving as needed. The highest probability of 2^{-24} was achieved by two paths, both having characteristic of 6-6-6 active S-boxes. In both cases, the difference of the first two rounds are a kernel and the first difference fits into a 1088 bit message. We chose the one which had less active bits one round later. This path stays valid for any translation in the z direction.

The best path: Let us define the following differences:

$$\Delta_1 : x = 0, y = 0, bit = 13, z = 0$$
$$x = 0, y = 1, bit = 8, \quad z = 0$$
$$x = 2, y = 1, bit = 10, z = 30$$
$$x = 2, y = 2, bit = 5, \quad z = 30$$
$$x = 1, y = 0, bit = 14, z = 63$$
$$x = 1, y = 2, bit = 4, \quad z = 63$$

$$\Delta_2 : x = 0, y = 0, bit = 13, z = 0$$
$$x = 0, y = 2, bit = 3, \quad z = 0$$
$$x = 2, y = 0, bit = 15, z = 9$$
$$x = 2, y = 3, bit = 25, z = 9$$
$$x = 1, y = 2, bit = 4, \quad z = 36$$
$$x = 1, y = 3, bit = 24, z = 36$$

$$\Delta_3 : x = 0, y = 0, bit = 13, z = 0$$
$$x = 2, y = 1, bit = 10, z = 3$$
$$x = 0, y = 4, bit = 18, z = 7$$
$$x = 3, y = 1, bit = 6, \quad z = 17$$
$$x = 3, y = 3, bit = 21, z = 24$$
$$x = 2, y = 3, bit = 25, z = 46$$

Then the path is the following (where we ignore ι since it is not important for the differences):

$$\Delta_1 \xoverbrace{\xrightarrow{\theta,\rho,\pi,} \Delta_2 \xrightarrow{\chi}}^{\text{round}} \Delta_2 \xoverbrace{\xrightarrow{\theta,\rho,\pi,} \Delta_3 \xrightarrow{\chi}}^{\text{round}} \Delta_3$$

3.4 Distinguisher on 4 Rounds of the Hash Function

Let $\mathcal{H}^{(4)}$ be the KECCAK-256 hash function reduced to 4 rounds and f_R one round the KECCAK$-f[1600]$ function. For a partial message M, we define by X_M the internal state of KECCAK after absorbing M.

In a first off-line step we find a padded message $M\|m$ such that $(X_M \oplus m, X_M \oplus m \oplus \Delta_1)$ will follow the differential path described in the previous section, *i.e.*

$$f_R^2(X_M \oplus m) \oplus f_R^2(X_M \oplus m \oplus \Delta_1) = \Delta_3 .$$

Such a message can be found by trying 2^{24} random messages. We recall that m and $m \oplus \Delta_1$ are the last message blocks including the correct padding.

Next we check how many *free bits* we have for our differential path within the range of the r message bits. We recall that by a free bit we mean a bit that we can change in m, so that the differential path is still verified. We find the following results:

- For the differential path described in Section 3.3, there are **81 free bits** within the $r = 1088$ bits of the message block. They are listed in Appendix A.1.

We can now define the vectors pace $A \subset \mathbb{F}_2^r$ containing all binary vectors of size r which are zero at all non-free bit positions, i.e. A has dimension 81. This means, for any $(X_M \oplus m, X_M \oplus m \oplus \Delta_1)$ which follows the differential path and any difference $\alpha \in A$, the pair of states $(X_M \oplus m \oplus \alpha, X_M \oplus m \oplus \Delta_1 \oplus \alpha)$ will also follow the differential path.

For our distinguisher we have to define what we mean by a bias. Let $\{0^r\}$ be the set containing the all-zero vector of size r bits.

Definition 1. *Let $\{\alpha_1, \ldots, \alpha_N\}$ be a set of N distinct differences, $\alpha_i \in A\backslash\{0^r\}$, and Δ a single difference. We denote by M the initial message with a length of some message blocks, and m be the final message part, such that m including the final padding fits into one message block. The concatenation of M and m is denoted as $M\|m$. Let $\mathcal{H}(M\|m)_i$ define the ith bit of the hash of $M\|m$ computed using the function $\mathcal{H}$. Then we define the bias of the ith bit as*

$$\epsilon_i^{\mathcal{H}} = \frac{\#\{1 \leq j \leq N : \mathcal{H}\Big(M\|(m \oplus \alpha_j)\Big)_i \oplus \mathcal{H}\Big(M\|(m \oplus \alpha_j \oplus \Delta)\Big)_i = 1\}}{N} - \frac{1}{2}$$

We will use the following property for our distinguisher:

Distinguishing property: For any M such that $(X_M \oplus m, X_M \oplus m \oplus \Delta_1)$ follows the differential path described in Section 3.3, any set of distinct differences $\{\alpha_1, \ldots, \alpha_N\}$, $\alpha_j \in A \backslash \{0^r\}$, $\Delta = \Delta_1$ and $\mathcal{H} = \mathcal{H}^{(4)}$ being the 4 round version of the KECCAK hash function, there are **18 positions i in the hash** where the absolute value of the bias is $|\epsilon_i^{\mathcal{H}^{(4)}}| = 2^{-1}$ for $N \geq 1$. The exact positions and their bias is listed in Appendix A.2

3.5   Implementation of the Distinguisher

In a first off-line step we search a message $M\|m$ such that $(X_M \oplus m, X_M \oplus m \oplus \Delta_1)$ follows the differential path. We have to try 2^{24} messages and for each of them compute first X_M by absorbing the message blocks in M and then two rounds of the round-function on $X_M \oplus m$ and on $X_M \oplus m \oplus \Delta_1$, to check if the differential path has been verified. Thus this step costs 2^{25} in time.

The next step is the online step, where we will determine if $\mathcal{H}$ is a random function or the 4 round version of KECCAK. We choose randomly a set of distinct differences $\{\alpha_1, \ldots, \alpha_N\}$, $\alpha_i \in A \setminus \{0^r\}$, and test the bias at the 18 predefined positions for Δ_1. In the case of KECCAK the bias will be correct for any $N \geq 1$, in the case of a random oracle this will happen with a probability of 2^{-18N}. Thus we can distinguish the 4-round version of KECCAK with an off-line precomputation complexity of 2^{25} and an on-line complexity of $2N$ where $N \geq 1$.

Remark 1. The previous test has a complexity of $2^{25} + 2N$ and searches in a precomputation phase for a suitable message M. Without this precomputation phase, we can still have a distinguisher with a cost of $2^{25}N$ where we need $N \geq 2$. For this we define a test $\mathcal{T}$, in the following way:

- A message M passes the test $\mathcal{T}$ if for Δ_1 and a set of distinct differences $\{\alpha_1, \ldots, \alpha_N\}$, $\alpha_i \in A \setminus \{0^r\}$, its bias has the correct value in the 18 positions defined in the previous section.

In the random case, a message M passes the test with probability 2^{-18N}, in the case of a 4-round KECCAK and a message where X_M follows the differential path, this happens with probability 1. We will find such an M with probability 2^{-24}, thus for the 4-round KECCAK the total probability of finding an M passing the test is $2^{-24} + 2^{-18N}$. For $N \geq 2$ the first term is dominating and by testing 2^{24} messages we are able to distinguish the 4-round KECCAK from a random oracle. The total time complexity is $2^{24} * 2N$.

4   Near-Collisions for 3 Rounds on the 256-bit Hash Function

In this section we show how to build near-collisions on the 3-round hash function by using the previous path. As we said before, for verifying the first two rounds and having 6 bit-differences after them, there are 24 conditions that need to be verified. This will happen with a probability of 2^{-24}. Thus, when inserting 2^{24} well padded message blocks from a fixed and chosen chaining value we can assume to find one that follows the differential path for two rounds.

We now study what happens one round later. For this we apply the linear part, θ, ρ and π, on the difference Δ_3. We call this new difference the state S.

The output of the hash function is given by the lanes of 64 bits at positions $y = 0$ and $x = 0, 1, 2, 3$. This means that if we have a difference in the hash value, we must have at least one difference with $y = 0$ in the corresponding slice before the final χ transformation. We have checked in state S which slices have differences in the middle lines, so in $y = 0$. This happens for 13 slices, and each of them contains just one difference in $y = 0$. Amongst them, four slices contain a difference in $x = 4$, which does not belong to the hash output. If such a difference passes χ in such a way that the 1-bit output difference equals the 1-bit input difference, we won't have a difference in the digest. If not, the 1-bit input difference might generate up to two bits of differences in the hash. The remaining $13 - 4 = 9$ slices have differences that will produce, each, at least a 1-bit difference in the hash value, so it will always have at least 9 differences after two rounds. Besides that, 6 amongst these differences can generate one more bit of difference, and 3 can generate up to two more difference bits, depending on χ.

At this stage, we can do two things:

1. Not increasing the complexity, and finding near-collisions for the bits that we know for sure won't have a difference. This number of bits is $256 - 4 * 2 - 6 * 2 - 3 * 3 = 227$. In this case, we can build near collisions of 227 bits with a complexity of 2^{24}, while the generic complexity would be $\left(\frac{2^{256}}{\binom{256}{227}} \right)^{1/2} = 2^{64}$. This has been verified experimentally, and an example is given in Appendix B.

2. Try to control the additional conditions and just have the inevitable 9 differences in the hash value. As we saw before, this means that we have to control $29 - 9 = 20$ additional bit conditions. These are the conditions given by χ so that it does not spread the already existing differences. The total complexity for having a near-collision on $256 - 9 = 247$ bits is $2^{24+20} = 2^{44}$, while the generic complexity is $\left(\frac{2^{256}}{\binom{256}{247}} \right)^{1/2} = 2^{101}$.

5 Hash Function Collisions on 2 Rounds

To find a collision on the hash function by means of a differential path we need to find a path that fits into the message and has no difference in the hash. This is not possible by a double kernel on three slices, however we found a suitable path considering double kernels on four slices.

$$
\begin{aligned}
\Delta_1 : x = 1, \, y = 2, \, bit = 4, \quad z = 0 \\
x = 1, \, y = 3, \, bit = 24, \, z = 0 \\
x = 0, \, y = 2, \, bit = 3, \quad z = 4 \\
x = 0, \, y = 3, \, bit = 23, \, z = 4 \\
x = 4, \, y = 0, \, bit = 12, \, z = 35 \\
x = 4, \, y = 2, \, bit = 2, \quad z = 35 \\
x = 1, \, y = 0, \, bit = 14, \, z = 61 \\
x = 1, \, y = 2, \, bit = 4, \quad z = 61
\end{aligned}
$$

$$\Delta_2 : x = 2, y = 1, bit = 10, z = 7$$
$$x = 2, y = 3, bit = 25, z = 7$$
$$x = 2, y = 3, bit = 25, z = 10$$
$$x = 2, y = 4, bit = 20, z = 10$$
$$x = 3, y = 1, bit = 6, \quad z = 45$$
$$x = 3, y = 4, bit = 16, z = 45$$
$$x = 0, y = 2, bit = 3, \quad z = 62$$
$$x = 0, y = 3, bit = 23, z = 62$$

$$\Delta_3 : x = 2, y = 1, bit = 10, z = 1$$
$$x = 4, y = 1, bit = 7, \quad z = 7$$
$$x = 1, y = 2, bit = 4, \quad z = 13$$
$$x = 3, y = 3, bit = 21, z = 22$$
$$x = 3, y = 3, bit = 21, z = 25$$
$$x = 1, y = 4, bit = 19, z = 36$$
$$x = 4, y = 3, bit = 22, z = 37$$
$$x = 3, y = 4, bit = 16, z = 39$$

We will use the following path:

$$\Delta_1 \xrightarrow{\overbrace{\theta,\rho,\pi,}^{\text{round}}} \Delta_2 \xrightarrow{\chi} \Delta_2 \xrightarrow{\overbrace{\theta,\rho,\pi,}^{\text{round}}} \Delta_3 \xrightarrow{\chi} \Delta_3$$

The differences Δ_2 and Δ_3 have each eight rows with a 1-bit difference in the input and in the output of χ. This lead to a total probability of 2^{-32} of following the differential characteristic. The difference in Δ_1 lies in the message block and difference in Δ_3 lies outside of the hash value. If we try random messages pairs where we introduce Δ_1 in the last message block, we have a probability of 2^{-32} that the last two rounds follows our differential path, which ensures that we have a collision in the final hash. Therefore, the complexity of this 2-round collision is 2^{33}. In practice, we find a collision much faster, after around 2^{13} steps. We give an example of such a collision in Appendix C.

6   Practical (Second) Preimages on 2 Rounds of the 256-bit Hash Function

In this section we present a (second) preimage attack for a reduced version with two rounds of KECCAK . The preimage works with a complexity of about 2^{33} in time and 2^{29} in memory. It also applies to several of the challenges with other parameters, but we present here the detailed case of 2 rounds of the recommended version for SHA-3. An example for a preimage can be found in Appendix D.

6.1   Main Scheme

Figure 2 gives a representation of the (second) preimage attack on 2 rounds. In it, we can see the slices at different states. In each slice, the square represents a 64 bit lane. The white lanes are known and the colored ones are not. For the sake of simplicity, we will omit the ι transformation in the explanation, as it does not affect the procedure of the attack, but it must be taken into account when implementing the attack.

We are given a hash value, which is 4 out of the 5 white lanes in the most right slice #4, in Fig. 2, that represents the final state after the permutation. The fifth lane is not known but we can choose a random value for it and fix it. What we want to find now is, given a chaining value, for example the initial one, a message block that produces the values of these five lanes, and so the initial given hash value corresponding to 4 out of these 5 lanes.

In Fig. 2, the gray lanes show into which lanes of the chaining value the message is xored. The lanes marked with a zero are the lanes of the message that we are going to fix to zero. The lanes marked with $(a_0, a_1, b_0, b_1, c_0, c_1, d_0, d_1, e_0, e_1)$ are the parts of the message that we do not fix until the end of the attack. The only condition we ask from them is that: $a_0 = a_1; \ldots; e_0 = e_1$. In a generic way we will say $x_0 = x_1$. These conditions are asked for so that the first operation θ will not change the unknown lanes. From the initial state #1 we can then compute the known lanes in #2 after θ, ρ and π, as well as the positions of the unknown lanes. Figure 3 shows the movement of the bits in detail. Imposing the previous conditions we still have $5 * 64$ degrees of freedom for the message, which is the same as the number of bits in which we want to collide in the end. We can then expect to find one solution.

In the backward direction, we can invert from #4 the known white line of five lanes in the final state with χ^{-1}. Then, we can apply the inverse of π and of ρ and obtain the values and positions of the 5 known lanes in #3.

Then, the issue is to find the values of the ten 64-bit words $(a_0, a_1, b_0, b_1, c_0, c_1, d_0, d_1, e_0, e_1)$ in #2 that make possible the transition by the operations χ and θ, from the state #2, that we have obtained computing forward, to the state

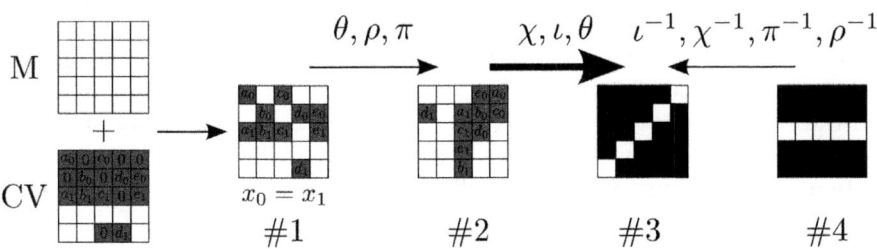

Fig. 2. Diagram of the 2-round preimage attack. Each square represents a 64 bit lane. Each white lane is a lane known and fixed, each colored one, a not-yet-fixed lane.

$$\theta, \rho, \pi$$

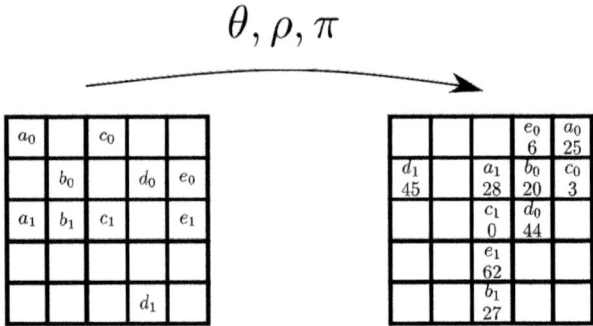

Fig. 3. Shows how the bits $a_0, \ldots, e_1$ get moved by θ, ρ, π. The number k under x_i means that x_i at slice z on the left side gets moved to slice $z + k$ (mod 64) after the transformation.

#3, that we have obtained computing backward. For this we are going to start by finding the bits that verify the relations for a few slices. The idea is, for example, to consider first groups of three slices where we guess all the involved bits of $a_0, \ldots, e_1$, and next we can do a sieving by just keeping the guesses ones that produce by χ and θ the values of the $5 \times 2 = 10$ known bits from #3 in the middle and last slices of the group of three slices. This is possible as for computing the output of θ in a specific slice, we need to know this same slice and the previous one in the input state. We say that one bit (from x_0 for example) is *repeated* in a group of slices when the bit from x_1, corresponding to the same slice in the initial difference, also appears in this group of slices.

6.2   Finding Partial Solutions

Partial solutions for 3 slices: We start by finding partial solutions for groups of three slices as previously described. In three slices there are 3×10 bit-variables from $a_0, \ldots, e_1$. If we consider the conditions $x_0 = x_1$ there are two variables that are repeated, so we have to guess in total $30 - 2 = 28$ variables. The two repeated values come from the values of ρ for d_0 and d_1, as there is a translation of 1 in between them. Out of the three bits of d_0 intervening in three consecutive slices, 2 of them will be equal to 2 of the bits of d_1 intervening in these same slices. For each guess we check if the 10 bits already fixed computing backwards on the two last slices collide, which will happen with a probability of 2^{-10}. In total, for each group of three slices that we try to find partial solutions for, we obtain $2^{28-10} = 2^{18}$ solutions. We will repeat this for 16 consecutive groups of 3 slices, leaving 16 out. By using the methods described in Section 6.4, the time complexity of building the list is given by the size of the list. Thus this step has a time and memory complexity of $16 * 2^{18} = 2^{22}$.

Partial solutions for 6 slices: We are going to merge here each two consecutive groups of three slices. We have 2^{18*2} possibilities, but, because of the conditions

$x_0 = x_1$, a group of three slices has 7 repeated variables regarding a consecutive group. Also, when we merge two groups, we can check if we obtain the wanted values on five more bits from the output (from the first slice of the second group). This way we obtain $2^{18*2-7-5} = 2^{24}$ solutions for each one of the 8 groups of 6 slices. The bottleneck of this step is the number of solutions, 8×2^{24} in both time and memory, as the merge of two lists can be done using the instant matching algorithm described in [9] by using the methods in 6.4. This algorithm can also be applied in the next steps, so in all the cases, the bottleneck will be the number of solutions obtained. This step has a time and memory complexity of $8*2^{24} = 2^{27}$.

Partial solutions for 12 slices: The same way, we merge here each two consecutive groups of 6 slices for generating 4 groups of 12 slices. In this case the number of repeated bits in the merge is 16, so the total number of solutions that we will obtain is $2^{24*2-16-5} = 2^{27}$. This step has a time and memory complexity of $4*2^{27} = 2^{29}$.

Partial solutions for 24 slices: We merge here each two consecutive groups of 12 slices for generating 2 groups of 24 slices. In this case the number of repeated bits in the merge is 22, so the total number of solutions that we will obtain is $2^{27*2-22-5} = 2^{27}$. We have a time and memory complexity of $2*2^{27} = 2^{28}$.

Partial solutions for 48 slices: Finally, we merge the 2 groups of 24 slices, for obtaining one group of solutions for 48 consecutive slices. In this case, the number of repeated bits due to conditions is also 22, so we obtain 2^{27} solutions. In these 48 slices we have determined 480 bit-variables from $a_0, \ldots, e_1$, and there are a total of

$$2*16 + 7*8 + 16*4 + 22*2 + 22 = 218$$

repeated variables amongst them. This means that there are $480 - 218*2 = 44$ bit-variables not repeated, that are then repeated in the 16 slices that we have yet to treat. This step has a time and memory complexity of 2^{27}.

Partial solutions for 16 slices: For finding solutions for 16 slices we first find solutions for the 12 rightmost slices the same way as before, and obtaining 2^{27} partial solutions. Let us remark here that in 12 slices there are 120 bit-variables fixed and 38 out of them are repeated ones. This means that there are 44 variables not repeated that must have their corresponding bit-variable in the 48-slice group or in the remaining 4-slice group.

Next, we can obtain solutions for the 4 remaining and consecutive slices, where we have 40 bit-variables, and 5 of them are repeated. Additionally, we have to collide on 5×3 bits that we can compute of the output. This leaves us with $2^{40-5-15} = 2^{20}$ solutions. As there are 40 bit-variables and 5 are repeated, there are $40 - 5*2 = 30$ variables not repeated.

We know that in the 48-slice group, there are 44 bit-variables that must be on either the independent group of 12 or on the groups of 4 slices. This means that out of the $44 + 30 = 74$ not-repeated variables in these last two groups, 44 will correspond to the ones in the 48-slices group. Then, of the 30 remaining variables, half must be in the groups of 12, and 15 must be in the group of 4.

This can also be seen as the following system of equations, where A represents the number of common bits between the 48-slice group and the 4-slice group, B is the number of common bits between the 48-slice group and the 12-slice group, and C is the number of common bits between the 4-slice group and the 12-slice group. From the previous numbers we know that $A + B = 44$, $B + C = 44$ and $A + C = 30$. Then we have $A = 15$, $B = 29$ and $C = 15$. The number of additional conditions determined by the repeated bits that we have for merging the group of 4 and the group of 12 slices is then $C = 15$.

We can then merge the group of 12 slices and the one of 4 obtaining

$$2^{27+20-15-5} = 2^{27} \text{ solutions.}$$

This step has a time complexity of $2^{27}+2^{20}+2^{27} \approx 2^{28}$ and a memory complexity of 2^{27}.

6.3   Matching 48 Slices with 16 Slices

Now, we can match the just obtained group of 16 slices with the one of 48. As we said, they have 44 variables in common where they have to collide, and there are $2*5$ bits of the output that will also be determined. This leaves us with

$$2^{27}2^{27}2^{-44}2^{2*5} = 1 \text{ solution,}$$

as expected. This step has a time complexity defined by the size of one input list and is thus 2^{27}. During the attack we keep only the latest generated lists in the memory, thus the the memory complexity is bounded by 2^{29}. The time complexity is given by $2^{22} + 2^{27} + 2^{29} + 2^{28} + 2^{27} + 2^{28} + 2^{27} \approx 2^{31}$ Thus, we can obtain a (second) preimage with a complexity of 2^{31} in time and 2^{29} in memory.

6.4   Implementation Remarks

Two methods can help in an efficient implementation of the attack. Let us assume we want to merge the block from slice i to j with the block from slice $j + 1$ to k. We first precompute a list containing all solutions for merging slice j and slice $j + 1$. We have 10 bits in each of the two slices, 1 repeated bit and 5 conditions from the output, thus we have in total 2^{14} solutions that we sort by the 2^{10} values in slice j. The costs of building this list is negligible in comparison to the remaining time complexities. Next, for each solution in the first block (i to j) we compute the values of the bits that will repeat in the second block. We will sort the solution in this first block by the value of the slice in j and the values of the repeated bits. We do the same thing for the second block ($j + 1$ to k) and sort it by the value of slice $j + 1$ and the values of the repeated bits. Now we can easily merge the two lists using the precomputed list of matches from slices j to $j + 1$.

6.5 Dealing with the Padding

In KECCAK, the message is not padded with the length, but with a simple padding where to the last message block we append a 1, a number of 0's and another 1 so that it completes the final block. For us to obtain a valid message block that fits into the last block, we need to have a 1 in the last position and a zero in the previous one. We have then a probability of 2^{-2} of obtaining a valid message. We can repeat the previous procedure for 4 different chaining values, and expect one to give a valid padded last block. Thus we have a time complexity of 2^{33} of finding a (second) preimage with a correct padding.

7 Conclusion

In this paper, we have presented new results in several directions on the security of KECCAK: A distinguisher on 4 rounds, preimage and collision attacks on 2 rounds, and near collisions on 3 rounds. These results apply to the 256 and to the 224 bits versions. First, these results concern the reduced round hash function rather than building blocks like the internal permutation or the compression function as considered in previous work. Next all our results are practical and have been implemented. The only known previous external cryptanalysis on the reduced round hash function setting was a marginally better than generic theoretical preimage attack. The number of rounds reached is far from the total, and the results do not present a threat against the whole hash function, but we believe they will contribute to a better understanding of the security of the KECCAK hash function.

References

1. Bernstein, D.J.: Second preimages for 6 (7 (8??)) rounds of Keccak? NIST mailing list (2010),http://ehash.iaik.tugraz.at/uploads/6/65/NIST-mailing-list_Bernstein-Daemen.txt
2. Bertoni, G., Daemen, J., Peeters, M., Assche, G.V.: The Keccak reference. Submission to NIST (Round 3) (2011), http://keccak.noekeon.org/Keccak-reference-3.0.pdf
3. Bertoni, G., Daemen, J., Peeters, M., Assche, G.V.: The Keccak SHA-3 submission. Submission to NIST (Round 3) (2011), http://keccak.noekeon.org/Keccak-submission-3.pdf
4. Bertoni, G., Daemen, J., Peeters, M., Van Assche, G.: On the Indifferentiability of the Sponge Construction. In: Smart, N.P. (ed.) EUROCRYPT 2008. LNCS, vol. 4965, pp. 181–197. Springer, Heidelberg (2008)
5. Biham, E., Chen, R.: Near-Collisions of SHA-0. In: Franklin, M. (ed.) CRYPTO 2004. LNCS, vol. 3152, pp. 290–305. Springer, Heidelberg (2004)
6. Boura, C., Canteaut, A., De Cannière, C.: Higher-Order Differential Properties of KECCAK and *Luffa*. In: Joux, A. (ed.) FSE 2011. LNCS, vol. 6733, pp. 252–269. Springer, Heidelberg (2011)

7. Knellwolf, S., Meier, W., Naya-Plasencia, M.: Conditional Differential Cryptanalysis of NLFSR-Based Cryptosystems. In: Abe, M. (ed.) ASIACRYPT 2010. LNCS, vol. 6477, pp. 130–145. Springer, Heidelberg (2010)
8. Morawiecki, P., Srebrny, M.: A SAT-based preimage analysis of reduced KECCAK hash functions. Cryptology ePrint Archive, Report 2010/285 (2010), http://eprint.iacr.org/2010/285.pdf
9. Naya-Plasencia, M.: How to Improve Rebound Attacks. In: Rogaway, P. (ed.) CRYPTO 2011. LNCS, vol. 6841, pp. 188–205. Springer, Heidelberg (2011)
10. Rechberger, C., Rijmen, V.: On Authentication with HMAC and Non-Random Properties. In: Dietrich, S., Dhamija, R. (eds.) FC 2007 and USEC 2007. LNCS, vol. 4886, pp. 119–133. Springer, Heidelberg (2007)
11. Sönmez Turan, M., Uyan, E.: Near-Collisions for the Reduced Round Versions of Some Second Round SHA-3 Compression Functions Using Hill Climbing. In: Gong, G., Gupta, K.C. (eds.) INDOCRYPT 2010. LNCS, vol. 6498, pp. 131–143. Springer, Heidelberg (2010)
12. Wang, X., Yin, Y.L., Yu, H.: Finding Collisions in the Full SHA-1. In: Shoup, V. (ed.) CRYPTO 2005. LNCS, vol. 3621, pp. 17–36. Springer, Heidelberg (2005)
13. Wang, X., Yu, H.: How to Break MD5 and Other Hash Functions. In: Cramer, R. (ed.) EUROCRYPT 2005. LNCS, vol. 3494, pp. 19–35. Springer, Heidelberg (2005)

A   Information for the Distinguisher

A.1   Free Bits Which Are in the 1088 Bit Message

In Table 3 we listed all the free bits from Section 3.

A.2   Bits in the Hash Output with a Bias

In the following, we give the list of bits in the hash with a bias of absolute value 2^{-1}. The numbering corresponds to the bit position in the 256-bit hash:

- For all $i \in \{22, 119, 126, 128, 138, 169, 205\}$ we have $\epsilon_i^{\mathcal{H}^{(4)}} = 2^{-1}$.
- For all $i \in \{56, 63, 98, 127, 149, 161, 162, 176, 195, 232, 252\}$ we have $\epsilon_i^{\mathcal{H}^{(4)}} = -2^{-1}$.

B   Hash Function Near-Collision Example for 3 Rounds

In this section we give an example for a near collision after two rounds. The two input messages collide in 234 out of 256 bits.

- input 1:

    0x09c2d45d03bae701a767c1b756e7e594c38ad4c618efc11dc32289
    31bb698feb072e3f9a6e9e8b414942e18102755b2e2faf545ac717
    402e12ac5f93ce54484955a870311867e2095b981797d778ee2e7e
    e3fa8fcb24e650ada1c4a07344f79ab8672027c502b240dda77eb9
    39c89134e778718ab86e39f75524e8a200c025ac0bdce3b246ddc5

Table 3. Free bits

$(x = 4, y = 0, z = 0)$ $(x = 4, y = 2, z = 0)$ $(x = 4, y = 1, z = 4)$ $(x = 4, y = 2, z = 4)$
$(x = 2, y = 0, z = 15)$ $(x = 2, y = 1, z = 15)$ $(x = 2, y = 2, z = 15)$ $(x = 1, y = 0, z = 20)$
$(x = 1, y = 2, z = 20)$ $(x = 1, y = 3, z = 20)$ $(x = 1, y = 0, z = 23)$ $(x = 1, y = 1, z = 23)$
$(x = 1, y = 2, z = 23)$ $(x = 0, y = 0, z = 24)$ $(x = 4, y = 0, z = 24)$ $(x = 0, y = 1, z = 24)$
$(x = 4, y = 1, z = 24)$ $(x = 0, y = 2, z = 24)$ $(x = 4, y = 2, z = 24)$ $(x = 0, y = 3, z = 24)$
$(x = 0, y = 0, z = 27)$ $(x = 0, y = 1, z = 27)$ $(x = 0, y = 2, z = 27)$ $(x = 3, y = 0, z = 28)$
$(x = 3, y = 1, z = 28)$ $(x = 3, y = 2, z = 28)$ $(x = 2, y = 0, z = 30)$ $(x = 2, y = 1, z = 30)$
$(x = 2, y = 2, z = 30)$ $(x = 0, y = 1, z = 31)$ $(x = 0, y = 2, z = 31)$ $(x = 0, y = 3, z = 31)$
$(x = 4, y = 0, z = 34)$ $(x = 4, y = 1, z = 34)$ $(x = 3, y = 0, z = 35)$ $(x = 4, y = 0, z = 35)$
$(x = 3, y = 1, z = 35)$ $(x = 4, y = 1, z = 35)$ $(x = 3, y = 2, z = 35)$ $(x = 4, y = 2, z = 35)$
$(x = 2, y = 0, z = 36)$ $(x = 2, y = 1, z = 36)$ $(x = 2, y = 2, z = 36)$ $(x = 3, y = 0, z = 37)$
$(x = 3, y = 1, z = 37)$ $(x = 3, y = 2, z = 37)$ $(x = 1, y = 0, z = 39)$ $(x = 1, y = 1, z = 39)$
$(x = 2, y = 0, z = 42)$ $(x = 2, y = 1, z = 42)$ $(x = 3, y = 0, z = 43)$ $(x = 3, y = 1, z = 43)$
$(x = 1, y = 0, z = 44)$ $(x = 1, y = 1, z = 44)$ $(x = 1, y = 2, z = 44)$ $(x = 1, y = 3, z = 44)$
$(x = 0, y = 1, z = 47)$ $(x = 0, y = 2, z = 47)$ $(x = 0, y = 3, z = 47)$ $(x = 0, y = 0, z = 54)$
$(x = 2, y = 0, z = 54)$ $(x = 0, y = 1, z = 54)$ $(x = 2, y = 1, z = 54)$ $(x = 0, y = 2, z = 54)$
$(x = 2, y = 2, z = 54)$ $(x = 1, y = 0, z = 56)$ $(x = 1, y = 1, z = 56)$ $(x = 1, y = 2, z = 56)$
$(x = 1, y = 3, z = 56)$ $(x = 1, y = 0, z = 57)$ $(x = 1, y = 1, z = 57)$ $(x = 1, y = 2, z = 57)$
$(x = 1, y = 3, z = 57)$ $(x = 0, y = 0, z = 60)$ $(x = 0, y = 1, z = 60)$ $(x = 0, y = 2, z = 60)$
$(x = 0, y = 3, z = 60)$ $(x = 0, y = 0, z = 61)$ $(x = 0, y = 1, z = 61)$ $(x = 0, y = 2, z = 61)$
$(x = 0, y = 3, z = 61)$

– input 2:

0x09c2d45d03bae701a767c1b756e7e594c38ad4c618efc11dc32289
31bb698feb072e3fda6e9e8bc14942e18102755b2e2faf545ac717
402e12ac5f93ce54484955a870311867e2095b9817d7d778ee2e7e
e3fa8fcb24e650ada1c4a07344f69ab8672027c502b240dda77eb9
39c89134e778718ab86e39775524e8a200c025ac0bdce3b246ddc4

we get the following difference in the hash value:

– output difference:

0x0000200001008009200000008140801020093000090000080000810
0000000001

C   Hash Function Collision Example for 2 Rounds

For the two inputs:

– input 1:

0x22458a902041831f3f7ffe62c58b16d1a3468df8f1e2c524499327
22458b17dab468d1983061c2c2850a1543860d1a860d1b37112346
8ca850a041a040800151a3468c52a44890993366ccca94295267cf
9f3e448811220c18306093264c988a142953be7cf9f22e5dbb7792
244992a4499327b265ca951f3e7cf81a3468d053a74f9fe8d1a244

– input 2:

0x22458a902041821f3f7ffe62c58b06d1a34685f8f1e2c524499327
22458b17dab468d1983061c2e2850a1543860d1b860d1b37112346
9ca850a041a040800151a3468c52a44890993366ccca94295267cf
9f3e448811220c18306093264c988a14295bbe7cf9f22e5dbb7792
244992a4499327b265ca953f3e7cf81a3468d053a74f9fe8d1a244

we get the same output, namely:

– output:

0xfa5f041d6152cc9b8f0747aa9f66b1be5365164ff14578436665f4
e828f1ea76

D Preimage for 2 Rounds of the 256-bit Hash Function

We chose as target output the all zero hash. The following message block, which has the correct final padding, fed to the 2 rounds version of the 256-bit KECCAK hash function, outputs a hash value of all zeros.

– input inclusive the padding:

0x8ed557e2ad30f0ee00000000000000000000000000000000e9a08c
fd1399f7ad000000000000000000000000000000000cb1dc30d0ad7
92214a0bb03f6aedcb230000000000000000a4a82caa2dfa4d388e
d557e2ad30f0ee00000000000000004a0bb03f6aedcb23e9a08cfd
1399f7ada4a82caa2dfa4d380000000000000000cb1dc30d0ad79221

Boomerang Distinguisher for the SIMD-512 Compression Function

Florian Mendel and Tomislav Nad

Institute for Applied Information Processing and Communications (IAIK)
Graz University of Technology, Inffeldgasse 16a, A-8010 Graz, Austria
Tomislav.Nad@iaik.tugraz.at

Abstract. In this paper, we present a distinguisher for the permutation of SIMD-512 with complexity $2^{226.52}$. We extend the attack to a distinguisher for the compression function with complexity $2^{200.6}$. The attack is based on the application of the boomerang attack for hash functions. Starting from the middle of the compression function we use techniques from coding theory to search for two differential characteristics, one for the backward direction and one for the forward direction to construct a second-order differential. Both characteristics hold with high probability. The direct application of the second-order differential leads to a distinguisher for the permutation. Based on this differential we extend the attack to distinguisher for the compression function.

Keywords: SHA-3, SIMD, cryptanalysis, higher-order differentials, hash function, distinguisher.

1 Introduction

Recently, the NIST hash function competition [21] has started. In this public competition to find an alternative hash function to replace the SHA-1 and SHA-2 hash functions, many new designs have been proposed. In November 2008, round one has started and in total 51 out of 64 submissions have been accepted. In December 2009 the 14 round 2 candidates and in December 2010 the final five were announced. During the competition distinguishing attacks on hash functions and their building blocks are getting more attention. In such attacks an adversary utilizes specific properties of a hash function to define a distinguishing property such that one can distinguish the output of a hash function from a random function. Usually, the existence of such properties is not intended by the designers. However, as shown in [4] for wide-pipe designs the impact of distinguishers is limited.

In this paper, we present a distinguisher for the compression function of SIMD-512. SIMD, designed by Leurent *et al.* [13], was submitted to the NIST competition and was one of the second round candidates. It is an iterative hash function based on the Merkle-Damgård design principle [5,18]. It is a wide-pipe design [14] producing a hash value up to 512 bits, denoted by SIMD-n, where n is the output length. The design of the compression function is similar to the MD4

D.J. Bernstein and S. Chatterjee (Eds.): INDOCRYPT 2011, LNCS 7107, pp. 255–269, 2011.

family. For the remainder of this paper wherever we mention SIMD we refer to SIMD-512.

We will show how one can use the boomerang attack on a hash function to construct a distinguisher with high probability. The first result is a distinguishing attack for the full permutation of SIMD-512 with complexity $\approx 2^{226.52}$. Next we show how this distinguisher can be extended to the full compression function of SIMD-512. with complexity $\approx 2^{200.6}$. The strategy to construct such second order differentials is based on the recently proposed cryptanalysis of reduced SHA-2 [12] and Blake [3].

The structure of this paper is as follows. In Section 2, we recall the basic definitions needed for the attack and give an overview how higher-order differentials can be used to attack hash functions. A short description of SIMD is given in Section 3. Section 4 presents the application on the permutation of SIMD-512. In Section 5, we show how the attack can be extended to the compression function of SIMD-512. Finally, we discuss the results in Section 6.

1.1   Related Work

The amount of available cryptanalysis of SIMD is low compared to other candidates. Mendel and Nad presented the first attack on the full SIMD-512 compression function [15]. They used techniques from coding theory to find a differential characteristic that holds with probability 2^{-507}. Based on this characteristic and the differential multicollision distinguisher introduced by Biryukov et al. [1] they constructed a distinguishing attack for the SIMD-512 compression function. Using IV/message modification the attack complexity was reduced to 2^{427}. The differential path used some unwanted properties in the permutation of SIMD. Therefore, the designers tweaked the hash function by changing the permutations and round constants of SIMD to prevent the attack.

A round reduced version of tweaked SIMD was attacked by Nikolić et al. [8]. They presented distinguishers for the compression function of SIMD-512 reduced to 24 round with a linearized message expansion and SIMD-512 reduced to 12 rounds with unmodified message expansion. Both are based on rotational properties of the compression function. The success probabilities for the distinguishers are 2^{-497} and 2^{-236}, respectively.

Later Yu and Wang [27] presented a free-start near-collision attack for SIMD-256 reduced to 20 rounds and for SIMD-512 reduced to 24 rounds. The attack complexities are 2^{107} and 2^{208}, respectively. Furthermore, they showed a distinguisher for the full compression function with complexity 2^{398}.

Finally, the designers [4] published a free-start distinguisher for the compression function exploiting the existence of symmetric states. Furthermore, they showed that distinguishers without differences in the message have only a minimal impact on the security of the hash function.

Higher-order differentials have been introduced by Lai in [11] and first applied to block ciphers by Knudsen in [10]. The application to stream ciphers was proposed by Dinur and Shamir in [6] and Vielhaber in [23].

Table 1. Notation

notation	description
$\neg X$	inversion of X
$X \oplus Y$	bit-wise XOR of X and Y
$X + Y$	modular addition of X and Y
$X \lll n$	bit-rotation of X by n positions to the left
$X \ggg n$	bit-rotation of X by n positions to the right
$X \ll n$	bit-shift of X by n positions to the left
$X \gg n$	bit-shift of X by n positions to the right

Recently, Lamberger and Mendel [12] showed how higher-order differentials can be used to attack SHA-256 and presented a distinguisher for 46 (out of 64) with practical complexity. The attack was recently extended to 47 steps in [2]. The attack stands between the *boomerang attack* and the *inside-out* attack which were both introduced by Wagner in the cryptanalysis of block ciphers [24]. A previous application of the boomerang attack to hash functions is due to Joux and Peyrin [9], who used the boomerang attack as a neutral bits tool to speed-up existing collision attacks. Another similar attack strategy for hash functions is the *rebound attack* introduced by Mendel *et al.* [17] and its extensions [7,16]. Furthermore, Biryukov et al. [3] presented a boomerang attack on the SHA-3 finalist Blake resulting in a distinguisher for 7 rounds of the Blake-32 compression function with a complexity of 2^{232}.

Notation. For the remainder of this paper we use the notation presented in Table 1.

2 Higher-Order Differentials and Hash Function

In order to find a distinguishing property we construct a second order differential collision for the compression function. In this section we recall the basic definitions and give a high level description of the attack strategy.

While a standard differential attack exploits the propagation of the difference between a pair of inputs to the corresponding outputs, a higher-order differential attack exploits the propagation of the difference between differences. Higher-order differential cryptanalysis was introduced by Lai in [11] and subsequently applied to block ciphers by Knudsen in [10]. We recall the basic definitions that we will need in the subsequent sections.

Definition 1. *Let $(S, +)$ and $(T, +)$ be Abelian groups. For a function $F : S \to T$, the derivative at a point $a \in S$ is defined as*

$$\Delta_a F(x) = F(x + a) - F(x). \tag{1}$$

The i-th derivative of F at $(a_1, a_2, \ldots, a_i)$ is then recursively defined as

$$\Delta^{(i)}_{a_1,\ldots,a_i} F(x) = \Delta_{a_i}(\Delta^{(i-1)}_{a_1,\ldots,a_{i-1}} F(x)). \tag{2}$$

When applying differential cryptanalysis to a hash function, a collision for the hash function corresponds to a pair of inputs with output difference zero. Similarly, when using higher-order differentials we define a higher-order differential collision for a function F as follows.

Definition 2. *An i-th order differential collision for a function F is an i-tuple $(a_1, a_2, \ldots, a_i)$ together with a value x_0 such that*

$$\Delta^{(i)}_{a_1,\ldots,a_i} F(x_0) = 0. \tag{3}$$

Note that the common definition of a collision for hash functions corresponds to a higher-order differential collision of order $i = 1$.

From (3) we see that we can freely choose $i+1$ of the input parameters, i.e. x_0 and $a_1, \ldots, a_i$, which then fix the remaining input. Hence, the expected number of solutions to (3) is one after choosing $2^{n/(i+1)}$ values for the inputs and the query complexity is:

$$\approx 2^{n/(i+1)} \tag{4}$$

In the following, we will only consider the case $i = 2$ for which the query complexity of the attack is $2^{n/3}$.

In order to construct a second-order differential collision for the function F, we use a strategy recently proposed in cryptanalysis of reduced SHA-2 in [12]. The idea of the attack is quite simple. Assume we are given two differentials for F_0 and F_1 with $F = F_1 \circ F_0$, where one holds in the forward direction and one in the backward direction. To be more precise, we have

$$F_0^{-1}(y + \beta) - F_0^{-1}(y) = \alpha$$

and

$$F_1(y + \gamma) - F_1(y) = \delta$$

where the differential in F_0^{-1} holds with probability p_0 and in F_1 holds with probability p_1. Using these two differentials, we can now construct a second order differential collision for F. This can be summarized as follows (see also Figure 1).

1. Choose a random value for X and compute $X^* = X + \beta$, $Y = X + \gamma$, and $Y^* = X^* + \gamma$.
2. Compute backward from X, X^*, Y, Y^* using F_0^{-1} to obtain P, P^*, Q, Q^*.
3. Compute forward from X, X^*, Y, Y^* using F_1 to obtain R, R^*, S, S^*.
4. Check if $P^* - P = Q^* - Q$ and $S - R = S^* - R^*$ is fulfilled.

Since

$$P^* - P = Q^* - Q = \alpha, \quad \text{resp. } S - R = S^* - R^* = \delta, \tag{5}$$

will hold with probability at least p_0^2 in the backward direction, resp. p_1^2 in the forward direction and assuming that the differentials are independent the attack succeeds with a probability of $p_0^2 \cdot p_1^2$. Hence, the expected number of solutions to (5) is 1, if we repeat the attack about $1/(p_0^2 \cdot p_1^2)$ times.

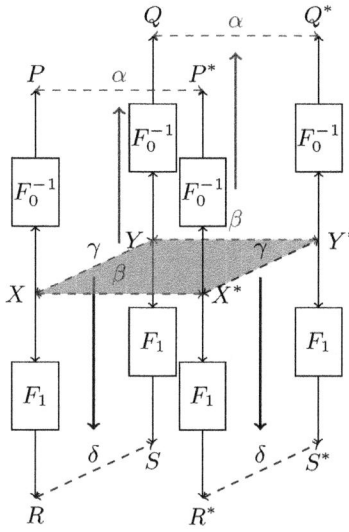

Fig. 1. Schematic view of the attack

3   Description of SIMD

SIMD is an iterative hash function that follows the Merkle-Damgård design. The main component of a Merkle-Damgård hash function is the compression function. In the case of SIMD-512 to compute the hash of a message M, it is first divided into k chunks of 1024 bits. By the use of a message expansion one block is expanded to 8192 bits. Then the compression function is used to compress the message chunks and the internal state. The padding rule to fill the last blocks is known as the Merkle-Damgård strengthening. The initial value of the internal state is called IV and is fixed in the specification of the hash function. The output of the hash function is given by computing a finalization function on the last internal state, which is a truncation for SIMD. The internal state of SIMD contains 32 32-bit words and is therefore twice as large as the output. SIMD consist of 4 rounds where each round consist of 8 steps. The feed-forward consists of four additional steps with the chaining value as message input. Since we inject differences only in the state variables and not in the message, our attack is independent from the message expansion and works for any given message. Therefore, we omit the description of the message expansion. For a detailed description of the hash function we refer to [13].

3.1   SIMD Step Function

The core part of SIMD is the step function of the state update. Figure 2 illustrates the step function at step t. The state update consists of eight step functions in parallel. To make the step function dependent from each other,

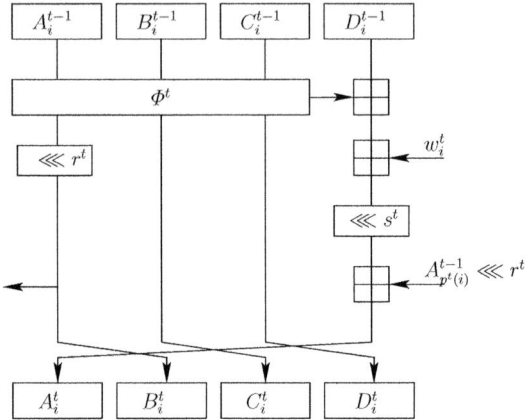

Fig. 2. Update function of SIMD at step t. $i = 0, \cdots, 7$

$(A_{p^t(i)}^{t-1} \lll r^t)$ is included in a modular addition, where $p^t(i)$ is a permutation, which is different for each step.

Equation (6) is the formal definition of the step function, where '+' denotes the addition modulo 2^{32}.

$$
\begin{aligned}
A_i^t &= (D_i^{t-1} + w_i^t + \Phi(A_i^{t-1}.B_i^{t-1}, C_i^{t-1})) \lll s^t + (A_{p^t(i)}^{t-1} \lll r^t)\\
B_i^t &= A_i^{t-1} \lll r^t\\
C_i^t &= B_i^{t-1}\\
D_i^t &= C_i^{t-1}
\end{aligned}
\tag{6}
$$

The permutation p for SIMD-512 is given by:

$$
\begin{aligned}
p^0(x) &= x \oplus 1\\
p^1(x) &= x \oplus 6\\
p^2(x) &= x \oplus 2\\
p^3(x) &= x \oplus 3\\
p^4(x) &= x \oplus 5\\
p^5(x) &= x \oplus 7\\
p^6(x) &= x \oplus 4
\end{aligned}
$$

The permutation used at step t is $p^{t \bmod 7}$. As mentioned before, the 32 steps of SIMD are divided into 4 rounds, each consisting of 8 steps. The boolean function Φ and the rotation constants (s and r) for a round are given in Table 2. The Boolean functions IF and MAJ are defined as follows:

$$
\begin{aligned}
f_{IF}(x, y, z) &= (x \wedge y)|(\neg x \wedge z)\\
f_{IF}(x, y, z) &= (x \wedge y)|(x \wedge z)|(y \wedge z).
\end{aligned}
$$

Table 2. Φ and rotation constants for a round

step	Φ	r	s
0	IF	π_0	π_1
1	IF	π_1	π_2
2	IF	π_2	π_3
3	IF	π_3	π_0
4	MAJ	π_0	π_1
5	MAJ	π_1	π_2
6	MAJ	π_2	π_3
7	MAJ	π_3	π_0

Table 3. Rotation constants for each round

round	π_0	π_1	π_2	π_3
0	3	23	17	27
1	28	19	22	7
2	29	9	15	5
3	4	13	10	25

Table 4. Φ and rotation constants for the feed-forward of SIMD

step	Φ	r	s
0	IF	4	13
1	IF	13	10
2	IF	10	25
3	IF	25	4

In Table 3 the rotation constants for each round are given. The feed-forward consist of four steps using the same step function. Table 4 lists the used Boolean function and the rotation constants for the feed-forward. In the feed-forward the chaining value is used as message input. In the first step A_i^0, in the second step B_i^0, in the third C_i^0 and in the fourth D_i^0 for $i = 0, \ldots, 7$ are used.

4   Application on SIMD-512

In this section we will show how to construct a second order differential collision which suits as a distinguishing property for the full permutation (compression function without feed-forward) of SIMD-512. For the permutation of SIMD-512 the attack strategy can be directly applied using a good differential characteristic for the forward and backward direction. We show how we construct such differential characteristic and compute the complexities. In contrast to the attack on SHA-256 [12], where the second-order collision for the internal block cipher immediately transfers to the compression function, we need to overcome the feed-forward which performs 4 additional steps with the chaining value as message input. In Section 5 we show how the attack can be extended to the compression function using a weaker attack scenario.

4.1   Searching for Characteristics

A common approach to construct differential characteristics, which have a high probability, is to use a linearized approximation of the attacked hash function. As observed by Rijmen and Oswald [22], all differential characteristics for a linearized hash function can be seen as the codewords of a linear code. To find good differential characteristics we used the same technique as Mendel and Nad in the cryptanalysis of the first version of SIMD [15]. The procedure can be described in the following way:

- Linearize the step function of SIMD, i.e. replace all nonlinear operations with linear ones.
- Construct a generator matrix.
- Use a probabilistic algorithm from coding theory to search for codewords with low Hamming weight.

The nonlinear parts of the step function are the modular additions and the Boolean function IF and MAJ. In the attack, we replace all modular additions by XORs. Since we aim for a characteristic with low Hamming weight, we replace the Boolean functions with the 0-function, *i.e.* we block each input difference in Φ such that the output difference is always zero. This has probability $1/2$ in most cases. Note that there is exactly one input difference for IF and one for MAJ where the output difference is always one. Such characteristics are discarded.

For the search we used the *CodingTool Library* [20], which is an open-source implementation of the needed coding theoretic algorithms and data structures. We searched for good differential characteristics for the backward and forward direction with no differences in the message. Moreover, we also searched for a good starting step. One would expect that starting from the exact middle (round 16) would result in the best probability, but it turns out that moving the starting step two steps further, results in a better overall probability.

Differential Characteristics. The complete differential characteristics are given in Appendix A. To describe the differential characteristics we used signed-bit differences introduced by Wang *et al.* [26] in the cryptanalysis of MD5. The advantage of using signed-bit differences is that there exists a unique mapping to both XOR and modular differences.

The characteristic for the backward direction consists of the first 18 steps of the permutation and has Hamming weight 72. The characteristic for the forward direction consists of the last 14 steps of the permutation and has Hamming weight 52.

To estimate the success probability of each characteristic we used the same heuristic as in [15]. The probability for blocking a difference in one bit at the input of IF or MAJ is $1/2$ or 0 for some cases, but then the characteristic is discarded. Hence, the total probability is determined by the sum of all differences

Table 5. Summary of the success probabilities

characteristic	Hamming weight	probability
backward	72	$2^{-72.04}$
forward	52	$2^{-51.4}$

at the input. Differences at the same bit positions are counted only once. For the modular additions carries are not prevented for each bit difference. By allowing carries in the first addition, one can compensate them at the second addition. However, the rotation after the first modular addition needs to be considered. Therefore, the probability in this part is slightly decreased, but results in a overall increase. Table 5 summarizes the overall probability of each characteristic.

4.2   Independency of the Characteristics

The assumption on independent characteristics is quite strong (cf. [19]). Nevertheless, one can check this property easily for few steps in both directions, which was done for the presented characteristics. Furthermore, the used characteristics have a low Hamming weight, which makes it very unlikely that they interfere with each other.

4.3   Complexity of the Attack

As described in Section 2 the generic complexity for the attack is $2^{n/3}$. For the SIMD compression function n is 1024 bits. Hence, the generic complexity is $\approx 2^{342}$. The total complexity of the attack based on the presented characteristic is $(2^{72.04} \cdot 2^{51.4})^2 \approx 2^{247}$ which can be improved by ignoring conditions at the end. As was already observed by Wang et $al.$ [25] in the cryptanalysis of SHA-1 conditions resulting from the modular addition in the last steps of the differential characteristic can be ignored, due to the fact that carries can be ignored since the modular difference at the output stays the same. This reduces the complexity by a factor $2^{8.24}$ in the backward direction and 2^2 in the forward direction which improves the overall complexity by a factor of $2^{2 \cdot 10.24}$ resulting in $2^{226.52}$.

$Remark$: Note that we also have the freedom to choose the actual values for the state (at the beginning of each characteristic) and for the message. Message/chaining input modification can be used to improve the attack complexities further.

5   Extending the Attack to the Compression Function

In contrast to SHA-2 it is not easy to extend the second-order differential collision to the compression function since the feed-forward of SIMD is non-linear. However, the first step of the feed-forward is almost linear and therefore we can show non-random properties in the output of the state variables D_i for $i = 0, \ldots, 7$.

In the feed-forward 4 additional steps with the initial value as message input are performed. This destroys the distinguishing property at the output of the permutation. However, the values of D_i^{36} for $i = 0, \ldots, 7$ (output of the feed-forward) are determined already in the first step of the feed-forward and not modified in the other three steps. By considering only D_i^{36} for $i = 0, \ldots, 7$ and accordingly only A_i^0 for $i = 0, \ldots, 7$ of the initial value the attack complexity is only slightly increased. Consequently, the dimension of the input and output space for the distinguisher is reduced to 256 bits $(8 \cdot 32)$. However, by fixing the differences in the rectangle in the middle of the second-order differential characteristic one can construct a distinguisher for the compression function.

5.1   Distinguisher for the Compression Function

For the feed-forward of SIMD we extend the scheme shown in Figure 1 to the one shown in Figure 3. The function F_2 takes two inputs, namely the state of the last step and the chaining value. As mentioned before we consider only A_i^0 in the initial value and D_i^{36} at the output which is denoted by the quartets $\{P_{A_i}, P_{A_i}^*, Q_{A_i}, Q_{A_i}^*\}$ and $\{\tilde{R}_{D_i}, \tilde{R}^*_{D_i}, \tilde{S}_{D_i}, \tilde{S}^*_{D_i}\}$, respectively.

So far we have considered the inputs X, β and γ to be unrelated. Due to the way we build the second-order collisions, we can see that they are the inputs to a rectangle, hence they are related in the middle of the rectangle (gray layer in Figure 3). Therefore, we can extend the attacks by fixing β and γ, since the complexity of the generic case for this type of attacks is 2^n (or 2^t) [3]. Since we show non-randomness only in part of the output, namely D_i for $i = 0, \ldots, 7$, the generic complexity of the attack becomes $2^t = 2^{8 \cdot 32} = 2^{256}$. Hence, by using the second-order differential characteristic from Section 4.1 one can construct a distinguisher for the compression function of SIMD. Note that the distinguisher becomes even more powerful if the attacker can find several of the above quartets with the same difference.

To summarize, the algorithm works as follows:

1. Use the differential from Section 4.1
2. Choose a random value for X and compute $X^* = X + \beta$, $Y = X + \gamma$, and $Y^* = X^* + \gamma$.
3. Compute backward from X, X^*, Y, Y^* using F_0^{-1} to obtain $P_{A_i}, P_{A_i}^*, Q_{A_i}, Q_{A_i}^*$.
4. Compute forward from X, X^*, Y, Y^* using F_1 and F_2 to obtain $\tilde{R}_{D_i}, \tilde{R}^*_{D_i}, \tilde{S}_{D_i}, \tilde{S}^*_{D_i}$.
5. Check if $P_{A_i}^* - P_{A_i} = Q_{A_i}^* - Q_{A_i}$ and $\tilde{S}_{D_i} - \tilde{R}_{D_i} = \tilde{S}^*_{D_i} - \tilde{R}^*_{D_i}$ and therefore $P_{A_i}^* - P_{A_i} - Q_{A_i}^* + Q_{A_i} + \tilde{S}_{D_i} - \tilde{R}_{D_i} - \tilde{S}^*_{D_i} + \tilde{R}^*_{D_i} = 0$ is fulfilled.

5.2   Complexity of the Attack

As mentioned before the attack complexity is increased slightly by the feed-forward. In fact using the backward and forward characteristics from Table 6

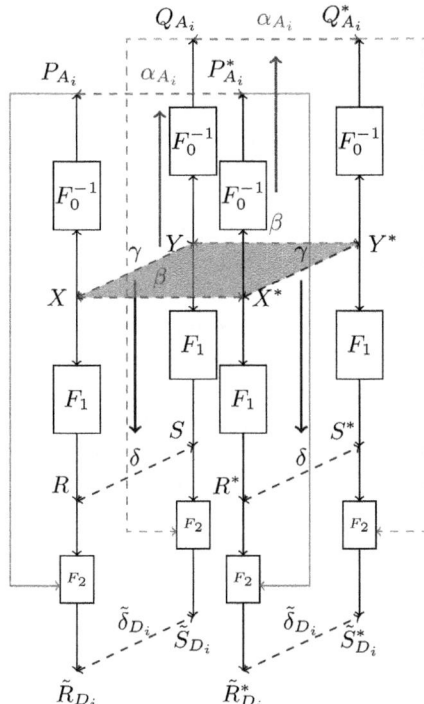

Fig. 3. Extending the attack to the compression function

and Table 7 the additional costs are negligible. In backward direction we have at the end only a difference in ΔA_6^{-1} which needs to be considered. This difference is rotated to the left by s bits. In the forward direction we have differences in ΔB_0^{31} and ΔA_3^{31}. Both are input to the Boolean IF function. Blocking each difference at the input of the IF function costs 2^2 for both differences. Additionally, ΔA_3^{31} is used to compute ΔA_6^{32} in the following way:

$$\Delta A_6^{32} = (\Delta D_6^{31} + \Delta A_6^{-1} + IF(\Delta A_6^{31}, \Delta B_6^{31}, \Delta C_6^{31})) \lll s^{32} + (\Delta A_3^{31} \lll r^{32}) \quad (7)$$

In Equation (7) only ΔA_6^{-1} and ΔA_3^{31} have differences. Only the rotation to the left by s^{32} bits adds a complexity about 2^1 [15].

Finally, we can ignore the costs of the last three steps in the backward $(2^{8.24+7+5})$ and forward $(2^{1+1.4+2})$ direction since we only consider the state variables A_i and D_i for $i = 0, \ldots, 7$ respectively. The differences in these variables do not change in the last three steps. Therefore, the total complexity is $(2^{72.04-20.24} \cdot 2^{51.4-4.4})^2 \cdot 2^1 \cdot 2^2 \approx 2^{200.6}$.

Hence, one can distinguish the compression function of SIMD from a random function with a complexity of about $2^{200.6}$. Note that the generic complexity for this attack is 2^{256}.

6   Conclusions and Discussion

In this paper, we present a distinguisher for the full permutation of SIMD-512 by an application of the boomerang attack on hash functions. Starting from the middle of the compression function we used techniques from coding theory to search for two differential characteristics, one for the backward direction and one for the forward direction, which hold with high probability. Then we construct a second-order differential and define a distinguishing property such that we can distinguish the permutation from a random permutation with a complexity of $2^{226.52}$.

Furthermore, we extend the attack to the full compression function of SIMD-512. By fixing the differences in the rectangle we can distinguish the output of the compression function from a random function with a complexity of $2^{200.6}$ compression function evaluations. This is a significant improvement to the current best known distinguisher with complexity 2^{398} [27].

However, our attack does not invalidate the security claims of the designers since it seems difficult to extend such an attack to the hash function and most of the security comes from the message expansion. In [4] the designers presented a more detailed analysis of SIMD regarding differential paths without differences in the message and are claiming that such characteristics does not affect the security of the SIMD hash function. Nevertheless, the results presented in this paper show how boomerang like attacks can be effectively used on compression functions. Furthermore, the results contribute to a better understanding of the security margin of SIMD.

Acknowledgments. The work in this paper has been supported by the European Commission under contract ICT-2007-216646 (ECRYPT II) and by the Austrian Science Fund (FWF, project P21936).

References

1. Biryukov, A., Khovratovich, D., Nikolić, I.: Distinguisher and Related-Key Attack on the Full AES-256. In: Halevi, S. (ed.) CRYPTO 2009. LNCS, vol. 5677, pp. 231–249. Springer, Heidelberg (2009)
2. Biryukov, A., Lamberger, M., Mendel, F., Nikolic, I.: Second-Order Differential Collisions for Reduced SHA-256. In: ASIACRYPT (to appear, 2011)
3. Biryukov, A., Nikolić, I., Roy, A.: Boomerang Attacks on BLAKE-32. In: Joux, A. (ed.) FSE 2011. LNCS, vol. 6733, pp. 218–237. Springer, Heidelberg (2011)
4. Bouillaguet, C., Fouque, P.-A., Leurent, G.: Security Analysis of SIMD. Cryptology ePrint Archive, Report 2010/323 (2010)
5. Damgård, I.B.: A Design Principle for Hash Functions. In: Brassard, G. (ed.) CRYPTO 1989. LNCS, vol. 435, pp. 416–427. Springer, Heidelberg (1990)
6. Dinur, I., Shamir, A.: Cube Attacks on Tweakable Black Box Polynomials. In: Joux, A. (ed.) EUROCRYPT 2009. LNCS, vol. 5479, pp. 278–299. Springer, Heidelberg (2009)
7. Gilbert, H., Peyrin, T.: Super-Sbox Cryptanalysis: Improved Attacks for AES-Like Permutations. In: Hong, S., Iwata, T. (eds.) FSE 2010. LNCS, vol. 6147, pp. 365–383. Springer, Heidelberg (2010)

8. Nikolić, P.S.I., Pieprzyk, J., Steinfeld, R.: Rotational Cryptanalysis of (Modified) Versions of BMW and SIMD (2010) Available online

9. Joux, A., Peyrin, T.: Hash Functions and the (Amplified) Boomerang Attack. In: Menezes, A. (ed.) CRYPTO 2007. LNCS, vol. 4622, pp. 244–263. Springer, Heidelberg (2007)

10. Knudsen, L.R.: Truncated and Higher Order Differentials. In: Preneel, B. (ed.) FSE 1994. LNCS, vol. 1008, pp. 196–211. Springer, Heidelberg (1995)

11. Lai, X.: Higher order derivatives and differential cryptanalysis. In: Blahut, R., Costello Jr., D., Maurer, U., Mittelholzer, T. (eds.) Communications and Cryptography, pp. 227–233. Kluwer (1992)

12. Lamberger, M., Mendel, F.: Higher-Order Differential Attack on Reduced SHA-256. Cryptology ePrint Archive, Report 2011/037 (2011)

13. Leurent, G., Bouillaguet, C., Fouque, P.-A.: SIMD Is a Message Digest. Submission to NIST (Round 2) (September 2009), http://csrc.nist.gov/groups/ST/hash/sha-3/Round2/submissions_rnd2.html

14. Lucks, S.: A Failure-Friendly Design Principle for Hash Functions. In: Roy, B. (ed.) ASIACRYPT 2005. LNCS, vol. 3788, pp. 474–494. Springer, Heidelberg (2005)

15. Mendel, F., Nad, T.: A Distinguisher for the Compression Function of SIMD-512. In: Roy, B., Sendrier, N. (eds.) INDOCRYPT 2009. LNCS, vol. 5922, pp. 219–232. Springer, Heidelberg (2009)

16. Mendel, F., Peyrin, T., Rechberger, C., Schläffer, M.: Improved Cryptanalysis of the Reduced Grøstl Compression Function, ECHO Permutation and AES Block Cipher. In: Jacobson Jr., M.J., Rijmen, V., Safavi-Naini, R. (eds.) SAC 2009. LNCS, vol. 5867, pp. 16–35. Springer, Heidelberg (2009)

17. Mendel, F., Rechberger, C., Schläffer, M., Thomsen, S.S.: The Rebound Attack: Cryptanalysis of Reduced Whirlpool and Grøstl. In: Dunkelman, O. (ed.) FSE 2009. LNCS, vol. 5665, pp. 260–276. Springer, Heidelberg (2009)

18. Merkle, R.C.: One Way Hash Functions and DES. In: Brassard, G. (ed.) CRYPTO 1989. LNCS, vol. 435, pp. 428–446. Springer, Heidelberg (1990)

19. Murphy, S.: The return of the cryptographic boomerang. IEEE Transactions on Information Theory 57(4), 2517–2521 (2011)

20. Nad, T.: The CodingTool Library. Workshop on Tools for Cryptanalysis 2010 (2010), http://www.iaik.tugraz.at/content/research/krypto/codingtool/

21. National Institute of Standards and Technology. Cryptographic Hash Algorithm Competition (November 2007), http://csrc.nist.gov/groups/ST/hash/sha-3/index.html

22. Rijmen, V., Oswald, E.: Update on SHA-1. In: Menezes, A. (ed.) CT-RSA 2005. LNCS, vol. 3376, pp. 58–71. Springer, Heidelberg (2005)

23. Vielhaber, M.: Breaking ONE.FIVIUM by AIDA an Algebraic IV Differential Attack. Cryptology ePrint Archive, Report 2007/413 (2007)

24. Wagner, D.: The Boomerang Attack. In: Knudsen, L.R. (ed.) FSE 1999. LNCS, vol. 1636, pp. 156–170. Springer, Heidelberg (1999)

25. Wang, X., Yin, Y.L., Yu, H.: Finding Collisions in the Full SHA-1. In: Shoup, V. (ed.) CRYPTO 2005. LNCS, vol. 3621, pp. 17–36. Springer, Heidelberg (2005)

26. Wang, X., Yu, H.: How to Break MD5 and Other Hash Functions. In: Cramer, R. (ed.) EUROCRYPT 2005. LNCS, vol. 3494, pp. 19–35. Springer, Heidelberg (2005)

27. Yu, H., Wang, X.: Cryptanalysis of the Compression Function of SIMD. In: Parampalli, U., Hawkes, P. (eds.) ACISP 2011. LNCS, vol. 6812, pp. 157–171. Springer, Heidelberg (2011)

A   Differential Characteristics for the Forward and Backward Direction

Table 6. Backward characteristic. The state at step -1 is the chaining value.

step	state	probability
-1	$\Delta D_0: -28, \Delta D_3: -7, \Delta B_5: -12, \Delta C_5: +4, \Delta A_6: +3, \Delta C_6: +25, \Delta C_7: +5, \Delta D_7: -15$	
0	$\Delta A_0: -19, \Delta A_3: -30, \Delta C_5: -12, \Delta D_5: +4, \Delta B_6: +6, \Delta D_6: +25, \Delta D_7: +5$	$2^{-8.24}$
1	$\Delta B_0: -10, \Delta B_3: -21, \Delta D_5: -12, \Delta C_6: +6, \Delta A_7: +22$	2^{-7}
2	$\Delta C_0: -10, \Delta C_3: -21, \Delta D_6: +6, \Delta B_7: +7$	2^{-5}
3	$\Delta D_0: -10, \Delta D_3: -21, \Delta A_6: +9, \Delta C_7: +7$	2^{-4}
4	$\Delta A_0: -1, \Delta B_6: +12, \Delta D_7: +7$	2^{-4}
5	$\Delta B_0: -24, \Delta C_6: +12$	2^{-3}
6	$\Delta C_0: -24, \Delta D_6: +12$	2^{-2}
7	$\Delta D_0: -24, \Delta A_6: +15$	2^{-2}
8	$\Delta B_6: +11$	2^{-2}
9	$\Delta C_6: +11$	2^{-1}
10	$\Delta D_6: +11$	2^{-1}
11	$\Delta A_6: +7$	2^{-1}
12	$\Delta A_1: +3, \Delta B_6: +3$	2^{-2}
13	$\Delta B_1: +22, \Delta A_5: +22, \Delta C_6: +3$	2^{-3}
14	$\Delta C_1: +22, \Delta A_4: +12, \Delta B_5: +12, \Delta D_6: +3$	2^{-4}
15	$\Delta D_1: +22, \Delta A_2: +19, \Delta B_4: +19, \Delta C_5: +12, \Delta A_6: +31$	$2^{-5.4}$
16	$\Delta A_0: +16, \Delta A_1: +31, \Delta B_2: +16, \Delta A_4: +28, \Delta C_4: +19, \Delta D_5: +12, \Delta B_6: +28$	$2^{-7.4}$
17	$\Delta B_0: +25, \Delta B_1: +8, \Delta A_2: +8, \Delta C_2: +16, \Delta A_3: +25, \Delta B_4: +5, \Delta D_4: +19, \Delta A_5: +27, \Delta C_6: +28, \Delta A_7: +5$	2^{-10}

Table 7. Forward characteristic

step	state	probability
17	$\Delta B_0 : +24, \Delta D_0 : +32, \Delta C_2 : -29, \Delta D_3 : +1, \Delta C_4 : -7, \Delta A_6 : -23, \Delta B_6 : +14, \Delta B_7 : +3, \Delta C_7 : -13$	
18	$\Delta A_0 : +5, \Delta C_0 : +24, \Delta D_2 : -29, \Delta D_4 : -7, \Delta B_6 : -6, \Delta C_6 : +14, \Delta C_7 : +3, \Delta D_7 : -13$	2^{-9}
19	$\Delta B_0 : +10, \Delta D_0 : +24, \Delta A_2 : -26, \Delta A_4 : -4, \Delta C_6 : -6, \Delta D_6 : +14, \Delta D_7 : +3$	2^{-8}
20	$\Delta C_0 : +10, \Delta B_2 : -23, \Delta B_4 : -1, \Delta D_6 : -6, \Delta A_7 : +12$	2^{-7}
21	$\Delta D_0 : +10, \Delta C_2 : -23, \Delta C_4 : -1, \Delta B_7 : +21$	2^{-5}
22	$\Delta A_0 : +15, \Delta D_2 : -23, \Delta D_4 : -1, \Delta C_7 : +21$	2^{-4}
23	$\Delta B_0 : +20, \Delta A_4 : -30, \Delta D_7 : +21$	2^{-4}
24	$\Delta C_0 : +20, \Delta B_4 : -2$	2^{-3}
25	$\Delta D_0 : +20, \Delta C_4 : -2$	2^{-2}
26	$\Delta A_0 : +13, \Delta D_4 : -2$	2^{-2}
27	$\Delta B_0 : +6$	2^{-2}
28	$\Delta C_0 : +6$	2^{-1}
29	$\Delta D_0 : +6$	2^{-1}
30	$\Delta A_0 : +31$	$2^{-1.4}$
31	$\Delta B_0 : +24, \Delta A_3 : +24$	2^{-2}

Lightweight Implementations of SHA-3 Candidates on FPGAs*

Jens-Peter Kaps, Panasayya Yalla, Kishore Kumar Surapathi, Bilal Habib, Susheel Vadlamudi, Smriti Gurung, and John Pham

ECE Department, George Mason University, Fairfax, VA 22030, U.S.A.
{jkaps,pyalla,ksurapat,bhabib,svadlamu,sgurung,jpham4}@gmu.edu
http://cryptography.gmu.edu

Abstract. The NIST competition for developing the new cryptographic hash algorithm SHA-3 has entered its third round. One evaluation criterion is the ability of the candidate algorithm to be implemented on resource-constrained platforms. This includes FPGAs for embedded and hand-held devices. However, there has not been a comprehensive set of lightweight implementations for FPGAs reported to date. We hope to fill this gap with this paper in which we present lightweight implementations of all SHA-3 finalists and all round-2 candidates with the exception of SIMD. All implementations were designed to achieve maximum throughput while adhering to an area constraint of 400-600 slices and one Block RAM on Xilinx Spartan-3 devices. We also synthesized them for Virtex-V, Altera Cyclone-II, and the new Xilinx Spartan-6 devices.

Keywords: SHA-3, FPGA, lightweight implementation, benchmarking.

1 Introduction and Motivation

The National Institute of Standards and Technology (NIST) started a public competition to develop a new cryptographic hash algorithm in November 2007. From the submitted 64 entries only 14 were selected for the second round of the competition and in December 2010, the 5 Secure Hash Algorithm-3 (SHA-3) finalists were announced. NIST is expected to announce the winner in 2012. In its decision which candidate algorithms should advance to the next round, NIST used the following criteria [40][3]: security, cost, and algorithm and implementation characteristics. The cost criterion describes the computational efficiency (speed) and memory requirements (gate counts for hardware implementations). One important implementation characteristic is the ability of the hash function to be "[...] implemented securely and efficiently on a wide variety of platforms, including constrained environments, such as smart cards"[3]. During the second phase of the SHA-3 competition many hardware implementations of the

---

* This work has been supported in part by NIST through the Recovery Act Measurement Science and Engineering Research Grant Program, under contract no. 60NANB10D004.

D.J. Bernstein and S. Chatterjee (Eds.): INDOCRYPT 2011, LNCS 7107, pp. 270–289, 2011.

candidates have been published. The first comprehensive analysis on FPGAs that included I/O overhead was done by Kobayashi et al. [32] on high throughput implementations of 8 round-2 candidates. The authors adapted an interface from [14] to the SASEBO [38]. Matsuo et al. [34] implemented all 14 round-2 candidates on FPGAs on SASEBO. Gaj et al.,[20] also implemented all 14 round-2 candidates on FPGAs optimized for throughput over area ratio with a different interface [15]. All implementations mentioned above only consider hash sizes of 256 bits. Homsirikamol et al. [27] and Baldwin et al [5] implemented all 14 round-2 candidates on FPGAs considering other hash sizes. Neither of these comprehensive implementations were done for resource-constrained applications. In a system-on-chip (SOC) on FPGAs, cryptographic functions such as encryption algorithms or hash algorithms are not necessarily the main purpose of the application but a part of it. Many other components such as soft-core processors, optimized signal processing algorithms, etc. are integrated in one chip. Furthermore, a space-constrained implementation could allow for using a smaller FPGA which in turn leads to cost and power savings. Recent developments in low-cost and low-power FPGAs [41] will increase their usage in battery powered devices which makes small implementations even more important.

Unfortunately, designing low-area implementations is not as straightforward as optimizing a design for best throughput over area. One has to go beyond merely reducing the datapath width and carefully evaluate the trade-off speed vs. area at every step of the design process. The control unit is an additional hurdle. Extensive component re-use in the datapath can lead to a very complex control logic which might negate the area savings in the datapath. There have been several publications that show low-area implementations of single SHA-3 candidates on FPGAs such as BLAKE [11], Grøstl [30], Keccak [9], and Skein [36][35]. Unfortunately, they are implemented on different FPGAs from different vendors and with different target sizes. This makes a fair comparison amongst these implementations impossible. Most recently Jungk [29] presented compact implementations of Grøstl, JH and Skein and Kerckhof et al. [31] of all five SHA-3 finalists at the "ECRYPT II Hash Workshop 2011". [31] shows results for 256-bit digest versions only on Virtex-6 devices. This device choice makes comparisons with previously reported results impossible. Furthermore, the authors did not formulate a clear design criterion other than "compact". Neither design is the smallest possible, yet the implementation area varies from 117 slices to 304 slices, the throughput from 105 Mbit/s to 960 Mbit/s, with the largest design (Grøstl) being the fastest. Only if one criterion is fixed (e.g. area or throughput) a meaningful comparison can be made.

Standardized interfaces have been proposed [14][15] for implementations of SHA-3 candidates and used by several comprehensive implementations in order to facilitate a fair comparison. Depending on the design of the hash function the interface can become a bottleneck. Furthermore, the interface protocol causes overhead and increases the size of the data path and control logic. Low area implementations will be particularly affected by the protocol overhead. Only the most recent publications, [29] and [31] use standardized interfaces.

In this paper we present low-area implementations of all five finalists and all round-2 candidates with the exception of SIMD[1], designed using the same criterion (space constraint), device, interface and optimization methods. This work is the most comprehensive analysis of lightweight implementations reported to date. In Sect. 2 we present the design methodology we used including clear assumptions and goals, interface description and performance metrics. Due to space constraints we describe only the datapaths of the five SHA-3 finalists in detail in Sect. 3. Our designs of the other algorithms are summarized in Table 1. Section 4 shows the results of our implementations and compares the 13 candidates with each other and other reported implementations.

2   Methodology

The primary target for our lightweight implementations are the low-cost Xilinx Spartan-3 FPGAs. We choose VHDL to describe our lightweight architectures. All implementations were designed at a low level for our main target FPGA family such that we can already obtain a rather precise estimate of the required area from detailed datapath diagrams. This approach allowed us to enforce a similar coding style across several designers and algorithms. Furthermore, we built a small VHDL library of elementary functions that was used by all designers.

2.1   Assumptions and Goals

Only SHA-3 variants with 256-bit digest have been implemented as these are the most likely variants to be used in area-constrained designs. Furthermore, we assume that padding is done in software. This assumption goes hand-in-hand with the application of hash functions to SOC designs. The salt values of all SHA-3 candidates who support them are set to zero. Typical optimization goals for hardware implementations are: maximum throughput, maximum throughput to area ratio, and minimum area. In order to compare lightweight implementations the minimum area target seems logical. However, optimizing the implementations for minimum area would yield a ranking of algorithms solely based on area, i.e. we would know which is the smallest and which is the largest irrespective of the throughput that is achieved by these implementations. This information is of not much use in practice. A different approach is to optimize for throughput given an area constraint. We believe that this is a much more realistic scenario. Additionally this optimization goal lets us determine how efficient an algorithm is in a constrained environment which is a factor of an algorithm's flexibility. This is a clearly stated evaluation criterion by NIST [3]. We choose to use an area range of 400 to 600 slices and 1 Block RAM on Xilinx Spartan-3 FPGAs as our constraint. The size of the range was chosen based on low-area implementation results published on the SHA-3 Zoo [2] website and our own analysis. Within this area constraint we try to achieve maximum throughput. Therefore, our final

---

[1] Our initial investigation has shown that it is unlikely that SIMD could be implemented within our area constraints, due to its complex underlying functions.

comparisons will be in terms of the ratio of throughput to area. The Block RAM was chosen due to the large storage requirements of some hash functions.

2.2 Tools and Result Generation

Even though all designs were targeted for Spartan-3 devices it is interesting to see how our implementations perform on low-cost devices from another vendor such as Altera Cyclone-II, newer devices such as Spartan-6 and on high speed devices such as Xilinx Virtex-V. Complete results are published in the ATHENa results database [1]. All designs were implemented using the vendor tools: Xilinx ISE 12.3 Web Pack and Altera Quartus II v. 10.0 Web Edition, and verified after place-and-route against known answer test files provided by the submissions packet of each hash function. All results were generated using the open source benchmarking tool ATHENa (Automated Tool for Hardware EvaluatioN) [21]. Other than simplifying the result generation, ATHENa also varies the vendor tool parameters to achieve optimal results.

2.3 Interface and Protocol

We based our hardware interface and I/O protocol (Fig. 1) on the one presented in [15] and updated in [20]. The SHA Core assumes that its inputs and outputs are connected to FIFOs. We believe that the FIFO interface model proposed in [15] is very suitable for lightweight implementations. In its simplest form a FIFO is a single w-bit wide register with minimal logic to support the handshake of read/write and ready. This can easily be interfaced to a microcontroller or other circuitry in an embedded system. Lightweight applications usually have smaller databus sizes than the 32 or 64 bits proposed in [15]. Therefore, we use a databus width w of 16 bits. The protocol supports two scenarios: 1) when the message length is known and 2) when the message length is not known. In case 1) the message is sent as a single segment starting with the message length after padding "msg_len_ap" in 32-bit words concatenated with a '1' followed by the message length before padding "msg_len_bp" in bits followed by the message. The "msg_len_bp" is needed by several algorithms even when the message is already padded. In case 2) the message can be processed in segments $seg_0, seg_1, \cdots, seg_{n-1}$. Each segment $seg_0, \cdots, seg_{n-2}$ is headed by the segment length after padding "seg_len_ap" concatenated with a '0' followed by the segment of the message. The last segment seg_{n-1} follows the format of case 1). It contain a block of the message and must contain all padding. The formulae to compute the total number of bits before padding and after padding are:

$$msg\_len\_ap = \sum_{i=0}^{n-1} seg\_len\_ap_i \cdot 32$$

$$msg\_len\_bp = \sum_{i=0}^{n-2} seg\_len\_ap_i \cdot 32 + seg\_len\_bp_{n-1}$$

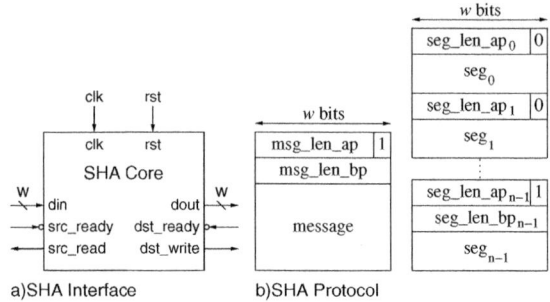

Fig. 1. Interface and protocol for our SHA cores

Furthermore in order to conserve logic resources needed for message counters we limit the total amount of data in a single message to 2^{32} bits i.e. 4 Gbits which we believe is sufficient for lightweight applications.

2.4   Area Minimization Techniques

Datapath: The most straightforward approach to reducing the area of the datapath is folding. Vertical folding reduces the datapath width while horizontal folding reduces the size of processing elements while maintaining the datapath width. How many times and in which direction a design can be folded depends on the algorithm. The extent to which folding can be applied to the SHA-3 candidates and how much it affects their throughput and throughput over area ratio has been examined by Homsirikamol et al. [28]. They show that only BLAKE can reach our area constraints through folding alone, Grøstl remains too large, JH area increases when folded, and Keccak as well as Skein cannot be folded at all and hence far exceed our area constraints. Another technique is reusing of processing elements. We heavily use this technique and additionally, we apply vertical folding at multiple levels down to single processing elements, not just the datapath as a whole as done in [28]. For example the Skein algorithm uses 4 Mix functions each using a 64-bit adder and a 64-bit XOR. We fold the 4 Mix functions into 1 and within the Mix function we reuse a 32-bit adder to perform 64-bit additions. The same adder is also reused for the key injections. Both folding and reuse of processing elements minimize the area consumption at the cost of an increased number of clock cycles. We reduced this increase to some extent by interleaving operations through pipelining.

Block RAM: Block RAMs (BRAMs) offer a large amount of memory space for storage but have a limited number of ports and I/O lines. Xilinx Spartan-3 BRAMs can be configured as single or dual port memories with a maximum data width of 64 bits or 32 bits per port, respectively. Each port is associated with a single address input. This limits the number of independent values and the number of bits that can be accessed in a single clock cycle. Our Grøstl design processes four 8-bit values in each clock cycle. Even though these are only 32 bits, a dual port BRAM does not allow reading of four independent values in one clock cycle. Hence, we store that data in 4 Distributed RAMs. The Spartan-3 BRAM data sheet specifies that data is written to the address applied in the current clock cycle, but read from the address of the previous clock cycle. Hence, computing $Mem[i] = Mem[i] + k$, where each element is a 64-bit word, requires 2 clock cycles per address location i, i.e. dedicated write cycles. These are not needed when computing $Mem[i + i] = Mem[i] + k$, i.e. when an address shift is acceptable. In our early Keccak design, this address shift increased the complexity of the control logic and with it the area consumption beyond our constraint. Hence it now uses dedicated write cycles. The new Xilinx Spartan-6 and Virtex-6 devices allow for independent read and write addresses for 64-bit data width.

Control Logic: The control logic of our implementations consists of a main finite state machine (FSM) with up-to 8 states, a single counter to count the clock cycles per state, and ROM-based FSMs for each state of the main FSM. ROM-based FSMs are more efficient in terms of area consumption and speed compared to conventional FSM [37], [39], [22], and their maximum frequency is independent of the complexity. However they are more complex to design. The area required to implement ROM-based FSMs is determined by the number of control signals and states. In order to reduce the number of control signals we try to use bits from the counter output, the main finite state machine, and simple boolean logic combinations thereof wherever possible. Furthermore, short sequences of control signals are placed in sub-controllers. The complexity of address generation for BRAMs can be reduced by placing datasets in memory locations starting at addresses which are a power of 2.

2.5   Performance Metrics

The number of clock cycles needed to hash N message blocks using our implementations can be computed from the number of clock cycles required to perform the following functions:

i	Initialization (if not precomputed)	p	Processing one block
h	Loading protocol header of message	z	Finalization
$l1$	Loading first block	o	Output of the hash value
l	Loading each subsequent block		

This results in the following formula for the number of clock cycles clk for hashing N blocks of data.

$$clk = i + h + l1 + l \cdot (N - 1) + p \cdot N + z + o$$

This formula can now be simplified to reflect the number of clock cycles needed for the initial steps before processing can begin $st = i + h + (l1 - l)$, loading and processing one block of data $l + p$, and finalization and output of the hash value $end = z + o$ resulting in (1).

$$clk = st + (l + p) \cdot N + end \tag{1}$$

Throughput is defined as the number of input bits processed per unit of time. The precise formula for throughput of a hash function is dependent on the number of message blocks N to be hashed, the block size b of the algorithm, the number of clock cycles needed to hash the message clk and the clock period T. We can derive the formula to compute the throughput from (1).

$$throughput(N) = \frac{b \cdot N}{clk \cdot T} = \frac{b \cdot N}{(st + (l + p) \cdot N + end) \cdot T} \tag{2}$$

Especially in embedded applications, messages can be very short. It is therefore important to also calculate the throughput for short messages. We use the empty message which after padding is one block long and therefore set $N = 1$ in (2) to compute the throughput.

When computing the throughput for very long messages, we can neglect st and end as their influence on the result goes to zero. This leads to the simplified equation (3).

$$throughput_{long} = \frac{b}{(l + p) \cdot T} \tag{3}$$

Resource Utilization of FPGAs is very difficult to define. All FPGAs contain configurable logic elements which contain flip-flops (Xilinx: slices, Altera: LE), BRAMs, multipliers and other resources. These resources have different features not only depending on the vendor but even on the FPGA family. Hence, we can compare implementations using the metric of throughput over area ratio only within a specific FPGA family and provided they use the same number of dedicated resources. As area in this formula we use solely *slices* for Xilinx and *LEs* for Altera devices as there is no direct mapping from BRAM utilization to slice or LE.

3   Implementations

Due to space constraints we only briefly describe our implementations of the SHA-3 finalists. A short list of implementation details of all 13 SHA-3 candidates evaluated in this paper is shown in Table 1. The throughput formulae for all implementations is shown in Table 2.

3.1   BLAKE

Our implementation of BLAKE-32 and BLAKE-256 (Fig. 2) uses the BRAM to store the message, constants, initial hash values, chaining hash values, salt, and a counter. It takes 16 clock cycles to initialize the internal state. The internal V-state is stored in four Distributed RAMs which can be accessed easily for each G-Function. We implemented 1/2 G-Function with interleaved pipeline stages such that it takes 20 clock cycles for computing 8 G-Functions. Additionally we need one extra clock cycle to store the registered value back into Distributed RAM leading to a total of 21 clock cycles for each round. The G-Function requires permuted values of constants and messages which are stored

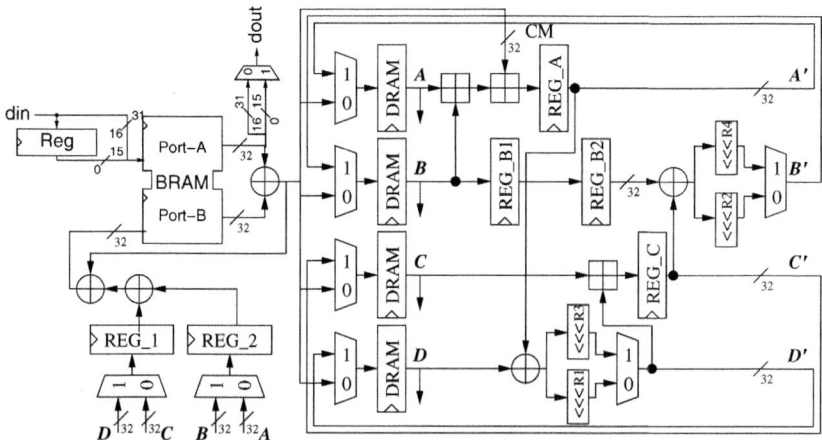

Fig. 2. Blockdiagram of BLAKE

in BRAM. This permutation doesn't have a repeatable pattern, therefore the BRAM addressing alone consumes 70% of the size of the controller. For round-3 of the SHA-3 competition a tweak was introduced for BLAKE which increases the number of rounds by four resulting in an increase in area consumption as the permutation function needs data values for 4 more rounds from BRAM. This version of BLAKE is called BLAKE-256.

3.2 Grøstl

Grøstl [23] is based on the AES round with the following sequence of operations: AddRoundConstant, SubBytes, ShiftBytes, and MixBytes. In our implementation (Fig. 3) the state, consisting of two 512-bit matrices P & Q, is stored in 16 4x8 Distributed RAMs. Each row is stored in one Distributed RAM. In order to get the first 64-bit column we access byte0 from RAM0, byte1 from RAM1... etc. This access scheme performs the ShiftBytes operation with which we start each round. SubBytes is implemented using 4 pipelined S-Boxes which are described as logic functions [19]. The multiplier takes a column from SubBytes and produces 32 bits of the new column in one clock cycle, the remaining 32 bits in the second clock cycle. It takes a total 3 clock cycles to produce a new column. Each round of P and Q computes 16 new columns which takes 48 clock cycles. We interleave the computations of P and Q through the pipeline. The XOR operation $(P \oplus Q \oplus h)$ takes 32 clock cycles. So a block of message is processed in 515 clock cycles ($48 \cdot 10 + 32 + 3$ clock cycles to fill the pipeline). BRAM is used in dual port mode and stores the initialization vector and the intermediate hash (h). For round-3 of the SHA-3 competition a tweak was introduced which changes the shifts in the ShiftBytes operation and introduces a different AddRoundConstant function. This has minimal effect on the area consumption and does not change the overall architecture. Grøstl from round-2 is now called Grøstl-0.

3.3   JH

Our implementation of JH (Fig. 4) stores the state and constants in BRAM. Two independent 32x8 Distributed RAMs store the state of the round constant generator. The BRAM is used as two independent memories to simplify the control logic and ease synchronization with the round constant generator. During initialization, which takes 35 cycles, the location of the state in BRAM is initialized with the precomputed starting value of $H^{(0)}$ from another address in the same BRAM. Grouping and de-grouping take advantage of the dual port memory and read two addresses from the BRAM simultaneously, retaining 4 bits from each address in registers and discarding the rest. This is repeated 4 times to write a full 32-bit value back into the BRAM and takes 160 cycles. The core round function is 32 times vertically folded and contains a pipelined permutation function. It needs a total of 34 cycles per round. Difficulties in creating this implementation were the memory access delays and the nonconsecutive read and write addresses. The tweak for round-3 of the SHA-3 competition increases the number of rounds to 42. This version of JH is called JH42.

3.4   Keccak

One round of Keccak [9] applies five functions, θ, ρ, π, χ, and ι to its state. In our implementation (Fig. 5), we store the state and the round constant in BRAM. The basic operations of Keccak use 64-bit data values which is also the maximum that we can read or write to BRAM in a single clock cycle. Therefore, in order to make the design more efficient we decided to quasi pipeline our functions. We have merged the θ and ρ functions. The later function uses a variable rotator. A barrel shifter consumes 192 slices on Spartan-3, hence we build a shifter that can only shift the 25 offsets Keccak needs. It uses on average 1.5 clock cycles per rotation and consumes only 128 slices. We use dedicated write cycles to accommodate the data rearrangement of the π function. These three functions take a total of 91 clock cycles. The χ function takes its operands from BRAM, applies a series of simple logical operations, and stores the result into BRAM. The ι operation combines a round constant with one 64-bit value of the new state. These operations take an additional 63 clock cycles. A single round operation thus takes a total of 154 clock cycles. Reducing the number of clock cycles would require more BRAM accesses which is not possible or more registers or Distributed RAMs, both would increase area consumption.

3.5   Skein

The basic building block of Skein [18] is a Mix function which consists of 64-bit ADD, XOR and rotate operations. Even though all main operations are of 64-bit size we chose a 32-bit datapath for our Skein-512-256 implementation (Fig. 6). The BRAM limits us to read two 32-bit values per clock cycle. Hence, we read the 32 LSB of two operands in one clock cycle and perform an addition, followed by the 32 MSB in the next clock cycle. The big advantage of this strategy is that

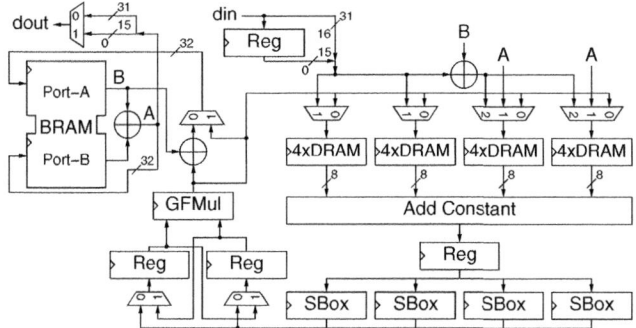

Fig. 3. Blockdiagram of Grøstl-0

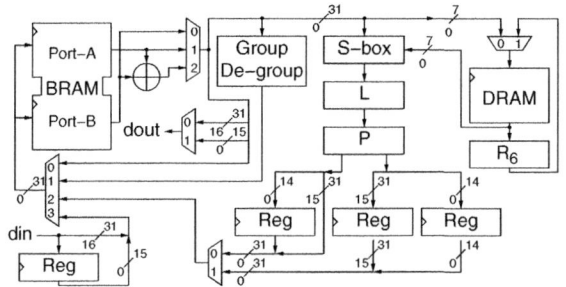

Fig. 4. Blockdiagram of JH

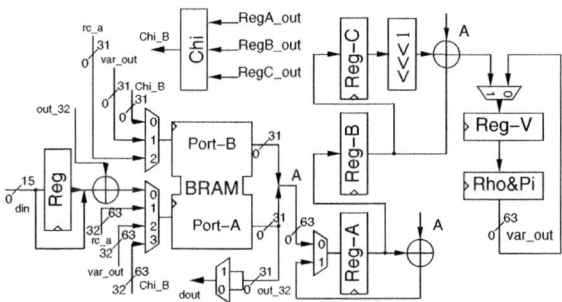

Fig. 5. Blockdiagram of Keccak

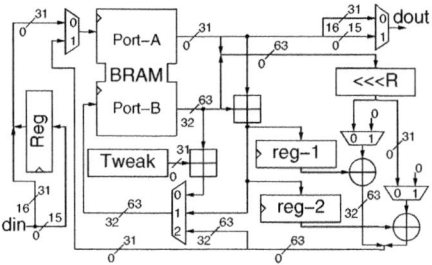

Fig. 6. Blockdiagram of Skein

Table 1. Implementation details of SHA-3 candidate implementations

Algorithm	Datapath Size (bits)	Rounds	Clock cycles per Round	Additional Clock cycles	Clock cycles per block p	Clock cycles per byte of message	Implementation Details
BLAKE-32	32	10	21	24	234	3.7	See description in Sect. 3.1
BLAKE-256	32	14	21	24	318	5.0	.
BMW	32	1	730	0	730	11.4	BRAM stores IV and state, shifter and rotator implemented separately to perform in parallel, the outputs of rotator, shifter and the adders are registered to reduce delay.
CubeHash	32	16	58	0	928	29.0	Processed IV and state is stored in BRAM. Finalization is equivalent to 10*Rounds. Distributed RAMs store immediate values to overcome swapping.
ECHO	64	8	290	129	2449	12.8	Message and state stored in BRAM. S-box implemented as logic. Round is generated using 32-bit adder and 32-bit register and stored in Distributed RAM.
Fugue	32	1	61	0	61	15.3	State is stored in BRAM. Super-Mix is created from 4 AES S-Boxes and fixed rotations.
Grøstl(-0)	32	10	48	35	515	7.8	See description in Sect. 3.2.
Hamsi	64	3	16	26	74	18.5	Message stored in register. Expansion function constants 32*128 bits, initialization constants, and state in BRAM. p/pf function constants in Distributed Ram. 64-bit p/pf function.
JH	32	36	34	384	1608	25.1	See description in Sect. 3.3
JH42	32	42	34	385	1813	28.3	
Keccak	64	24	154	0	3696	27.2	See description in Sect. 3.4.
Luffa	32	8	66	78	606	18.9	Message injection through serialized XOR. Tweak uses shift register. The SubCrumb is implemented as ROM. Constants, IVs and state stored in BRAM
Shabal	32	48	1	16	64	1.0	Design from [17] with IVs and C-state located in BRAM.
SHAvite-3	32	12	38	288	744	11.6	BRAM stores state and IV. 4 S-Boxes implemented as ROM. 4x32-bit shift register is tapped at 32 bit positions to provide data for MixColumns multiplication. Key generation uses same datapath and takes 288 clock cycles.
Skein	32	72	20	967	2407	37.6	See description in Sect. 3.5.

Table 2. Throughput formulae for our implementations of SHA-3 candidates

Algorithm	Speci-fication	Block Size (bits) b	Clock Cycles to hash N blocks $clk =$ $st + ($ $l +$ $p) \cdot N +$ end	Throughput $\dfrac{b}{(l+p) \cdot T}$
BLAKE-32	[4]	512	$2 + (32 + 234) \cdot N + 17$	$512/(266 \cdot T)$
BLAKE-256	[4]	512	$2 + (32 + 318) \cdot N + 17$	$512/(350 \cdot T)$
BMW	[25]	512	$2 + (32 + 730) \cdot N + 757$	$512/(762 \cdot T)$
CubeHash	[7]	256	$2 + (16 + 928) \cdot N + 9312$	$256/(944 \cdot T)$
ECHO	[6]	1536	$18 + (96 + 2449) \cdot N + 17$	$1536/(2545 \cdot T)$
Fugue	[26]	32	$33 + (2 + 61) \cdot N + 990$	$32/(63 \cdot T)$
Grøstl-0, Grøstl	[23], [24]	512	$2 + (32 + 515) \cdot N + 532$	$512/(547 \cdot T)$
Hamsi	[33]	32	$2 + (2 + 74) \cdot N + 65$	$32/(76 \cdot T)$
JH	[42]	512	$35 + (32 + 1608) \cdot N - 15$	$512/(1640 \cdot T)$
JH42	[43]	512	$35 + (32 + 1813) \cdot N - 15$	$512/(1845 \cdot T)$
Keccak	[9], [10]	1088	$2 + (68 + 3696) \cdot N + 17$	$1088/(3764 \cdot T)$
Luffa	[16]	256	$2 + (16 + 606) \cdot N + 647$	$256/(622 \cdot T)$
Shabal	[13]	512	$32 + (32 + 64) \cdot N + 208$	$512/(96 \cdot T)$
SHAvite-3	[12]	512	$18 + (32 + 744) \cdot N + 17$	$512/(776 \cdot T)$
Skein	[18]	512	$5 + (32 + 2407) \cdot N + 2423$	$512/(2439 \cdot T)$

we can use a 32-bit adder which has a much shorter critical path than a 64-bit adder. The variable rotator is realized as a 64-bit barrel shifter with a single pipeline stage. It is the single largest block in our design. We use the BRAM to store the state and the processed IV. This allows us to skip the initialization. Hashing a single block of data takes 72 rounds and 19 key injections. Within each round the Mix function is used 4-times which takes 20 clock cycles. The first key injection takes 48 clock cycles and all following 45, resulting in 858 clock cycles per block. After the round function completes a new chaining value has to be generated which is used to generate the new key for the next message block. Permutations take an additional 109 clock cycles. When all message blocks are processed the message finalization starts. This finalization is equivalent to processing a message block, except no new key has to be generated. For round-3 of the SHA-3 competition only a single constant got changed. This neither affects the size of our implementation nor its speed.

4   Results and Conclusions

4.1   Implementation Results

The results of our implementations are summarized by the graph shown in Fig. 7. It shows the area consumption of each implementation on the x-axis and the throughput on the y-axis. Hash functions where the implementations did not change between round 2 and round 3 of the SHA competition are marked as

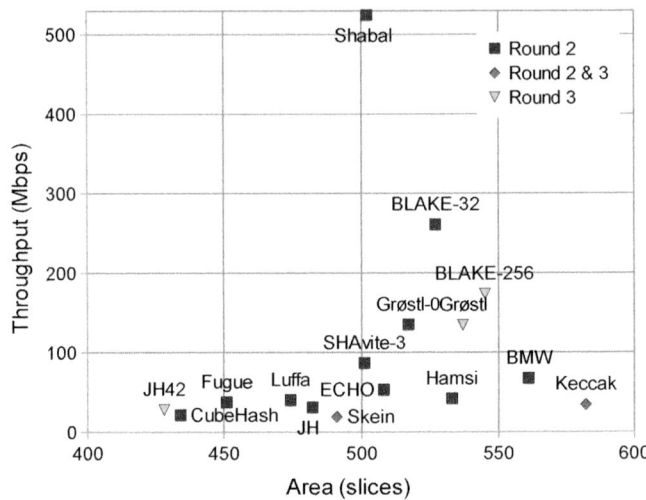

Fig. 7. Throughput over area of our SHA-3 implementations on Xilinx Spartan-3

"Round 2 & 3". Otherwise they are grouped by competition rounds. It can be seen that all implementations fall within our target range of 400 to 600 slices. Each algorithm was optimized for maximum throughput without violating the area constraint. The throughput over area ratio of each implementation on a Xilinx Spartan-3 and, due to page limitations, only of the finalists on other FPGA devices is shown in Fig. 8. Each graph is sorted by throughput over area for long messages according to (3) in red. The results for short messages of one block only are computed according to (2) and shown in light-blue. The order of the algorithms differs slightly depending on the implementation platform. Shabal outperforms all other hash functions for long messages. Of the five finalists, BLAKE-256 performs better on Xilinx devices, Grøstl on the Altera device. It can clearly be seen, that algorithms that have a lengthy finalization step do not perform well for short messages. It is interesting to note, that all round-3 candidates which introduced a tweak after round-2 that requires a change in the datapath perform slightly worse after the tweak with the exception of JH42. Its tweak leads to an increase in the number of rounds, however, this penalty is compensated for by the simpler datapath resulting from dropping the half round. The detailed results of our implementations on Xilinx Spartan-3, Spartan-6, Virtex-V and Altera Cyclone-II devices are summarized in Tables 3 and 4. Both tables show first the results of the SHA-3 finalists followed by the remaining round-2 candidates.

One challenge when implementing the hash functions for low-area, is the trade-off between simplicity of the datapath versus complexity of the control logic. Reuse of components in the datapath for several different functions, clock cycle optimized usage of the BRAM, pipelining complex functions to achieve low critical path delay, and interleaving elementary functions of a hash algorithm all lead to a better throughput over area ratio of the datapath. However, all of

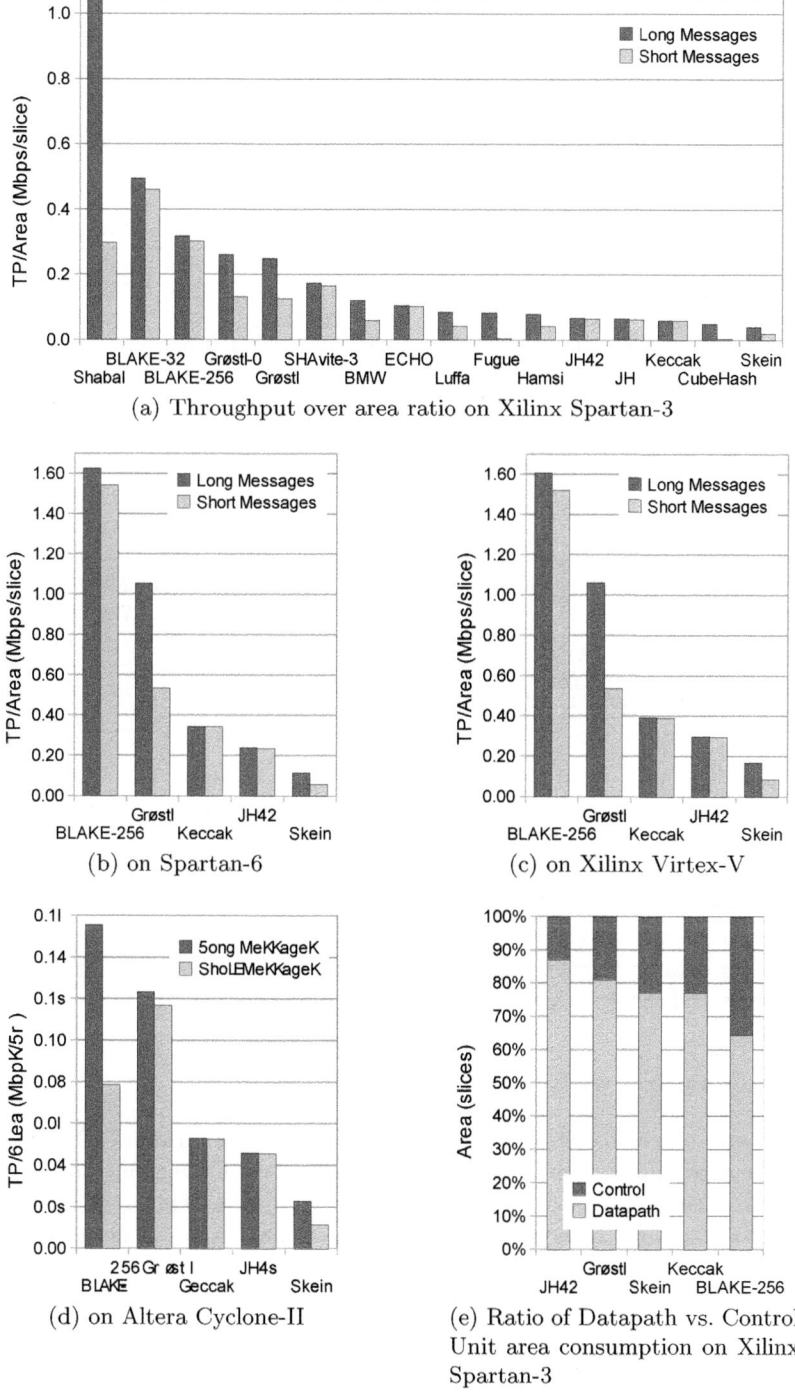

(a) Throughput over area ratio on Xilinx Spartan-3

(b) on Spartan-6

(c) on Xilinx Virtex-V

(d) on Altera Cyclone-II

(e) Ratio of Datapath vs. Control Unit area consumption on Xilinx Spartan-3

Fig. 8. Lightweight SHA-3 implementations results

Table 3. Implementation results of our implementations of SHA-3 candidates

Algorithm	Xilinx xc3s50-5 Area (slices)	Block RAMs	Maximum Delay (ns) T	Xilinx xc6slx4csg-3 Area (slices)	Block RAMs	Maximum Delay (ns) T	Xilinx xc5vlx20-2 Area (slices)	Block RAMs	Maximum Delay (ns) T	Altera ep2c5f256c6 Area (LEs)	Memory Bits	Maximum Delay (ns) T
BLAKE-256	545	1	8.42	139	1	6.47	212	1	4.30	1,365	2,048	8.70
Grøstl	537	1	6.95	163	1	5.44	234	1	3.77	1,026	2,560	5.86
JH42	428	1	9.74	142	1	8.16	176	1	5.28	702	8,704	8.59
Keccak	582	1	8.30	142	1	5.91	196	1	3.74	996	8,192	5.48
Skein	491	1	10.68	227	1	8.07	215	1	5.74	930	4,096	9.89
BLAKE-32	527	1	7.38	138	1	7.23	238	1	4.37	1,262	2,048	8.71
BMW	561	1	9.99	183	1	7.15	233	1	5.16	1,104	8,192	9.45
CubeHash	434	1	12.58	131	1	8.11	231	1	6.05	2,761	16,384	9.18
ECHO	508	1	11.33	155	1	9.72	232	1	5.34	1,069	16,512	10.14
Fugue	451	1	13.48	269	1	12.84	209	1	6.43	940	16,384	7.81
Grøstl-0	517	1	6.93	163	1	5.23	232	1	3.61	1,020	2,560	5.93
Hamsi	533	1	9.97	162	1	12.83	208	1	5.28	687	10,240	8.87
JH	482	1	9.99	180	1	8.35	161	1	4.97	702	8,704	8.59
Luffa	474	1	10.17	107	1	6.86	176	1	5.08	946	8,192	7.66
Shabal	502	1	10.17	165	1	7.66	231	1	4.86	2,093	1,760	9.16
Shavite-3	501	1	7.60	120	1	5.24	136	1	3.37	471	16,384	7.01

them also lead to a complex control logic. Figure 8e shows how many percent of the total area consumption of each hash function was used for the datapath and for the control unit. A small control unit indicates that the control signals needed for an algorithm are very regular, i.e. the algorithm is very regular and can be easily scaled down for lightweight implementations. A large control unit might indicate the opposite. On the other hand, BLAKE-256 has a very good throughput over area ratio, yet, due to its permutation schedule with 210 entries it requires a relatively large control unit.

4.2 Comparison with Other Reported Results

We compare our results with previously reported ones in Table 5. Due to space limitations we concentrate only on the SHA-3 finalists. Even though our primary design target is Xilinx Spartan-3, we synthesized our implementations for other devices to match the devices of reported results. This puts our designs at a disadvantage as we could not take full advantage of their features. Most notably, pipeline stages might become unbalanced when synthesizing a design for a device with 4-input LUTs on a 6-input LUT device. The compact BLAKE-32 design reported in [11] uses two BRAMs, one for the controller and one to store the message, constant, internal states, hash values and counters. In order to have a fair comparison we moved the control unit of our BLAKE-32 design to a second

Table 4. Throughput results of our lightweight implementations of SHA-3 candidates. First are round-3 results followed by round-2 results.

Message Algorithm	Xilinx xc3s50-5				Xilinx xc6slx4csg225-3			
	Long		Short		Long		Short	
	Throughput (Mbps)	TP/Area (Mbps/slice)	Throughput (Mbps)	TP/Area (Mbps/slice)	Throughput (Mbps)	TP/Area (Mbps/slice)	Throughput (Mbps)	TP/Area (Mbps/slice)
BLAKE-256	173.8	0.32	164.8	0.302	226.0	1.63	214.4	1.542
Grøstl	134.6	0.25	68.1	0.127	171.9	1.05	87.0	0.534
JH-42	28.5	0.07	28.2	0.066	34.0	0.24	33.6	0.237
Keccak	34.8	0.06	34.7	0.060	48.9	0.34	48.6	0.343
Skein	19.7	0.04	9.9	0.020	26.0	0.11	13.0	0.057
BLAKE-32	261.0	0.50	243.6	0.462	266.4	1.93	248.6	1.802
BMW	67.3	0.12	33.7	0.060	93.9	0.51	47.1	0.257
CubeHash	21.6	0.05	2.0	0.005	33.5	0.26	3.1	0.024
ECHO	53.3	0.10	52.5	0.103	62.1	0.40	61.2	0.395
Fugue	37.7	0.08	2.2	0.005	39.6	0.15	2.3	0.009
Grøstl-0	135.1	0.26	68.4	0.132	179.0	1.10	90.6	0.556
Hamsi	42.2	0.08	22.4	0.042	32.8	0.20	17.4	0.108
JH	31.2	0.06	30.9	0.064	37.4	0.21	37.0	0.205
Luffa	40.5	0.09	19.8	0.042	60.0	0.56	29.4	0.275
Shabal	524.7	1.05	149.9	0.299	687.5	4.17	196.4	1.190
Shavite-3	86.8	0.17	83.1	0.166	126.0	1.05	120.6	1.005

Message Algorithm	Xilinx xc5vlx20-2				Altera ep2c5f256c6			
	Long		Short		Long		Short	
	TP (Mbps)	(Mbps /slice)	TP (Mbps)	(Mbps /slice)	TP (Mbps)	(Mbps /LE)	TP (Mbps)	(Mbps /LE)
BLAKE-256	340.4	1.61	322.8	1.523	168.1	0.12	159.4	0.117
Grøstl	248.5	1.06	125.7	0.537	159.7	0.16	80.8	0.079
JH-42	52.5	0.30	52.0	0.295	32.3	0.05	31.4	0.046
Keccak	77.2	0.39	76.8	0.392	52.7	0.05	52.5	0.053
Skein	36.6	0.17	18.3	0.085	21.2	0.02	10.6	0.011
BLAKE-32	440.8	1.85	411.4	1.728	220.9	0.18	206.2	0.163
BMW	130.3	0.56	65.3	0.280	71.1	0.06	35.6	0.032
CubeHash	44.8	0.19	4.1	0.018	29.6	0.01	2.7	0.001
ECHO	113.0	0.49	111.5	0.481	59.5	0.06	58.7	0.055
Fugue	79.0	0.38	4.6	0.022	65.0	0.07	3.8	0.004
Grøstl-0	259.6	1.12	131.4	0.566	157.9	0.15	79.9	0.078
Hamsi	79.7	0.38	42.4	0.204	47.5	0.07	25.2	0.037
JH	62.8	0.39	62.0	0.385	36.3	0.05	35.9	0.051
Luffa	81.1	0.46	39.7	0.225	53.8	0.06	26.3	0.028
Shabal	1097.8	4.75	313.7	1.358	582.5	0.28	166.4	0.080
Shavite-3	196.1	1.44	187.6	1.380	94.1	0.20	90.1	0.191

Table 5. Comparison of lightweight implementations of SHA-3 finalists on Xilinx FPGAs ([TW] – this work)

Algorithm	Reference	Device	I/O Width	Datapath Width	Clock Cycles $(l+p)$	Area (slices)	Block RAMs	Maximum Delay (ns)	Throughput (Mbps)	TP/Area (Mbps/slice)
BLAKE-32	[11]	xc3s50-5	32	32	846	124	2	5.26	115.0	0.927
BLAKE-32	[TW]	xc3s50-5	16	32	220	360	2	7.38	315.6	0.877
BLAKE-256	[31]	xc6vlx75t-1	64	64	1,336	117	0	3.65	105.0	0.897
BLAKE-256	[TW]	xc6vlx75t-1	16	32	350	146	1	5.27	277.7	1.902
Grøstl-0	[30]	xc3s200	64	64	160	1,276	0	16.67	192.0	0.150
Grøstl-0	[TW]	xc3s200-5	16	32	547	529	1	7.15	131.0	0.248
Grøstl	[31]	xc6vlx75t-1	64	64	176	285	0	3.57	815.0	2.860
Grøstl	[TW]	xc6vlx75t-1	16	32	547	179	1	4.13	226.7	1.266
Grøstl	[29]	xc5v	32	64	160	470	0	2.82	1,132.0	2.409
Grøstl	[TW]	xc5v	16	32	547	234	1	3.77	248.5	1.062
JH42	[31]	xc6vlx75t-1	64	64	689	240	0	3.47	214.0	0.892
JH42	[TW]	xc6vlx75t-1	16	32	1,845	164	1	5.39	51.5	0.314
JH42	[29]	xc5v	32	8	6,466	205	0	2.93	27.0	0.132
JH42	[TW]	xc5v	16	32	1,845	176	1	5.28	52.5	0.299
Keccak	[8]	xc5vlx50-3	64	64	5,492	448	1	3.77	52.5	0.117
Keccak	[TW]	xc5vlx50-3	16	64	3,764	192	1	3.54	81.8	0.426
Keccak	[31]	xc6vlx75t-1	64	64	2,125	144	0	4.00	128.0	0.889
Keccak	[TW]	xc6vlx75t-1	16	64	3,764	154	1	3.69	78.3	0.509
Skein	[31]	xc6vlx75t-1	64	64	458	240	0	6.25	179.0	0.746
Skein	[TW]	xc6vlx75t-1	16	32	2,439	162	1	6.02	34.9	0.215
Skein	[29]	xc5v	32	64	585	555	0	3.69	237.0	0.427
Skein	[TW]	xc5v	16	32	2,439	215	1	5.74	36.6	0.170

BRAM. This enabled us to reduce the number of clock cycles. The designs are quite comparable in terms of throughput to area ratio. Furthermore, the designers of [11] did not include the clock cycles needed for loading a message block which would bring our designs even closer. Our BLAKE-256 design performs significantly better than the design by Kerckhof [31]. Our result for Keccak on Virtex-V compares favorably with the design reported in [8]. For comparison we are assuming that we can equate their system memory with one BRAM. The reason for the difference in clock cycles is that we chose to use a quasi pipelined design whereas [8] has implemented each of the functions separately. However, Kerckhof's result for Keccak [31] is better than ours even though their area is slightly smaller. One reason is, that we have to use dedicated write cycles for the BRAM and that our rotator requires 1.5 clock cycles on average. The Grøstl-0 implementation reported in [30] is more than twice as large as our design, yet our throughput to area ratio is 1.5 times better. Their design processes the data in fewer clock cycles leading to a higher throughput. However, as they are using 8 S-boxes with no pipeline register in between, their delay is high and the area is comparatively large. The Grøstl implementations by Kerckhof [31] and Jungk [29] outperform our implementation. However, both are significantly larger, have a wider I/O, and a 64-bit datapath. We can make the same observation also for the JH implementation in [31] and Skein in [31] and [29]. Our area constraint and the BRAM restricts us to a 32-bit datapath. It is interesting to note though, that the throughput over area ratio is highly non-linear with the area of implementations of the same algorithm, i.e. the more area is available, the disproportional more the throughput improves. Jungk's implementation of JH [29] has a worse performance than ours due to its 8-bit datapath.

4.3   Conclusions

In this paper we presented the first comprehensive comparison of lightweight FPGA implementations of all SHA-3 finalists and all round-2 candidates with the exception of SIMD. All algorithms were implemented using the same assumptions, goals, tools, interface, and the same area optimization techniques. The lightweight implementations were evaluated with regards to their throughput over area ratio. The resulting ranking of algorithms is very different from implementations for best throughput over area reported in the literature [34], [20], [5]. The finalists with the best throughput over area ratio on Xilinx devices is BLAKE-256 followed by Grøstl. On the Altera Cyclone-II Grøstl is followed by BLAKE-256. JH42 and Keccak are very close to each other. The finalist with the lowest ratio is Skein. However, this ranking might change, when we change the available area or the BRAM requirement. As future work we would like to explore how much more area is needed for each of these algorithms in order to achieve a significant increase in throughput over area. This will give us an even better understanding of the scalability of the algorithms.

References

1. ATHENa results database. Automated Tool for Hardware EvaluatioN project, http://cryptography.gmu.edu/athenadb/
2. The SHA-3 Zoo. ECRYPT, Information Societies Technology (IST) Programme of the European Commission, http://ehash.iaik.tugraz.at/wiki/The_SHA-3_Zoo
3. Announcing request for candidate algorithm nominations for a new cryptographic hash algorithm (SHA-3) family. Federal Register 72(212), notices 62212 (November 2007)
4. Aumasson, J.P., Henzen, L., Meier, W., Phan, R.C.W.: SHA-3 proposal BLAKE. Submission to NIST (Round 3) (2010), http://131002.net/blake/blake.pdf
5. Baldwin, B., Hanley, N., Hamilton, M., Lu, L., Byrne, A., O'Neill, M., Marnane, W.P.: FPGA implementations of the round two SHA-3 candidates. Tech. rep., Second SHA-3 Candidate Conference (2010)
6. Benadjila, R., Billet, O., Gilbert, H., Macario-Rat, G., Peyrin, T., Robshaw, M., Seurin, Y.: SHA-3 proposal: ECHO. Submission to NIST (updated) (February 2009), http://crypto.rd.francetelecom.com/echo/
7. Bernstein, D.J.: CubeHash specification (2.b.1). Submission to NIST (Round 2) (2009), http://cubehash.cr.yp.to/
8. Bertoni, G., Daemen, J., Peeters, M., Gilles, V.A.: Keccak function version 2.0 (September 2009)

9. Bertoni, G., Daemen, J., Peeters, M., Van Assche, G.: Keccak sponge function family main document. version 1.2 (April 2009), http://keccak.noekeon.org
10. Bertoni, G., Daemen, J., Peeters, M., Van Assche, G.: The Keccak SHA-3 submission. Submission to NIST (Round 3) (2011), http://keccak.noekeon.org/Keccak-submission-3.pdf
11. Beuchat, J.L., Okamoto, E., Yamazaki, T.: Compact implementations of BLAKE-32 and BLAKE-64 on FPGA. Cryptology ePrint Archive, Report 2010/173 (2010)
12. Biham, E., Dunkelman, O.: The SHAvite-3 hash function. Submission to NIST (Round 2) (2009), http://www.cs.technion.ac.il/~orrd/SHAvite-3/Spec.15.09.09.pdf
13. Bresson, E., et al.: Shabal, a submission to NISTs cryptographic hash algorithm competition. Submission to NIST (October 2008), http://ehash.iaik.tugraz.at/uploads/6/6c/Shabal.pdf
14. Chen, Z., Morozov, S., Schaumont, P.: A hardware interface for hashing algorithms. Cryptology ePrint Archive, Report 2008/529 (2008), http://eprint.iacr.org/
15. Cryptographic Engineering Research Group, George Mason University: Hardware Interface of a Secure Hash Algorithm (SHA), v. 1.4 edn. (January 2010)
16. De Cannière, C., Sato, H., Watanabe, D.: Hash function Luffa: Specification. Submission to NIST (Round 2) (October 2009), http://www.sdl.hitachi.co.jp/crypto/luffa/Luffa_v2_Specification_20091002.pdf
17. Detrey, J., Gaudry, P., Khalfallah, K.: A low-area yet performant FPGA implementation of Shabal. Cryptology ePrint Archive, Report 2010/292 (2010)
18. Ferguson, N., Lucks, S., Schneier, B., Whiting, D., Bellare, M., Kohno, T., Callas, J., Walker, J.: The Skein hash function family. Submission to NIST (Round 3) (2010), http://www.skein-hash.info/sites/default/files/skein1.3.pdf
19. Gaj, K., Chodowiec, P.: FPGA and ASIC Implementations of AES. In: Cryptographic Engineering, pp. 235–294. Springer, Heidelberg (2009)
20. Gaj, K., Homsirikamol, E., Rogawski, M.: Fair and Comprehensive Methodology for Comparing Hardware Performance of Fourteen Round two SHA-3 Candidates Using FPGA. In: Mangard, S., Standaert, F.-X. (eds.) CHES 2010. LNCS, vol. 6225, pp. 264–278. Springer, Heidelberg (2010)
21. Gaj, K., Kaps, J.P., Amirineni, V., Rogawski, M., Homsirikamol, E., Brewster, B.Y.: ATHENa – Automated Tool for Hardware EvaluatioN: Toward fair and comprehensive benchmarking of cryptographic hardware using FPGAs. In: FPL 2010, pp. 414–421. IEEE (2010)
22. García-Vargas, I., Senhadji-Navarro, R., Jiménez-Moreno, G., Civit-Balcells, A., Guerra-Gutiérrez, P.: ROM-based finite state machine implementation in low cost FPGAs. In: Int. Symposium on Industrial Electronics, ISIE 2007, pp. 2342–2347. IEEE Press (June 2007)
23. Gauravaram, P., Knudsen, L.R., Matusiewicz, K., Mendel, F., Rechberger, C., Schäffer, M., Thomsen, S.S.: Grøstl – a SHA-3 candidate. Submission to NIST (October 2008), http://www.groestl.info/
24. Gauravaram, P., Knudsen, L.R., Matusiewicz, K., Mendel, F., Rechberger, C., Schäffer, M., Thomsen, S.S.: Grøstl – a SHA-3 candidate. Submission to NIST (Round 3) (2011), http://www.groestl.info/Groestl.pdf
25. Gligoroski, D., Klima, V., Knapskog, S.J., El-Hadedy, M., Amundsen, J., Mjølsnes, S.F.: Cryptographic hash function Blue Midnight Wish. Submission to NIST (Round 2) (September 2009), http://people.item.ntnu.no/~danilog/Hash/BMW-SecondRound/Supporting_Documentation/BlueMidnightWishDocumentation.pdf

26. Halevi, S., Hall, W.E., Jutla, C.S.: The hash function Fugue. Submission to NIST (updated) (September 2009),
   http://domino.research.ibm.com/comm/
   research_projects.nsf/pages/fugue.index.html
27. Homsirikamol, E., Rogawski, M., Gaj, K.: Comparing hardware performance of fourteen round two SHA-3 candidates using FPGAs. Cryptology ePrint Archive, Report 2010/445 (2010), http://eprint.iacr.org/
28. Homsirikamol, E., Rogawski, M., Gaj, K.: Throughput vs. Area Trade-Offs Architectures of Five Round 3 SHA-3 Candidates Implemented Using Xilinx and Altera FPGAs. In: Preneel, B., Takagi, T. (eds.) CHES 2011. LNCS, vol. 6917, pp. 491–506. Springer, Heidelberg (2011)
29. Jungk, B.: Compact implementations of Grøstl, JH and Skein for FPGAs. In: ECRYPT II Hash Workshop 2011 (May 2011)
30. Jungk, B., Reith, S.: On FPGA-based implementations of Grøstl. Cryptology ePrint Archive, Report 2010/260 (2010)
31. Kerckhof, S., Durvaux, F., Veyrat-Charvillon, N., Regazzoni, F., de Dormale, G.M., Standaert, F.X.: Compact FPGA implementations of the five SHA-3 finalists. In: ECRYPT II Hash Workshop 2011 (May 2011)
32. Kobayashi, K., Ikegami, J., Matsuo, S., Sakiyama, K., Ohta, K.: Evaluation of hardware performance for the SHA-3 candidates using SASEBO-GII. (January 2010), http://eprint.iacr.org/2010/010
33. Küçük, Ö.: The hash function Hamsi. Submission to NIST (updated) (2009), http://www.cosic.esat.kuleuven.be/publications/article-1203.pdf
34. Matsuo, S., Knežević, M., Schaumont, P., Verbauwhede, I., Satoh, A., Sakiyama, K., Ota, K.: How can we conduct "fair and consistent" hardware evaluation for SHA-3 candidate? Tech. rep., Second SHA-3 Candidate Conference (2010)
35. Namin, A., Hasan, M.: Hardware implementation of the compression function for selected SHA-3 candidates. Tech. Rep. 28, Centre for Applied Cryptographic Research (CACR), University of Waterloo (July 2009)
36. Namin, A., Hasan, M.: Implementation of the compression function for selected SHA-3 candidates on FPGA. In: International Parallel Distributed Processing Symposium, Workshops and Phd Forum (IPDPSW), pp. 1–4. IEEE (2010)
37. Rawski, M., Selvaraj, H., Luba, T.: An application of functional decomposition in ROM-based FSM implementation in FPGA devices. J. Syst. Archit. 51(6-7), 424–434 (2005)
38. Research Centre for Information Security (RCIS), National Institute of Advanced Industrial Science and Technology (AIST): Side-channel Attack Standard Evaluation Board SASEBO-GII Specification, version 1.01 edn. (November 2009)
39. Sklyarov, V.: Synthesis and Implementation of RAM-Based Finite State Machines in fPGAs. In: Grünbacher, H., Hartenstein, R.W. (eds.) FPL 2000. LNCS, vol. 1896, pp. 718–728. Springer, Heidelberg (2000)
40. Sönmez Turan, M., Perlner, R., Bassham, L.E., Burr, W., Chang, D., jen Chang, S., Dworkin, M.J., Kelsey, J.M., Paul, S., Peralta, R.: Status report on the second round of the SHA-3 cryptographic hash algorithm competition. In: NIST Interagency Report 7764, NIST, Gaithersburg (2011)
41. Tuan, T., Kao, S., Rahman, A., Das, S., Trimberger, S.: A 90nm low-power FPGA for battery-powered applications. In: FPGA 2006, ACM/SIGDA, pp. 3–11. ACM, New York (2006)
42. Wu, H.: The hash function JH. Submission to NIST (updated) (September 2009), http://icsd.i2r.a-star.edu.sg/staff/hongjun/jh/
43. Wu, H.: The hash function JH. Submission to NIST (round 3) (2011), http://www3.ntu.edu.sg/home/wuhj/research/jh/jh_round3.pdf

Publicly Verifiable Secret Sharing
for Cloud-Based Key Management

Roy D'Souza[1], David Jao[2,*], Ilya Mironov[1,3], and Omkant Pandey[1,**]

[1] Microsoft Corporation, Redmond WA, USA
{royd,omkantp}@microsoft.com
[2] University of Waterloo, Waterloo ON, Canada
djao@math.uwaterloo.ca
[3] Microsoft Research Silicon Valley Center, Mountain View CA, USA
mironov@microsoft.com

Abstract. Running the key-management service of cryptographic systems in the cloud is an attractive cost saving proposition. Supporting key-recovery is an essential component of every key-management service. We observe that to verifiably support key-recovery in a public cloud, it is essential to use publicly verifiable secret-sharing (PVSS) schemes. In addition, a *holistic* approach to security must be taken by requiring that running the key-management service in the (untrusted) cloud does not violate the security of the cryptographic system at hand.

This paper takes such a holistic approach for the case of public-key encryption which is one of the most basic cryptographic tasks. The approach boils down to formalizing the security of public-key encryption *in the presence of* PVSS. We present such a formalization and observe that the PVSS scheme of Stadler [29] can be shown to satisfy our definition, albeit in the Random Oracle Model.

We construct a new scheme based on pairings which is much more efficient than Stadler's scheme. Our scheme is noninteractive and can support any monotone access structure. In addition, it is proven secure in the *standard* model under the Bilinear Diffie-Hellman (BDH) assumption. Interestingly, our PVSS scheme is actually the *first* non-interactive scheme proven secure in the *standard* model; all previous non-interactive PVSS schemes assume the existence of a Random Oracle. Our scheme is simple and efficient; an implementation of our scheme demonstrates that our scheme compares well with the current fastest known PVSS schemes.

1 Introduction

Today, there is a huge emphasis on cloud computing. The "cloud" can be thought of as an infrastructure which is available to everyone at all times and provides reliable data storage and computational power at low cost. More and more applications are now moved from private machines to run in clouds to reduce capital

---

[*] Partially supported by NSERC.
[**] Also affiliated with Microsoft Research India, Bangalore, India.

D.J. Bernstein and S. Chatterjee (Eds.): INDOCRYPT 2011, LNCS 7107, pp. 290–309, 2011.

and operational expense. Leveraging the cloud infrastructure is an increasingly popular value proposition for IT companies [10,17]. The cloud can either be operated *privately* by a trusted party, or *publicly* by an untrusted third party. In this paper, we will be concerned only with public clouds which cannot be trusted for cryptographic purposes.

Key-management is an essential component of systems that deploy cryptographic techniques. Since availability and reliability are two crucial requirements of a good key-management service, running it in the cloud is a natural proposition for cost reduction [18]. In this work, we shall focus on one of the most basic cryptographic tasks: public-key encryption. A similar formal treatment can be easily provided for other cryptographic tasks as well such as signature schemes.

Suppose that (PK, SK) is the public and secret key-pair of an employee U of a business B. The key-management service, run by the business (administrator) B, has two fundamental tasks: securely store SK from where *only* U can access SK (for example, by providing his passphrase); and support *key-recovery*. That is, if U loses SK (or his passphrase), it should be possible for some authorized party (such as U's manager) to recover SK for U. Typically, who can recover SK for U is determined by an *access policy* $\mathbb{A}$ which represents authorized sets of parties which must collude to recover SK for U. This set may or may not include the system administrator B. At the time of registration when (PK, SK) are generated for U, strict guidelines must be followed to *verifiably* ensure that SK can indeed be recovered as defined by the policy $\mathbb{A}$. Failure to recover SK in legitimate circumstances can result in data loss which can lead to severe financial damage and perhaps legal consequences.

We would like to bring attention to a subtle yet crucial point here. The verification steps to ensure that key-recovery can indeed be performed when needed, is usually performed by an automatized procedure controlled by the system administrator B. Such a process usually needs one-time access to the generated key SK to successfully complete the verification task. As a result, in principle, administrator B might be able to look at SK even if ideally he should not be allowed to do so. The usual "fix" around this problem is that B is bound by a legal contract trusted entity. Therefore, it is not considered the part of the adversary trying to break the security of the encryption scheme. That is, in the mathematical formalizations of security of PKE (such as the IND-CPA game), B is not modeled as a separate entity.

However, when the key-management service is moved to the cloud, this becomes a severe problem. The cloud, unlike B, must be treated as an untrusted entity and hence the adversary. There is really no alternative but to use cryptographic methods to ensure that key-recovery can be successfully performed when needed. Therefore, we need a cryptographic mechanism which ensures that the cloud must be able to store SK in some "encrypted" form (from where U can access it), verify that SK can be successfully recovered by authorized set of parties as defined by the policy $\mathbb{A}$, and *yet be unable to break* the IND-CPA security of messages encrypted under PK.

Securely storing SK in "encrypted" form in the cloud so that it does not compromise IND-CPA security of encrypted messages is quite easy. Simply use an appropriate Key Encapsulation Mechanism (KEM) [28] to encrypt SK under U's password (or some other appropriate key stored, e.g., in U's employee smartcard). This allows U to access SK, and security can be argued using standard techniques almost automatically (see [28]). In this paper, we focus on the key-recovery part: how to allow the cloud to (non-interactively) verify that SK can be recovered by the legitimate parties and yet ensure that allowing the cloud to do so does not compromise the IND-CPA security of the associated public-key encryption scheme. The cryptographic tool which allows such a public verification is known as *publicly verifiable secret-sharing* (PVSS) scheme. However, plain PVSS schemes do not explicitly consider supporting public-key encryption. Rather, their goal is to simply ensure that there is a *unique* and well defined secret value s that will be recovered by (all) authorized sets of parties. It is not *explicitly* required that $s = SK$ where SK is a legitimate secret-key for PK. This is because such a guarantee is not really needed in cryptographic tasks which use PVSS, e.g., secure function evaluation [9,11], electronic voting applications [25], and so on. In our context, however, we need to ensure that $s = SK$ without compromising IND-CPA security of messages encrypted under PK.

To do this, we first present a formal security model for public-key encryption schemes that support publicly verifiable secret-sharing schemes. We then observe that even though traditional PVSS schemes do not satisfy all our requirements, the scheme of Stadler [29] can be shown to satisfy these requirements in the Random Oracle Model [4]. The public-key encryption scheme supported by Stadler's construction is the ElGamal cryptosystem [12]. The scheme, however, relies on parallel repetition of zero-knowledge proofs for proving relations about double discrete logarithms [29]. Such parallel repetitions are necessary to reduce the soundness error to an acceptable level, and makes the scheme quite inefficient in practice. Indeed, this was later addressed by Schoenmakers [25] who presented a much more efficient scheme based on simple discrete logarithms (as opposed to the double discrete logarithms). However, Schoenmakers' scheme is only a plain PVSS, i.e., it does not support a public-key encryption scheme.

To address this problem, we construct a new PVSS scheme and a corresponding PKE scheme using pairing-based techniques [16,8]. Our construction is highly efficient: asymptotically, it is optimal just like Schoenmakers' construction; in addition, our implementation shows that even in practice, despite the use of pairings, the system performs very well when compared to Schoenmakers system (which, to the best of our knowledge, is the fastest known PVSS). The implementation details, and our test results can be found in Section 4.

Our construction has an added theoretical benefit. It is proven secure in the *standard* model, under the standard bilinear Diffie-Hellman (BDH) assumption. All previous (non-interactive) PVSS constructions assume the existence of a Random Oracle. While not our main motivation, this actually resolves an open question in the area of publicly-verifiable secret-sharing schemes: ours is the first construction of a non-interactive PVSS scheme proven secure in the standard model under standard assumptions.

Related Work. To our knowledge, a formal treatment of key-management from the point of view of running key-management services on untrusted computing facilities (such as the cloud) has not previously appeared. Indeed, our work also only focuses on one of the crucial aspects of key-recovery, and does not aim to explicitly provide a full treatment to key-management in the cloud.

Nevertheless, we have been able to focus on the aspect of key-recovery which is common to almost all good key-management services. Our work relies heavily on the techniques of secret-sharing schemes, in particular the standard extension of Blakley-Shamir secret-sharing scheme [5,27] to access trees [14,24]. The idea of verifiability in secret-sharing schemes (VSS) was introduced by Chor et al. [9]. Efficient non-interactive versions were presented by Feldman [11], where verifiability of the secret is information-theoretic but secrecy relies on computational assumptions, and by Pedersen [23], where verifiability is only guaranteed computationally while secrecy is unconditional. Publicly verifiable secret-sharing schemes most relevant to our work are those of Stadler [29] and Schoenmakers [25]; both schemes hide the secret computationally.

The idea of using PVSS schemes to enforce verifiability of shares for a secret key is not new in its own, and has been used in a closely related goal of verifiable *key-escrow*. Key-escrow were initially designed with the purpose of allowing the government to recover DES keys to monitor suspected activities [21]. This was followed by extensive research considering various issues to partial key-escrow [20], verifiability in key-escrow [3,19], and so on (see [3] for a good exposure). In summary, adding verifiability to the key-escrow problem naturally brought the usage of PVSS schemes and variations of the same were developed as needed. We note that the focus of these works is different, and as such a holistic approach to security of PKE was never considered by any of these works. In addition, our scheme is the first non-interactive PVSS proven secure in the standard model.

2 Preliminaries

We assume familiarity with public key encryption scheme [13]. Unless stated otherwise, $\kappa \in \mathbb{N}$ will denote the security parameter. All parties, and mechanisms are assumed to have the security parameter as an implicit input in the form 1^κ, and run in time polynomial in κ. A function is called *negligible* if it approaches zero faster than the inverse of every polynomial.

2.1 Definitions

We first recall the definition of an Access Structure involving parties $P_1, \ldots, P_n$.

Definition 2.1 (Access Structure [2]). *Let $\{P_1, \ldots, P_n\}$ be a set of parties. A collection $\mathbb{A} \subseteq 2^{\{P_1,\ldots,P_n\}}$ is monotone if $\forall B, C$: if $B \in \mathbb{A}$ and $B \subseteq C$ then $C \in \mathbb{A}$. An* access structure *(resp., monotone access structure) is a collection (resp., monotone collection) $\mathbb{A}$ of non-empty subsets of $\{P_1, \ldots, P_n\}$; i.e., $\mathbb{A} \subseteq 2^{\{P_1,\ldots,P_n\}} \setminus \{\emptyset\}$. The sets in $\mathbb{A}$ are called* authorized sets, *and the sets not in $\mathbb{A}$ are called* unauthorized sets.

In our context, these parties will receive encrypted shares of a secret key. Each party P_i will be defined by its public parameters PP_i to be fixed by the scheme. Description of $\mathbb{A}$, defined over P_i, then includes the public parameters PP_i of relevant parties P_i. Unless stated otherwise, we shall only deal with monotone access structures in this paper.

Let PKE $= \{\mathcal{K}, \mathcal{E}, \mathcal{D}\}$ be a public key encryption scheme. We assume that there exists an efficient algorithm VALID such that VALID$(PK, SK) = 1$ if and only if (PK, SK) is in the range of $\mathcal{K}(1^\kappa)$ for some $\kappa \in \mathbb{N}$.

We now formally define PKE schemes that support publicly verifiable secret sharing (PVSS).

PKE Supporting Public-VSS. A public-key encryption scheme supporting publicly verifiable secret sharing for an access structure $\mathbb{A}$ consists of seven algorithms $\{\mathcal{K}, \mathcal{E}, \mathcal{D}, \texttt{Setup}, \texttt{GenShare}, \texttt{Verify}, \texttt{Reconst}\}$ such that the triplet PKE $= \{\mathcal{K}, \mathcal{E}, \mathcal{D}\}$ is an (ordinary) public-key encryption scheme and:

$\texttt{Setup}(1^\kappa, n)$. This is a randomized algorithm. For every $i \in [n]$, it computes a public-value PP_i (defining the party P_i) and a corresponding secret-value SK_i. It outputs the vector of pairs $\{(PP_i, SK_i), \dots, (PP_n, SK_n)\}$.

$\texttt{GenShare}(PK, SK, \mathbb{A})$. This is a randomized algorithm for generating encrypted shares. It takes as input a public-secret key-pair (PK, SK) and an access structure $\mathbb{A}$; it outputs a string π. Recall that the description of $\mathbb{A}$ includes public parameters PP_j of relevant parties.

$\texttt{Verify}(PK, \pi, \mathbb{A})$. This is a deterministic (verification) algorithm. On input $(PK, \pi, \mathbb{A})$, the algorithm either outputs 1 or 0. We require that for every $\kappa \in \mathbb{N}$ and for every (valid[1]) $\mathbb{A}$:

$$\Pr\left[\texttt{Verify}(PK, \pi, \mathbb{A}) = 1 : (PK, SK) \leftarrow \mathcal{K}(1^\kappa) \wedge \pi \leftarrow \texttt{GenShare}(PK, SK, \mathbb{A})\right]$$
$$= 1.$$

This requirement is known as the *correctness* condition.

$\texttt{Reconst}(PK, \pi, \mathbb{A}, SK_S)$. This is a deterministic algorithm for reconstructing the secret key SK from (encrypted shares in) π. Formally, let $S \in \mathbb{A}$ be an authorized set, and let $SK_S = \{SK_j\}_{j:P_j \in S}$ be the set of secret keys of parties $P_j \in S$. Algorithm $\texttt{Reconst}$ takes as input $(PK, \pi, \mathbb{A}, SK_S)$ and outputs a string SK'.

Informally, we require that the no polynomial time adversary can produce a (PK^*, π^*) which will be accepted by $\texttt{Verify}$ but $\texttt{Reconst}$ will fail to recover a valid secret key SK' for PK^*. This requirement is known as the *soundness* condition. The formal definition follows.

---

[1] $\mathbb{A}$ is defined over P_i, which in turn are defined over PP_i; $\mathbb{A}$ is valid if it satisfies Definition 2.1 and every PP_i is an output of $\texttt{Setup}(1^\kappa, n)$.

Soundness. Formally, we require that there exists a negligible function $\text{negl}(\cdot)$ such that for every valid $\mathbb{A}$, every $S \in \mathbb{A}$, every non-uniform PPT algorithm U^*, and every sufficiently large $\kappa \in \mathbb{N}$:

$$\Pr\left[\begin{array}{l}(PK^*, \pi^*) \leftarrow U^*(\mathbb{A}); SK' \leftarrow \text{Reconst}(PK^*, \pi^*, \mathbb{A}, SK_S); \\ \text{Verify}(PK^*, \pi^*, \mathbb{A}) = 1 \bigwedge \text{VALID}(PK^*, SK') = 0\end{array}\right] \le \text{negl}(\kappa).$$

Security Game for PKE supporting Public-VSS. The security is defined by considering a game played between the challenger and the adversary. We shall give the adversary flexibility to choose the public parameters of the parties it wishes to corrupt. However, we shall only consider *static* corruptions where the adversary chooses these parameters before the challenge phase. The game proceeds in the following phases:

Setup. The challenger runs the Setup algorithm to obtain system parameters $\{PP_i, SK_i\}_{i=1}^n$; it then samples ("user") keys $(PK, SK) \leftarrow \mathcal{K}(1^\kappa)$. Public parameters PK and $\{PP_i\}_{i=1}^n$ are sent to the adversary.

Corruption. The adversary "corrupts" a set of parties by sending the following to the challenger: a set $C \subset [n]$ of indices and a public parameter PP_i^* for every $i \in C$. The new public parameters for the system are: $PK, \{PP_i^*\}_{i \in C} \cup \{PP_i\}_{i \in [n] \setminus C}$.

Phase 1. The adversary sends a (valid) access structure $\mathbb{A}^*$ to the challenger such that set C of corrupted parties does not satisfy $\mathbb{A}^*$; that is, $C \notin \mathbb{A}^*$. The challenger runs the GenShare algorithm on inputs $(PK, SK, \mathbb{A}^*)$ and sends the resulting output to the adversary.

Challenge. The adversary sends two distinct and equal length messages m_0 and m_1. The challenger samples a random bit b and computes the challenge ciphertext $CT^* \leftarrow \mathcal{E}_{PK}(m_b)$. Adversary receives CT^*.

Phase 2. Phase 1 is repeated.

Guess. Adversary outputs a guess bit b'.

The advantage of an adversary in this game is defined to be $\Pr[b' = b] - \frac{1}{2}$.

Definition 2.2. *A public-key encryption scheme supporting publicly verifiable secret-sharing is said to be secure in the (static) corruption model if all (non-uniform) polynomial time algorithms have at most a negligible advantage in the security game.*

2.2   Access Trees

We will consider access structures that are representable by a tree of threshold gates. This is a very large class of access structures and have been used in many previous works including attribute-based encryption and verifiable secret sharing. To facilitate working with them, the basic framework of access trees is recalled here.

Access Tree $\mathcal{T}$. Let $\mathcal{T}$ be a tree representing an access structure. Each non-leaf node of the tree represents a threshold gate, described by its children and a threshold value. If num_x is the number of children of a node x and k_x is its threshold value, then $0 \le k_x \le num_x$ When $k_x = 1$, the threshold gate is an OR gate and when $k_x = num_x$, it is an AND gate. Each leaf node x of the tree is described by a party P_i and a threshold value k_x.

To facilitate working with the access trees, we define a few functions. We denote the parent of the node x in the tree by parent(x). The access tree $\mathcal{T}$ defines an ordering between the children of every node. That is, the children of a node x are numbered from 1 to num_x. The function id(z) returns such a number associated with the node z. (We assume that there is a publicly known method to assign such index values so that they are unique for every node x). Note that by definition, the leaf nodes do not have any children; instead they are associated with a party in $\{P_1, \dots, P_n\}$. If x is a leaf node, function id(x) returns the index $i \in [n]$ of the party associated with x.

Satisfying an Access Tree. Let $\mathcal{T}$ be an access tree with root r. Denote by $\mathcal{T}_x$ the subtree of $\mathcal{T}$ rooted at the node x. Hence $\mathcal{T}$ is the same as $\mathcal{T}_r$. If a set $\gamma \subseteq [n]$ of indices satisfies the access tree $\mathcal{T}_x$, we denote it as $\mathcal{T}_x(\gamma) = 1$. We compute $\mathcal{T}_x(\gamma)$ recursively as follows. If x is a non-leaf node, evaluate $\mathcal{T}_{x'}(\gamma)$ for all children x' of node x. $\mathcal{T}_x(\gamma)$ returns 1 if and only if at least k_x children return 1. If x is a leaf node, then $\mathcal{T}_x(\gamma)$ returns 1 if and only if id$(x) \in \gamma$.

2.3   Cryptographic Assumptions

Bilinear Diffie-Hellman (BDH) Assumption. We assume familiarity with bilinear maps (see [16,8]). Let $\mathbb{G}_1$ be bilinear group of prime order p and generator g. In addition, let $e \colon \mathbb{G}_1 \times \mathbb{G}_1 \to \mathbb{G}_2$ be the bilinear map with the target group $\mathbb{G}_2$. Let $a, b, c, d \in \mathbb{Z}_p$ be chosen at random and g be a generator of $\mathbb{G}_1$. The BDH assumption [6] states that no (non-uniform) probabilistic polynomial time algorithm $\mathcal{B}$ can distinguish the tuple $(g^a, g^b, g^c, e(g,g)^{abc})$ from the tuple $(g^a, g^b, g^c, e(g,g)^d)$ with more than a negligible advantage. Here, the advantage of $\mathcal{B}$ is defined by:

$$\left| \Pr \left[\mathcal{B}(g^a, g^b, g^c, e(g,g)^{abc}) = 1 \right] - \Pr \left[\mathcal{B}(g^a, g^b, g^c, e(g,g)^d) = 1 \right] \right|.$$

3   An Efficient Scheme without Random Oracles

In this section we shall present a public-key encryption scheme which will support public-VSS for access trees. This construction is based on bilinear pairings, and is proven secure in the standard model under the (standard) BDH assumption. As noted before, we will first present an encryption scheme, and then present a publicly verifiable secret sharing for the specific purpose of sharing the decryption keys of this encryption scheme. While our encryption schemes is new, the secret-sharing scheme will follow standard approaches. Nevertheless, since this is the first time, for completeness we will present the secret sharing part in full detail.

Let $\mathbb{G}_1$ be a bilinear group of prime order p, and let g be a randomly chosen generator of $\mathbb{G}_1$. In addition, let $e\colon \mathbb{G}_1 \times \mathbb{G}_1 \to \mathbb{G}_2$ denote the bilinear map with target group $\mathbb{G}_2$. Note that the security parameter κ determines the size of the groups, and parameters $g, \mathbb{G}_1, \mathbb{G}_2, p$ are available to all parties.

The Encryption Scheme. Our new encryption scheme, PKE $= \{\mathcal{K}, \mathcal{E}, \mathcal{D}\}$, is a variant of the ElGamal encryption scheme. It encrypts messages in $\mathbb{G}_2$. The description can be found in Figure 1.

Key Generation $\mathcal{K}$: $h \xleftarrow{\$} \mathbb{G}_1$. $SK = h$, and $PK = e(g, h)$.

Encryption $\mathcal{E}_{PK}(m \in \mathbb{G}_2)$: $R \xleftarrow{\$} \mathbb{Z}_p$, output: $\langle g^R, m \cdot PK^R \rangle$.

Decryption $\mathcal{D}(\langle C_1, C_2 \rangle, SK)$: Output $C_2/e(C_1, SK)$.

Fig. 1. Encryption scheme PKE

Observe that for correctly generated ciphertexts $\langle C_1, C_2 \rangle$: $C_2/e(C_1, SK) = m \cdot PK^R/e(g^R, h) = m$, since the denominator $e(g^R, h) = e(g, h)^R = PK^R$. Also observe that corresponding to every public key, there is a *unique* secret key, and it is possible to efficiently test if a proposed secret key SK^* is valid for a given public PK by testing that: $e(g, SK^*) = PK$.

Supporting Public-VSS Property for Access Trees. We complete the description of the remaining four algorithms $\{\texttt{Setup}, \texttt{GenShare}, \texttt{Verify}, \texttt{Reconst}\}$ in our system. Recall that our access structure $\mathbb{A}$ is represented by an access tree $\mathcal{T}$.

For $i \in \mathbb{Z}_p$ and a set S consisting of elements in $\mathbb{Z}_p$, we define the Lagrange coefficient $\Delta_{i,S}(X) = \prod_{j \in S \setminus \{i\}} \frac{X-i}{i-j}$.

$\texttt{Setup}(1^\kappa, n)$. For every $i \in [n]$: sample $y_i \xleftarrow{\$} \mathbb{Z}_p$; output $SK_i = y_i$ and $PP_i = g^{y_i}$.

$\texttt{GenShare}(PK, SK, \mathcal{T})$. Recall that $SK = h$ and $PK = e(g, h)$. Let $s \in \mathbb{Z}_p$ such that $h = g^s$. For clarity, we break the algorithm in three steps.

1. *Define polynomials*: Choose a polynomial q_x for every node x (including the leaves) in the $\mathcal{T}$. These polynomials are chosen in the following way in a top-down manner, starting from the root node r.
   For each node x in the tree, set the degree d_x of the polynomial q_x to be one less than the threshold value k_x of that node; that is, $d_x = k_x - 1$. Now, for the root node r, set $q_r(0) = s$. That is, the constant term of q_r is set

to s. Choose d_r more points randomly to completely fix the polynomial q_r. For every other node x, set $q_x(0) = q_{\text{parent}(x)}(\text{id}(x))$; i.e., the constant term of q_x is set to $q_{\text{parent}(x)}(\text{id}(x))$. Choose the remaining d_x points randomly to completely define the polynomial q_x.

2. *Encapsulate shares*: For every *leaf* node x, the share of node x is defined by: $\lambda_x = g^{q_x(\text{id}(x))}$. This value can be computed by using polynomial interpolation since all points are known (recall that $\text{id}(x)$ returns the index $i \in [n]$ of the party P_i at the leaf node x). Now, choose a random value $R_x \in \mathbb{Z}_p$; the encapsulation of λ_x is $\langle B_x, C_x \rangle$, where:

$$B_x = g^{R_x}, \qquad C_x = \lambda_x \cdot PP_{\text{id}(x)}^{R_x}.$$

Observe that the encapsulation of λ_x is simply an ElGamal encryption of λ_x under the public parameter $PP_{\text{id}(x)}$.

3. *Proof*: Finally, to enable public verification, the algorithm will "commit" to polynomials of every node x in the target group $\mathbb{G}_2$. For every node x and every $0 \le i \le d_x$, define the following values:

$$A_{x,i} = g^{q_x(i)} \quad \text{and} \quad \widehat{A}_{x,i} = e(g, A_{x,i}) = e(g,g)^{q_x(i)}.$$

The output string π consists of the following:

1. For every node x (including the leaf nodes), the "committed polynomial": $\{\widehat{A}_{x,i}\}_{i=1}^{d_x}$;
2. For every leaf node, the encapsulations: $\langle B_x, C_x \rangle$.

Verify$(PK, \pi, \mathcal{T})$. The algorithms proceeds in following steps:

1. For every node x in $\mathcal{T}$, parse π to obtain the committed points $\{\widehat{A}_{x,i}\}_{i=1}^{d_x}$ of polynomial q_x. For every leaf node x in $\mathcal{T}$, parse π to obtain the encapsulations $\langle B_x, C_x \rangle$ of secrets λ_x. (Note that λ_x is not publicly known).

2. For the root node, verify that $\widehat{A}_{r,0} = PK$. For every other node x, verify that:

$$\widehat{A}_{x,0} = \prod_{i=0}^{d_z} \left(\widehat{A}_{z,i} \right)^{\Delta_{i,\gamma_z}(w)}, \tag{1}$$

where $z = \text{parent}(x)$, $w = \text{id}(x)$, and the set $\gamma_z = \{0, 1, \ldots, d_z\}$.

3. For every leaf node x, verify that:

$$\widehat{A}_{x,0} = \frac{e(g, C_x)}{e(B_x, PP_i)}, \tag{2}$$

where $i = \text{id}(x)$.

If all tests pass, output 1; otherwise output 0. We quickly note that for correctly generated values all tests do pass, because:

$$\text{RHS of (1)} = \prod_{i=0}^{d_z} \left(\widehat{A}_{z,i}\right)^{\Delta_{i,\gamma_z}(w)} = e(g,g)^{\sum_{i=0}^{d_z} q_z(i) \cdot \Delta_{i,\gamma_z}(w)}$$

$$= e(g,g)^{q_z(w)} = e(g,g)^{q_x(0)} = \widehat{A}_{x,0},$$

$$\text{RHS of (2)} = \frac{e(g, C_x)}{e(B_x, PP_i)} = \frac{e(g, \lambda_x \cdot PP_i^{R_x})}{e(g^{R_x}, PP_i)}$$

$$= \frac{e(g, \lambda_x) \cdot e(g, PP_i^{R_x})}{e(g, PP_i)^{R_x}} = e(g, \lambda_x) = \widehat{A}_{x,0}.$$

$\mathtt{Reconst}(PK, \pi, \mathcal{T}, SK_S)$. Informally, the reconstruction procedure works as follows. First "decrypt" the shares λ_x for relevant leaf nodes. Then, apply the standard polynomial interpolation in the exponent recursively (e.g., see [24,14]). The formal description follows.

For every node x in $\mathcal{T}$, parse π to obtain the committed coefficients $\{\widehat{A}_{x,i}\}_{i=1}^{d_x}$ of polynomial q_x. For every leaf node x in $\mathcal{T}$, parse π to obtain the encapsulations $\langle B_x, C_x \rangle$ of secrets λ_x.

Now, we define a recursive algorithm $\mathrm{DecryptNode}(\pi, SK_S, x)$ that takes as input the string π and the secret key set SK_S (we assume that S is included in SK_S), and a node x in the tree. It outputs an element in $\mathbb{G}_1$ or $\perp$.

Let $i = \mathrm{id}(x)$. If x is a leaf node then let $y_i \in SK_S$ be the secret key corresponding to PP_i. The algorithm is defined as follows: if $i \in S$,

$$\mathrm{DecryptNode}(\pi, SK_S, x) = \frac{C_x}{B_x^{y_i}} = \frac{\lambda_x \cdot PP_i^{R_x}}{g^{R_x \cdot y_i}} = \lambda_x = g^{q_x(0)}.$$

If $i \notin S$, then we define $\mathrm{DecryptNode}(\pi, SK_S, x) = \perp$.

We now consider the case when x is not a leaf node. In this case, the algorithm $\mathrm{DecryptNode}(\pi, SK_S, x)$ proceeds as follows: for all nodes z that are *children* of x, it calls $\mathrm{DecryptNode}(\pi, SK_S, z)$ and stores the output as F_z. Let γ_x be an arbitrary k_x-sized set of child nodes z such that $F_z \neq \perp$. If no such set exists then the node was not satisfied and the algorithm returns $\perp$.

Otherwise, compute:

$$F_x = \prod_{z \in \gamma_x} F_z^{\Delta_{i,\gamma_x'}(0)}, \qquad \text{where } \begin{cases} i = \mathrm{id}(z) \\ \gamma_x' = \{\mathrm{id}(z) : z \in \gamma_x\} \end{cases}$$

$$= \prod_{z \in \gamma_x} g^{q_z(0) \cdot \Delta_{i,\gamma_x'}(0)}$$

$$= \prod_{z \in \gamma_x} g^{q_{\mathrm{parent}(z)}(\mathrm{id}(z)) \cdot \Delta_{i,\gamma_x'}(0)} \qquad \text{(by construction)}$$

$$= \prod_{z \in \gamma_x} g^{q_x(i) \cdot \Delta_{i,\gamma_x'}(0)}$$

$$= g^{q_x(0)} \qquad \text{(using polynomial interpolation)}$$

and return the result.

Having defined the recursive algorithm DecryptNode, our reconstruction algorithm Reconst simply calls the function DecryptNode on the root node r of the tree with inputs (π, SK_S). Observe that: DecryptNode$(\pi, SK_S, r) = g^{q_r(0)} = g^s = SK$ if and only if $\mathcal{T}(S) = 1$ (as desired).

Efficiency. Note that for ease of exposition, we have defined the simplest form of reconstruction algorithm. There are several optimizations possible. See the discussion in [14] on how to minimize the number of exponentiations (ignoring the pairing computations). Note that the Reconst algorithm does not perform any pairing computations; the computation cost is thus dominated by number of exponentiations.

On supporting every LSSS-realizable $\mathbb{A}$. Our construction is only described for access trees. However, it can be easily extended to suppose every access structure $\mathbb{A}$ which can be realized by a *linear secret-sharing scheme* (LSSS, see [2]). Such access structures $\mathbb{A}$ are represented by a *monotone span program*. Our construction will commit to the randomness of such secret-sharing scheme instead of committing to the coefficients. The full construction can be obtained by following the details of construction in Section A of [14].

4 System Implementation

Recall that our access structures are composed of a tree of threshold gates. For the purposes of evaluating performance, it suffices to consider a single (k, n)-threshold gate. For such a gate, it is easy to calculate that the theoretical cost of our scheme is $O(n)(T_1 + T_3)$ for GenShare, $O(k^2)T_2 + O(n)T_3$ for Verify, and $O(k^2 + n)T_1$ for Reconst, where T_1, T_2, and T_3 are the costs of a $\mathbb{G}_1$-exponentiation, a $\mathbb{G}_2$-exponentiation, and a pairing respectively. Hence, in terms of asymptotic cost complexity, our scheme has performance similar to [25]. In order to compare the performance of the two schemes in more detail, we implemented the two schemes and measured their running times empirically.

Our implementation is based on Mike Scott's MIRACL library [26]. For the pairing, our protocol requires a type 1 (symmetric) cryptographic pairing. We used the Tate pairing on supersingular elliptic curves over $\mathbb{F}_p$ of embedding degree 2. Although other type 1 pairings lead to a sizable performance improvement [1], we chose the Tate pairing implementation built into MIRACL because it has the advantages of public availability and integration with the supporting MIRACL API. We evaluated the performance of both schemes at the 80, 112, 128, and 256-bit security levels. Following the guidance of [22, Table 1], we used corresponding group sizes of 160, 224, 256, and 512 bits, and field sizes of 1024, 2048, 3072, and 15360 bits respectively. For the pairing-based implementation, the field size is the size of $\mathbb{G}_2$ in bits; the size of $\mathbb{G}_1$ in bits is half that of $\mathbb{G}_2$ (since the embedding degree is 2). All tests were run on an AMD 2.4GHz Opteron in 64-bit mode.

The results of our tests are presented in Appendix B. We observe that, in general, the performance of the two schemes on GenShare is comparable. Our scheme is slower for `GenShare` at the 256-bit security level because pairing operations over such large curves are slow. For `Verify`, our scheme is slower than [25] for the smallest measured values of k and faster for the largest values. We expect such a performance improvement in asymptotic terms since our scheme avoids the double exponentiation step of [25, p. 154]. For `Reconst`, our scheme is slower by about a factor of 2, in this case because group operations on large elliptic curves are slow. As mentioned above, one possible strategy for improving performance would be to use pairings on supersingular curves over fields of small characteristic with larger embedding degrees [1]. We mention, however, that execution of `Reconst` is normally needed only in unforeseen circumstances such as the loss of a key, and will not be performed simultaneously for too many users.

5   Security Proof for Our Construction

In this section, we provide a full proof of security of our pairing based scheme. First note that the proof of soundness (of the `Reconst` procedure) is straightforward. Further details can be found in Appendix A. We move on to prove the security of encryption (in the presence of public-VSS). The security of our scheme is proven by reduction to the BDH assumption. We show that if an adversary can win the security game for PKE supporting Public-VSS with non-negligible advantage, then one can construct a simulator to break the BDH assumption.

Theorem 5.1. *If a polynomial time adversary $\mathcal{A}$ wins the security for PKE scheme supporting publicly verifiable secret-sharing scheme, then there exists a polynomial time simulator $\mathcal{B}$ to break the Bilinear Diffie-Hellman Assumption.*

Proof. Suppose that $\mathcal{A}$ can succeed in the security game for PKE supporting public-VSS with advantage ϵ. We construct a simulator $\mathcal{B}$ that succeeds in the decisional BDH game with advantage $\epsilon/2$ or more. The simulation proceeds as follows.

We first let the challenger set the groups $\mathbb{G}_1, \mathbb{G}_2$ of prime order p with an efficient bilinear map e and a generator g. The challenger flips a fair coin μ outside the view of $\mathcal{B}$. If $\mu = 0$ the challenger sets $(A, B, C', D) = (g^a, g^b, g^c, e(g, g)^{abc})$; otherwise, it sets $(A, B, C', D) = (g^a, g^b, g^c, e(g, g)^d)$ for random (a, b, c, d). Now, the simulator initiates the adversary $\mathcal{A}$ interacting with it through various phases as follows.

Setup. $\mathcal{B}$ prepares the following values. First it sets $PK = e(A, B) = e(g, g)^{ab}$. Next, for every $i \in [n]$, it chooses a random value $\beta_i \in \mathbb{Z}_p$ and sets $PP_i = B^{\beta_i} = g^{b\beta_i}$. Adversary receives $(PK, \{PP_i\}_{i \in [n]})$.

Corruption. The adversary corrupts a set $C \subset [n]$ of parties by fixing public parameter PP_i^* of its own choice for every $i \in C$. The new public parameters for the system are: $PK, \{PP_i^*\}_{i \in C} \cup \{PP_i\}_{i \in [n] \setminus C}$.

Phase 1. The adversary sends an access tree $\mathcal{T}$ to the simulator such that $\mathcal{T}(C) = 0$. The simulator needs to respond with a string π as its response to the public VSS query. It proceeds as follows.

Let $s = ab$ so that $PK = e(g, g)^s$ and $y_i = b\beta_i$ so that $PP_i = g^{y_i}$ for every $i \in [n] \setminus C$. The simulator first needs to define a polynomial q_x of degree d_x for every node x. We define the following two procedures to be executed by the $\mathcal{B}$ later: PolySat and PolyUnsat. These are recursive procedures, and append values to the output string π (initially empty).

PolySat$(\mathcal{T}_x, C, \delta_x)$ This procedure sets up the polynomials for all nodes of an access *sub*-tree whose root node is *satisfied* by parties in C; that is $\mathcal{T}_x(C) = 1$. The inputs to the procedure are: the subtree $\mathcal{T}_x$ rooted at node x of $\mathcal{T}$, the set C, and an integer $\delta_x \in \mathbb{Z}_p$.

The procedure starts by defining a polynomial q_x for node x; it sets $q_x(0) = \delta_x$. It then sets the remaining points of q_x randomly to completely fix the polynomial q_x. For $0 \leq i \leq d_x$, values $\widehat{A}_{x,i} = e(g, g)^{q_x(i)}$ are then appended to π.

Now, for every child node x' of x, we call PolySat$(\mathcal{T}_{x'}, C, q_x(\mathrm{id}(x')))$. This fixes the polynomials for every node z in the access sub-tree $\mathcal{T}_x$ and appends relevant values $\widehat{A}_{x,i}$ to π. Note that by construction, all nodes satisfy the constraint that: $q_z(0) = q_{\mathrm{parent}(z)}(\mathrm{id}(z))$.

PolyUnsat$(\mathcal{T}_x, C, e(g, g)^{\delta_x})$ This procedure sets up the polynomials for all nodes of an unsatisfied access sub-tree $\mathcal{T}_x$; that is $\mathcal{T}_x(C) = 0$. The inputs to the procedure are: the subtree $\mathcal{T}_x$ rooted at node x of $\mathcal{T}$, the set C, and an element $e(g, g)^{\delta_x} \in \mathbb{G}_2$ where $\delta_x \in \mathbb{Z}_p$.

It first defines a polynomial q_x of degree d_x for the root node x such that $q_x(0) = \delta_x$. Since $\mathcal{T}_x(C) = 0$, at most $h_x \leq d_x$ children of x are satisfied. For each satisfied child x' of x, the procedure chooses a random value $\delta_{x'}$ and sets $q_x(\mathrm{id}(x')) = \delta_{x'}$. It then fixes the remaining $d_x - h_x$ points of q_x randomly to completely fix the polynomial. Let γ_x be the set of these d_x points where the value of the polynomial is chosen. That is, except for $i = 0$, value of $q(i)$ is known to $\mathcal{B}$ for every $i \in \gamma_x$.

Now the algorithm recursively defines polynomials for the rest of the nodes in the tree as follows. For each child node x' of x, the algorithm calls:

- PolySat$(\mathcal{T}_x, C, \delta_{x'})$, if x' is a satisfied child node. Note that the value $\delta_{x'} = q_x(\mathrm{id}(x'))$ is chosen by $\mathcal{B}$ in this case.
- PolyUnsat$(\mathcal{T}_x, C, e(g, g)^{q_x(\mathrm{id}(x'))})$, if x' is an unsatisfied child node. The unknown value $e(g, g)^{q_x(\mathrm{id}(x'))}$ is computed by polynomial interpolation. To see this, we obtain a general formula as follows. First note that:

$$q_x(X) = \sum_{i \in \gamma_x} q_x(i) \Delta_{i, \gamma_x}(X)$$

$$= q_x(0) \Delta_{0, \gamma_x}(X) + \underbrace{\sum_{i \in \gamma_x \setminus \{0\}} q_x(i) \Delta_{i, \gamma_x}(X)}_{\xi_x(X) \quad (=\text{known})}$$

$$= \delta_x \cdot \Delta_{0, \gamma_x}(X) + \xi_x(X).$$

Then, the following function is computable by $\mathcal{B}$:

$$e(g,g)^{q_x(X)} = \left(e(g,g)^{\delta_x}\right)^{\Delta_{0,\gamma_x}(X)} \cdot e(g,g)^{\xi_x(X)}. \tag{3}$$

Hence, the procedure can compute the input $e(g,g)^{q_x(\mathrm{id}(x'))}$ as needed above.

Before finishing the execution, the procedure computes the values $\widehat{A}_{x,i} = e(g,g)^{q_x(i)}$ for every $0 \le i \le d_x$ using formula (3) and appends it to the output π.

Having defined the two procedures, the simulator runs $\mathrm{PolyUnsat}(\mathcal{T}, C, PK)$. The procedure returns a partially complete output π which includes the committed polynomials corresponding to every node x in $\mathcal{T}$. To complete the output, $\mathcal{B}$ needs to compute the encapsulations corresponding to every leaf node x. These are computed as follows and appended to the string π:

1. If x is a satisfied leaf node (i.e., $\mathrm{id}(x) \in C$), then value $\lambda_x = g^{q_x(0)}$ is known to $\mathcal{B}$. In this case, $\mathcal{B}$ generates the encapsulation as usual; choose $R_x \xleftarrow{\$} \mathbb{Z}_p$ and output $\langle B_x, C_x \rangle$ where $B_x = g^{R_x}$ and $C_x = \lambda_x \cdot PP_i^{R_x} = \lambda_x \cdot B^{\beta_i R_x}$.
2. If x is an unsatisfied leaf node (i.e., $\mathrm{id}(x) \notin C$), then value $\lambda_x = g^{q_x(0)}$ is not known to $\mathcal{B}$. However, by construction of $\mathrm{PolyUnsat}$, we have that:

$$\lambda_x = g^{\xi_1 + ab \cdot \xi_2},$$

where both ξ_1 and ξ_2 are known values (computed recursively using interpolations and functions $\xi_x(X)$ defined above). The simulator sets $R_x = -a\xi_2/\beta_{\mathrm{id}(x)} + R'_x$ for a randomly chosen $R'_x \in \mathbb{Z}_p$. Let $i = \mathrm{id}(x)$, then the encapsulation of λ_x includes:

$$B_x = g^{R_x} = g^{-a\xi_2/\beta_i + R'_x} = g^{R'_x} \cdot A^{-\xi_2/\beta_i};$$
$$C_x = \lambda_x \cdot PP_i^{R_x} = g^{\xi_1 + ab \cdot \xi_2} \cdot \left(g^{b\beta_i}\right)^{-a\xi_2/\beta_i + R'_x} = g^{\xi_1} \cdot B^{\beta_i R'_x}.$$

Hence, $\mathcal{B}$ can compute encapsulations for all unsatisfied nodes as well.

Therefore, the simulator is able to answer the Phase 1 queries of $\mathcal{A}$. Furthermore, these queries are distributed identically to that in the original scheme.

Challenge. $\mathcal{A}$ sends two distinct equal length messages $m_0, m_1 \in \mathbb{G}_2$. The simulator chooses a random bit ν and responds by sending the following values: $\langle C', m_\nu \cdot D \rangle$.

If $\nu = 0$, then $D = e(g,g)^{abc} = PK^c$. In this case, $\langle C', m_\nu \cdot D \rangle$ is a valid encryption of m_ν and distributed identical to the original scheme. Whereas if $\nu = 1$, D is a random element of $\mathbb{G}_2$ and hence the ciphertext $\langle C', m_\nu \cdot D \rangle$ contains no information about m_ν.

Phase 2. The simulator acts exactly as it did in Phase 1.

Guess. $\mathcal{A}$ will submit a guess ν' of ν. If $\nu' = \nu$ the simulator will output $\mu' = 0$ to indicate that it was given a valid BDH-tuple otherwise it will output $\mu' = 1$ to indicate it was given a random 4-tuple.

As shown in the construction, the simulator's generation of public parameters and answers to the queries of $\mathcal{A}$ in all stages are identical to that of the actual scheme.

In the case where $\mu = 1$ the adversary gains no information about ν. Therefore, we have $\Pr[\nu \neq \nu' \mid \mu = 1] = \frac{1}{2}$. Since the simulator guesses $\mu' = 1$ when $\nu \neq \nu'$, we have $\Pr[\mu' = \mu \mid \mu = 1] = \frac{1}{2}$.

If $\mu = 0$ then the adversary sees an encryption of m_ν. The adversary's advantage in this situation is ϵ by definition. Therefore, we have $\Pr[\nu = \nu' \mid \mu = 0] \geq \frac{1}{2} + \epsilon$. Since the simulator guesses $\mu' = 0$ when $\nu = \nu'$, we have $\Pr[\mu' = \mu \mid \mu = 0] \geq \frac{1}{2} + \epsilon$.

The overall advantage of the simulator in the Decisional BDH game is equal to $\frac{1}{2}\Pr[\mu' = \mu \mid \mu = 0] + \frac{1}{2}\Pr[\mu' = \mu \mid \mu = 1] - \frac{1}{2} \geq \frac{1}{2}(\frac{1}{2} + \epsilon) + \frac{1}{2} \cdot \frac{1}{2} - \frac{1}{2} = \frac{1}{2}\epsilon$.

References

1. Aranha, D.F., López, J., Hankerson, D.: High-Speed Parallel Software Implementation of the η_t Pairing. In: Pieprzyk, J. (ed.) CT-RSA 2010. LNCS, vol. 5985, pp. 89–105. Springer, Heidelberg (2010)
2. Beimel, A.: Secure Schemes for Secret Sharing and Key Distribution. PhD thesis, Israel Institute of Technology, Technion, Haifa, Israel (June 1996)
3. Bellare, M., Goldwasser, S.: Verifiable partial key escrow. In: ACM Conference on Computer and Communications Security, pp. 78–91 (1997)
4. Bellare, M., Rogaway, P.: Random oracles are practical: A paradigm for designing efficient protocols. In: ACM Conference on Computer and Communications Security, pp. 62–73 (1993)
5. Blakley Jr., G.R.: Safeguarding cryptographic keys. In: AFIPS 1979, National Computer Conference, vol. 48, pp. 313–317 (1979)
6. Boneh, D., Boyen, X.: Efficient Selective-ID Secure Identity-Based Encryption Without Random Oracles. In: Cachin, C., Camenisch, J.L. (eds.) EUROCRYPT 2004. LNCS, vol. 3027, pp. 223–238. Springer, Heidelberg (2004)
7. Boneh, D., Franklin, M.: Identity-Based Encryption from the Weil Pairing. In: Kilian, J. (ed.) CRYPTO 2001. LNCS, vol. 2139, pp. 213–229. Springer, Heidelberg (2001)
8. Boneh, D., Franklin, M.K.: Identity-based encryption from the Weil pairing. SIAM J. Comput. 32(3), 586–615 (2003); Ealier version in [7]
9. Chor, B., Goldwasser, S., Micali, S., Awerbuch, B.: Verifiable secret sharing and achieving simultaneity in the presence of faults (extended abstract). In: 26th Annual Symposium on Foundations of Computer Science (FOCS), pp. 383–395. IEEE (1985)
10. Creeger, M.: Cloud computing: An overview. Queue 7, 2:3–2:4 (2009)
11. Feldman, P.: A practical scheme for non-interactive verifiable secret sharing. In: 28th Annual Symposium on Foundations of Computer Science (FOCS), pp. 427–437. IEEE (1987)
12. El Gamal, T.: A Public Key Cryptosystem and a Signature Scheme Based on Discrete Logarithms. In: Blakely, G.R., Chaum, D. (eds.) CRYPTO 1984. LNCS, vol. 196, pp. 10–18. Springer, Heidelberg (1985)

13. Goldwasser, S., Micali, S.: Probabilistic encryption. J. Comput. Syst. Sci. 28(2), 270–299 (1984)
14. Goyal, V., Pandey, O., Sahai, A., Waters, B.: Attribute-based encryption for fine-grained access control of encrypted data. In: Juels, A., Wright, R.N., De Capitani di Vimercati, S. (eds.) ACM Conference on Computer and Communications Security, pp. 89–98. ACM (2006)
15. Joux, A.: A one round protocol for tripartite Diffie-Hellman. In: Bosma, W. (ed.) ANTS 2000. LNCS, vol. 1838, pp. 385–394. Springer, Heidelberg (2000)
16. Joux, A.: A one round protocol for tripartite Diffie-Hellman. J. Cryptology 17(4), 263–276 (2004); Earlier version in [15]
17. Klien, M.: Six Benefits of Cloud Computing (2010), http://resource.onlinetech.com/the-six-benefits-of-cloud-computing/
18. Martin, L.: Federated Key Management for Secure Cloud Computing. Presentation by Voltage Security, Inc. (May 2010), http://storageconference.org/2010/Presentations/KMS/17.Martin.pdf%
19. Micali, S.: Fair Public-Key Cryptosystems. In: Brickell, E.F. (ed.) CRYPTO 1992. LNCS, vol. 740, pp. 113–138. Springer, Heidelberg (1993)
20. Micali, S., Shamir, A.: Partial key-escrow (1996) (manuscript)
21. Escrowed encryption standard (EES). FIPS PUB 185, National Institute of Standards and Technology (February 1994)
22. National Institute of Standards and Technology. NIST Special Publication 800-57: Recommendation for Key Management — Part 1: General (revised) (2007)
23. Pedersen, T.P.: Non-Interactive and Information-Theoretic Secure Verifiable Secret Sharing. In: Feigenbaum, J. (ed.) CRYPTO 1991. LNCS, vol. 576, pp. 129–140. Springer, Heidelberg (1992)
24. Sahai, A., Waters, B.: Fuzzy Identity-Based Encryption. In: Cramer, R. (ed.) EUROCRYPT 2005. LNCS, vol. 3494, pp. 457–473. Springer, Heidelberg (2005)
25. Schoenmakers, B.: A Simple Publicly Verifiable Secret Sharing Scheme and Its Application to Electronic Voting. In: Wiener, M. (ed.) CRYPTO 1999. LNCS, vol. 1666, pp. 148–164. Springer, Heidelberg (1999)
26. Scott, M.: MIRACL—A Multiprecision Integer and Rational Arithmetic C/C++ Library. Shamus Software Ltd, Dublin, Ireland (2010), http://www.shamus.ie/
27. Shamir, A.: How to share a secret. Commun. ACM 22(11), 612–613 (1979)
28. Shoup, V.: Encryption algorithms—part 2: Asymmetric ciphers. Final Committee Draft 18033-2, ISO/IEC (December 2004), http://www.shoup.net/iso/std6.pdf
29. Stadler, M.: Publicly Verifiable Secret Sharing. In: Maurer, U.M. (ed.) EUROCRYPT 1996. LNCS, vol. 1070, pp. 190–199. Springer, Heidelberg (1996)

A   Proof of Soundness

Proof of Soundness. Let U^* be an arbitrary non-uniform PPT adversary. Let T be an arbitrary access tree (representing an access structure $\mathbb{A}$), and let S be such that $T(S) = 1$. Let (PK^*, π^*) be the output of U^* on input T (we assume that the advice string is in-built in the description of U^*). If $\texttt{Verify}(PK^*, \pi^*, T) = 1$ then we have the following:

1. For a node x, let $\{\widehat{A}^*_{x,i}\}_{i=1}^{d_x}$ be the values parsed by $\texttt{Verify}$. From the uniqueness of the discrete logarithm in the prime-order groups, we have that for every x and every $0 \le i \le d_x$, there exists a *unique* value $\alpha^*_{x,i}$ such that

$\widehat{A}^*_{x,i} = e(g,g)^{\alpha^*_{x,i}}$. This fixes a *unique* polynomial of degree d_x for the node x. Also the algorithm tests that for the root node, $\widehat{A}^*_{r,0} = PK = e(g,g)^s$; we have that $q^*_r(0) = s$.

2. Let x be a leaf node, and let $\langle B^*_x, C^*_x \rangle$ be the parsed encapsulated values. First, we claim that $\text{DecryptNode}(\pi^*, SK_S) = g^{q^*_x(0)}$ for every x such that $\text{id}(x) \in S$.

   From the test in (2), we have that there exist unique $R^*_x \in \mathbb{Z}_p$ and $\lambda^*_x \in \mathbb{G}_1$) such that: $B^*_x = g^{R^*_x}$, $C^*_x = \lambda^*_x \cdot PP^{R^*_x}_{\text{id}(x)}$, and $\widehat{A}^*_{x,0} = e(g, \lambda^*_x)$. This implies that $\lambda^*_x = g^{q^*_x(0)}$. Observe that the output of $\text{DecryptNode}(\pi^*, SK_S)$ is λ^*_x if $\text{id}(x) \in S$. This proves the claim.

3. Finally, from the test (1), we have for every node x: $q^*_x(0) = q^*_{\text{parent}(x)}(\text{id}(x))$. It follows that for every node x, the value F_x computed by the Reconst algorithm returns $g^{q^*_x(0)}$. As a result, the output of the Reconst algorithm is: $F_r = g^{q^*_r(0)} = SK$.

This establishes the soundness of the protocol.

B   Empirical Benchmarks

In the following tables, we list the observed timings of our implementation of the GenShare, Verify, and Reconst algorithms. For comparison, we also implemented and measured the performance of the corresponding algorithms in Schoenmakers' scheme [25], and both sets of timings are provided in the tables below. For further details regarding our implementation, see Section 4.

80 bit	k = 1	5	10	15	20	25	30	35	40	45	50
n = 10	70 / 110	80 / 110	80 / 110								
15	110 / 160	110 / 160	120 / 160	110 / 170							
20	160 / 210	150 / 210	150 / 230	150 / 220	160 / 220						
25	190 / 270	190 / 260	200 / 280	190 / 270	200 / 280	190 / 270					
30	230 / 310	230 / 320	230 / 320	220 / 310	230 / 320	230 / 320	230 / 330				
35	270 / 360	260 / 370	270 / 360	270 / 370	260 / 370	270 / 370	270 / 380	260 / 390			
40	300 / 410	300 / 420	310 / 420	300 / 430	300 / 420	290 / 430	300 / 440	300 / 430	310 / 440		
45	350 / 460	340 / 470	330 / 470	350 / 470	340 / 470	350 / 480	350 / 480	340 / 480	340 / 490	340 / 520	
50	380 / 520	380 / 520	380 / 520	370 / 520	370 / 520	380 / 540	380 / 530	390 / 540	380 / 540	370 / 560	370 / 540

112 bit	k = 1	5	10	15	20	25	30	35	40	45	50
n = 10	320 / 360	320 / 370	320 / 410								
15	480 / 540	480 / 560	480 / 560	470 / 580							
20	630 / 740	640 / 730	670 / 740	630 / 750	650 / 770						
25	790 / 910	790 / 910	790 / 930	810 / 930	800 / 950	800 / 960					
30	980 / 1080	970 / 1090	960 / 1100	960 / 1120	990 / 1130	970 / 1130	980 / 1170				
35	1120 / 1270	1120 / 1260	1110 / 1290	1120 / 1290	1120 / 1300	1140 / 1310	1110 / 1320	1120 / 1350			
40	1280 / 1440	1290 / 1440	1300 / 1450	1280 / 1460	1370 / 1480	1280 / 1490	1270 / 1500	1280 / 1520	1280 / 1540		
45	1450 / 1620	1450 / 1640	1440 / 1620	1470 / 1630	1440 / 1660	1440 / 1670	1430 / 1840	1460 / 1690	1450 / 1730	1450 / 1720	
50	1590 / 1810	1600 / 1790	1580 / 1830	1590 / 1810	1590 / 1850	1590 / 1850	1600 / 1880	1610 / 1880	1590 / 1880	1600 / 1890	1600 / 1900

128 bit	k = 1	5	10	15	20	25	30	35	40	45	50
n = 10	760 / 830	760 / 830	770 / 870								
15	1150 / 1210	1140 / 1260	1140 / 1270	1140 / 1280							
20	1530 / 1600	1520 / 1630	1520 / 1640	1560 / 1670	1520 / 1750						
25	1880 / 2010	1890 / 2020	1900 / 2050	1900 / 2080	1890 / 2120	1890 / 2120					
30	2290 / 2400	2260 / 2410	2290 / 2440	2250 / 2480	2260 / 2520	2280 / 2810	2270 / 2560				
35	2700 / 2830	2650 / 2830	2680 / 2880	2650 / 2880	2660 / 2900	2650 / 2940	2670 / 2990	2700 / 3020			
40	3100 / 3180	3030 / 3220	3030 / 3280	3060 / 3300	3020 / 3500	3170 / 3330	3060 / 3360	3050 / 3410	/ 3430		
45	3440 / 3630	3470 / 3650	3380 / 3650	3420 / 3650	3410 / 3690	3450 / 3740	3400 / 3760	3400 / 3780	3450 / 3860	3400 / 3840	
50	3800 / 4000	3800 / 4040	3810 / 4090	3810 / 4070	3790 / 4120	3780 / 4430	3780 / 4150	3790 / 4250	3780 / 4240	3770 / 4230	3940 / 4290

256 bit	k = 1	5	10	15	20	25	30	35	40	45	50
n = 10	44140 / 30430	44480 / 31570	44740 / 32570								
15	66920 / 45660	66620 / 46530	66790 / 47870	66410 / 48840							
20	88360 / 60290	88190 / 61160	89870 / 62580	88760 / 64060	92600 / 70620						
25	111120 / 75400	110910 / 76410	111830 / 77460	111600 / 78900	110720 / 80530	110660 / 81100					
30	133120 / 90520	133640 / 91350	133580 / 92510	132860 / 94550	133010 / 95400	133350 / 96240	132520 / 97790				
35	155400 / 105040	155310 / 106440	157830 / 108020	155500 / 108950	167470 / 110310	154880 / 111290	155760 / 113630	155460 / 114240			
40	178990 / 120650	180610 / 121830	193690 / 122900	176750 / 123580	179140 / 125150	180810 / 126980	177330 / 127510	177860 / 129210	178090 / 129470		
45	199700 / 136270	198990 / 135750	198270 / 137610	199030 / 138630	202850 / 139540	200080 / 142110	199290 / 142130	201540 / 162080	201210 / 144850	220940 / 146050	
50	223930 / 151200	222360 / 150740	221820 / 152370	222500 / 154600	220730 / 155340	226480 / 156660	221660 / 158420	221080 / 158790	224450 / 159460	220410 / 162610	221810 / 170520

Fig. 2. Time in milliseconds for GenShare, at various security levels, for selected values of k and n. Top numbers in each cell are for our scheme; bottom numbers are for [25].

80 bit

n	k = 1	5	10	15	20	25	30	35	40	45	50
10	100 / 70	100 / 80	130 / 120								
15	140 / 110	150 / 130	170 / 180	220 / 230							
20	180 / 140	190 / 150	210 / 200	260 / 280	330 / 370						
25	230 / 180	240 / 200	260 / 240	310 / 320	380 / 440	480 / 580					
30	270 / 210	290 / 230	310 / 280	360 / 370	420 / 480	520 / 620	630 / 800				
35	320 / 250	330 / 270	360 / 320	390 / 410	480 / 550	560 / 690	670 / 870	830 / 1100			
40	370 / 280	380 / 300	400 / 360	450 / 460	520 / 580	610 / 760	710 / 940	850 / 1180	1020 / 1450		
45	420 / 320	420 / 340	450 / 400	500 / 510	570 / 640	650 / 810	770 / 1020	910 / 1300	1090 / 1610	1250 / 1860	
50	470 / 380	470 / 380	490 / 440	540 / 550	620 / 690	700 / 880	820 / 1100	970 / 1390	1110 / 1640	1330 / 1980	1490 / 2310

112 bit

n	k = 1	5	10	15	20	25	30	35	40	45	50
10	410 / 300	430 / 330	530 / 480								
15	610 / 430	640 / 570	720 / 630	940 / 900							
20	830 / 600	840 / 630	950 / 790	1150 / 1160	1390 / 1470						
25	1020 / 730	1040 / 780	1130 / 960	1310 / 1310	1610 / 1680	1940 / 2210					
30	1210 / 880	1230 / 940	1320 / 1130	1520 / 1420	1800 / 1880	2160 / 2430	2600 / 3180				
35	1420 / 1170	1450 / 1090	1540 / 1280	1710 / 1610	1990 / 2070	2350 / 2650	2790 / 3400	3350 / 4300			
40	1600 / 1170	1700 / 1230	1820 / 1440	1930 / 1780	2220 / 2260	2560 / 2890	2990 / 3670	3700 / 4580	4310 / 5710		
45	1810 / 1330	1840 / 1600	1910 / 1610	2120 / 1980	2450 / 2500	2740 / 3140	3240 / 3940	3730 / 4910	4390 / 6080	5110 / 7290	
50	2010 / 1450	2040 / 1550	2170 / 1770	2340 / 2150	2600 / 2730	2950 / 3390	3530 / 4220	3940 / 5210	4560 / 6390	5280 / 7700	6060 / 9130

128 bit

n	k = 1	5	10	15	20	25	30	35	40	45	50
10	990 / 690	1050 / 780	1280 / 1120								
15	1510 / 1050	1550 / 1150	1770 / 1510	2170 / 2130							
20	1980 / 1390	2040 / 1490	2310 / 1880	2700 / 2510	3280 / 3510						
25	2470 / 1760	2530 / 1860	2740 / 2230	3190 / 2930	3770 / 3900	4590 / 5230					
30	3020 / 2090	3020 / 2230	3240 / 2620	3640 / 3340	4250 / 4410	5040 / 5680	6060 / 7430				
35	3520 / 3020	3560 / 2600	3780 / 3030	4200 / 3750	4760 / 4830	5570 / 6220	6560 / 7940	8380 / 10060			
40	4030 / 2770	4070 / 2910	4340 / 3410	4670 / 4210	5280 / 5350	6140 / 6740	7030 / 8550	8290 / 10800	9640 / 13550		
45	4480 / 3150	4520 / 3300	4720 / 3860	5160 / 4600	5790 / 5800	6870 / 7300	7550 / 9210	8730 / 11350	10210 / 14000	11700 / 16990	
50	4960 / 3480	5140 / 3670	5410 / 4200	5610 / 5200	6220 / 6270	7020 / 7930	8030 / 9810	9210 / 12260	10580 / 14930	12200 / 17960	14240 / 21640

256 bit

n	k = 1	5	10	15	20	25	30	35	40	45	50
10	67990 / 28930	70720 / 31970	79360 / 45790								
15	101660 / 43570	105260 / 47030	112520 / 60520	129060 / 86390							
20	136150 / 57780	138440 / 61270	146660 / 75580	166150 / 101290	189490 / 140420						
25	173420 / 72300	173300 / 76110	182370 / 90900	200640 / 117560	222070 / 156970	265640 / 207800					
30	205330 / 87170	207350 / 90880	215660 / 108450	231710 / 133560	258000 / 172690	297980 / 249110	328120 / 292050				
35	240600 / 101080	241060 / 106180	262750 / 122500	266270 / 153180	290560 / 207970	321650 / 244380	361750 / 311540	411490 / 392170			
40	274870 / 115490	276150 / 121170	288990 / 137970	299010 / 168420	325440 / 207840	355200 / 262240	397230 / 330940	448800 / 413090	496840 / 509530		
45	306960 / 130950	307270 / 134880	316040 / 151680	333910 / 182950	374740 / 223810	389810 / 318080	428760 / 349830	491120 / 436530	535690 / 532520	594150 / 656010	
50	400070 / 144880	342270 / 149700	352770 / 167680	368720 / 199020	392030 / 243940	425850 / 341040	464270 / 372970	516150 / 458320	565050 / 556530	630380 / 670890	704710 / 798930

Fig. 3. Time in milliseconds for Verify, at various security levels, for selected values of k and n. Top numbers in each cell are for our scheme; bottom numbers are for [25].

80 bit	$k = 1$	5	10	15	20	25	30	35	40	45	50
$n = 10$	0	20	110								
	0	10	50								
15	10	30	120	280							
	0	10	40	110							
20	0	30	120	270	490						
	0	20	50	110	200						
25	0	30	120	270	480	850					
	0	10	40	100	190	300					
30	10	30	110	270	480	770	1160				
	0	10	50	100	190	320	430				
35	10	30	110	270	500	780	1150	1600			
	0	10	50	110	190	290	420	590			
40	0	30	120	270	490	780	1130	1560	2080		
	0	10	50	110	190	290	420	590	770		
45	0	30	110	260	490	780	1140	1580	2170	2640	
	0	10	40	100	180	290	420	580	760	990	
50	0	30	110	260	490	790	1130	1590	2080	2790	3300
	0	10	50	100	190	290	430	630	760	970	1200

112 bit	$k = 1$	5	10	15	20	25	30	35	40	45	50
$n = 10$	10	100	460								
	0	40	190								
15	0	100	440	1100							
	10	50	180	430							
20	0	110	460	1090	2000						
	0	40	180	420	770						
25	10	110	450	1070	1960	3190					
	0	40	180	410	780	1220					
30	10	100	440	1060	1960	3140	4660				
	0	40	180	410	780	1200	1840				
35	0	110	460	1080	1970	3160	4630	6410			
	0	50	180	420	830	1200	1760	2460			
40	10	90	460	1060	1970	3160	4590	6430	8490		
	10	40	180	410	760	1200	1760	2440	3260		
45	0	110	440	1120	1960	3130	4820	6370	8420	10900	
	0	40	180	410	760	1200	1770	2420	3240	4120	
50	10	100	460	1070	2070	3160	4610	6350	8400	10730	13570
	10	40	180	410	770	1210	1760	2410	3250	4110	5100

128 bit	$k = 1$	5	10	15	20	25	30	35	40	45	50
$n = 10$	20	220	1020								
	10	90	440								
15	20	220	980	2420							
	10	100	410	1010							
20	10	220	1000	2370	4410						
	10	100	420	990	1890						
25	20	220	1000	2390	4330	7080					
	0	110	420	990	1850	2930					
30	10	220	990	2350	4350	6930	10360				
	10	90	420	980	1850	2910	4300				
35	10	250	1020	2400	4460	6990	10360	14230			
	0	100	430	990	1820	2900	4250	5920			
40	20	210	1000	2350	4340	6960	10190	14120	18830		
	10	90	430	1000	1840	2900	4230	5880	7960		
45	10	230	980	2360	4330	7120	10150	14110	18680	24030	
	10	110	410	1000	1800	2890	4230	5830	7710	9900	
50	10	240	990	2380	4350	6980	10240	14050	18620	23920	30170
	0	100	430	980	1830	2930	4220	5900	7820	9900	12510

256 bit	$k = 1$	5	10	15	20	25	30	35	40	45	50
$n = 10$	510	8330	39510								
	260	3880	18260								
15	520	8670	37310	93260							
	260	4120	17440	43280							
20	530	8720	38750	90380	171110						
	250	3970	17800	42160	79370						
25	510	8670	38060	90820	168180	288340					
	250	4090	17660	42440	78020	126620					
30	510	8310	37160	92560	166860	268980	408860				
	250	3890	17320	42160	77490	124620	185240				
35	510	9400	38790	91470	169770	269710	402200	563480			
	250	4320	17880	42550	78280	145900	183820	256290			
40	520	7930	38490	91830	167840	268260	395790	552200	731230		
	260	3720	17840	42180	78040	125120	183340	253250	339000		
45	520	9140	37070	90830	166960	268720	394180	548450	726240	938150	
	250	4070	17250	42350	77050	124730	182300	254770	336120	430460	
50	510	9080	38140	90530	169500	272900	393910	547660	731220	929300	1169760
	250	4150	17670	42880	78680	125350	183860	253640	333800	429270	536560

Fig. 4. Time in milliseconds for `Reconst`, at various security levels, for selected values of k and n. Top numbers in each cell are for our scheme; bottom numbers are for [25].

On Constructing Families of Pairing-Friendly Elliptic Curves with Variable Discriminant[*]

Robert Dryło

Institute of Mathematics, Polish Academy of Sciences,
ul. Śniadeckich 8, 00-956 Warszawa, Poland
Instytut Matematyki, Uniwersytet Humanistyczno-Przyrodniczy w Kielcach,
ul. Świetokrzyska 15, 25-406 Kielce, Poland
r.drylo@impan.gov.pl

Abstract. In [10] Freeman, Scott and Teske consider three types of families: complete, sparse and complete with variable discriminant. A general method for constructing complete families is due to Brezing and Weng. In this note we generalize this method to construct families of the latter two types. As an application, we find variable-discriminant families for a few embedding degrees, which improve the previous best ρ-values of families given in [10].

1 Introduction

Elliptic curves with small embedding degrees are used for implementing pairing-based cryptosystems (see, e.g., [3,4,8,11,15]). For details on such curves and their constructions we refer to an exhaustive paper of Freeman, Scott and Teske [10].

Recall that the *embedding degree* of an elliptic curve over a finite field $\mathbb{F}_q$ with respect to its $\mathbb{F}_q$-rational subgroup of prime order r is defined to be the smallest integer k such that $r|(q^k - 1)$.

To construct an ordinary elliptic curve E with a given embedding degree k one finds parameters r, t, q of E, where t is the trace of E, and then makes use of the CM method to find its equation. For efficiency of the CM method the discriminant D of E must be sufficiently small, where D is the square-free part of $4q - t^2 > 0$. Furthermore, elliptic curves preferred for applications should have the bit size of r close to the size of the order $|E(\mathbb{F}_q)|$; in other words, the parameter $\rho = \log q / \log r$ should be close to 1. Parameters of such curves are usually obtained as values of the following polynomials.

Definition 1. ([10, Def.2.7]) Let $r(x), t(x), q(x)$ be nonzero polynomials with rational coefficients.
(i) The triple (r, t, q) *parametrizes a family of elliptic curves with embedding degree k and discriminant D*, where $D > 0$ is a square-free integer, if the following conditions are satisfied:
(1) $q(x) = p(x)^d$, where $p(x)$ represents primes and $d \geq 1$.

---

[*] Research supported by the Polish Minister of Science as project O R00 0111 12.

D.J. Bernstein and S. Chatterjee (Eds.): INDOCRYPT 2011, LNCS 7107, pp. 310–319, 2011.
© Springer-Verlag Berlin Heidelberg 2011

(2) $r(x)$ is irreducible, integer-valued, and has positive leading coefficient.

(3) $r(x)$ divides $q(x) + 1 - t(x)$.

(4) $r(x)$ divides $\Phi_k(t(x) - 1)$, where Φ_k is the kth cyclotomic polynomial.

(5) The equation $4q(x) - t(x)^2 = Dy^2$ has infinitely many integer solutions (x, y).

(ii) We say that a family (r, t, q) is *complete* if there exists $y(x) \in \mathbb{Q}[x]$ such that $4q(x) - t(x)^2 = Dy(x)^2$; otherwise we say that the family is *sparse*.

(iii) We say that a family (r, t, q) is *potential* if it satisfies conditions (2)-(5).

The parameter ρ of a family is defined as

$$\rho = \frac{\deg q(x)}{\deg r(x)};$$

if $r(x), t(x), q(x)$ for some $x \in \mathbb{Z}$ are parameters of elliptic curves, then their ρ-values tend to ρ.

For constructions of families with $\rho = 1$ and embedding degrees $k = 3, 4, 6, 10, 12$ see [2,9,12,14,16]. A general method for constructing complete families with a given embedding degree k and discriminant D is due to Brezing and Weng [6] (see also [1]). For higher security it may be desired to use curves with sufficiently large discriminant. Therefore Freeman, Scott and Teske [10, Sec.6.4] (see also [5,16]) considered *complete families with variable discriminant*, which here we define as follows.

Definition 2. *A triple of polynomials $(r(x), q(x), t(x))$ parametrizes a complete family of elliptic curves with embedding degree k and variable discriminant if it satisfies conditions (1)-(4) of Definition 1 and $4q(x) - t(x)^2 = xh(x)^2$ for some $h(x) \in \mathbb{Q}[x]$. Then substitution $x \leftarrow Dx^2$, where $D > 0$ is a square-free integer, yields a complete family $r(Dx^2), t(Dx^2), q(Dx^2)$ with discriminant D, whenever conditions (1), (2) of Definition 1 are satisfied.*

In [10, Th.6.19] variable-discriminant families are obtained from complete families (r, t, q) with discriminant D such that there exists polynomials $r', t', q', y' \in \mathbb{Q}[x]$ satisfying $r(x) = r'(x^2)$, $t(x) = t'(x^2)$, $q(x) = q'(x^2)$, and $4q(x) - t(x)^2 = Dx^2(y'(x^2))^2$. Then the family (r', t', q') satisfies Definition 2 after substituting $x \leftarrow x/D$. However, it is not clear how to construct such complete families, although a few remarkable cyclotomic families were found in [10].

The main contribution of this paper is an explicit algorithm given in Algorithm 5 for constructing complete families with variable discriminant, which generalizes the Brezing-Weng method [6]. As an application, we construct sporadic families (i.e., $r(x)$ is not a cyclotomic polynomial) with variable discriminant, which improve the previous best ρ-values of families given in [10], and we interpret constructions of cyclotomic families.

To obtain elliptic curves with variable discriminant one can also use sparse families, which for some embedding degrees may improve ρ-values. We also generalize the Brezing-Weng method to construct these families, however in this case the algorithm is less efficient and finding suitable families is more difficult, nevertheless we give some examples.

At the end of the paper we collect the families that improve on ρ-values in [10] in a table.

2 Constructing Complete Families with Variable Discriminant

We start by recalling the Brezing-Weng method.

Algorithm 3. Input: A number field K containing kth roots of unity and $\sqrt{-D}$, where $D > 0$ is a square-free integer.
Output: A complete potential family with embedding degree k and discriminant D.

1. Find a polynomial $r(x) \in \mathbb{Q}[x]$ such that $K = \mathbb{Q}[x]/(r(x))$.
2. Choose a kth primitive root of unity $\zeta_k \in K$.
3. Let $t(x) \in \mathbb{Q}[x]$ be a lift of $\zeta_k + 1$.
4. Let $y(x) \in \mathbb{Q}[x]$ be a lift of $(\zeta_k - 1)/\sqrt{-D}$.
5. Let $q(x) = \frac{1}{4}(t(x)^2 + Dy(x)^2)$.
6. Return $(r(x), t(x), q(x))$.

The following simple fact allows us to extend the Brezing-Weng algorithm to construct complete families with variable discriminant.

Lemma 4. *If (r, t, q) is a complete family with variable discriminant such that $4q(x) - t(x)^2 = xh(x)^2$, then $-\bar{x}$ is a square in $K = \mathbb{Q}[x]/(r(x))$ and $\bar{h} = \pm(\zeta_k - 1)/\sqrt{-\bar{x}}$, where the bar denotes the residue class mod r.*

Proof. Reducing the equation $4q(x) - t(x)^2 = xh(x)^2$ mod $r(x)$ and using the fact that $\bar{q} = \bar{t} - 1 = \zeta_k$, we have $\bar{x}\bar{h}^2 = 4\bar{q} - \bar{t}^2 = 4\zeta_k - (\zeta_k + 1)^2 = -(\zeta_k - 1)^2$, which implies the assertion.

Therefore to construct families in question we need for a number field K containing kth roots of unity to find a defining polynomial $r(x)$ such that $\sqrt{-\bar{x}} \in K$. It can be obtained as the minimal polynomial of $-z^2$, where $z \in K$ and z^2 should be a primitive element of K. Hence we have the following algorithm.

Algorithm 5. Input: A number field K containing kth roots of unity.
Output: A complete potential family with embedding degree k and variable discriminant.

1. Choose $z \in K$ such that $a = -z^2$ is a primitive element of K.
2. Let $r(x)$ be a minimal polynomial of a, and write $K = \mathbb{Q}[x]/(r(x))$.
3. Choose a primitive kth root of unity $\zeta_k \in K$.
4. Let $t(x)$ and $h(x)$ be lifts of $\zeta_k + 1$ and $(\zeta_k - 1)/\sqrt{-\bar{x}}$, respectively.
5. Let $q(x) = \frac{1}{4}(t(x)^2 + xh(x)^2)$.
6. Return (r, t, q).

Note that we obtain families with parameter $\rho = \max\{2 \deg t, 1 + 2 \deg h\}/\deg r$, which is generically equal only to $(2 \deg r - 1)/\deg r$, whenever $\deg t, \deg h < \deg r$.

First we explain from the point of view of the above algorithm constructions of cyclotomic families given by Freeman, Scott and Teske [10, Constructions 6.2, 6.3, 6.20, 6.24].

Example 6. If k is odd, then $\sqrt{\zeta_k} = \pm\zeta_k^{(k+1)/2} \in K = \mathbb{Q}(\zeta_k)$. Let us take $a = -\zeta_k = \zeta_{2k}$ in Algorithm 5 and $r(x) = \Phi_{2k}(x)$. Then a family depends on the choice of a kth root of unity ζ_k^u, where $0 < u < k$ is an integer relatively prime to k. Let $t(x) \to \zeta_k^u + 1$ and $h(x) \to (\zeta_k^u - 1)/\sqrt{-a} = (\zeta_k^u - 1)/\sqrt{\zeta_k} = (\zeta_k^u - 1)\zeta_k^{(k-1)/2} = \zeta_k^{u+(k-1)/2} - \zeta_k^{(k-1)/2}$. The ρ-value depends on the degree of $\zeta_k^u + 1$ and $\zeta_k^{u+(k-1)/2} - \zeta_k^{(k-1)/2}$ with respect to ζ_k.

(i) For $u = 1$ we obtain the family with an odd embedding degree k and $\rho = \frac{k+2}{\varphi(k)}$

$$r(x) = \Phi_{2k}(x),$$

$$t(x) = -x + 1,$$

$$q(x) = \tfrac{1}{4}(x^{k+2} + 2x^{k+1} + x^k + x^2 - 2x + 1).$$

(ii) For $u = (k+1)/2$ we have $\zeta_k^{u+(k-1)/2} - \zeta_k^{(k-1)/2} = 1 - \zeta_k^{(k-1)/2}$. Hence we obtain the family with an odd embedding degree k and $\rho = \frac{k+1}{\varphi(k)}$

$$r(x) = \Phi_{2k}(x),$$

$$t(x) = (-x)^{(k+1)/2} + 1,$$

$$q(x) = \tfrac{1}{4}(x^{k+1} + x^k + 4(-x)^{(k+1)/2} + x + 1).$$

However, if $k \equiv 1 \pmod 4$, then $q(1) = 0$, so $q(x)$ is reducible and does not represent primes. Using Magma in the remaining cases one can check as in [10] that $q(x)$ represents primes for $k \le 1000$.

Example 7. To obtain cyclotomic families with embedding degree $2k$, where k be odd, let us take as above $a = \zeta_{2k} = -\zeta_k$ in $K = \mathbb{Q}(\zeta_{2k})$, and $r(x) = \Phi_{2k}(x)$. For an integer $0 < u < 2k$ relatively prime to $2k$, let $t(x) \to \zeta_{2k}^u + 1$ and $h(x) \to (\zeta_{2k}^u - 1)/\sqrt{-\zeta_{2k}} = (\zeta_{2k}^u - 1)(-\zeta_{2k})^{(k-1)/2} = (-1)^{(k-1)/2}(\zeta_{2k}^{u+(k-1)/2} - \zeta_{2k}^{(k-1)/2})$.

(i) For $u = 1$ we obtain the family with embedding degree $2k$ and $\rho = (k+2)/\varphi(k)$

$$r(x) = \Phi_{2k}(x)$$

$$t(x) = x + 1$$

$$q(x) = \tfrac{1}{4}(x^{k+2} - 2x^{k+1} + x^k + x^2 + 2x + 1)$$

(ii) If $k \equiv 1 \pmod 4$, then $u = (k+1)/2$ is relatively prime to $2k$, and $h \to -(1 + \zeta_{2k}^{(k-1)/2})$. Hence we have the following family with embedding degree $2k$ and $\rho = (k+1)/\varphi(k)$

$$r(x) = \Phi_{2k}(x)$$

$$t(x) = x^{(k+1)/2} + 1$$

$$q(x) = \tfrac{1}{4}(x^{k+1} + x^k + 4x^{(k+1)/2} + x + 1).$$

Now we construct a few sporadic families with variable discriminant, which slightly improve ρ-values of families given in [10]. Similarly as in the paper of Kachisa, Schaefer and Scott [13], we generate polynomials $r(x)$ as minimal polynomials of elements $-z^2$, where z has small coefficients in the cyclotomic basis.

Remark 8. If (r, t, q) is a variable-discriminant complete family such that its polynomials evaluated at $a + bx$, $a, b \in \mathbb{Z}$, are integer-valued, and $r(a+bx)/c, q(a+bx)$ represent primes for some $c \in \mathbb{Z}$, then we can obtain a complete family with discriminant $D \equiv 1, a \pmod{b}$ by evaluating r, t, q at $D(a + bx)^2$ or $D(1 + bx)^2$, respectively. Since the values of the last two polynomials are congruent to a $\pmod{b}$ for $x \in \mathbb{Z}$, we obtain families whose polynomials are integer-valued, so it remains to check whether conditions (1) and (2) of Definition 1 are satisfied.

Example 9. The following family has embedding degree $k = 8$ and $\rho = 1.75$,

$$r(x) = x^4 - 4x^3 + 8x^2 + 8x + 4,$$

$$t(x) = \tfrac{1}{12}(-x^3 + 5x^2 - 16x + 14),$$

$$q(x) = \tfrac{1}{576}(4x^7 - 39x^6 + 170x^5 - 311x^4 + 52x^3 + 716x^2 - 384x + 196).$$

We obtain r as the minimal polynomial of $1 - 2\zeta_8 + \zeta_8^2 = -(-\zeta_8^2 + \zeta_8^3)^2$. These polynomials evaluated at $11 + 24x$ have integer coefficients, and $q(11 + 24x)$, $r(11 + 24x)/9$ represent primes. Hence we can obtain a complete family with discriminant $D \equiv 1, 11 \pmod{24}$ by evaluating this family at $D(11 + 24x)^2$ or $D(1 + 24x)^2$, respectively. Note that $4q - t^2 = \tfrac{1}{36}x(x^3 - 5x^2 + 10x + 4)^2$.

We have noticed that in the examples below families are obtained by taking $r(x)$ to be the minimal polynomial of ζ_k/a such that $\sqrt{-\zeta_k/a} \in \mathbb{Q}(\zeta_k)$ for some $a \in \mathbb{Z}$. Note that then we could also use $a\zeta_k$, but families determined by ζ_k/a have smaller denominators.

Example 10. $k = 9$, $\rho \approx 1.666$,

$$r(x) = 729x^6 + 27x^3 + 1,$$

$$t(x) = 243x^5 + 1,$$

$$q(x) = \tfrac{1}{4}(59049x^{10} + 6561x^9 + 8748x^8 + 2916x^7 + 972x^6 + 1296x^5$$
$$+ 108x^4 + 36x^3 + 12x^2 + x + 1).$$

We obtain $\frac{1}{729}r$ as the minimal polynomial of $\frac{1}{3}\zeta_9 = -\frac{1}{9}(\zeta_9^5 + 2\zeta_9^2)^2$. Both q and r evaluated at $1 + 2x$ have integer coefficients and represent primes. Thus we can obtain a complete family with an odd discriminant D by evaluating r, t, q at $D(1 + 2x)^2$. This family improves the construction with $\rho = 1.833$ given in [10].

Example 11. $k = 15$, $\rho = 1.625$,

$$r(x) = 6561x^8 - 2187x^7 + 243x^5 - 81x^4 + 27x^3 - 3x + 1,$$

$$t(x) = 9x^2 + 1,$$

$$q(x) = \tfrac{1}{4}(531441x^{13} - 236196x^{11} + 39366x^{10} + 39366x^9 - 8748x^8 - 729x^7$$

$$+486x^6 - 243x^5 + 135x^4 + 18x^3 + 18x^2 + x + 1).$$

We obtain $\frac{1}{6561}r$ as the minimal polynomial of

$$\frac{1}{3}\zeta_{15} = -\frac{1}{9}(-\zeta_{15}^7 + \zeta_{15}^5 - \zeta_{15}^4 - \zeta_{15}^3 - \zeta_{15} + 1)^2.$$

Both q and r evaluated at $1 + 2x$ have integer coefficients and represent primes. Hence we obtain a complete family with odd discriminant D by evaluating r, t, q at $D(1 + 2x)^2$. This improves the construction in [10] with $\rho = 1.75$ and $2 \deg r(x) = 32$.

Example 12. $k = 28$, $\rho = 1.5$,

$$r(x) = 4096x^{12} - 1024x^{10} + 256x^8 - 64x^6 + 16x^4 - 4x^2 + 1,$$

$$t(x) = 512x^9 + 1,$$

$$q(x) = \tfrac{1}{4}(262144x^{18} + 65536x^{17} - 32768x^{15} + 16384x^{14} + 12288x^{13}$$

$$-3072x^{11} + 2816x^9 - 192x^7 + 48x^5 + 16x^4 - 8x^3 + x + 1).$$

Then $\frac{1}{4096}r$ is the minimal polynomial of $\frac{1}{2}\zeta_{28} = -\frac{1}{4}(\zeta_{28}^{11} + \zeta_{28}^4)^2$. Evaluating q and r at $3 + 4x$ we get polynomials with integer coefficients, which represent primes. Thus we can obtain a complete family with an odd discriminant D by evaluating the family at $D(3 + 4x)^2$ or $D(1 + 4x)^2$ for $D \equiv 1, 3 \pmod 4$, respectively. This improves the construction in [10] with $\rho = 1.917$.

Example 13. $k = 30$, $\rho = 1.625$,

$$r(x) = 390625x^8 + 78125x^7 - 3125x^5 - 625x^4 - 125x^3 + 5x + 1,$$

$$t(x) = -25x^2 + 1,$$

$$q(x) = \tfrac{1}{4}(244140625x^{13} + 195312500x^{12} + 78125000x^{11} + 19531250x^{10} + 2343750x^9$$

$$-140625x^7 - 43750x^6 - 6875x^5 - 125x^4 + 150x^3 - 50x^2 + 9x + 1).$$

We obtain $\frac{1}{390625}r$ as the minimal polynomial of

$$\frac{1}{5}\zeta_{30} = -\frac{1}{25}(\zeta_{30}^7 + 2\zeta_{30}^6 + \zeta_{30}^5 - \zeta_{30}^4 - \zeta_{30}^3 - \zeta_{30} + 1)^2.$$

Both q and r evaluated at $1 + 2x$ have integer coefficients and represent primes, so we obtain complete families with odd discriminant D by evaluating r, t, q at $D(1 + 2x)^2$. This family improves the construction in [10] with $\rho = 1.813$ and $2 \deg r(x) = 32$.

3   Constructing Sparse Families

Sometimes one can obtain variable-discriminant families with better ρ-value using sparse families. We generalize the Brezing-Weng method to construct these families, however the method is much less efficient than the previous one.

Note that in order the CM equation $4q(x) - t(x)^2 = Dy^2$ of a family (r, t, q) to have infinitely many integer solutions (x, y) its left-hand side must be of the form

$$4q(x) - t(x)^2 = g(x)h(x)^2, \text{ where } g, h \in \mathbb{Q}[x], \deg g \le 2, \text{ and } \mathrm{lc}(g) > 0. \quad (1)$$

This fact was observed by Freeman [9], and is a consequence of Siegel's theorem. Write $4q(x) - t(x)^2 = g(x)h(x)^2$, where $g(x) \in \mathbb{Q}[x]$ has no multiple roots, and multiply the CM equation $g(x)h(x)^2 = Dy^2$ by a non-zero integer $c = ab^2$, where $a, b \in \mathbb{Z}$ and $g_1 = ag$, $h_1 = bh$ have integer coefficients. Then the equation $g_1(x)h_1(x)^2 = D'y^2$ has infinitely many integer solutions, where D' is a square-free part of Dab^2, and hence so does the equation $g_1(x) = D'y^2$. From Siegel's theorem [17, Th.IX.4.3] it follows that $\deg g_1 \le 2$.

Similarly as in Lemma 4, reducing the equation (1) mod r we find that $-\bar{g}$ is a square in $K = \mathbb{Q}[x]/(r)$ and $\bar{h} = \pm(\zeta_k - 1)/\sqrt{-\bar{g}}$. Hence we have the following algorithm for constructing families satisfying conditions (1) and (2)-(4) of Definition 1, which by abuse of notation we will also call potential.

Algorithm 14. Input: A number field K containing kth roots of unity. Output: A potential family with embedding degree k.

1. Find a polynomial $r(x) \in \mathbb{Q}[x]$ such that $K = \mathbb{Q}[x]/(r(x))$.
2. Find $g(x) \in \mathbb{Q}[x]$ with $\deg g \le 2$ and positive leading coefficient such that $-g \bmod r$ is a square in K.
3. Choose a kth primitive root of unity $\zeta_k \in K$.
4. Let $t(x)$ and $h(x)$ be lifts of $\zeta_k + 1$ and $(\zeta_k - 1)/\sqrt{-\bar{g}}$, respectively.
5. Let $q(x) = \frac{1}{4}(t(x)^2 + g(x)h(x)^2)$.
6. Return (r, t, q).

This algorithm is much more time consuming than Algorithm 5, because for each polynomial r we must look for new polynomials g (only if g is constant, it does not depend on r; then the family is complete). Given r, the polynomials

g in step 2 can be obtained as follows. Let $n = \deg r$, and $G_i \in \mathbb{Q}[X_1, \ldots, X_n]$, $i = 0, \ldots, n-1$, be quadratic polynomials satisfying mod r

$$\left(\sum_{i=1}^{n} x_i \bar{x}^{i-1} \right)^2 = \sum_{i=0}^{n-1} G_i(x_1, \ldots, x_n) \bar{x}^i \text{ for } x_1, \ldots, x_n \in \mathbb{Q}.$$

Then g's are lifts of $-(G_0(\mathbf{x}) + G_1(\mathbf{x})\bar{x} + G_2(\mathbf{x})\bar{x}^2)$ for some $\mathbf{x} \in S = \{G_3 = \cdots = G_{n-1} = 0\} \cap \mathbb{Q}^n$. Note that g and $u^2 g$ for $u \in \mathbb{Q} \backslash 0$ give the same family, so each family is determined by a point with integer coordinates in S. To save some work throughout looking for such points, one can enumerate part of variables $x_1, \ldots, x_m \in \mathbb{Z}$ and determine the remaining coordinates $x_{m+1}, \ldots, x_n \in \mathbb{Q}$ such that $(x_1, \ldots, x_n) \in S$ by solving the system $G_i(x_1, \ldots, x_m, X_{m+1}, \ldots, X_n) = 0$, $i = 3, \ldots, n-1$. We expect that for $m = 3$ this system will have generically finitely many solutions. Of course, this part of the algorithm is very time consuming. Furthermore, the resulting families have $\rho = \max\{2 \deg t, \deg g + 2 \deg h\} / \deg r$, which is generically equal to 2, whenever $\deg t, \deg h < \deg r$. We give the following examples.

Example 15. We first explain Freeman's construction [9] of a family with $k = 10$ and $\rho = 1$,

$$r(x) = 25x^4 + 25x^3 + 15x^2 + 5x + 1,$$

$$t(x) = 10x^2 + 5x + 3,$$

$$q(x) = 25x^4 + 25x^3 + 25x^2 + 10x + 3.$$

The minimal polynomial of $\frac{1}{5}(-2\zeta_{10}^2 + \zeta_{10} - 2) \in \mathbb{Q}(\zeta_{10})$ times the common denominator of its coefficients is equal to r. Then we take $g = 15x^2 + 10x + 3 \equiv -(10x^2 + 5x + 1)^2 \pmod{r}$, $\zeta_{10} \to 10x^2 + 5x + 2$, and $h = 1$.

Example 16. $k = 8$, $\rho = 1.5$,

$$r(x) = x^4 - 2x^2 + 9,$$

$$t(x) = \tfrac{1}{12}(-x^3 + 3x^2 + 5x + 9),$$

$$q(x) = \tfrac{1}{576}(x^6 - 6x^5 + 7x^4 - 36x^3 + 135x^2 + 186x - 63).$$

We obtain r as the minimal polynomial of $-\zeta_8^3 + \zeta_8^2 + \zeta_8 \in \mathbb{Q}(\zeta_8)$. Then we take $\zeta_8 \to \tfrac{1}{12}(-x^3 + 3x^2 + 5x - 3)$, $g = 8x^2 - 16 \equiv -(\bar{x}^2 - 5)^2 \pmod{r}$, and $h = \tfrac{1}{12}(-x + 3)$. Note that $4q - t^2 = \tfrac{1}{18}(x^2 - 2)(x - 3)^2$. The polynomials r, t, q evaluated at $3 + 12x$ have integer coefficients, and $q(3 + 12x)$, $r(3 + 12x)/72$ represent primes.

Example 17. $k = 12, \rho = 1.5$,

$$r(x) = x^4 - 2x^3 - 3x^2 + 4x + 13,$$

$$t(x) = \tfrac{1}{15}(-x^3 + 4x^2 + 5x + 6),$$

$$q(x) = \tfrac{1}{900}(x^6 - 8x^5 + 18x^4 - 56x^3 + 202x^2 + 258x - 423).$$

We find r as the minimal polynomial of $-\zeta_{12}^3 + \zeta_{12}^2 + 2\zeta_{12} \in \mathbb{Q}(\zeta_{12})$. Then we choose $\zeta_k \to \tfrac{1}{15}(-x^3 + 4x^2 + 5x - 9)$, $g = 12x^2 - 12x - 51 \equiv -(x^3 - x - 8)^2$ (mod r), and $h = \tfrac{1}{15}(-x + 3)$. Note that $4q - t^2 = \tfrac{4}{75}(x^2 - x - 17/4)(x - 3)^2$. Evaluating r, t, q at $3 + 30x$ or $23 + 30x$, we obtain polynomials with integer coefficients such that $q(3 + 30x)$, $r(3 + 30x)/25$, $q(23 + 30x)$, $r(23 + 30x)/225$ represent primes. This family improves the construction with $\rho = 1.75$ given in [10].

4   Conclusion

We have generalized the Brezing-Weng method to construct complete families with variable discriminant and sparse families. This allows us to find families for a few embedding degrees, which improve on ρ-values given by Freeman, Scott and Teske [10]. Our improvements are summarized in the table below. If (r, t, q) is a variable-discriminant family, then "degree" means $2 \deg r$ if the family is complete, and $\deg r$ if the family is sparse.

k	ρ	D	Degree	Constr.
8	1.500	some	4	16
9	1.666	odd	12	10
12	1.500	some	4	17
15	1.625	odd	16	11
28	1.500	odd	24	12
30	1.625	odd	16	13

Acknowledgements. The author would like to thank Michael Naehrig and the referees for their helpful comments, which allow to improve an earlier draft of this paper.

References

1. Barreto, P.S.L.M., Lynn, B., Scott, M.: Constructing Elliptic Curves with Prescribed Embedding Degrees. In: Cimato, S., Galdi, C., Persiano, G. (eds.) SCN 2002. LNCS, vol. 2576, pp. 257–267. Springer, Heidelberg (2003)
2. Barreto, P.S.L.M., Naehrig, M.: Pairing-Friendly Elliptic Curves of Prime Order. In: Preneel, B., Tavares, S. (eds.) SAC 2005. LNCS, vol. 3897, pp. 319–331. Springer, Heidelberg (2006)

3. Boneh, D., Franklin, M.: Identity-Based Encryption from the Weil Pairing. In: Kilian, J. (ed.) CRYPTO 2001. LNCS, vol. 2139, pp. 213–229. Springer, Heidelberg (2001)

4. Boneh, D., Lynn, B., Shacham, H.: Short Signatures from the Weil Pairing. In: Boyd, C. (ed.) ASIACRYPT 2001. LNCS, vol. 2248, pp. 514–532. Springer, Heidelberg (2001); Full version: J. Cryptol. 17, 297–319 (2004)

5. Bisson, G., Satoh, T.: More Discriminants with the Brezing-Weng Method. In: Chowdhury, D.R., Rijmen, V., Das, A. (eds.) INDOCRYPT 2008. LNCS, vol. 5365, pp. 389–399. Springer, Heidelberg (2008)

6. Brezing, F., Weng, A.: Elliptic curves suitable for pairing based cryptography. Des. Codes Cryptogr. 37, 133–141 (2005)

7. Cocks, C., Pinch, R.G.E.: Identity-based cryptosystems based on the Weil pairing (2001) (unpublished manuscript)

8. Cha, J.C., Cheon, J.H.: An Identity-Based Signature from Gap Diffie-Hellman Groups. In: Desmedt, Y.G. (ed.) PKC 2003. LNCS, vol. 2567, pp. 18–30. Springer, Heidelberg (2002)

9. Freeman, D.: Constructing Pairing-Friendly Elliptic Curves with Embedding Degree 10. In: Hess, F., Pauli, S., Pohst, M. (eds.) ANTS 2006. LNCS, vol. 4076, pp. 452–465. Springer, Heidelberg (2006)

10. Freeman, D., Scott, M., Teske, E.: A taxonomy of pairing-friendly elliptic curves. J. Cryptol. 23, 224–280 (2010)

11. Joux, A.: A one round protocol for tripartite DiffieHellman. In: Bosma, W. (ed.) ANTS 2000. LNCS, vol. 1838, pp. 385–393. Springer, Heidelberg (2000)

12. Galbraith, S., McKee, J., Valença, P.: Ordinary abelian varieties having small embedding degree. Finite Fields Appl. 13, 800–814 (2007)

13. Kachisa, E.J., Schaefer, E.F., Scott, M.: Constructing Brezing-Weng Pairing-Friendly Elliptic Curves Using Elements in the Cyclotomic Field. In: Galbraith, S.D., Paterson, K.G. (eds.) Pairing 2008. LNCS, vol. 5209, pp. 126–135. Springer, Heidelberg (2008)

14. Miyaji, A., Nakabayashi, M., Takano, S.: New explicit conditions of elliptic curves traces for FR-reduction. IEICE Trans. Fundam. E84-A, 1234–1243 (2001)

15. Sakai, R., Ohgishi, K., Kasahara, M.: Cryptosystems based on pairings. In: 2000 Symposium on Cryptography and Information Security, SCIS 2000, Okinawa, Japan (2000)

16. Scott, M., Barreto, P.S.L.M.: Generating more MNT elliptic curves. Des. Codes Cryptogr. 38, 209–217 (2006)

17. Silverman, J.: The Arithmetic of Elliptic Curves. Springer, Berlin (1986)

Attractive Subfamilies of BLS Curves for Implementing High-Security Pairings

Craig Costello[1,2,*], Kristin Lauter[2], and Michael Naehrig[2,3]

[1] Information Security Institute
Queensland University of Technology, GPO Box 2434, Brisbane QLD 4001, Australia
craig.costello@qut.edu.au
[2] Microsoft Research
One Microsoft Way, Redmond, WA 98052, USA
klauter@microsoft.com
[3] Department of Mathematics and Computer Science
Technische Universiteit Eindhoven, P.O. Box 513, 5600 MB Eindhoven, Netherlands
michael@cryptojedi.org

Abstract. Barreto-Lynn-Scott (BLS) curves are a stand-out candidate for implementing high-security pairings. This paper shows that particular choices of the pairing-friendly search parameter give rise to four subfamilies of BLS curves, all of which offer highly efficient and implementation-friendly pairing instantiations.

Curves from these particular subfamilies are defined over prime fields that support very efficient towering options for the full extension field. The coefficients for a specific curve and its correct twist are automatically determined without any computational effort. The choice of an extremely sparse search parameter is immediately reflected by a highly efficient optimal ate Miller loop and final exponentiation. As a resource for implementors, we give a list with examples of implementation-friendly BLS curves through several high-security levels.

Keywords: Pairing-friendly, high-security pairings, BLS curves.

1 Introduction

Current public-key security recommendations have influenced a concentrated effort from the pairing-based community towards optimizing the implementation of pairings on Barreto-Naehrig (BN) curves [4]. Indeed, aside from an array of many other attractive properties (see [20,23] for more details), BN curves are perfectly suited to the security level of 128 bits (cf. [2], [32], and [12, §1.1]), since they achieve an optimal balance between the necessary sizes of the three groups involved in the pairing $e : \mathbb{G}_1 \times \mathbb{G}_2 \to \mathbb{G}_T$. The BN pairing speed record of 10

* Acknowledges funding from the Australian-American Fulbright Commission, the Queensland Government Smart State PhD Scholarship, and an Australian Postgraduate Award.

D.J. Bernstein and S. Chatterjee (Eds.): INDOCRYPT 2011, LNCS 7107, pp. 320–342, 2011.

million cycles set by Hankerson *et al.* in 2008 [14] stood until mid 2010, when three papers appeared in rapid succession [22,7,1], each one shaving more time off the previous record and pushing the limit of efficiency at this security level. This work was pinnacled by Aranha *et al.* [1], who applied a combination of improvements to accelerate the entire pairing computation to less than 2 million cycles. As implementors seemingly converge towards a "satisfaction asymptote" at the 128-bit security level, the focus is now beginning to shift to optimizing pairings at higher security levels.

As Scott details [25], scaling security in pairing-based cryptography is fundamentally different than doing so in traditional public-key protocols that only require one group definition. Whilst increasing the security of other number theoretic protocols usually requires an increase in the size of the modulus, an optimized scaling in the context of pairing-based cryptography can be achieved by increasing the embedding degree. Utilizing the flexibility of a higher embedding degree allows an implementor to inflate the size of the finite field target group $\mathbb{G}_T$ at a greater rate than the corresponding inflation in the elliptic curve groups $\mathbb{G}_1$ and $\mathbb{G}_2$, paying respect to the faster subexponential attacks that must be resisted in $\mathbb{G}_T$. Stepping up to a significantly higher security level therefore calls for a different family of pairing-friendly elliptic curves altogether, and although many of the previous optimizations (e.g. for BN curves) can be immediately or easily transferred to computing pairings on the new curves, there are naturally new issues that arise when advancing towards a thorough optimization.

This work focuses on pairings that employ the Barreto-Lynn-Scott (BLS) family with $k = 24$ [3] (see also [12, §6.6]). The BLS family has already been identified as the prime candidate for 256-bit secure pairings by Scott [27], who now holds the current software speed record at this level. The aim of this paper is to provide some of the finer details that will begin to pave the way for implementors who may wish to further accelerate the state-of-the-art timings on BLS curves. The bulk of our discussion is motivated by Pereira *et al.*'s work in the case of BN curves [23], where they detail a very simple method of generating highly optimal instantiations of *implementation-friendly* BN curves. With the same intent, we point out four highly attractive subfamilies of BLS curves that facilitate very efficient instantiations of high-security pairings.

The proposed curves are found by restricting the search parameter x in the polynomial representation to any one of four specific congruency classes, namely $x_0 \equiv 7, 16, 31, 64 \pmod{72}$. These choices result in prime fields of characteristic $p \equiv 19 \pmod{24}$, which in turn leads to three very efficient towering options for the full extension field $\mathbb{F}_{p^{24}}$. We show that all curves found in any of the proposed subfamilies can immediately be given by the same short Weierstraß equation over $\mathbb{F}_p$, and the unique sextic twist E' of correct order for use in the ate pairing setting is automatically determined. Not only is there no computational effort necessary to write down the curve equations once the prime p is found, both E and E' can be represented very compactly by the polynomial parameter x_0 alone.

We give some details on line function computation and discuss an efficient version of the hard part in the final exponentiation. As a resource for implementors, we give elaborate lists of example curves covering a range of high-security levels between 192- and 320-bit security, including the dedicated BLS security level of 256 bits. We have chosen the examples with a very sparse parameter x which leads to a low number of addition steps in the Miller loop and simultaneously very efficient exponentiations by x, which are needed in the final exponentiation.

The remainder of this paper is organized as follows. Section 2 gives a brief background on the BLS family for embedding degree $k = 24$ and on computing optimal pairings on BLS curves. Section 3 describes the proposed implementation-friendly subfamilies of BLS curves, whilst Section 4 details some choices that can facilitate more simple pairing code. In Section 5 we apply the work of Scott *et al.* [29] to give details on the final exponentiation routine. Lists of example curves at several security levels are presented in Section 6. Finally, Appendix A provides timing results of a C implementation of the optimal ate pairing on BLS curves.

2   Background

The BLS family with embedding degree 24. In [3], Barreto, Lynn and Scott propose polynomial parametrizations for certain complete pairing-friendly curve families for specific fixed embedding degrees. All curves belonging to one of these so-called cyclotomic families have CM discriminant $D = 3$ (i.e. j-invariant $j = 0$) and can be given by a short Weierstraß equation $E : y^2 = x^3 + b$. For embedding degree $k = 24$, the Barreto-Lynn-Scott (BLS) family is given by the following parametrization (see also [12, Construction 6.6]):

$$p(x) = (x-1)^2(x^8 - x^4 + 1)/3 + x, \qquad r(x) = x^8 - x^4 + 1,$$
$$n(x) = (x-1)^2(x^8 - x^4 + 1)/3, \qquad t(x) = x + 1,$$
$$f(x) = (x-1)(2x^4 - 1)/3. \tag{1}$$

Finding a specific BLS curve is achieved by running through integer values $x_0 \equiv 1 \pmod{3}$ until $p(x_0)$ and $r(x_0)$ are both prime (note that $x_0 \equiv 1 \pmod 3$ leads to all involved parameters being integers). For each set of parameters, there exists an elliptic curve E over $\mathbb{F}_p$ such that $\#E(\mathbb{F}_p) = n(x_0)$. The correct curve E can be found by trying different values for b (i.e. different twists) and checking for the right group order. Another alternative is to compute the coefficient by the algorithm described in [24]. Since $r(x_0) \mid n(x_0)$, there is a subgroup of $E(\mathbb{F}_p)$ of prime order $r(x_0)$. The CM discriminant is $D = 3$ because $4p(x_0) - t(x_0)^2 = 3f(x_0)^2$. The family has a ρ-value of $\rho = \deg(p)/\deg(r) = 1.25$.

BLS curves achieve the smallest ρ-value for $k = 24$ (see [12, Section 8]); they have twists of degree 6 and allow for many optimizations when computing pairings, similar to Barreto-Naehrig (BN) curves. A particularly nice property is that the ate pairing [15] already provides an optimal pairing [33]. The number of iterations in Miller's algorithm to compute the ate pairing is $\log(t(x_0) - 1) = \log(x_0) \approx \log(r(x_0))/\varphi(k) = \log(r(x_0))/8$.

According to key size recommendations in [2], [32], and [12, Section 1], BLS curves are a good choice for pairings at the high-security 256-bit level. Indeed, Scott [27] recently demonstrated an efficient implementation of pairings on BLS curves at that level.

An optimal ate pairing. We now briefly recall the ate pairing on BLS curves and the arithmetic that is involved in its computation. Let $E/\mathbb{F}_p$ be a BLS curve with parameters constructed as above given an integer $x_0 \equiv 1 \pmod 3$, i.e. $p = p(x_0)$, $r = r(x_0)$ and so forth. Then there exists a unique twist $E'/\mathbb{F}_{p^4}$ with $r \mid \#E'(\mathbb{F}_{p^4})$ with twisting isomorphism $\psi : E' \to E$ over $\mathbb{F}_{p^4}$. Let us define the usual groups $G_1 = E(\mathbb{F}_p)[r]$ and $G_2 = \ker(\phi_p - [p]) \subset E(\mathbb{F}_{p^{24}})[r]$ as the 1- and p-eigenspaces of the Frobenius endomorphism ϕ_p on $E[r]$. Let the group $G_2' = \psi^{-1}(G_2)$ be the preimage of G_2 under ψ.

Let $T = t - 1$. As outlined in [9], we can either compute the original ate pairing

$$a_T : G_2' \times G_1 \to \mathbb{F}_{p^{24}}^*, \ (Q', P) \mapsto f_{T,\psi(Q')}(P)^{\frac{p^{24}-1}{r}},$$

by "untwisting" Q' for the computation of $f_{T,\psi(Q')}$; or we can compute

$$a_T' : G_2' \times G_1 \to \mathbb{F}_{p^{24}}^*, \ (Q', P) \mapsto f_{T,Q'}(\psi^{-1}(P))^{\frac{p^{24}-1}{r}},$$

i.e. compute entirely on the twist by "twisting" P to E'.

A function $f_{T,R}(S)$ for points R, S on E or E' as above is computed with Miller's algorithm [19]. It involves curve arithmetic in G_2', i.e. doubling and addition of points on the curve E' over the field $\mathbb{F}_{p^4}$ and the computation of line functions from points in G_2' evaluated at a point in G_1 or its preimage under ψ. The computations in $\mathbb{F}_{p^4}$ for point doubling and addition and for obtaining the corresponding line function coefficients share partial results and are thus usually optimized together. Depending on the context in which the pairing is used and the computing platform, one needs to choose between affine and different variants of projective coordinates to find the best formulas for these operations.

The function $f_{T,R}(S)$ is built up by accumulating the line function values. This requires efficient squaring and multiplication-by-line-function operations in the full extension field $\mathbb{F}_{p^{24}}$. The final exponentiation on elements in $\mathbb{F}_{p^{24}}^*$ has a fixed exponent and it is possible to use specialized more efficient squaring operations as well as other optimizations [13,17,1]. Full extension field arithmetic needs to be particularly efficient as pairing efficiency strongly depends on it. It is therefore important to work with a well-chosen extension field tower [6].

3   Particularly Friendly Subfamilies

In this section we show that specializing the congruency classes of the curve-finding search parameter gives rise to four subfamilies of $k = 24$ BLS curves that are highly efficient in terms of all operations required in a pairing computation. Specifically, we show that rather than searching with $x_0 \equiv 1 \pmod 3$, searching with any of $x_0 \equiv 7, 16, 31, 64 \pmod{72}$ guarantees that the curves found offer (among other things) the following advantages.

- **The curve constant b is immediately determined** (see Proposition 3 below). This saves performing expensive computations that test different values of b and the corresponding group order until the correct twist is found; or the checks for quadratic and cubic residuosity and root computations for the algorithm in [24].
- **Highly efficient field tower options are available** (see Proposition 2 below). This facilitates very efficient field arithmetic in the full extension field (and all intermediate subfields).
- **The correct twist is immediately determined** (see Proposition 4 below). Among other savings, having an automated and general representation for the *correct* twist saves group arithmetic on curves over the quartic extension field $\mathbb{F}_{p^4}$ in the generation phase. Also, the representations of the twists are always simple and facilitate nice back-and-forth isomorphisms between E and E'.

In addition to the above efficiency benefits, generating curves with identical or consistent parameters also offers the advantage of code reusability across different instantiations and security levels. This allows an implementor the flexibility of scaling parameter sizes to better match a neighboring security level without changing any of the pairing code.

Furthermore, having fixed coefficients for the curve and its twist leads to a very compact way of representing the curve data. Note that in this setting, the knowledge of the generating parameter x_0 uniquely determines all information about both curves. What only remains is to give generators for the groups G_1 and G'_2, except for the case of $x_0 \equiv 64 \pmod{72}$ for which a compact generator in G_1 is always available as $[h](3,5)$ with high probability, where h is the cofactor, i.e. $[h]$ maps elements of $E(\mathbb{F}_p)$ into G_1.

Table 1. Four attractive subfamilies of BLS curves

x_0 (mod 72)	$p(x_0)$ (mod 72) (eq. (1))	$n(x_0)$ (mod 72) (eq. (1))	efficient tower (Prop. 2)	E (Prop. 3)	E' (Prop. 4)
7	19	12	✓	$y^2 = x^3 + 1$	$y^2 = x^3 \pm 1/v$
16	19	3	✓	$y^2 = x^3 + 4$	$y^2 = x^3 \pm 4v$
31	43	12	✓	$y^2 = x^3 + 1$	$y^2 = x^3 \pm v$
64	19	27	✓	$y^2 = x^3 - 2$	$y^2 = x^3 \pm 2/v$

We split the rest of this section into two subsections. The first subsection is dedicated to proving the claims in Table 1 and showing that taking $x_0 \equiv 7, 16, 31, 64 \pmod{72}$ will always give rise to highly efficient BLS instantiations. The intention of the second subsection is to detail why $x_0 \equiv 7, 16, 31, 64 \pmod{72}$ reign supreme over the other possible congruence classes.

3.1   Using the Four Classes $x_0 \equiv 7, 16, 31, 64 \pmod{72}$

We start with a lemma that is instrumental in some of the proofs that follow. For a prime p, let $\mathrm{QR}(p)$ denote the set of quadratic residues modulo p.

Lemma 1. *Let $x_0 \in \mathbb{Z}$ be any of $x_0 \equiv 7, 16, 31, 64 \pmod{72}$, and let $p = p(x_0)$ with p given by (1) be a prime. Then 2 is neither a quadratic nor a cubic residue modulo p.*

Proof. A simple calculation shows that for $x_0 \equiv 7, 16, 64 \pmod{72}$, we have $p(x_0) \equiv 19 \pmod{72}$. For $x_0 \equiv 31 \pmod{72}$, we have $p(x_0) \equiv 43 \pmod{72}$. In both cases, it is easy to deduce that $2 \notin \mathrm{QR}(p)$.

It remains to show that 2 is not a cube modulo p for each of the x_0 values. For all four cases, we use [16, Prop. 9.6.2], which states that for $p \equiv 1 \pmod{3}$, 2 is a cubic residue modulo p if and only if there exist integers C, D such that $p = C^2 + 27D^2$. According to [16, Prop. 8.3.2], for $p \equiv 1 \pmod{3}$ there always exist integers A and B such that $4p = A^2 + 27B^2$ and A, B are unique up to sign. Thus, if A and B in this unique representation are both even, 2 is a cube modulo p, otherwise it is not. In each of the four cases $x_0 \equiv 7, 16, 31, 64 \pmod{72}$, we examine A and B in terms of the polynomials from (1). The CM norm equation for the BLS family is $4p = t^2 + 3f^2$, where $f = (x_0 - 1)/3 \cdot (2x_0^4 - 1)$, see (1).

For $x_0 \equiv 64 \pmod{72}$, we have $f(x_0) \equiv 0 \pmod{3}$ which allows us to write $4p(x_0) = t(x_0)^2 + 27 \cdot (f(x_0)/3)^2$, so $A = t(x_0) = x_0 + 1 \equiv 65 \pmod{72}$ and $B = f(x_0)/3 = (x_0 - 1)/9 \cdot (2x_0^4 - 1) \equiv 25 \pmod{72}$ are both odd, meaning that for $x_0 \equiv 64 \pmod{72}$, 2 is not a cubic residue modulo $p(x_0)$.

For the other three cases $x_0 \equiv 7, 16, 31 \pmod{72}$, it is easy to show that $f(x_0) \not\equiv 0 \pmod{3}$, so the CM equation does not directly yield A and B. The two cases $x_0 \equiv 7, 16 \pmod{72}$ can be handled by considering a transformation of the CM norm as

$$4p = \left(\frac{3f + t}{2}\right)^2 + 27 \left(\frac{t - f}{6}\right)^2.$$

For $x_0 \equiv 7 \pmod{72}$, $t(x_0), f(x_0) \equiv 2 \pmod{6}$, so that both $A = (3f + t)/2$ and $B = (t - f)/6$ are integers. Furthermore, $f(x_0) \equiv 2 \mod 4$ and $t(x_0) \equiv 0 \pmod{4}$ reveal that $3f + t \equiv 2 \pmod{4}$ so that A is odd, from which it follows that 2 is not a cubic residue modulo $p(x_0)$ for $x_0 \equiv 7 \pmod{72}$.

For $x_0 \equiv 16 \pmod{72}$, we have $t(x_0), f(x_0) \equiv 1 \pmod{2}$ and $t(x_0), f(x_0) \equiv 2 \pmod{3}$. Furthermore, since $t(x_0) \equiv 1 \pmod{4}$ and $f(x_0) \equiv 3 \pmod{4}$, again we conclude that $3f + t \equiv 2 \pmod{4}$, meaning that $A = (t - 3f)/2$ is odd and 2 is not a cube modulo $p(x_0)$ for $x_0 \equiv 16 \pmod{72}$.

Finally, for $x_0 \equiv 31 \pmod{72}$, we require a slightly different transformation of the CM equation as

$$4p = \left(\frac{t - 3f}{2}\right)^2 + 27 \left(\frac{t + f}{6}\right)^2.$$

In this case $t(x_0) \equiv 2 \pmod{6}$ and $f(x_0) \equiv 4 \pmod{6}$ so that $A = (t - 3f)/2$ and $B = (t + f)/6$ are integers. Since $t(x_0) \equiv 0 \pmod{4}$ and $f(x_0) \equiv 2 \pmod{4}$, it follows that $A \equiv 1 \pmod 4$ is odd and 2 is not cube modulo $p(x_0)$ for $x_0 \equiv 31$ $\pmod{72}$. □

The ideal way to form a quadratic extension field arises when $p \equiv 3 \pmod 4$ which allows us to take $\mathbb{F}_{p^2} = \mathbb{F}_p(u)$, $u^2 + 1 = 0$. Operations in $\mathbb{F}_p[u]/(u^2 + 1)$ are cheaper than operations in $\mathbb{F}_p[u]/(u^2 - \alpha)$ for any other non-residue $\alpha \in \mathbb{F}_p$, since multiplication by $\alpha \neq \pm 1$ costs additions in $\mathbb{F}_p$ (or see [10]). Since we have $p \equiv 19 \pmod{24}$ in all four cases, we always have $p \equiv 3 \pmod 4$.

For the extension from $\mathbb{F}_{p^2}$ to $\mathbb{F}_{p^4} = \mathbb{F}_{p^2}(v)$, the ideal irreducible binomial in terms of simplicity would be $v^2 + u$. Unfortunately the following proposition shows that if we form $\mathbb{F}_{p^2}$ as above, then this binomial cannot be used to define $\mathbb{F}_{p^4}$ (and this statement is true for any quartic extension fields, whether in the context of pairings or not).

Proposition 1. *If $p \equiv 3 \pmod 4$, and $\mathbb{F}_{p^2}$ is constructed as $\mathbb{F}_p(u)$, $u^2 + 1 = 0$, then the polynomial $x^2 + su$ with $s \in \mathbb{F}_p$ is reducible over $\mathbb{F}_{p^2}$. In particular, $\mathbb{F}_{p^4}$ cannot be constructed over $\mathbb{F}_{p^2}$ using a binomial of the above form.*

Proof. Since $-1 \notin \mathrm{QR}(p)$, precisely one of $s/2$ or $-s/2$ is a quadratic residue modulo p. In the first case, write $x^2 + su = x^2 + 2a^2u = (x + au - a)(x - au + a)$ for $s = 2a^2$, $a \in \mathbb{F}_p$. In the second case, taking $s = -2a^2$ for some $a \in \mathbb{F}_p$ gives $x^2 + su = x^2 - 2a^2u = (x + au + a)(x - au - a)$. Thus, $x^2 + su$ with $s \in \mathbb{F}_p$ is reducible over $\mathbb{F}_{p^2}$. □

Nevertheless, a binomial that is almost as attractive in terms of efficiency is $v^2 + (u + 1)$, since multiplications by $u + 1$ in $\mathbb{F}_{p^2}$ also come almost for free. The following proposition shows that the proposed subfamilies of BLS curves always allow $\mathbb{F}_{p^4}$ to be constructed using this binomial. Furthermore, we also show that the rest of the tower up to $\mathbb{F}_{p^{24}}$ can be constructed in three different ways; all three of which employ optimal binomials, but may be preferred by implementors depending on various factors, such as the nature of previous pairing code, or whether compression is desired.

Proposition 2. *Let $x_0 \in \mathbb{Z}$ be any of $x_0 \equiv 7, 16, 31, 64 \pmod{72}$. If $p = p(x_0)$ given by the polynomial in (1) is prime, then the extension field $\mathbb{F}_{p^{24}}$ can be constructed using any of the following towering options T_1, T_2, T_3:*

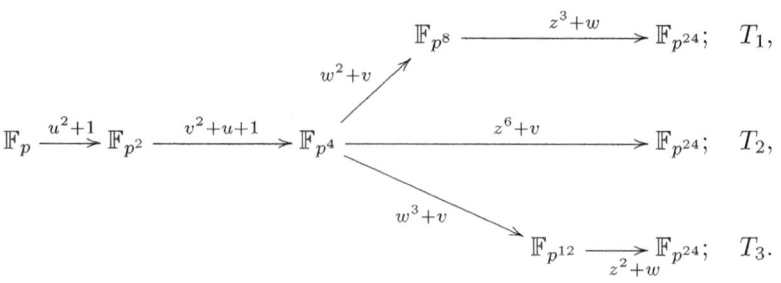

Proof. For all four congruency classes $x_0 \equiv 7, 16, 31, 64 \pmod{72}$, we have $p(x_0) \equiv 19 \pmod{24}$, so that $p \equiv 3 \pmod 4$ and we can use $\mathbb{F}_{p^2} = \mathbb{F}_p(u) = \mathbb{F}_p[u]/(u^2+1)$. For the remaining irreducibility arguments, we make use of a theorem due to Benger and Scott [6, Thm. 4]. We immediately note that $2, 3 \mid p-1$.

Let us compute $N_{\mathbb{F}_{p^2}/\mathbb{F}_p}(-(u+1)) = N_{\mathbb{F}_{p^2}/\mathbb{F}_p}(-1)N_{\mathbb{F}_{p^2}/\mathbb{F}_p}(u+1) = (u+1)^{p+1} = (u+1)^p(u+1) = (1-u)(u+1) = 2$. Since 2 is not a quadratic residue modulo p, Theorem 4 of [6] ensures that $v^2 + u + 1$ is irreducible in $\mathbb{F}_{p^2}[v]$ and we can construct $\mathbb{F}_{p^4} = \mathbb{F}_{p^2}[v]/(v^2 + u + 1)$.

Next, we compute $N_{\mathbb{F}_{p^4}/\mathbb{F}_p}(-v) = N_{\mathbb{F}_{p^4}/\mathbb{F}_p}(-1)N_{\mathbb{F}_{p^4}/\mathbb{F}_p}(v) = N_{\mathbb{F}_{p^4}/\mathbb{F}_p}(v) = v^{1+p+p^2+p^3} = (v^{1+p^2})^{1+p} = ((-(u+1))^{1+p^2})^{(1+p)/2} = ((u+1)^2)^{(1+p)/2} = (u+1)^{p+1} = 2$. Lemma 1 tells us that 2 is neither a quadratic nor a cubic residue modulo p, and thus it follows from [6, Thm. 4] that $w^2 + v$, $w^3 + v$ and $z^6 + v$ are all irreducible over $\mathbb{F}_{p^4}$, giving rise to the tower T_2, to T_1 up to $\mathbb{F}_{p^8}$ and T_3 up to $\mathbb{F}_{p^{12}}$.

For the remaining parts of T_1 and T_3, we similarly compute $N_{\mathbb{F}_{p^{12}}/\mathbb{F}_p}(-w) = 2^{(1+p^4+p^8)/3}$ and $N_{\mathbb{F}_{p^8}/\mathbb{F}_p}(-w) = 2^{(1+p^4)/2}$. Since $p \equiv 1 \pmod 6$ for all cases, it follows that $(1 + p^4 + p^8)/3$ is odd and so $N_{\mathbb{F}_{p^{12}}/\mathbb{F}_p}(-w)$ is not a square in $\mathbb{F}_p$ since 2 is not a square by Lemma 1. Similarly, $(p^4 + 1)/2 \equiv 1 \pmod 3$ and so $N_{\mathbb{F}_{p^8}/\mathbb{F}_p}(-w)$ is not a cube in $\mathbb{F}_p$. Therefore, [6, Thm. 4] ensures irreducibility of the remaining polynomials and completes the proof. □

We have now shown that once a BLS prime p is found using $x_0 \equiv 7, 16, 31, 64 \pmod{72}$, a highly efficient tower is immediately available. The following two propositions show that the curve constant and twisted curve are also immediate in all four cases.

Proposition 3. *If $x_0 \equiv 7, 31 \pmod{72}$ in (1) produces a prime $p = p(x_0)$, then the curve $E/\mathbb{F}_p : y^2 = x^3 + 1$ is always such that $r = r(x_0) \mid n = \#E(\mathbb{F}_p)$. Similarly, for $x_0 \equiv 16 \pmod{72}$, the desired curve is always $E/\mathbb{F}_p : y^2 = x^3 + 4$. Finally, for $x_0 \equiv 64 \pmod{72}$, the desired curve is always $E/\mathbb{F}_p : y^2 = x^3 - 2$.*

Proof. It is well known that if g is neither a square nor a cube in $\mathbb{F}_p$, then all possible group orders an elliptic curve $E : y^2 = x^3 + b$ can have over $\mathbb{F}_p$ occur as the order of one of the 6 twists with $b \in \{1, g, g^2, g^3, g^4, g^5\}$. Specifically, choosing b as exactly one of $\{1, g, g^2, g^3, g^4, g^5\}$ will give the correct number of points (cf. [31, §X.5]), i.e. the curve with $r \mid n = \#E(\mathbb{F}_p)$. Lemma 1 shows that we can take $g = 2$, so that in all four cases the correct b is exactly one of $\{1, 2, 4, 8, 16, 32\}$.

For both $x_0 \equiv 7 \pmod{72}$ and $x_0 \equiv 31 \pmod{72}$, we have $n(x_0) = (x_0 - 1)^2(x_0^8 - x_0^4 + 1)/3 \equiv 12 \pmod{72}$ from (1). Thus, both cases have $2, 3 \mid n$, meaning that the correct curves $E/\mathbb{F}_p$ necessarily contain points of order 2 and points of order 3. This implies that b is both a quadratic and cubic residue modulo p, from which it follows that $b = 1$ is the only option.

For $x_0 \equiv 16 \pmod{72}$, observe that $n(x_0) \equiv 3 \pmod{72}$ and thus the correct curve E has a point of order 3, but not a point of order 2. This rules out $b = 1, 8$, since the points $(-1, 0)$ and $(-2, 0)$ have order 2 on the respective curves. The curve $E/\mathbb{F}_p : y^2 = x^3 + b$ has a point of order 3 if and only if b is a square in

$\mathbb{F}_p$, which rules out $b = 2$, and therefore $b = 32$ as well. To rule out $b = 16$, we first observe that since $n = n(x_0) \equiv 3 \pmod{72}$, $9 \nmid n$ and $E/\mathbb{F}_p$ has at most three points of order 3. If $b = 16$, then two such points are $(0, -4)$ and $(0, 4)$. It is easy to see that $-3 \in \mathrm{QR}(p)$, so let $\nu^2 = -3$ for $\nu \in \mathbb{F}_p$, and write $P = (-4, 4\nu) \in E(\mathbb{F}_p) : y^2 = x^3 + 16$. An easy calculation (e.g. using point doubling formulas) shows that $[2]P = (-4, -4\nu) = -P$, so that P has order 3, and similarly for $-P$. Thus, there are at least four points of order 3 in $\mathbb{F}_p$ if $b = 16$ contradicting $9 \nmid n$, which leaves $b = 4$ as the only option.

For $x_0 \equiv 64 \pmod{72}$, we can make use of Algorithm 3.5 in [24], where in our case $U = t/2$ and $V = f/2$. Since $2V = f(x_0) \equiv 0 \bmod 3$ and $2U = t(x_0) = 2 \bmod 3$, we immediately have $E/\mathbb{F}_p : y^2 = x^3 + 16$ as the correct curve. Lastly, since $-2 \in \mathrm{QR}(p)$, write $\mu^2 = -2$ for $\mu \in \mathbb{F}_p$, so that $\mu^6 = -8$. Since $E/\mathbb{F}_p : y^2 = x^3 + 16$ is isomorphic to $\tilde{E}/\mathbb{F}_p : y^2 = x^3 + 16/\mu^6$ over $\mathbb{F}_p$, we can take $b = 16/ -8 = -2$ as the curve constant instead.   □

Proposition 4. *If $x_0 \equiv 7 \pmod{72}$ produces the BLS curve $y^2 = x^3 + 1$ described in Proposition 3, and $\mathbb{F}_{p^4}$ is constructed as in Proposition 2, then the correct sextic twist with $r = r(x_0) \mid \#E'(\mathbb{F}_{p^4})$ can be obtained as both $E'/\mathbb{F}_{p^4} = y^2 = x^3 + 1/v$ and $E'/\mathbb{F}_{p^4} : y^2 = x^3 - 1/v$. Similarly, $x_0 \equiv 16 \bmod 72$ gives rise to the correct twist as both $E'/\mathbb{F}_{p^4} : y^2 = x^3 + 4v$ or $E'/\mathbb{F}_{p^4} : y^2 = x^3 - 4v$; $x_0 \equiv 31 \bmod 72$ gives rise to the correct twist as both $E'/\mathbb{F}_{p^4} : y^2 = x^3 + v$ or $E'/\mathbb{F}_{p^4} : y^2 = x^3 - v$; and finally, $x_0 \equiv 64 \bmod 72$ gives rise to the correct twist as both $E'/\mathbb{F}_{p^4} : y^2 = x^3 + 2/v$ or $E'/\mathbb{F}_{p^4} : y^2 = x^3 - 2/v$.*

Proof. We first note that $-1 = u^6$ is a sixth power in $\mathbb{F}_{p^2}$. Therefore, for $b' \in \mathbb{F}_{p^4}$, the curves given by $y^2 = x^3 + b'$ and $y^2 = x^3 - b'$ are isomorphic over $\mathbb{F}_{p^4}$.

In each case the correct twist E' is unique with the property that $r \mid \#E'(\mathbb{F}_{p^4})$ and it has degree 6 (see [15]). Since there are exactly two twists of E of degree 6 over $\mathbb{F}_{p^4}$, there are only two possible group orders. We have $\#E(\mathbb{F}_{p^4}) = p^4 + 1 - t_4$, where t_4 can be computed from t and f (notation as before) as $t_4 = (t^4 - 18t^2 f^2 + 9f^2)/8$ (see [31, §V.2]). If we define f_4 by $4p^4 = t_4^2 + 3f_4^2$, the possible group orders for the sextic twist are given in [15] as

$$ n_{4,1} = p^4 + 1 - (3f_4 - t_4)/2, \qquad n_{4,2} = p^4 + 1 - (-3f_4 - t_4)/2. $$

We compute both $n_{4,1}$ and $n_{4,2}$ as polynomials in terms of the parametrization (1), and evaluate them at each of the congruency classes which reveals opposing parities each time. The remainder of the proof is essentially the same for all four congruencies classes, so we demonstrate completely with $x_0 \equiv 16 \pmod{72}$. Taking $x_0 \equiv 16 \pmod{72}$ gives $n_{4,1} \equiv 28 \pmod{72}$ and $n_{4,2} \equiv 49 \pmod{72}$. In particular, $n_{4,1}$ is even and $n_{4,2}$ is odd in this case. From the polynomial parametrization, it is easy to check that $r \mid n_{4,1}$. Therefore, the unique sextic twist we are looking for in this case has an even group order over $\mathbb{F}_{p^4}$.

Proposition 2 shows that v is neither a square nor a cube in $\mathbb{F}_{p^4}$, so the correct twist can be given as $y^2 = x^3 + 4v$ or as $y^2 = x^3 + 4v^5$. We have $4v = (N_{\mathbb{F}_{p^4}/\mathbb{F}_p}(v))^2 v = v^{3+2p+2p^2+2p^3}$ as in the proof of Proposition 2. Since the exponent on the right hand side of the last equation is divisible by 3, we conclude that $4v$ is a cube in $\mathbb{F}_{p^4}$. Similarly, one can show that $4v^5$ is not a cube. Since $4v$ is a cube, the curve $y^2 = x^3 + 4v$ has a point of order 2, namely $(-c, 0)$ with $c^3 = 4v$. Hence its order is even and we have found the correct twist for $x_0 \equiv 16$ (mod 72). The other three cases are proven analogously.   □

3.2   The Other Congruency Classes

We now show why the four congruency classes $x_0 \equiv 7, 16, 31, 64$ (mod 72) stand out over the other congruency classes. After all, restricting $x_0 \equiv 1$ (mod 3) to any or all of the proposed classes essentially discards 20 out of 24 other congruency classes modulo 72. Of course, there will always be examples where some of the discarded congruency classes produce curves that also perform highly efficiently. However, we argue that $x_0 \equiv 7, 16, 31, 64$ (mod 72) are the only classes for which we can *always* simultaneously guarantee the propositions in the previous subsection.

Quadratic extension to $\mathbb{F}_{p^2}$: We start by eliminating the classes of x_0 mod 72 which do not facilitate the quadratic extension as $\mathbb{F}_{p^2} = \mathbb{F}_p(u)$, $u^2 + 1 = 0$. Of the 24 possible x_0 values modulo 72 in $\{1, 4, 7, ..., 67, 70\}$, 12 have $x_0 \equiv 1, 10$ (mod 12), which always (undesirably) produce $p \equiv 1$ (mod 12). The remaining 12 are $\{4, 7, 16, 19, 28, 31, 40, 43, 52, 55, 64, 67\}$ with $x_0 = 4, 7$ (mod 12), which always produce $p(x_0) \equiv 7$ (mod 12).

Quadratic extension to $\mathbb{F}_{p^4}$: If $\mathbb{F}_{p^4}$ is to be constructed as $\mathbb{F}_{p^4} = \mathbb{F}_{p^2}(v)$, $v^2 + u + 1 = 0$, then (refer back to the proof of Proposition 2) we can only guarantee this if $N_{\mathbb{F}_{p^2}/\mathbb{F}_p}(-(u+1)) = 2$ is not a quadratic residue in $\mathbb{F}_p$. Substituting the remaining 12 candidates for x_0 (mod 72) into (1) reveals only 4 possibilities for p (mod 72), those being $p \equiv 7, 19, 43, 55$ (mod 72). It is easy to check that only two of these have $2 \notin QR(p)$, namely $p \equiv 19, 43$ (mod 72). These correspond to 6 of the remaining x_0 congruencies, shrinking the pool of preferred candidates to $\{7, 16, 31, 40, 55, 64\}$.

Sextic extension to $\mathbb{F}_{p^{24}}$: The proposed sextic extensions in Proposition 2 that employ simple binomials to form $\mathbb{F}_{p^{24}}$ over $\mathbb{F}_{p^4}$ require that 2 is not a cube in $\mathbb{F}_p$. This does not always happen for $x_0 \in \{40, 55\}$. For the sake of counter examples, $x_0 = 12856 \equiv 40$ (mod 72) and $x_0 = 1135 \equiv 55$ (mod 72) produce BLS curves where 2 is a cube modulo p, and therefore fields which can not use the tower in Proposition 2. Even if alternative binomials can be found in such cases, the fact that 2 is a cubic residue also affects the ease of guaranteeing the smallest identical curve constant b for all curves in the subfamily, as we were able to do for the proposed four congruency classes in Proposition 3. For example, the smallest curve constant for the curve found with $x_0 = 12856 \equiv 40$ (mod 72) is $b = -3$, whilst the smallest constant for the curve found with $x_0 = 25312 \equiv 40$ (mod 72) is $b = 9$.

4   Choosing Simple Lines: Twisting vs. Untwisting

The aim of this short section is to detail some choices that can facilitate more simple (and theoretically faster) pairing code.

For 128-bit security BN implementations, the complexity of a single Miller loop is higher than complexity of a single final exponentiation (see any of [22,7,1]). However, as the security level and embedding degree increases, a final exponentiation becomes much more costly than a Miller loop [11] (or see our timings in Appendix A). This could influence implementors paying less attention to more complicated subtleties within the Miller loop, like adopting projective coordinates, and instead focusing on speedups within the exponentiation routine. On the other hand, a large number of recent pairing protocols like attribute-based encryption (ABE) require many Miller loops for each exponentiation (see Scott's recent work [27] for an in depth look), and therefore in these scenarios savings within the loop again become more significant overall. In any case, a thoroughly optimized implementation will make use of the fastest formulas inside the Miller loop, so even though our implementation slightly favored affine coordinates, and indeed Scott's current 256-bit record [27] also employed affine coordinates, it could well be that a thoroughly optimized projective routine (like [1] for $k = 12$) ends up outperforming affine formulas at this level too. We refer to [8,9,18] for ways to find the most efficient coordinate system depending on the specific protocol and implementation situation.

As we mentioned briefly in the previous section, the choice of which tower (T_1, T_2, T_3 in Proposition 2) to use in an implementation could be influenced by a number of factors. For example, if low bandwidth requirements favored maximum compression techniques [28], then T_1 would seem most appropriate. On the other hand, if the major priority is raw speed, then one could employ the field arithmetic presented in [10, §6] and favor the (slightly faster) quadratic over cubic extension offered by T_3. Or perhaps most commonly, if the implementor is adopting BLS curves to scale a prior (say BN) pairing implementation to a higher security level, then the towered code for the BN sextic extension from $\mathbb{F}_{p^2}$ to $\mathbb{F}_{p^{12}}$ could be easily updated to BLS code for the extension from $\mathbb{F}_{p^4}$ to $\mathbb{F}_{p^{24}}$. In any case, there are efficient formulas available for all three of the towering choices [10,9], so we will treat all three cases in parallel and highlight the differences that arise.

One such difference lies in the sparse doubling and addition lines that are used to update the pairing function. For each of the proposed congruency classes, Table 2 follows the exposition in [26, §5] and details the correct placing of the line function coefficients for the three tower choices T_1, T_2, T_3 and the two twist choices in Proposition 4, both of which have the same group order but result in different looking line functions.

The point P is always kept in affine coordinates (x_P, y_P). For the affine formulas the line simply is $\ell(P) = y_P - \lambda x_P - c$, where $\lambda \in \mathbb{F}_{p^4}$ is the slope of the line as usual and $c \in \mathbb{F}_{p^4}$ is the constant coefficient. Projective formulas for these line functions usually output three coefficients $L_{0,0}, L_{1,0}, L_{0,1} \in \mathbb{F}_{p^4}$ [9] that define the evaluated update $\ell(P) = L_{0,0} + L_{1,0} x_{\psi^{-1}(P)} + L_{0,1} y_{\psi^{-1}(P)} \in \mathbb{F}_{p^{24}}$

by being attached to different algebraic elements in the representation of $\mathbb{F}_{p^{24}}$ over $\mathbb{F}_{p^4}$ depending on the twisting isomorphism ψ. In all cases $\ell(P)$ is a sparse element of $\mathbb{F}_{p^{24}}$, with the only difference being the places that the ℓ_i occupy as a result of the different algebraic towerings and maybe a sign change.

To explain the different lines in Table 2, let us briefly look at the conversion between representations of an element in the different towering options T_1, T_2, and T_3. Let us start with an element a in T_2 given by the coefficients $a_0, \ldots, a_5 \in \mathbb{F}_{p^4}$, i.e. $a = a_0 + a_1 z + a_2 z^2 + a_3 z^3 + a_4 z^4 + a_5 z^5$. Converting to T_1, we take $z^3 \mapsto w$ and thus the same element is represented as $a = (a_0 + a_3 w) + (a_1 + a_4 w)z + (a_2 + a_5 w)z^2$. To go to T_3, we use $z \mapsto -uz$ (i.e. $z^2 \mapsto w$) and obtain $a = (a_0 + a_2 w + a_4 w^2) + (-a_1 u - a_3 uw - a_5 uw^2)z$.

An optimized routine must take advantage of the sparse nature of $\ell(P)$ and tailor make a specialized multiplication routine to exploit the presence of zero entries. Roughly speaking, the options in Table 2 will give rise to similar speeds, but it is obvious to see which twist constant an implementor would choose (all other things being equal) if their choice of tower is already concrete. For example, if T_1 is the chosen tower, then using $b' = -4v$ gives a slightly easier line function to code than using $b' = 4v$.

A more important difference arises when choosing whether to leave $P \in E$ and "untwist" $Q' \in E'$ to $Q \in E$ for the line function computation, or choosing to put both points on the twist for the entire routine [9]. The nature of the proposed tower actually means that there is a significant difference between the simplicity of the twisting and untwisting isomorphisms, and one option will be more desirable to implement than the other. For example, when using T_1 and $b' = -4v$, the untwisting isomorphism ψ is $\psi : (x', y') \mapsto (x/z^2, y/w) = (wvz(u-1)/2 \cdot x', wv(1-u)/2 \cdot y')$, which is more annoying to code (and theoretically slightly slower) than using $\psi^{-1} : (x, y) \mapsto (z^2 x, wy)$ to twist the second argument P instead. On the other hand, if the twist is given as $E' : y^2 = x^3 \pm b/v$, as is the cases when $x_0 \equiv 7, 64 \pmod{72}$, then it is the untwisting isomorphism that is clearly preferable.

5   The Final Exponentiation

An optimized final exponentiation routine is critical for fast pairings at high-security levels. Scott et $al.$ [29] propose the most efficient algorithm to date, which exploits the polynomial representations of p and r to reduce the work encountered after the Miller loop. The first step is to split the exponent into two parts by factoring $(p^k - 1)/r$ and exploiting the Frobenius operator which raises elements to the power of p almost for free. In our case the exponent splits as:

$$(p^{24} - 1)/r = [(p^{12} - 1) \cdot (p^4 + 1)] \cdot \underbrace{[(p^8 - p^4 + 1)/r]}_{\text{hard part}}.$$

After exploiting Frobenius operations to quickly raise the output of the Miller loop to the exponent in the left square parentheses, we then face the major bottle-neck in the exponentiation: raising a value $m \in \mathbb{F}_{p^{24}}$ to the power of "the hard part".

Table 2. Details of the Miller line function depending on the choice of tower (and twist constant)

cong. class	twist b'	tower choice	twist isomorphism $\psi^{-1}: E \to E'$	Miller lines — Affine	Miller lines — Projective
16,31	bv	T_1	$(x,y) \mapsto (-z^2x, uwy)$	$\left[-c : y_P\cdot u \,\middle|\, \underset{z}{0} : \underset{z^2}{0} \,\middle|\, \lambda \cdot x_P : \underset{w}{0}\right]$	$\left[L_{0,0} : L_{0,1}\cdot y_P u \,\middle|\, \underset{z}{0} : \underset{z^2}{0} \,\middle|\, -L_{1,0}\cdot x_P : \underset{w}{0}\right]$
		T_2	$(x,y) \mapsto (-z^2x, uz^3y)$	$\tfrac{1}{-c}\left[\underset{z}{0} : \lambda\cdot x_P : y_P\cdot u : \underset{z^3}{0} : \underset{z^4}{0} : \underset{z^5}{0}\right]$	$\tfrac{1}{L_{0,0}}\left[\underset{z}{0} : -L_{1,0}\cdot x_P : L_{0,1}\cdot y_P u : \underset{z^3}{0} : \underset{z^4}{0} : \underset{z^5}{0}\right]$
		T_3	$(x,y) \mapsto (-wx, wzy)$	$\left[-c : \lambda\cdot x_P : \underset{w}{0} : \underset{w^2}{y_P} : \underset{z}{0} : \underset{z^2}{0}\right]$	$\left[L_{0,0} : -L_{1,0}\cdot x_P : \underset{w}{0} : \underset{w^2}{0} : L_{0,1}\cdot y_P : \underset{z}{0}\right]$
	$-bv$	T_1	$(x,y) \mapsto (z^2x, wy)$	$\left[-c : \underset{w^2}{y_P} \,\middle|\, \underset{z}{0} : \underset{z^2}{0} \,\middle|\, -\lambda \cdot x_P : \underset{w}{0}\right]$	$\left[L_{0,0} : L_{0,1}\cdot y_P \,\middle|\, \underset{z}{0} : \underset{z^2}{0} \,\middle|\, L_{1,0}\cdot x_P : \underset{w}{0}\right]$
		T_2	$(x,y) \mapsto (z^2x, z^3y)$	$\tfrac{1}{-c}\left[\underset{z}{0} : -\lambda\cdot x_P : \underset{w}{y_P} : \underset{z^3}{0} : \underset{z^4}{0} : \underset{z^5}{0}\right]$	$\tfrac{1}{L_{0,0}}\left[\underset{z}{0} : L_{1,0}\cdot x_P : L_{0,1}\cdot y_P : \underset{z^3}{0} : \underset{z^4}{0} : \underset{z^5}{0}\right]$
		T_3	$(x,y) \mapsto (wx, uwzy)$	$\left[-c : -\lambda\cdot x_P : \underset{w}{0} : \underset{w^2}{0} : -y_P\cdot u : \underset{z}{0}\right]$	$\left[L_{0,0} : L_{1,0}\cdot x_P : \underset{w}{0} : \underset{w^2}{0} : -L_{0,1}\cdot y_P u : \underset{z}{0}\right]$

cong. class	twist b'	tower choice	untwist isomorphism $\psi': E' \to E$	Miller lines — Affine	Miller lines — Projective
7,64	b/v	T_1	$(x',y') \mapsto (-z^2x', uwy')$	$\left[y_P : -c\cdot u \,\middle|\, \underset{z}{0} : \underset{z^2}{0} \,\middle|\, \lambda\cdot x_P \cdot u : \underset{w}{0}\right]$	$\left[L_{0,1}\cdot y_P : L_{0,0}\cdot u \,\middle|\, \underset{z}{0} : \underset{z^2}{0} \,\middle|\, L_{1,0}\cdot x_P\cdot u : \underset{w}{0}\right]$
		T_2	$(x',y') \mapsto (-z^2x', uz^3y')$	$\tfrac{1}{y_P}\left[\underset{z}{0} : \lambda\cdot u \cdot x_P : \underset{z^2}{0} : -c\cdot u : \underset{z^4}{0} : \underset{z^5}{0}\right]$	$\tfrac{1}{L_{0,1}\cdot y_P}\left[\underset{z}{0} : -L_{1,0}\cdot u\cdot x_P : \underset{z^2}{0} : L_{0,0}\cdot u : \underset{z^4}{0} : \underset{z^5}{0}\right]$
		T_3	$(x',y') \mapsto (-wx', wzy')$	$\left[y_P : \underset{w}{0} : \lambda\cdot x_P : -c\cdot u : \underset{w^2}{0} : \underset{z}{0}\right]$	$\left[L_{0,1}\cdot y_P : \underset{w}{0} : -L_{1,0}\cdot x_P : L_{0,0} : \underset{w^2}{0} : \underset{z}{0}\right]$
	$-b/v$	T_1	$(x',y') \mapsto (z^2x', wy')$	$\left[y_P : -c \,\middle|\, \underset{z}{0} : \underset{z^2}{0} \,\middle|\, -\lambda\cdot x_P : \underset{w}{0}\right]$	$\left[y_P : L_{0,0} \,\middle|\, \underset{z}{0} : \underset{z^2}{0} \,\middle|\, -L_{1,0}\cdot x_P : \underset{w}{0}\right]$
		T_2	$(x',y') \mapsto (z^2x', z^3y')$	$\tfrac{1}{y_P}\left[\underset{z}{0} : -\lambda\cdot x_P : \underset{z^2}{0} : -c : \underset{z^4}{0} : \underset{z^5}{0}\right]$	$\tfrac{1}{y_P}\left[\underset{z}{0} : L_{1,0}\cdot x_P : \underset{z^2}{0} : L_{0,0} : \underset{z^4}{0} : \underset{z^5}{0}\right]$
		T_3	$(x,y) \mapsto (wx, uwzy')$	$\left[y_P : \underset{w}{0} : \underset{w^2}{0} : -\lambda\cdot u\cdot x_P : -c\cdot u : \underset{z}{0}\right]$	$\left[y_P : \underset{w}{0} : \underset{w^2}{0} : L_{1,0}\cdot u\cdot x_P : L_{0,0}\cdot u : \underset{z}{0}\right]$

One helpful observation which aids the remaining computations is that, after exponentiation to the power $p^{12} - 1$, the value $m \in \mathbb{F}_{p^{24}}$ is now such that $N_{\mathbb{F}_{p^{24}}/\mathbb{F}_{p^{12}}}(m) = 1$. This allows any inversions in $\mathbb{F}_{p^{24}}$ to be computed for free using a simple conjugation [28,21,29], and any squarings in $\mathbb{F}_{p^{24}}$ to be computed more efficiently than standard squarings [13,17,1]. To apply the algorithm in [29], we use the parameterizations in (1) to write the hard part as

$$(p(x)^8 - p(x)^4 + 1)/r(x) = \sum_{i=0}^{7} \lambda_i(x)p(x)^i.$$

In an appendix of her thesis, Benger [5] computed the λ_i for a range of curve families, including BLS curves with $k = 24$, giving $\lambda_i = \nu_i/3$, where

$$\nu_7(x) = x^2 - 2x + 1,$$
$$\nu_6(x) = x^3 - 2x^2 + x = x \cdot \nu_7(x),$$
$$\nu_5(x) = x^4 - 2x^3 + x^2 = x \cdot \nu_6(x),$$
$$\nu_4(x) = x^5 - 2x^4 + x^3 = x \cdot \nu_5(x),$$
$$\nu_3(x) = x^6 - 2x^5 + x^4 - x^2 + 2x - 1 = x \cdot \nu_4(x) - \nu_7(x),$$
$$\nu_2(x) = x^7 - 2x^6 + x^5 - x^3 + 2x^2 - x = x \cdot \nu_3(x),$$
$$\nu_1(x) = x^8 - 2x^7 + x^6 - x^4 + 2x^3 - x^2 = x \cdot \nu_2(x),$$
$$\nu_0(x) = x^9 - 2x^8 + x^7 - x^5 + 2x^4 - x^3 + 3 = x \cdot \nu_1(x) + 3.$$

This representation reveals another nice property exhibited by $k = 24$ BLS curves: namely, a very convenient way to compute the ν_i with essentially just multiplications by x. Letting $\mu_i = m^{\nu_i(x_0)}$, this structure allows us to write the hard part of the final exponentiation as

$$m^{(p^8 - p^4 + 1)/r} = \mu_0 \cdot \mu_1^p \cdot \mu_2^{p^2} \cdot \mu_3^{p^3} \cdot \mu_4^{p^4} \cdot \mu_5^{p^5} \cdot \mu_6^{p^6} \cdot \mu_7^{p^7},$$

where the μ_i can be computed using the following sequence of operations:

$$\mu_7 = (m^{x_0})^{x_0} \cdot (m^{x_0})^{-2} \cdot m, \; \mu_6 = (\mu_7)^{x_0}, \; \mu_5 = (\mu_6)^{x_0}, \; \mu_4 = (\mu_5)^{x_0},$$
$$\mu_3 = (\mu_4)^{x_0} \cdot (\mu_7)^{-1}, \; \mu_2 = (\mu_3)^{x_0}, \; \mu_1 = (\mu_2)^{x_0}, \mu_0 = (\mu_1)^{x_0} \cdot m^2 \cdot m.$$

The computation of $m^{(p^8 - p^4 + 1)/r}$ requires 9 exponentiations by x_0, 12 multiplications in $\mathbb{F}_{p^{24}}$, 2 special squarings, 2 conjugations to compute the inverses and 7 p-power Frobenius operations. We detail a possible scheduling for the full exponentiation routine in Table 3. Note that we can simply forget about the difference between the λ_i and the ν_i; by leaving away the 3 in the denominators, we just compute the third power of the pairing.

By far the most costly stage of the final exponentiation is the nine exponentiations by x_0, which are performed using a standard square-and-multiply routine. This is where the BLS pairing computation benefits most from the faster squarings in $\mathbb{F}_{p^{24}}$, and also from an x_0 value that has low hamming-weight which

Table 3. The final exponentiation for BLS curves with $k = 24$

FinalExp	Input: $f_{r,Q}(P) \in \mathbb{F}_{p^{24}}$ and loop parameter x_0

Initialize $f \leftarrow f_{r,Q}(P)$,

$t_0 \leftarrow 1/f,$ $m \leftarrow \bar{f},$ $m \leftarrow m \cdot t_0, \; t_0 \leftarrow \pi_p^4(m), \; m \leftarrow m \cdot t_0,$

$m_1 \leftarrow m^x,$ $m_2 \leftarrow m_1^x$ $m_1 \leftarrow m_1^2, \; m_1 \leftarrow \overline{m_1}, \; \mu_7 \leftarrow m_2 \cdot m_1, \mu_7 \leftarrow \mu_7 \cdot m,$

$\mu_6 \leftarrow \mu_7^x,$ $\mu_5 \leftarrow \mu_6^x,$ $\mu_4 \leftarrow \mu_5^x, \; \mu_7' \leftarrow \overline{\mu_7}, \quad \mu_3 \leftarrow \mu_4^x, \quad \mu_3 \leftarrow \mu_3 \cdot \mu_7',$

$\mu_2 \leftarrow \mu_3^x,$ $\mu_1 \leftarrow \mu_2^x,$ $\mu_0 \leftarrow \mu_1^x, \quad m' \leftarrow m^2, \quad \mu_0 \leftarrow \mu_0 \cdot m', \; \mu_0 \leftarrow \mu_0 \cdot m,$

$f \leftarrow \pi_p(\mu_7), \; f \leftarrow f \cdot \mu_6, \quad f \leftarrow \pi_p(f), \; f \leftarrow f \cdot \mu_5, \quad f \leftarrow \pi_p(f), \quad f \leftarrow f \cdot \mu_4,$

$f \leftarrow \pi_p(f), \; f \leftarrow f \cdot \mu_3, \quad f \leftarrow \pi_p(f), \; f \leftarrow f \cdot \mu_2, \; f \leftarrow \pi_p(f), \quad f \leftarrow f \cdot \mu_1,$

$f \leftarrow \pi_p(f), \; f \leftarrow f \cdot \mu_0,$

Return $f_{r,Q}(P)^{(p^{24}-1)/r} \leftarrow f.$

	Output: $f_{r,Q}(P)^{(p^{24}-1)/r}$

reduces the $\mathbb{F}_{p^{24}}$ multiplications encountered. In the following section we give several examples of very low hamming-weight x_0 values that give rise to curves in the proposed implementation-friendly BLS subfamily.

6   Example Curves

This section provides four lists of implementation-friendly BLS curves at security levels where the entire BLS family is either competitive across all families, or is clearly the current outright favorite. Each list (Table 4 through to Table 7) corresponds to one of the four proposed subfamilies.

We start with 192-bit security since our timings of a C implementation of the optimal ate pairing at this level (see Appendix A) agreed with Scott's comment [27] that the stand-out candidate curve family at this level is not yet as obvious as the 128- and 256-bit levels. In our implementation both a $k = 12$ BN curve and $k = 24$ BLS curve outperformed a $k = 18$ KSS curve at this level.

Each table lists curves where x_0 is very sparse in signed binary representation, meaning here that it has weight 3, 4 or 5. Working with signed binary representation for the Miller loop parameter and the powerings in the final exponentiation can be considered standard and extends the space of nice curves compared to just using plain binary representation. Nevertheless, we also included many x_0 values which have the same plain binary representation as the signed binary representation; these are the x_0 values which share the same sign for each power of 2.

All curves given have the implementation-friendly properties outlined in the previous sections. In particular, curves in Table 4 and Table 6 have $x_0 \equiv 7$ (mod 72) and $x_0 \equiv 31$ (mod 72) respectively, and are given by $E : y^2 = x^3 + 1$; curves in Table 5 have $x_0 \equiv 16$ (mod 72) and are given by $E : y^2 = x^3 + 4$, and curves in Table 7 have $x_0 \equiv 64$ (mod 72) and are given by $y^2 = x^3 - 2$. In all cases all parameters are uniquely defined by the short value x_0.

Table 4. BLS curves with low-weight parameter $x_0 \equiv 7 \pmod{72}$ aiming at several security levels given in the first column. The columns "words for p" and "words for r" give the necessary number of 32- or 64-bit words to store the values for p and r, respectively. The last column provides the estimated actual security by the formula in [32].

security level	$x_0 \equiv 7 \pmod{72}$	weight	p (bits)	words for p	r (bits)	words for r	security (bits)
192	$-1 - 2^8 + 2^{38} + 2^{45}$	4	449	15×32	361	6×64	181
	$-1 + 2^3 - 2^5 - 2^{19} + 2^{46}$	5	459		368		184
	$-1 - 2^{11} - 2^{26} - 2^{35} - 2^{47}$	5	469		377		189
	$-1 + 2^{19} - 2^{24} + 2^{27} - 2^{48}$	5	479		384		192
	$-1 - 2^{11} - 2^{28} - 2^{35} - 2^{49}$	5	489	8×64	393	13×32	197
	$-1 - 2^4 - 2^{21} - 2^{50}$	4	499		401		201
	$-1 + 2^{11} - 2^{28} - 2^{51}$	4	509		409		205
	$-1 - 2^{22} - 2^{26} - 2^{36} - 2^{52}$	5	519	17×32	417	7×64	209
	$-1 + 2^{44} + 2^{51} + 2^{53}$	4	532		427		214
224	$-1 - 2^3 - 2^{29} - 2^{38} - 2^{55}$	5	549	9×64	441	7×64	221
	$-1 + 2^3 - 2^{11} - 2^{51} + 2^{56}$	5	558		448		224
	$-1 - 2^{15} - 2^{22} - 2^{56}$	4	559		449	15×32	225
	$-1 - 2^{16} + 2^{23} - 2^{28} + 2^{57}$	5	569		456		228
	$-1 - 2^{28} + 2^{51} + 2^{58}$	4	579	19×32	465		233
	$-1 - 2^{12} - 2^{28} - 2^{50} - 2^{58}$	5	579		465		233
256	$-1 + 2^{15} + 2^{43} - 2^{61}$	4	609	10×64	488	8×64	244
	$-1 + 2^{49} + 2^{55} + 2^{62}$	4	619		497		249
	$-1 - 2^{19} - 2^{23} - 2^{26} - 2^{63}$	5	629		505		253
	$-1 - 2^8 - 2^{35} - 2^{61} - 2^{63}$	5	632		507		254
	$-1 + 2^{17} - 2^{54} + 2^{61} - 2^{64}$	5	637		511		256
	$-1 + 2^{35} + 2^{60} - 2^{64}$	4	638		512		256
	$-1 + 2^{10} + 2^{14} - 2^{18} + 2^{64}$	5	639		512		256
	$-1 - 2^3 - 2^{37} - 2^{50} - 2^{65}$	5	649	21×32	521	17×32	261
288	$-1 - 2^{52} + 2^{59} - 2^{81}$	4	809	13×64	648	21×32	282
	$-1 - 2^{26} - 2^{74} + 2^{82}$	4	819		656		283
	$-1 + 2^{21} - 2^{73} - 2^{82}$	4	819		657		283
	$-1 - 2^7 - 2^{13} - 2^{27} - 2^{82}$	5	819		657		283
	$-1 - 2^{11} - 2^{23} - 2^{32} - 2^{83}$	5	829		665		285
	$-1 - 2^{48} - 2^{52} - 2^{72} - 2^{84}$	5	839	27×32	673	11×64	286
	$-1 + 2^8 - 2^{12} + 2^{16} - 2^{85}$	5	849		680		287
	$-1 - 2^3 + 2^{31} - 2^{86}$	4	859		688		289
	$-1 - 2^{16} - 2^{20} - 2^{87}$	4	869	14×64	697		290
	$-1 + 2^{53} - 2^{56} + 2^{88}$	4	879		704		292
	$-1 + 2^{23} + 2^{67} + 2^{90}$	4	899	29×32	721	23×32	295
320	$-1 - 2^{12} - 2^{93} - 2^{95} - 2^{107}$	5	1069	17×64	857	27×32	317
	$-1 + 2^{65} - 2^{75} + 2^{109}$	4	1089	35×32	872	14×64	319
	$-1 - 2^{64} - 2^{100} + 2^{110}$	4	1099		880		321
	$-1 + 2^{15} + 2^{93} - 2^{111}$	4	1109		888		322
	$-1 - 2^{13} + 2^{57} - 2^{112}$	4	1119		896		323

Table 5. BLS curves with low-weight parameter $x_0 \equiv 16 \pmod{72}$ aiming at several security levels given in the first column. The columns "words for p" and "words for r" give the necessary number of 32- or 64-bit words to store the values for p and r, respectively. The last column provides the estimated actual security by the formula in [32].

security level	$x_0 \equiv 16 \pmod{72}$	weight	p (bits)	words for p	r (bits)	words for r	security (bits)
192	$2^{47} + 2^{16} - 2^5$	3	469	15×32	377	12×32	188
	$2^{47} + 2^{43} + 2^{36} + 2^3$	4	470		377		188
	$2^{47} + 2^{44} - 2^{32} + 2^7$	4	471		378		189
	$-2^{47} - 2^{45} + 2^{32} + 2^{28}$	4	472		379		189
	$-2^{48} + 2^{45} + 2^{31} - 2^7$	4	477		383		191
	$2^{48} - 2^{14} - 2^{12} - 2^4$	4	479		384		192
	$-2^{50} + 2^{21} + 2^{17} - 2^{13}$	4	499	8×64	400	7×64	200
	$2^{51} - 2^{48} + 2^{46} - 2^{16}$	4	507		407		203
	$-2^{51} + 2^{47} - 2^{22} + 2^{15}$	4	508		408		204
	$-2^{51} - 2^8 - 2^6 - 2^4$	4	509		409		204
	$-2^{51} - 2^{48} + 2^{45} + 2^{39}$	4	510		410		205
224	$2^{56} - 2^{53} - 2^{31} - 2^9$	4	557	9×64	447	7×64	223
	$-2^{56} + 2^{40} - 2^{26} - 2^6$	4	559		448		224
	$2^{56} + 2^{40} - 2^{20}$	3	559		449	15×32	224
	$2^{57} + 2^{25} + 2^{18} + 2^{11}$	4	569		457		228
	$2^{57} + 2^{54} + 2^{51} + 2^{39}$	4	571		458		229
256	$2^{63} - 2^{47} + 2^{38}$	3	629	10×64	504	8×64	252
	$2^{63} + 2^{59} + 2^{45} - 2^{17}$	4	630		505		252
	$-2^{63} - 2^{60} - 2^{44} - 2^{16}$	4	631		506		253
	$-2^{64} + 2^{61} - 2^{35} + 2^3$	4	637		511		254
	$2^{64} - 2^{46} + 2^{15} + 2^9$	4	639		512		255
288	$2^{83} - 2^{78} + 2^{60} - 2^{22}$	4	828	13×64	664	11×64	284
	$-2^{83} - 2^{46} + 2^{24}$	3	829		665		285
	$2^{83} + 2^{81} + 2^{12} + 2^7$	4	832		667		285
	$-2^{86} + 2^{82} + 2^{71} - 2^{24}$	4	858	27×32	688	22×32	289
	$2^{86} + 2^{77} - 2^{54} + 2^{27}$	4	859		689		289
	$-2^{89} + 2^{86} + 2^{28} - 2^{15}$	4	887	14×64	711	12×64	293
	$2^{89} - 2^{84} - 2^{50} + 2^{10}$	4	888		712		293
	$2^{89} + 2^{33} - 2^{29} - 2^6$	3	889		713		293
320	$2^{108} + 2^{66} - 2^{42}$	3	1079	17×64	865	14×64	318
	$-2^{108} - 2^{105} + 2^{55} + 2^{11}$	4	1081		866		318
	$2^{109} - 2^{106} + 2^{71} + 2^{22}$	4	1087		871		319
	$2^{111} + 2^{70} + 2^{66}$	3	1109	35×32	889	28×32	322
	$2^{111} + 2^{109} - 2^{100} - 2^{83}$	4	1112		891		322
	$2^{111} + 2^{110} + 2^{103} + 2^{43}$	4	1115		893		322
	$-2^{112} + 2^{107} - 2^{57} + 2^{33}$	4	1118		896		323
	$2^{112} - 2^{66} + 2^{42}$	3	1119		896		323

Table 6. BLS curves with low-weight parameter $x_0 \equiv 31 \pmod{72}$ aiming at several security levels given in the first column. The columns "words for p" and "words for r" give the necessary number of 32- or 64-bit words to store the values for p and r, respectively. The last column provides the estimated actual security by the formula in [32].

security level	$x_0 \equiv 31 \pmod{72}$	weight	p (bits)	words for p	r (bits)	words for r	security (bits)
192	$-1+2^{16}+2^{21}+2^{45}$	4	449	15 × 32	361	6 × 64	181
	$-1-2^{17}+2^{20}-2^{36}+2^{46}$	5	459		368		184
	$-1-2^{28}-2^{37}+2^{47}$	4	469		376		188
	$-1-2^{6}-2^{16}-2^{47}$	4	469		377		189
	$-1-2^{13}-2^{25}-2^{30}+2^{48}$	5	479		384		192
	$-1-2^{15}-2^{32}-2^{48}$	4	479		385	13 × 32	193
	$-1+2^{27}-2^{43}-2^{48}$	4	479		385		193
	$-1-2^{15}-2^{19}-2^{31}-2^{48}$	5	479		385		193
	$-1-2^{8}+2^{15}+2^{17}-2^{50}$	5	499	8 × 64	400		200
	$-1-2^{8}-2^{15}+2^{51}$	4	509		408		204
	$-1+2^{18}-2^{28}-2^{52}$	4	519	17 × 32	417	7 × 64	209
224	$-1-2^{37}+2^{40}-2^{43}+2^{55}$	5	549	9 × 64	440	7 × 64	220
	$-1-2^{18}+2^{29}+2^{35}-2^{56}$	5	559		448		224
	$-1+2^{14}-2^{22}+2^{57}$	4	569		456	15 × 32	228
	$-1+2^{17}+2^{27}-2^{57}$	4	569		456		228
	$-1-2^{8}-2^{34}-2^{50}-2^{58}$	5	579	19 × 32	465		233
	$-1+2^{13}-2^{30}-2^{59}$	4	589		473		237
256	$-1+2^{16}+2^{20}-2^{24}+2^{62}$	5	619	10 × 64	496	8 × 64	248
	$-1+2^{45}-2^{49}+2^{63}$	4	629		504		252
	$-1-2^{9}+2^{11}-2^{27}+2^{64}$	5	639		512		256
	$-1-2^{10}-2^{22}-2^{24}-2^{64}$	5	639		513	17 × 32	257
	$-1-2^{3}-2^{36}-2^{57}-2^{65}$	5	649	21 × 32	521		261
	$-1+2^{20}-2^{43}+2^{58}+2^{65}$	5	649		521		261
288	$-1+2^{25}+2^{49}-2^{81}$	4	809	13 × 64	648	21 × 32	282
	$-1+2^{3}-2^{74}-2^{81}$	4	809		649		282
	$-1-2^{11}+2^{57}-2^{82}$	4	819		656		283
	$-1-2^{18}-2^{39}+2^{83}$	4	829		664		285
	$-1-2^{18}-2^{39}+2^{83}$	4	829		664		285
	$-1+2^{31}-2^{77}-2^{84}$	4	839	27 × 32	673	11 × 64	286
	$-1-2^{20}+2^{71}+2^{86}$	4	859		689		289
320	$-1-2^{39}-2^{54}+2^{107}$	4	1069	17 × 64	856	27 × 32	317
	$-1+2^{3}+2^{10}+2^{18}-2^{109}$	5	1089	35 × 32	872	14 × 64	319
	$-1+2^{26}+2^{36}+2^{57}-2^{111}$	5	1109		888		322
	$-1-2^{8}-2^{24}+2^{37}+2^{111}$	5	1109		889		322

Table 7. BLS curves with low-weight parameter $x_0 \equiv 64 \pmod{72}$ aiming at several security levels given in the first column. The columns "words for p" and "words for r" give the necessary number of 32- or 64-bit words to store the values for p and r, respectively. The last column provides the estimated actual security by the formula in [32].

security level	$x_0 \equiv 64 \pmod{72}$	weight	p (bits)	words for p	r (bits)	words for r	security (bits)
192	$-2^{16}-2^{27}+2^{46}$	3	459	15 × 32	368	6 × 64	184
	$2^8+2^{12}+2^{40}+2^{46}$	4	459		369		185
	$-2^{14}+2^{19}+2^{21}-2^{47}$	4	469		376		188
	$-2^{12}-2^{30}-2^{35}-2^{48}$	4	479		385	13 × 32	193
	$-2^{10}-2^{14}-2^{17}-2^{48}$	4	479		385		193
	$2^{15}+2^{22}+2^{25}+2^{49}$	4	489	8 × 64	393		197
	$2^{12}-2^{17}-2^{31}-2^{49}$	4	489		393		197
	$2^{31}-2^{33}-2^{49}$	3	489		393		197
224	$2^{10}+2^{21}-2^{28}+2^{55}$	4	549	9 × 64	440	7 × 64	220
	$-2^4-2^{30}+2^{32}-2^{56}$	4	559		448		224
	$2^5-2^{10}+2^{27}+2^{56}$	4	559		449	15 × 32	225
	$2^8+2^{10}+2^{16}-2^{57}$	4	569		456		228
	$2^{31}+2^{35}-2^{41}+2^{57}$	4	569		456		228
	$-2^7+2^{10}+2^{16}+2^{58}$	4	579	19 × 32	465		233
	$2^7-2^{25}+2^{31}-2^{60}$	4	599		480		240
256	$2^{19}-2^{26}-2^{37}-2^{62}$	4	619	10 × 64	497	8 × 64	249
	$2^{16}-2^{42}-2^{60}-2^{62}$	4	622		499		250
	$2^9-2^{38}-2^{56}-2^{63}$	4	629		505		253
	$-2^{14}+2^{39}-2^{56}-2^{63}$	4	629		505		253
	$-2^{23}-2^{42}-2^{44}-2^{64}$	4	639		513	17 × 32	257
	$-2^{12}-2^{21}-2^{60}-2^{64}$	4	640		513		257
	$2^{17}-2^{52}-2^{54}-2^{65}$	4	649	21 × 32	521		261
	$-2^{11}-2^{35}-2^{55}-2^{65}$	4	649		521		261
288	$-2^{40}-2^{49}-2^{81}$	3	809	13 × 64	649	21 × 32	282
	$-2^7-2^{48}-2^{70}-2^{82}$	4	819		657		283
	$-2^7-2^{46}-2^{54}-2^{82}$	4	819		657		283
	$-2^{15}-2^{17}+2^{83}$	3	829		664		285
	$2^{41}+2^{47}+2^{68}+2^{83}$	4	829		665		285
	$2^{17}+2^{21}+2^{29}+2^{84}$	4	839	27 × 32	673	11 × 64	286
	$2^9+2^{13}+2^{34}+2^{85}$	4	849		681		287
	$-2^{31}-2^{66}+2^{86}$	3	859		688		289
	$2^8+2^{26}+2^{34}+2^{86}$	4	859		689		289
	$2^{34}-2^{82}-2^{87}$	3	869	14 × 64	697		290
	$2^{13}+2^{17}-2^{27}-2^{88}$	4	879		705	23 × 32	292
	$-2^{10}-2^{12}-2^{14}-2^{89}$	4	889		713		293
	$-2^6-2^{45}+2^{90}$	3	899	29 × 32	720		295
	$2^{60}-2^{67}+2^{91}$	3	909		728		296
320	$2^{14}+2^{17}-2^{38}+2^{107}$	4	1069	17 × 64	856	27 × 32	317
	$2^6-2^{32}+2^{39}+2^{107}$	4	1069		857		317
	$2^5-2^{18}-2^{25}-2^{108}$	4	1079		865	14 × 64	318
	$2^{20}+2^{49}+2^{71}+2^{111}$	4	1109	35 × 32	889		322

The curves in all four tables were found by trying all possibilities for the signed binary representation of x_0 with a fixed weight such that x_0 belongs to the right congruence class modulo 72. In our search, we did not find any curves in the considered range of parameter sizes where x_0 is plus or minus a power of 2 (i.e. weight 1) or where it is a binomial, a sum of two such powers (i.e. weight 2). In this sense, our search indicates that weights 3, 4 and 5 are optimal for the security levels considered in this paper. The even congruencies ($x_0 \equiv 16, 64 \pmod{72}$) gain the slight advantage over the odd congruencies ($x_0 \equiv 7, 31 \pmod{72}$), since the last bit of the binary representation of odd congruencies is obviously forced to be 1. Thus, curves in the even congruency classes commonly have weights 3 and 4 whilst curves in the odd congruency classes commonly have weights 4 and 5. On the other hand, the odd congruencies both give rise to curves with $b = 1$ which would make for slightly faster point operations, but (all other things being equal) one would probably achieve a faster implementation by taking the x_0 value with the lowest weight possible, since one less bit in x_0 saves over 10 full $\mathbb{F}_{p^{24}}$ multiplications per single pairing.

Restricting x_0 to sparse values only results in a certain inflexibility when adjusting the parameter sizes to exact values, for example certain multiples of word sizes on a target implementation platform. However, recent high-speed implementations of pairings at the 128-bit security level have shown that lazy reduction techniques give significant improvements in the field tower arithmetic and thus the overall pairing computation [7,1]. Such techniques can be employed efficiently when the bit size of the prime characteristic p is a few bits less than a multiple of the word size, which provides a certain space for delaying reductions for field arithmetic. In Tables 4, 5, 6 and 7 we have tried to account for this (as far as possible), by including different choices of curves at each security level that have a varying gap between the prime field size and multiples of standard word sizes 32 and 64, which are also given in the table. We believe that most implementors of pairings in software will find a suitable curve at the desired security level in our tables, or else will be able to find a suitable curve themselves with similar properties.

References

1. Aranha, D.F., Karabina, K., Longa, P., Gebotys, C.H., López, J.: Faster Explicit Formulas for Computing Pairings Over Ordinary Curves. In: Paterson, K.G. (ed.) EUROCRYPT 2011. LNCS, vol. 6632, pp. 48–68. Springer, Heidelberg (2011)
2. Barker, E., Barker, W., Burr, W., Polk, W., Smid, M.: Recommendation for key management - part 1: General (revised). Technical report, NIST National Institute of Standards and Technology, Published as NIST Special Publication 800–57 (2007), http://csrc.nist.gov/groups/ST/ toolkit/documents/SP800-57Part1_3-8-07.pdf
3. Barreto, P.S.L.M., Lynn, B., Scott, M.: Constructing Elliptic Curves with Prescribed Embedding Degrees. In: Cimato, S., Galdi, C., Persiano, G. (eds.) SCN 2002. LNCS, vol. 2576, pp. 257–267. Springer, Heidelberg (2003)

4. Barreto, P.S.L.M., Naehrig, M.: Pairing-Friendly Elliptic Curves of Prime Order. In: Preneel, B., Tavares, S. (eds.) SAC 2005. LNCS, vol. 3897, pp. 319–331. Springer, Heidelberg (2006)
5. Benger, N.: Cryptographic Pairings: Efficiency and DLP Security. PhD thesis, Dublin City University (May 2010)
6. Benger, N., Scott, M.: Constructing Tower Extensions of Finite Fields for Implementation of Pairing-Based Cryptography. In: Hasan, M.A., Helleseth, T. (eds.) WAIFI 2010. LNCS, vol. 6087, pp. 180–195. Springer, Heidelberg (2010)
7. Beuchat, J.-L., González-Díaz, J.E., Mitsunari, S., Okamoto, E., Rodríguez-Henríquez, F., Teruya, T.: High-Speed Software Implementation of the Optimal Ate Pairing Over Barreto–Naehrig Curves. In: Joye, M., Miyaji, A., Otsuka, A. (eds.) Pairing 2010. LNCS, vol. 6487, pp. 21–39. Springer, Heidelberg (2010)
8. Costello, C., Hişil, H., Boyd, C., Nieto, J.M.G., Wong, K.K.-H.: Faster pairings on special Weierstrass curves. In: Shacham and Waters [30], pp. 89–101 (2009)
9. Costello, C., Lange, T., Naehrig, M.: Faster Pairing Computations on Curves with High-Degree Twists. In: Nguyen, P.Q., Pointcheval, D. (eds.) PKC 2010. LNCS, vol. 6056, pp. 224–242. Springer, Heidelberg (2010)
10. Devegili, A.J., hÉigeartaigh, C.Ó., Scott, M., Dahab, R.: Multiplication and squaring on pairing-friendly fields. Cryptology ePrint Archive, Report 2006/471 (2006), http://eprint.iacr.org/
11. Dominguez Perez, L.J., Scott, M.: Private communication (November 2010)
12. Freeman, D., Scott, M., Teske, E.: A taxonomy of pairing-friendly elliptic curves. J. Cryptology 23(2), 224–280 (2010)
13. Granger, R., Scott, M.: Faster Squaring in the Cyclotomic Subgroup of Sixth Degree Extensions. In: Nguyen, P.Q., Pointcheval, D. (eds.) PKC 2010. LNCS, vol. 6056, pp. 209–223. Springer, Heidelberg (2010)
14. Hankerson, D., Menezes, A.J., Scott, M.: Software implementation of pairings. In: Joye, M., Neven, G. (eds.) Identity-Based Cryptography, pp. 188–206. IOS Press (2008)
15. Heß, F., Smart, N.P., Vercauteren, F.: The eta pairing revisited. IEEE Transactions on Information Theory 52, 4595–4602 (2006)
16. Ireland, K., Rosen, M.: A Classical Introduction to Modern Number Theory. Graduate texts in mathematics, vol. 84. Springer, Heidelberg (1990)
17. Karabina, K.: Squaring in cyclotomic subgroups. Cryptology ePrint Archive, Report 2010/542 (2010), http://eprint.iacr.org/
18. Lauter, K., Montgomery, P.L., Naehrig, M.: An Analysis of Affine Coordinates for Pairing Computation. In: Joye, M., Miyaji, A., Otsuka, A. (eds.) Pairing 2010. LNCS, vol. 6487, pp. 1–20. Springer, Heidelberg (2010)
19. Miller, V.S.: The Weil pairing, and its efficient calculation. Journal of Cryptology 17, 235–261 (2004)
20. Naehrig, M.: Constructive and computational aspects of cryptographic pairings. PhD thesis, Eindhoven University of Technology (May 2009)
21. Naehrig, M., Barreto, P.S.L.M., Schwabe, P.: On Compressible Pairings and Their Computation. In: Vaudenay, S. (ed.) AFRICACRYPT 2008. LNCS, vol. 5023, pp. 371–388. Springer, Heidelberg (2008)
22. Naehrig, M., Niederhagen, R., Schwabe, P.: New Software Speed Records for Cryptographic Pairings. In: Abdalla, M., Barreto, P.S.L.M. (eds.) LATINCRYPT 2010. LNCS, vol. 6212, pp. 109–123. Springer, Heidelberg (2010)
23. Pereira, G.C.C.F., Simplício Jr., M.A., Naehrig, M., Barreto, P.S.L.M.: A family of implementation-friendly BN elliptic curves. Journal of Systems and Software 84(8), 1319–1326 (2011), http://cryptojedi.org/papers/#fast-bn

24. Rubin, K., Silverberg, A.: Choosing the correct elliptic curve in the CM method. Mathematics of Computation 79, 545–561 (2010)
25. Scott, M.: Scaling security in pairing-based protocols. Cryptology ePrint Archive, Report 2005/139 (2005), http://eprint.iacr.org/
26. Scott, M.: A note on twists for pairing friendly curves (February 2009), Personal webpage ftp://ftp.computing.dcu.ie/pub/resources/crypto/twists.pdf
27. Scott, M.: On the efficient implementation of pairing-based protocols. Cryptology ePrint Archive, Report 2011/334 (2011), http://eprint.iacr.org/
28. Scott, M., Barreto, P.S.L.M.: Compressed Pairings. In: Franklin, M. (ed.) CRYPTO 2004. LNCS, vol. 3152, pp. 140–156. Springer, Heidelberg (2004)
29. Scott, M., Benger, N., Charlemagne, M., Dominguez Perez, L.J., Kachisa, E.J.: On the final exponentiation for calculating pairings on ordinary elliptic curves. In: Shacham and Waters [30], pp. 78–88 (2009)
30. Shacham, H., Waters, B. (eds.): Pairing 2009. LNCS, vol. 5671. Springer, Heidelberg (2009)
31. Silverman, J.H.: The Arithmetic of Elliptic Curves. Graduate texts in mathematics, vol. 106. Springer, Heidelberg (1986)
32. Smart, N. (ed.): ECRYPT II yearly report on algorithms and keysizes (2009-2010). Technical report, ECRYPT II – European Network of Excellence in Cryptology, EU FP7, ICT-2007-216676, Published as deliverable D.SPA.13 (2010), http://www.ecrypt.eu.org/documents/D.SPA.13.pdf
33. Vercauteren, F.: Optimal pairings. IEEE Transactions on Information Theory 56(1), 455–461 (2010)

A Timings

This section provides timings of a plain C implementation of the (optimal) ate pairing on BLS curves with embedding degree $k = 24$ and parameter $x_0 \equiv 16$ (mod 72). We give timings for field operations of base field and all extension fields. Timings for pairings are split up into timing for the Miller loop, the final exponentiation, and the complete pairing. The number denoted "Product" refers to the time per pairing in a product of 20 pairings.

	add cyc	add μs	sub cyc	sub μs	M cyc	M μs	S cyc	S μs	I cyc	I μs
$\mathbb{F}_p$	320	0.10	248	0.08	1112	0.34	1026	0.36	20320	6.83
$\mathbb{F}_{p^2}$	540	0.18	487	0.16	4627	1.52	4414	1.44	52334	17.16
$\mathbb{F}_{p^4}$	1068	0.35	942	0.32	21722	7.21	20532	6.76	131246	43.65
$\mathbb{F}_{p^{12}}$	3510	1.19	2750	0.93	163570	54.83	154325	51.21	744451	247.18
$\mathbb{F}_{p^{24}}$	6269	2.15	5536	1.85	539026	179.70	506029	168.13	2326049	777.80

Pairings	Miller loop	Final exp.	Single pairing	Product
cyc	30,335,982	97,561,935	127,897,917	24,124,734
ms	10.00	32.62	42.62	8.05

(a) Cycle counts and timings: pfc-bls384-p478-k24a

	add cyc	add μs	sub cyc	sub μs	M cyc	M μs	S cyc	S μs	I cyc	I μs
$\mathbb{F}_p$	367	0.12	393	0.10	1475	0.52	1465	0.49	28511	9.70
$\mathbb{F}_{p^2}$	659	0.22	587	0.20	6272	2.09	5973	1.99	71990	23.96
$\mathbb{F}_{p^4}$	1272	0.42	1146	0.38	29253	9.70	27512	9.24	181534	60.34
$\mathbb{F}_{p^{12}}$	3884	1.36	3443	1.15	218255	72.04	209100	68.05	1010078	337.78
$\mathbb{F}_{p^{24}}$	7644	2.68	6982	2.32	708585	236.46	665836	221.82	3127211	1041.84

Pairings	Miller loop	Final exp.	Single pairing	Product
cyc	53,827,736	168,824,048	222,651,784	41,951,965
ms	17.98	56.29	74.27	13.99

(b) Cycle counts and timings: pfc-bls513-p639-k24a

Fig. 1. Timings for two BLS curves of security roughly 192-bits (pfc-bls384-p478-k24a, group size 384 bits and base field size 478 bits) and 256-bits (pfc-bls513-p639-k24a). All measurements were done on an Intel Core 2 Duo CPU (E6850) running 64-bit Windows 7 at 3.00 GHz with 4GB RAM.

Stone Knives and Bear Skins: Why Does the Internet Run on Pre-historic Cryptography?

Eric Rescorla

RTFM, Inc.
ekr@rtfm.com

Abstract. While cryptography has advanced greatly since since 2001, Internet security protocols have not. Here is a list of the algorithms that are used in common SSL/TLS stacks:

- RSA in PKCS#1 1.5 mode (1993)
- MD5 (1982)
- SHA-1 (1993)
- DES (1976) and AES (2001) in CBC mode (with chained IVs)
- RC4 (1987, leaked 1994)

The situation is similar for other protocols such as IPsec and S/MIME. Without exception, all of these algorithms have known deficiencies, and in many cases these deficiencies have led to practical or semi-practical attacks. Despite this, implementors and users have responded either by ignoring these issues or by adding layers of countermeasures to the attacks which are presently known. Even when new protocols are designed – for instance the IETF's new JSON-based secure messaging effort – designers often select older algorithms over newer, more secure ones. In this talk, we explore how we got into this situation, if we can get out, and if we even want to.

D.J. Bernstein and S. Chatterjee (Eds.): INDOCRYPT 2011, LNCS 7107, p. 343, 2011.
© Springer-Verlag Berlin Heidelberg 2011

The Limits of Common Coins: Further Results

Hemanta K. Maji[1,*] and Manoj Prabhakaran[2]

[1] University of California, Los Angeles
hmaji2@illinois.edu
[2] University of Illinois, Urbana-Champaign
mmp@illinois.edu

Abstract. In [8] it was shown that the coin-tossing functionality $\mathcal{F}_{coin}$ has limited use in 2-party secure function evaluation (SFE) in the computationally unbounded (a.k.a information-theoretic) setting. Further it was shown that for $\mathcal{F}_{coin}$ to be useful in securely realizing any one in a *a large class* of symmetric SFE (SSFE) functionalities, a certain computational assumption (namely the existence of a semi-honest secure OT protocol) is necessary and sufficient. In this work, we close a gap in the class of SSFE functionalities for which this result was proven in [8]: we show that $\mathcal{F}_{coin}$ can be used to securely realize *any* SSFE functionality that cannot be realized in the computationally unbounded setting, if and only if there exists a semi-honest secure OT protocol.

1 Introduction

Multi-party computation is a central problem in modern cryptography. An important question regarding secure function evaluation has been to understand the relative "cryptographic complexity" of the different functions that are evaluated – i.e., secure evaluation of which functions can be reduced to that of which other functions. While several aspects of this problem have been well-studied (e.g. [4,5,9,7,6] to name a few) a large number of problems remain open. In particular, we do not fully understand how, in the probabilistic polynomial time (PPT) setting, the cryptographic complexity of various functionalities relates to computational intractability assumptions, though again several works have resolved this question in various special cases (see for e.g. [1,10]).

In this work, we follow up on [8], which studied the power of the coin-tossing functionality $\mathcal{F}_{coin}$ in the 2-party setting. Our focus will be on randomized symmetric secure function evaluation (SSFE) functionalities, in which both parties get the same output. In [8], the following was shown (along with other results):

- A 2-party SSFE functionality with "bi-directional influence" reduces to $\mathcal{F}_{coin}$ in the PPT setting if and only if there exists a semi-honest secure OT protocol.

The assumption that there exists a semi-honest secure OT protocol (in the PPT setting) is a classic assumption in the context of secure function evaluation, comparable in generality to the existence of a one-way function, but stronger. The

---

* This work was carried out when the author was at University of Illinois, Urbana-Champaign.

D.J. Bernstein and S. Chatterjee (Eds.): INDOCRYPT 2011, LNCS 7107, pp. 344–358, 2011.
© Springer-Verlag Berlin Heidelberg 2011

notion of reduction here is that of secure reduction in the Universal Composition framework.

Here, "bi-directional influence" refers to the case that the output distribution depends on both parties' inputs. The class of SSFE functionalities without bi-directional influence can be "less complex," and may indeed reduce to $\mathcal{F}_{\mathsf{coin}}$ unconditionally. On the other hand, some other SSFE functionalities without bi-directional influence may be complex enough that they require some computational complexity assumption to reduce to $\mathcal{F}_{\mathsf{coin}}$. This raises the question if such reductions are equivalent to *other* computational intractability assumptions. In this paper we answer this question in the negative:

- A 2-party SSFE functionality $\mathcal{F}$ reduces to $\mathcal{F}_{\mathsf{coin}}$ in the PPT setting, if and only if,
  - either $\mathcal{F}$ reduces to $\mathcal{F}_{\mathsf{coin}}$ in the computationally unbounded setting, or
  - there exists a semi-honest secure OT protocol.

Since [8] explicitly characterized the class of SSFE functionalities that reduce to $\mathcal{F}_{\mathsf{coin}}$ in the computationally unbounded setting, our result completely resolves the question of which SSFE functionalities reduce to $\mathcal{F}_{\mathsf{coin}}$ under what computational assumptions. Indeed, *our result shows that there is only one relevant complexity assumption in this case, namely, the existence of a semi-honest secure OT protocol.*

1.1 Outline

The main components in proving our main result (Theorem 1) are the following:

- In Section 2 we define a new class of SSFE functionalities with "uni-directional influence" called *oblivious sampling functionalities* (Definition 1) that is crucial to our new result.
- Oblivious Sampling functionalities are essentially the ones that the result in [8] mentioned above, did not cover. This is formalized as Lemma 2.
- Then we go on to show that if an oblivious sampling functionality reduces to $\mathcal{F}_{\mathsf{coin}}$, then there must exist a semi-honest secure OT protocol (Lemma 3).

In Section 3 we provide an immediate corollary of our main result (Corollary 2) as well as a generalization (Theorem 3) which replaces $\mathcal{F}_{\mathsf{coin}}$ with a larger class of SSFE functionalities (called publicly-selectable source, originally defined in [8] and extended in Section 2).

2   Preliminaries

In this section we introduce important definitions (including some from [8]) that are required to state and prove our results.

Secure function evaluation functionalities. A (randomized) 2-party symmetric secure function evaluation (SSFE) functionality $\mathcal{F}_f$ is specified by a function $f : X \times Y \times R \to Z$.[1] The functionality takes inputs $x \in X$ from Alice, $y \in Y$ from Bob, uniformly samples $r \in R$ and outputs $f(x, y, r)$ to both Alice and Bob. (If a party is actively corrupt, it can obtain its own output first and decide whether the output should be delivered to the other party.) We shall write $f(x, y)$ to denote the *distribution* of $f(x, y, r)$ when r is sampled uniformly from R. An example is the common randomness functionality, denoted by $\mathcal{F}_{\text{coin}}$, with $X = Y = \{0\}$, $R = \{0, 1\}$), $f(x, y, r) = r$.

We shall partition SSFE functionalities into three classes. We say that in $\mathcal{F}_f$ Alice has influence on the output if there exist $x', x'' \in X$, and $y \in Y$ such that the distributions $f(x', y) \neq f(x'', y)$ (and similarly we define Bob having influence on the output).

- *Uninfluenced functionalities.* In this case neither Alice nor Bob has influence on the output; i.e., the output is from a constant distribution and hence we can set $f(x, y, r)$ to be $f(r)$.
- *Functionalities with unidirectional influence.* In this case exactly one party, say Alice (or Bob), has influence on the output. We shall denote the output distribution as $f(x)$ (or $f(y)$) in this case. By abuse of notation, we will use the same notation to denote a random variable drawn from this distribution.
- *Functionalities with bidirectional influence.* In this case both parties have influence on the output.

Oblivious Sampling. In order to prove our results, we define a new class of SSFE functionalities with unidirectional influence called oblivious sampling functionalities. Intuitively, in these functionalities, Alice decides the distribution from which the common output is sampled, but Bob does not fully learn which distribution the output comes from.

Towards formally defining this class, we require the notion of a non-redundant input. Let $\mathcal{F}_f$ be a 2-party SSFE functionality with unidirectional influence. Then x is a *redundant input* if there exists a set $X' \subseteq X$ such that for each $x' \in X'$, $f(x') \neq f(x)$, but $f(x)$ is a convex combination of $\{f(x') | x' \in X'\}$ (i.e., $f(x) = \sum_{x' \in X'} \alpha_{x'} f(x')$ for $\alpha_{x'} > 0$ such that $\sum_{x' \in X'} \alpha_{x'} = 1$).

Definition 1. *A 2-party SSFE functionality $\mathcal{F}_f$ with unidirectional influence (say Alice has influence) is called an* Oblivious Sampling *functionality if there exist two non-redundant inputs x_0, x_1 such that*

---

[1] In this work, unless otherwise specified, we allow functionalities to be randomized by default. A "symmetric" functionality is one which gives the same output to both parties, and we restrict our general definition to such functionalities. Finally, as in [8] and related works, we shall always consider X, Y, R, Z to be finite and of constant size independent of the security parameter that appears in the definition of security; also the probabilities involved are constant. However, the results do extend to the case when these sets are polynomially large in the security parameter, and the probabilities are such that the outcome for a "non-redundant" input (x, y) cannot be approximated as a convex combination of other inputs within an inverse polynomial statistical distance.

- $\exists z' \Pr[f(x_0) = z'] \neq \Pr[f(x_1) = z']$ *(i.e., $f(x_0) \neq f(x_1)$), and*
- $\exists z \Pr[f(x_0) = z], \Pr[f(x_1) = z] > 0$ *(i.e., supports of $f(x_0)$ and $f(x_1)$ intersect).*

From a characterization in [8], it follows that Oblivious Sampling functionalities are not UC-securely reducible to $\mathcal{F}_{\text{coin}}$ in the computationally unbounded setting (publicly-selectable sources, as defined below, are the only SSFE functionalities which are). A modification of the proof there could be used to show that one-way functions must exist for such a reduction to be possible in the PPT setting. What we shall show is that, in fact, a semi-honest OT protocol must exist for such a reduction.

We also note that Oblivious Sampling functionalities are not complex enough to be complete in the computationally unbounded setting. Since the characterization in [5] of complete SSFE functions in the *semi-honest security setting* requires both parties to have more than one input value,[2] there is no semi-honest secure reduction of say, OT to an Oblivious Sampling functionality; and therefore there is no UC-secure of reduction of OT to an Oblivious Sampling functionality (as OT is deviation-revealing [12]). This makes it non-trivial to prove that such a reduction in the PPT setting implies the existence of a semi-honest OT protocol.

Publicly-selectable sources. [8] defined an SSFE functionality to be a publicly-selectable source if it is of the form $\mathcal{F}_f$ where $f(x, y, r) = (g(x), h(g(x), r))$ (possibly relabeled using an output alphabet), for some functions g and h (or with Alice's and Bob's roles interchanged). That is, the function's output distribution for different values of x must either be identical (when $g(x) = g(x')$) or have disjoint supports (when $g(x) \neq g(x')$). We slightly extend this definition so that $\mathcal{F}_f$ is called a publicly-selectable source if $\mathcal{F}'_f$ obtained by restricting $\mathcal{F}_f$ to non-redundant inputs has the same property.

Definition 2. *A 2-party SSFE functionality $\mathcal{F}_f$ with unidirectional influence (say Alice has influence) is called a* publicly-selectable source *if, for every two non-redundant inputs x_0, x_1 such that $f(x_0) \neq f(x_1)$, supports of $f(x_0)$ and $f(x_1)$ are disjoint (i.e., $\nexists z \Pr[f(x_0) = z], \Pr[f(x_1) = z] > 0$).*

Note that with this modification in the definition, an oblivious sampling functionality could alternately be defined as an SSFE functionality with unidirectional influence that is *not* a publicly-selectable source (as redefined here).

We point out that any publicly-selectable source functionality UC-securely reduces to $\mathcal{F}_{\text{coin}}$ by a protocol in which the party with the influence specifies one of the non-redundant inputs, and then the two parties use $\mathcal{F}_{\text{coin}}$ to sample an outcome from that distribution. Indeed, it follows from the characterization in [8] that these are the only SSFE functionalities which can be UC-securely reduced to $\mathcal{F}_{\text{coin}}$.

---

[2] [5] showed that an SSFE functionality is semi-honest complete iff there exist x_0, x_1, y_0, y_1, z, such that $\Pr[z|x_0, y_0], \Pr[z|x_0, y_1] > 0$ and $\Pr[z|x_0, y_0] \Pr[z|x_1, y_1] \neq \Pr[z|x_0, y_1] \Pr[z|x_1, y_0]$.

Secure Reductions. We use standard security notions, that are summarized in Appendix A. We say that a functionality $\mathcal{F}$ *UC-securely reduces* (or simply, reduces) to a functionality $\mathcal{G}$ if there exists a universally composable protocol that securely realizes $\mathcal{F}$, in which the parties are allowed access to ideal instances of the functionality $\mathcal{G}$. We distinguish between security in the probabilistic polynomial time (PPT) setting — in which the adversaries and the environment — are restricted to be PPT, and the computationally unbounded setting.

A functionality is called *trivial* if it can be UC-securely realized by a protocol in which the parties use only a plain communication functionality to interact with each other. For finite functionalities as we consider, the class of trivial functionalities remains the same in the PPT and computationally unbounded settings. In particular, based on the characterization in [12] it is easy to see that the 2-party SSFE functionalities that are trivial are unidirectional functionalities in which one party, say Alice, can determine the outcome as a deterministic function of its input (and may in addition have redundant inputs).

Oblivious Transfer. We shall refer to the oblivious transfer or OT functionality which takes two bits (x_0, x_1) as input from Alice, a single bit b from Bob as input, and outputs x_b to Bob, but nothing to Alice. (Note that this is not an SSFE functionality, because it is not symmetric.) The only computational intractability assumption that is referred to by our results is that there exists a protocol that is a secure realization of the OT functionality against semi-honest adversaries in the PPT setting. It is known that this is equivalent to the existence of an OT protocol that is secure against active adversaries as well, if restricted to the standalone security (as opposed to UC security) case (and even in a "black-box" sense [3]).

3   Our Results

In this section we describe our main results, which are proven in the subsequent sections.

Theorem 1. *If an SSFE functionality $\mathcal{F}$ reduces to $\mathcal{F}_{\mathrm{coin}}$ in the PPT setting, then either $\mathcal{F}$ is reducible to $\mathcal{F}_{\mathrm{coin}}$ in the computationally unbounded setting or there exists a semi-honest secure OT protocol.*

Our main contribution is to recognize the gap left behind by the results in [8] which only addressed the case of SSFE functionalities $\mathcal{F}$ with bi-directional influence. We fill this gap by identifying *oblivious sampling functionalities* (Definition 1) as an interesting class of SSFE functionalities, and showing that if such a functionality reduces to $\mathcal{F}_{\mathrm{coin}}$, then there exists a semi-honest secure OT protocol (Lemma 3).

We note that if a semi-honest secure OT protocol exists, then it is known that $\mathcal{F}_{\mathrm{coin}}$ is complete in the PPT setting and a hence $\mathcal{F}_{\mathrm{coin}}$ is useful for realizing a functionality like OT [11]. Hence we have the following corollary of the above theorem.

Corollary 2. *The following statements are equivalent:*

1. *Some 2-party SSFE functionality $\mathcal{F}$ that is not UC-reducible to $\mathcal{F}_{\text{coin}}$ in the computationally unbounded setting, reduces to $\mathcal{F}_{\text{coin}}$ in the PPT setting.*
2. *There exists a semi-honest secure OT protocol (in the PPT setting).*
3. *Every 2-party SSFE functionality reduces to $\mathcal{F}_{\text{coin}}$ in the PPT setting.*

Finally, we can extend the above results to a wider class of functionalities than just $\mathcal{F}_{\text{coin}}$. In particular, we show the following.

Theorem 3. *If $\mathcal{G}$ is a publicly-selectable source, and an SSFE functionality $\mathcal{F}$ reduces to $\mathcal{G}$ in the PPT setting, then either $\mathcal{F}$ is reducible to $\mathcal{G}$ in the computationally unbounded setting or there exists a semi-honest secure OT protocol.*

To prove this we show that any non-trivial publicly-selectable source $\mathcal{G}$ is "equivalent" to $\mathcal{F}_{\text{coin}}$ in that either functionality can be reduced to the other (Lemma 1).

4 Proofs

In this section we prove Theorem 1 and Theorem 3. But first we prove the following simple lemma that will be useful in both proofs.

Lemma 1. *For any publicly-selectable source $\mathcal{G}$, in the computationally unbounded (as well as PPT) setting, $\mathcal{G}$ reduces to $\mathcal{F}_{\text{coin}}$; also, $\mathcal{F}_{\text{coin}}$ reduces to $\mathcal{G}$, unless $\mathcal{G}$ is trivial.*

Proof. W.l.o.g, let Alice be the party whose input may have influence on the output in $\mathcal{G}$. Let $\mathcal{D}$ denote the set of output distributions for non-redundant inputs for Alice. Note that since $\mathcal{G}$ is a publicly-selectable source, the distributions in $\mathcal{D}$ have disjoint supports.

$\mathcal{G}$ reduces to $\mathcal{F}_{\text{coin}}$: this follows from a simple protocol for $\mathcal{G}$ as follows (we omit the routine security analysis):

- On input x, Alice determines the unique convex combination of distributions in $\mathcal{D}$ that equals the output distribution for x. The uniqueness is a consequence of those distributions having disjoint supports.
- Alice samples an element from $\mathcal{D}$ according to its weight in the above convex combination, and announces it. (We remark that a cheating Alice could use any strategy to choose an element from $\mathcal{D}$; however it can be mapped to simply choosing an input and then following the protocol honestly.)
- Alice and Bob obtain coins from $\mathcal{F}_{\text{coin}}$, and use them to sample an outcome from the announced distribution.

$\mathcal{F}_{\text{coin}}$ reduces to $\mathcal{G}$, unless $\mathcal{G}$ is trivial: $\mathcal{G}$ is trivial iff every distribution in $\mathcal{D}$ has zero entropy. Otherwise the following is a secure protocol[3] for $\mathcal{F}_{\text{coin}}$ using $\mathcal{G}$

---

[3] Note that in a secure realization for $\mathcal{F}_{\text{coin}}$ (without guaranteed output delivery) either party is allowed to abort the protocol, possibly after seeing the outcome of the protocol. This is the standard UC security guarantee for 2-party functionalities (and more generally, when there is no honest majority assumption).

(again, we omit the standard security analysis). Briefly, in this protocol the parties apply a von Neumann extractor to the outcome sampled from $\mathcal{G}$, to obtain a fair coin.

- Let x be a fixed non-redundant input for Alice such that the output distribution for x is in $\mathcal{D}$ and has positive entropy. Let $Z_0 \subseteq Z$ be a subset of the outcomes so that for input x, the probability that the outcome is in Z_0 is p, $0 < p < 1$. Let $Z_1 = Z \backslash Z_0$.
- Alice and Bob repeat the following until they are "satisfied":
  - Alice sends x to $\mathcal{G}$ twice.
  - If in either instance, the output from $\mathcal{G}$ is not from the support of the distribution corresponding to x, Bob aborts the protocol. Note that since $\mathcal{G}$ is a publicly-selectable source, this essentially forces Alice to either send an input equivalent to x or probabilistically abort.
  - Else, if exactly one of the outputs is from Z_0 and one from Z_1 then the parties are satisfied
- If the first and second outputs in the last pair of invocations of $\mathcal{G}$ were in Z_0 and Z_1 respectively, the common output is 0; else (the outputs where in Z_1 and Z_0 respectively) the common output is 1.

Since p is a constant independent of the security parameter, this protocol runs in expected constant number of rounds, and except with negligible probability, ends in a polynomial number of rounds.   □

4.1   Proof of Theorem 1

We start with the following classification of 2-party (randomized) SSFE functionalities.

Lemma 2. *Every 2-party SSFE functionality falls into one of the following categories.*

1. *UC-reduces to $\mathcal{F}_{\mathsf{coin}}$ (in the computationally unbounded setting).*
2. *An oblivious sampling functionality.*
3. *A functionality with bi-directional influence.*

Proof sketch: This follows from the partitioning of SSFE functionalities into (a) uninfluenced functionalities, (b) functionalities with unidirectional influence, and (c) those with bidirectional influence (see Section 2). As noted in Section 2, an uninfluenced SSFE functionality amounts to sampling from a fixed distribution, and this readily UC-reduces to $\mathcal{F}_{\mathsf{coin}}$. Hence functionalities of type (a) fall into Category 1. For a functionality $\mathcal{F}$ of type (b), if $\mathcal{F}$ is a publicly-selectable source, then again it falls into Category 1, by the first part of Lemma 1. On the other hand if $\mathcal{F}$ of type (b) is not a publicly-selectable source, then as pointed out after Definition 2, it is an oblivious sampling functionality and hence falls into Category 2. Finally (c) is the same as Category 3.   □

Given the above classification we prove Theorem 1 by considering functionalities in each of the above categories separately.

- *Category 1.* Since $\mathcal{F}$ in this category is UC-reducible to $\mathcal{F}_{\text{coin}}$ in the computationally unbounded setting, the condition in the theorem is satisfied.
- *Category 2.* If $\mathcal{F}_{\text{coin}}$ is useful for UC-securely realizing a functionality $\mathcal{F}$ in this category, and therefore in particular $\mathcal{F}$ UC-securely reduces to $\mathcal{F}_{\text{coin}}$, then below we shall give a semi-honest secure OT protocol.
- *Category 3.* In [8] it has already been shown that if a functionality in this category reduces to $\mathcal{F}_{\text{coin}}$, then there exists a semi-honest secure OT protocol.

Thus to complete the proof of Theorem 1 it remains to show the following.

Lemma 3. *If an oblivious sampling functionality $\mathcal{F}$ has a UC-secure protocol in the $\mathcal{F}_{\text{coin}}$-hybrid model, then there exists a semi-honest secure OT protocol.*

Proof. Since $\mathcal{F}$ is an oblivious sampling functionality, it is an SSFE functionality $\mathcal{F}_f$ with unidirectional influence (w.l.o.g, assume that Alice's input influences Bob's output) such that there exist two non-redundant inputs $x_0, x_1 \in X$ and an output $z \in Z$, such that the distributions $f(x_0) \neq f(x_1)$ and z falls in the intersection of the supports of $f(x_0)$ and $f(x_1)$.

Suppose π is a protocol in $\mathcal{F}_{\text{coin}}$-hybrid that securely realizes $\mathcal{F}$. Before we specify and analyze our protocol, we elaborate on what it means for π to securely realize $\mathcal{F}$: there exists a simulator $\mathcal{S}_\pi^A$, such that for any environment and corrupt Alice, it will be indistinguishable whether Alice is taking part in an execution of π or Alice is taking part in an execution simulated by $\mathcal{S}_\pi^A$. (Similarly, there is a simulator $\mathcal{S}_\pi^B$ for corrupt Bob.) This simulator $\mathcal{S}_\pi^A$ behaves as follows: it interacts with corrupt Alice simulating to her Bob's messages in π, while also interacting with the ideal functionality $\mathcal{F}$ playing Alice's role. At some point $\mathcal{S}_\pi^A$ would send an input to $\mathcal{F}$ on behalf of Alice, and obtain an outcome (which Bob also obtains and outputs to the environment). We use the following observation about the input that $\mathcal{S}_\pi^A$ sends to $\mathcal{F}$, when corrupt Alice follows the protocol π honestly. Here, two inputs x and x' are called *equivalent* if the distributions $f(x)$ and $f(x')$ are identical.

Claim. Consider the ideal execution involving a corrupt Alice, $\mathcal{S}_\pi^A$ and the ideal functionality $\mathcal{F}$. If corrupt Alice follows π honestly using a non-redundant input x, then the input that $\mathcal{S}_\pi^A$ sends to $\mathcal{F}$ is, except with negligible probability, equivalent to x.

Proof. Let $\alpha_{x'}$ be the probability with which $\mathcal{S}_\pi^A$ sends the input x' to $\mathcal{F}$. Then the resulting output distribution is $\sum_{x' \in X} \alpha_{x'} f(x')$. However, for the simulation to be good, we require this to be negligibly different from $f(x)$. Consider the set X' of all inputs not equivalent to x. Since x is not redundant, $f(x)$ lies outside the convex hull of the set of distributions $\{f(x') | x' \in X'\}$. Since the probabilities are constant (independent of the security parameter), the Euclidean distance between $f(x)$ (considered a point in the space $\mathbb{R}^{|Z|}$) and this convex hull is some constant, say ℓ. Then, the distribution $\sum_{x' \in X} \alpha_{x'} f(x')$ has a Euclidean distance of at least $\ell(\sum_{x' \in X'} \alpha_{x'})$ from $f(x)$. Since this distance must be negligible (as the Euclidean distance is at most twice the statistical distance), and ℓ is constant, it must be that $\sum_{x' \in X'} \alpha_{x'}$ is negligible. In other words, except with negligible probability $\mathcal{S}_\pi^A$ sends an input equivalent to x, completing the proof of the claim.

$\square$

To show that there exists a semi-honest secure protocol for OT, we shall show that there is such a protocol for the functionality $\mathcal{F}_{\text{uni-AND}}$, which takes a bit each from Alice and Bob and outputs their logical AND to Bob (Alice gets an empty output). (This is enough since it is easy to see that in the semi-honest case OT reduces to $\mathcal{F}_{\text{uni-AND}}$.) Consider the following protocol for $\mathcal{F}_{\text{uni-AND}}$.

Let Alice's input be $x^* \in \{0,1\}$ and Bob's input be $y^* \in \{0,1\}$. Let π be a UC-secure protocol for an oblivious sampling functionality $\mathcal{F}$, with two non-redundant inputs x_0 and x_1 such that $f(x_0) \neq f(x_1)$ and z is in the support of both $f(x_0)$ and $f(x_1)$.

For $i = 1$ to k

    Until Alice and Bob are "satisfied"

        Alice picks $b_i \leftarrow \{0,1\}$, and executes π with Bob, with Bob implementing $\mathcal{F}_{\text{coin}}$, with input $x^i := x_{b_i}$

        If $y^* = 0$, then

            Bob executes the protocol π with Alice, implementing $\mathcal{F}_{\text{coin}}$ himself, and obtains output $\hat{z}$.

        Else ($y^* = 1$),

            Bob runs the simulator $\mathcal{S}_\pi^A$ for a corrupt Alice in π, until the simulator extracts an input $\hat{x}^i$; the simulator expects a response from $\mathcal{F}$ on sending this input to it.

            Bob samples $\hat{z}$ from $f(\hat{x}^i)$, and feeds this back to the simulator as the output from $\mathcal{F}$.

            Bob continues executing the simulator until the end of the protocol.

        If $\hat{z} = z$ then Alice and Bob are satisfied, else not.

Alice sends $w = x^* \oplus b_1 \oplus b_2 \oplus \ldots \oplus b_k$ to Bob.

If $y^* = 0$ Bob outputs 0, else he outputs $w \oplus \hat{b}_1 \oplus \hat{b}_2 \oplus \ldots \oplus \hat{b}_k$, where the bit $\hat{b}_i$ is 0 iff $\hat{x}^i = x_0$.

We shall argue that if π is a secure protocol for $\mathcal{F}$, then this protocol is a semi-honest secure protocol for $\mathcal{F}_{\text{uni-AND}}$ in the PPT setting.

Firstly, we show that the protocol is correct: for any pair of inputs, the outputs of the protocol is the same as that of the ideal functionality $\mathcal{F}_{\text{uni-AND}}$. Alice produces an empty output in the protocol and in the ideal execution. When $y^* = 0$, Bob's output is 0 in both cases. It only remains to analyze the case when $y^* = 1$. For this case, we argue that in the protocol, $\hat{x}^i = x^i$ for all i (except with negligible probability), so that Bob's ouput is indeed x^* as it will be in the ideal execution. This follows from the above claim regarding the correctness of the input x extracted by the simulator $\mathcal{S}_\pi^A$, and a union bound over $i = 1$ to k.

It remains to consider the case when exactly one of Alice and Bob is passively corrupt (the case when both are corrupt being trivial). In each case, we need to show that the view of the corrupt party can be simulated based on the corrupt party's input and output (and given those, independent of the input of the other party).

If Alice is corrupt, consider a simulator which simply runs our protocol with Bob's input set to (say) 0, and sends Alice's input to $\mathcal{F}_{\text{uni-AND}}$. By the correctness of the protocol, we need only argue that the view of Alice is nearly the same as in the simulation for $y^* = 0$ and $y^* = 1$. Clearly this is true when $y^* = 0$. On the other hand, Alice's view is nearly identical when $y^* = 1$ and $y^* = 0$ by the indistinguishability guarantee of the simulator $\mathcal{S}_\pi^A$.

If Bob is corrupt, consider the following (semi-honest) simulation. If $y^* = 1$, then the simulator sends 1 to $\mathcal{F}_{\text{uni-AND}}$ and obtains x^* in response; then it faithfully runs our protocol with Alice's input set to x^*. If $y^* = 0$, then the simulator obtains no information from $\mathcal{F}_{\text{uni-AND}}$; in this case it simply picks an arbitrary input for Alice, say 0, and runs our protocol faithfully. Note that this has the effect that the last message sent in the protocol when $x^* = 1$ could be wrongly distributed. However we argue that the last message when $x^* = 1$ is nearly identically distributed as when $x^* = 0$, conditioned on Bob's view in the rest of the protocol. For this, we first replace each execution of π in our protocol as well as in our simulation with a simulation using $\mathcal{S}_\pi^B$ interacting with an instance of the ideal functionality $\mathcal{F}$. This causes a negligible change in the two distributions. Then, for an execution of π, conditioned on Bob's view (in which the only information about each b_i is the fact that the response from the ideal functionality $\mathcal{F}$ is z), $p := \Pr[b_i = 0] = \Pr[f(x_0) = z]/(\Pr[f(x_0) = z] + \Pr[f(x_1) = z])$, and $\Pr[b_i = 1] = 1 - p$ (independently for each i), for some constant (i.e., independent of k) p, with $0 < p < 1$. Then $|\Pr[\bigoplus_{i=1}^{k} b_i = 0] - \Pr[\bigoplus_{i=1}^{k} b_i = 1]| = |(p - (1 - p))^k|$ is negligible, or in other words $\bigoplus_{i=1}^{k} b_i$ is close to a uniformly distributed bit. Thus the last message sent out by Alice is nearly identically distributed for $x^* = 0$ and $x^* = 1$.   □

4.2   Proof of Theorem 3

Theorem 3 extends Theorem 1 to allow any publicly-selectable source $\mathcal{G}$ in place of $\mathcal{F}_{\text{coin}}$. We show that it follows from Theorem 1 and Lemma 1. If $\mathcal{F}$ reduces to a publicly-selectable source $\mathcal{G}$ in the PPT setting, then using Lemma 1, $\mathcal{F}$ reduces to $\mathcal{F}_{\text{coin}}$ in the PPT setting. By Theorem 1, then either there is a semi-honest secure protocol for OT, or $\mathcal{F}$ reduces to $\mathcal{F}_{\text{coin}}$ in the computationally unbounded setting. In the former case we are done. In the latter case, if $\mathcal{G}$ is not trivial then by Lemma 1 again, we have that $\mathcal{F}$ reduces to $\mathcal{G}$ in the computationally unbounded setting. On the other hand, if $\mathcal{G}$ is trivial, then so must $\mathcal{F}$ be (since only trivial functionalities can be reduced to trivial functionalities, even in the PPT setting [12]), and then again $\mathcal{F}$ reduces to $\mathcal{G}$ in the computationally unbounded setting.

5   Conclusions and Open Problems

This work closes a gap left in the recent work of [8], thereby characterizing the computational assumption necessary and sufficient for reducing *any* SSFE functionality to $\mathcal{F}_{\mathsf{coin}}$ (or to any publicly-selectable source). The main technical contribution in this work is to identify a new class of functionalities called oblivious sampling functionalities, and to provide a semi-honest secure protocol for OT, assuming that an oblivious sampling functionality reduces to $\mathcal{F}_{\mathsf{coin}}$.

Despite our complete understanding regarding the question of reduction to $\mathcal{F}_{\mathsf{coin}}$ and other publicly-selectable sources, several other aspects of randomized SSFE functionalities remain relatively less understood. In particular, the question of reduction to functionalities other than publicly-selectable sources remains unexplored. Also, we leave open the question of extending our current characterization to functionalities beyond SSFE functionalities. Finally, by considering reductions under other security notions (like semi-honest security), we come across more open problems for randomized functionalities. In particular, it is not known which SSFE functionalities are trivial in the semi-honest security model.

We hope that our techniques – especially, the identification of oblivious sampling functionalities, and the resulting classification of SSFE functionalities – will add to the tools that will aid in resolving these questions.

Acknowledgments. We would like to thank Pichayoot Ouppaphan for being part of this work in its initial stages. This material is based upon work supported by the National Science Foundation, USA under grant CNS 07-47027. Any opinions, findings, and conclusions or recommendations expressed in this material are those of the authors and do not necessarily reflect the views of the National Science Foundation.

References

1. Beimel, A., Malkin, T., Micali, S.: The All-or-Nothing Nature of Two-Party Secure Computation. In: Wiener, M. (ed.) CRYPTO 1999. LNCS, vol. 1666, pp. 80–97. Springer, Heidelberg (1999)
2. Canetti, R.: Universally composable security: A new paradigm for cryptographic protocols. In: Electronic Colloquium on Computational Complexity (ECCC) TR01-016 (2001); Previous version "A unified framework for analyzing security of protocols" availabe at the ECCC archive TR01-016. Extended abstract in FOCS 2001
3. Haitner, I.: Semi-Honest to Malicious Oblivious Transfer—the Black-Box Way. In: Canetti, R. (ed.) TCC 2008. LNCS, vol. 4948, pp. 412–426. Springer, Heidelberg (2008)
4. Kilian, J.: Founding cryptography on oblivious transfer. In: STOC, pp. 20–31. ACM (1988)
5. Kilian, J.: More general completeness theorems for secure two-party computation. In: Proc. 32th STOC, pp. 316–324. ACM (2000)

6. Kraschewski, D., Müller-Quade, J.: Completeness Theorems with Constructive Proofs for Finite Deterministic 2-Party Functions. In: Ishai, Y. (ed.) TCC 2011. LNCS, vol. 6597, pp. 364–381. Springer, Heidelberg (2011)
7. Künzler, R., Müller-Quade, J., Raub, D.: Secure Computability of Functions in the IT Setting with Dishonest Majority and Applications to Long-Term Security. In: Reingold, O. (ed.) TCC 2009. LNCS, vol. 5444, pp. 238–255. Springer, Heidelberg (2009)
8. Maji, H.K., Ouppaphan, P., Prabhakaran, M., Rosulek, M.: Exploring the limits of Common Coins Using Frontier Analysis of Protocols. In: Ishai, Y. (ed.) TCC 2011. LNCS, vol. 6597, pp. 486–503. Springer, Heidelberg (2011)
9. Maji, H.K., Prabhakaran, M., Rosulek, M.: Complexity of Multi-Party Computation Problems: The Case of 2-Party Symmetric Secure Function Evaluation. In: Reingold, O. (ed.) TCC 2009. LNCS, vol. 5444, pp. 256–273. Springer, Heidelberg (2009)
10. Maji, H.K., Prabhakaran, M., Rosulek, M.: Cryptographic complexity classes and computational intractability assumptions. In: Yao, A.C.-C. (ed.) ICS, pp. 266–289. Tsinghua University Press (2010)
11. Maji, H.K., Prabhakaran, M., Rosulek, M.: A Zero-One Law for Cryptographic Complexity with Respect to Computational UC Security. In: Rabin, T. (ed.) CRYPTO 2010. LNCS, vol. 6223, pp. 595–612. Springer, Heidelberg (2010)
12. Prabhakaran, M., Rosulek, M.: Cryptographic Complexity of Multi-Party Computation Problems: Classifications and Separations. In: Wagner, D. (ed.) CRYPTO 2008. LNCS, vol. 5157, pp. 262–279. Springer, Heidelberg (2008)

A Security Definitions

We follow standard conventions and terminology for defining security of protocols for multi-party computation tasks. For easy reference we reproduce these definitions as given in [8], whose results we improve up on.

A protocol is secure if for every adversary in the real world (in which parties execute a protocol), there is an adversary, or *simulator*, in the ideal world (in which the task is carried out on behalf of the parties by a trusted third party called a *functionality*) that achieves the same effect in every environment. Depending on the nature or adversary/simulator and the environment, we consider three different kinds of security notions.

- A semi-honest or passive adversary (in the real or ideal execution) is one which is not allowed to deviate from the (real or ideal) protocol. *Semi-honest* or *passive security* is achieved if for every semi-honest adversary in the real world there is a semi-honest simulator in the ideal world as above.
- A standalone environment is one which does not interact with the adversary during the execution of the protocol. *Standalone security* is achieved if we restrict the security requirement to standalone environments; in this case the simulator can rewind the adversary without the environment detecting it. In this work we do not consider this notion of security.
- *Universally composable (UC) security* [2] is achieved when the security requirement is met against all adversaries (possibly active) and all environments (possibly not standalone); the simulator is allowed to be an active

adversary. In this case there must exist a straight-line blackbox simulation (i.e., the simulator internally runs the adversary as a blackbox and never rewinds it).

In this work, we exclusively consider *static* adversaries, who do not adaptively corrupt honest parties during the execution of a protocol.

PPT vs. computationally unbounded setting. In the PPT setting we restrict all entities – the environment, the adversary and simulator – to probabilistic polynomial time computation. In the computationally unbounded setting all these entities can be computationally unbounded. (However, for the purpose of the results in this work, one could require the simulator in the computationally unbounded setting to be efficient (PPT) with blackbox access to the adversary. Then, if a protocol is secure in the computationally unbounded setting, it will be secure in the PPT setting too.

Hybrids. The *plain model* is a real world in which protocols only have access to a simple communication channel; a *hybrid model* is a real world in which protocols can additionally use a particular trusted functionality. While hybrid worlds are usually considered only for UC security, we also use the terminology in the setting of standalone security. We note that protocols for *non-reactive* functionalities (i.e., those which receive input from all parties, then give output, and then stop responding) do securely compose even in the standalone security setting.

Reduction. We say that a functionality $\mathcal{F}$ *reduces* to a functionality $\mathcal{G}$ if $\mathcal{F}$ can be UC-securely realized in the $\mathcal{G}$-hybrid. In the real world protocol, that parties have access to a trusted implementation of $\mathcal{G}$ in addition to the secure point-to-point communication channel to securely realize $\mathcal{F}$. Suppose π is a UC-secure protocol for $\mathcal{F}$ in the $\mathcal{G}$-hybrid. Then, parties generate a transcript based on their local views and π can also call the trusted $\mathcal{G}$ implementation. The functionality $\mathcal{G}$ can be any arbitrary functionality, i.e. it need not be a two party function, parties need not play fixed roles while calling $\mathcal{G}$ and, in fact, both parties can provide multiple inputs while performing a call to $\mathcal{G}$.

UC-security in Hybrid worlds. As mentioned earlier, we shall only consider static corruption of parties, i.e. at the beginning of an execution the adversary announces which party it wants to corrupt and cannot corrupt any further party during the execution of the protocol. To show that a protocol π is a UC-secure realization of $\mathcal{F}$ in the $\mathcal{G}$ hybrid, we need to show that for every adversarial strategy in the $\mathcal{G}$-hybrid there exists a simulator in the Ideal world such that any environment is unable to distinguish the Real execution from the Ideal execution. In this work, we shall restrict ourselves to reductions where both $\mathcal{F}$ and $\mathcal{G}$ are (at most) two party functionalities. Henceforth, we present the definition restricted to this particular case. Suppose Alice is corrupted by the adversary and Bob is honest. The simulator S_{π}^{A} for Alice in the Ideal execution, interacts with the adversarial Alice so that no environment can distinguish the Real from the Ideal execution. The simulator also forwards communication between adversarial Alice and the environment. Note, in particular, it implies that the simulator cannot be rewinding in this case. During this execution, the calls to the $\mathcal{G}$ functionality made by the adversarial Alice is answered by the simulator S_π^A. At some point during the interaction with adversarial Alice, the simulator sends an input x to the ideal functionality $\mathcal{F}$ and receives and answer z. The simulator continues the execution with the adversarial Alice and terminates after generating a complete transcript (we can assume that the adversarial Alice strategy always completes a protocol).

If there exists an efficient S_π^A which can make the Ideal execution indistinguishable from the Real execution to any environment, then $\mathcal{F}$ is secure in the $\mathcal{G}$-hybrid when Alice is corrupt. Additionally, if there exists an efficient simulator S_π^B which shows that $\mathcal{F}$ is secure in the $\mathcal{G}$-hybrid when Bob is corrupt then π is a UC-secure protocol for $\mathcal{F}$ in the $\mathcal{G}$-hybrid. Intuitively, the existence of a simulator shows that any effect achieved by the adversarial party could be reflected in the Ideal world itself. The additional power of the simulator lies in the fact that it receives the calls to $\mathcal{G}$, i.e. it gets to see the input sent by the adversarial party to $\mathcal{G}$, and it decides the reply to this call. So, for example, when $\mathcal{G}$ is $\mathcal{F}_{\text{coin}}$, the simulator can determine all the coin outcomes at the beginning of the execution and this could provide additional power to the simulator over the parties in the $\mathcal{G}$-hybrid. Another example is when $\mathcal{G}$ is an oblivious sampling functionality, so it is not possible to be always certain of the input to $\mathcal{G}$ just from the output given by $\mathcal{G}$, the simulator gets additional information when it sees the query made to $\mathcal{G}$.

B   Representative Functionalities

It is instructive to consider some representative examples of SSFE functionalities to form an intuition about the classes of functionalities under consideration. The functions will be represented by a matrix where the (i, j)-th entry represents the output distribution when Alice uses input i and Bob uses input j. A distribution is represented as a vector where the k-th entry represents the probability of the k-th output symbol. In most of the examples below, Bob has a single possible input, and hence the matrix has a single column.

Influence of Inputs. Below we list three functionalities with no influence, and two functionalities with uni- and bi- directional influence respectively. The first function corresponds to $\mathcal{F}_{\text{coin}}$– it is an inputless function that outputs an unbiased coin. The second function is a communication channel which sends Alice's input to Bob. Finally, the third function represents the function in which Alice and Bob each has a bit as input and the functionality provides them the XOR of the bits.

$$\left(\langle 1/2, 1/2\rangle\right) \qquad \begin{pmatrix} \langle 1,0\rangle \\ \langle 0,1\rangle \end{pmatrix} \qquad \begin{pmatrix} \langle 1,0\rangle & \langle 0,1\rangle \\ \langle 0,1\rangle & \langle 1,0\rangle \end{pmatrix}$$

Publicly-selectable sources. If the output of the function uniquely determines the input of the function (after we have already removed redundant inputs), then the function is a publicly-selectable source. In other words, in a publicly-selectable source the output distributions of the non-redundant inputs should have disjoint supports.

$$\begin{pmatrix} \langle 1, 0 \rangle \\ \langle 0, 1 \rangle \end{pmatrix} \qquad\qquad \begin{pmatrix} \langle 1/2, 1/2, 0, 0 \rangle \\ \langle 0, 0, 3/4, 1/4 \rangle \end{pmatrix}$$

Oblivious sampling. Following are some oblivious sampling functionalities.

$$\begin{pmatrix} \langle 1/2, 1/2 \rangle \\ \langle 1/4, 3/4 \rangle \end{pmatrix} \qquad \begin{pmatrix} \langle 1/2, 1/2 \rangle \\ \langle 1, 0 \rangle \end{pmatrix} \qquad \begin{pmatrix} \langle 1/2, 1/2, 0 \rangle \\ \langle 0, 1/2, 1/2 \rangle \end{pmatrix}$$

Redundancies. Consider the following function $\mathcal{F}$ with unidirectional influence:

$$\begin{pmatrix} \langle 1, 0 \rangle \\ \langle 1/2, 1/2 \rangle \\ \langle 0, 1 \rangle \end{pmatrix}$$

This function is not an oblivious sampling function, even though the output distributions from different inputs have overlapping support. This is because the second row of this matrix represents a redundant input as it can be expressed as a convex linear combination of the first and the last rows. So, after removal of the redundant input, the function is seen to be a publicly-selectable source.

Secure Message Transmission in Asynchronous Directed Graphs

Shashank Agrawal, Abhinav Mehta, and Kannan Srinathan

Center for Security, Theory and Algorithmic Research (C-STAR),
International Institute of Information Technology, Hyderabad, 500032, India
{shashank.agrawal@research.,abhinav_mehta@research.,srinathan@}iiit.ac.in

Abstract. We study the problem of secure message transmission (SMT) in asynchronous directed graphs, where an unbounded Byzantine adversary can corrupt some subset of nodes specified via an adversary structure. We focus on the particular variant $(0, \delta)$-SMT, where the message remains perfectly private, but there is a small chance that the receiver $\mathbf{R}$ may not obtain it. This variant can be of two kinds: Monte Carlo - where $\mathbf{R}$ may output an incorrect message with small probability; and Las Vegas - where $\mathbf{R}$ never outputs an incorrect message. For a Monte Carlo $(0, \delta)$-SMT protocol to exist in an asynchronous directed graph, we show that the minimum connectivity required in the network does not decrease even when privacy of the message being transmitted is not required. In the case of Las Vegas $(0, \delta)$-SMT, we show that the minimum connectivity required matches exactly with the minimum connectivity requirements of the zero-error variant of SMT – $(0, 0)$-SMT. For a network that meets the minimum connectivity requirements, we provide a protocol efficient in the size of the graph and the adversary structure. We also provide a protocol efficient in the size of the graph for an important family of graphs, when the adversary structure is threshold.

Keywords: directed graph, asynchronous network, information-theoretic security, Byzantine adversary.

1 Introduction

In a network where secure channels are available for communication between certain pairs of nodes, a node $\mathbf{S}$ (the sender) wishes to send a message securely to another node $\mathbf{R}$ (the receiver). If there is a secure channel from $\mathbf{S}$ to $\mathbf{R}$, the node $\mathbf{S}$ simply puts the message on this channel. However, if no secure channel is available, the message transmission protocol must involve other nodes. When some of the involved nodes are corrupt, protocol-designing becomes complex, and it is is not immediately obvious whether a protocol indeed exists. In short, the problem of secure message transmission (SMT) is to design a protocol which when run by the nodes in a network leads to the secure transmission of a message from a sender node to a receiver node, even when some subset of the nodes may be corrupt.

D.J. Bernstein and S. Chatterjee (Eds.): INDOCRYPT 2011, LNCS 7107, pp. 359–378, 2011.
© Springer-Verlag Berlin Heidelberg 2011

The standard way of modelling corruption in a network is via a fictitious entity known as *adversary*. Secure message transmission has been widely studied in literature for an adversary which has unbounded computing power, in a variety of settings [6,7,3,9,25,10,2]. Protocols designed to withstand such a strong adversary provide security in an information-theoretic sense – no information about the message being transmitted from sender to receiver is revealed to the adversary. In this work, we consider an unbounded adversary which corrupts nodes in Byzantine fashion, i.e., nodes under the control of adversary deviate arbitrarily from the protocol. The possible subset of nodes which an adversary can corrupt is captured via an adversary structure [8].

We model the underlying communication network as a directed graph. For several real-life networks where a node can communicate with another node but not the other way round, undirected graphs are not a suitable model. For instance, in a sensor network where different nodes have different transmission power, communication links tend to be uni-directional: a node u can hear v but v cannot hear u [27]. Note that our network model is essentially the same as in [26,23], where every node in the network can perform computations, but different from another popular way of modelling directed networks [3,17], where nodes besides the sender and receiver can only forward received messages but not perform computations on them.

A wide body of work on SMT [7,3,26] has focussed on synchronous networks where an upper bound is known *a priori* on the delay in delivery of messages. While synchronous network is an appealing model to work on, it is hard to achieve synchrony in practice. In a real-life network, messages can be arbitrarily delayed over a channel, or may not arrive at all. Hence, protocols which promise guaranteed delivery of messages must cope with such inconsistent behaviour. This motivates the study of secure communication over asynchronous networks where no assumption is made regarding the relative speed of processes running at individual nodes or the delay in delivery of messages.

The general version of secure message transmission problem is denoted by (ϵ, δ)-SMT, where ϵ denotes the error in privacy and δ denotes the error in reliability. The particular case of $(0, \delta)$-SMT, where the message transmitted by the sender remains perfectly hidden from the adversary (0-privacy or perfect privacy) and the receiver outputs this message with at least $1 - \delta$ probability (δ-reliability), can have two variants: Monte Carlo and Las Vegas. In the former variant, receiver can output an incorrect message (with probability at most δ), but in the latter variant it cannot (it may output a special symbol, or simply abort, again with probability at most δ). We study the minimum connectivity required in an asynchronous directed graph, which is under the influence of a unbounded Byzantine adversary, for a Las Vegas $(0, \delta)$-SMT protocol to exist; and then, for a Monte Carlo $(0, \delta)$-SMT protocol to exist. A similar study of the Monte Carlo and Las Vegas variants of reliable message transmission $((1, \delta)$-SMT) have been recently done in [13].

Desmedt et. al. [3] as well as Choudhary et. al. [2] have shown that if a protocol providing only reliability (Monte Carlo $(1, \delta)$-SMT) exists in their respective network models, one can design a protocol that also achieves perfect privacy (Monte Carlo $(0, \delta)$-SMT). We show that this holds true in our more general network model as well. On a more interesting note, in an asynchronous directed graph, we show that a protocol for Las Vegas $(0, \delta)$-SMT exists if and only if a protocol for $(0, 0)$-SMT exists. This is an interesting result connecting two seemingly different problems. Furthermore, our protocols for Las Vegas $(0, \delta)$-SMT are efficient in the size of the adversary structure and the size of the graph. For a t-threshold adversary structure, which has all subsets of size at most t (where t is less than the number of nodes in the network), we provide protocols – efficient in the size of the graph – for an important class of graphs.

1.1 Related Work

In [4], Dolev et al. initiate the study of SMT through its perfect variant, i.e. PSMT or $(0, 0)$-SMT. They model the network as a collection of channels between two synchronized non-faulty processors **S** and **R**. These channels can be either all 1-way (allowing information to flow from **S** to **R** only) or all 2-way (allowing information to flow in both directions). They provide several solutions to the problem of PSMT, which vary in the number of phases required, communication complexity and fault-tolerance. The problem of PSMT in Dolev et al.'s model has been quite popular, and there have been several results in literature optimizing one or the other aspect of its solution [24,1,10].

In [3], Desmedt and Wang introduce a better way of modelling networks where not every communication channel is bi-directional. In their model, there are some channels from **S** and **R**, and some other channels from **R** to **S**. The two variants of PSMT – $(0, \delta)$-SMT and $(0, 0)$-SMT – have been extensively studied in this model [18,19,16,17]. The case of non-threshold adversary has also been dealt with [28]. While Desmedt and Wang's way of modelling directed networks is suitable when the intermediate nodes act as routers (can forward received messages but not perform any computation on them), Srinathan and Pandu Rangan [26] propose that directed graphs should be used to model directed networks when every node in the network can do computations. Several variants of SMT have also been studied in this model [23,13,15].

An important thing to note is that all the results discussed above are for synchronous networks. Sayeed and Abu-Amara [21] are the first to tackle the problem of SMT in asynchronous networks. They provide protocols for both 1-way and 2-way PSMT. However, Choudhary et al. [2] show that Sayeed and Abu-Amara's protocols do not provide perfect security. They prove lower bounds on the connectivity required in an asynchronous network for the existence of a $(0, 0)$-SMT protocol, and then for the existence of a $(0, \delta)$-SMT protocol, when the adversary is threshold Byzantine. They also provide protocols for networks which have sufficient connectivity.

While the two papers discussed above work with Dolev et al.'s network model [4], we initiate the study of SMT in asynchronous networks in the directed graph model [26]. A natural questions arises here: does it really make a difference which model we work with. We address this concern in the next section where we show that when every node in a network can compute, modelling the network as a collection of channels may fail to yield a protocol, when there actually exists one!

1.2   A Motivating Example

Consider the simple asynchronous network $\mathcal{N}$ shown in Figure 1. Here $\mathbf{S}$ and $\mathbf{R}$ are the sender and receiver respectively; they are assumed to be non-faulty. One among the nodes b_1, b_2 and b_3 may be Byzantine corrupt, i.e., the adversary structure is given by $\mathbb{A} = \{\{b_1\}, \{b_2\}, \{b_3\}\}$. If we model the network as a collection of channels (which is suitable when intermediate nodes are routers), then since there are only 3 node disjoint paths from $\mathbf{S}$ to $\mathbf{R}$, PSMT is impossible – see Theorem 6 in [2]. However, modelling the network as a directed graph (which is suitable when every node in the network can compute), we show that a protocol Π exists for PSMT in this network.

Let m be the secret $\mathbf{S}$ wants to send. We give an informal description of Π here. In Π, nodes do the following:

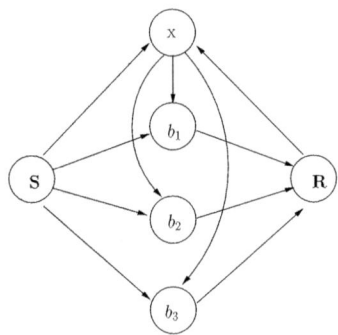

Fig. 1. The network $\mathcal{N}$

- Node $\mathbf{S}$: Apply a $(2,4)$ secret sharing scheme [22] to m and get 4 shares m_0, m_1, m_2, m_3. Send m_0 to node x and m_i to node b_i $(1 \le i \le 3)$.
- Node x: Wait for the share m_0 to arrive from $\mathbf{S}$, and random number ρ to arrive from $\mathbf{R}$. Send $l = m_0 + \rho$ to each b_i.
- Node b_i: Wait for the share m_i to arrive from $\mathbf{S}$ and send it to $\mathbf{R}$. Wait for l to arrive from node x and forward it to $\mathbf{R}$.
- Node $\mathbf{R}$: Choose a random number ρ and send it to node x. Simultaneously wait for:
  - A share of the message m to arrive from b_i; call this share m_i' $(1 \le i \le 3)$.
  - Two identical values of l to arrive from different b_i. Add $-\rho$ to l to obtain m_0'.

  Wait till three consistent shares $m_\alpha', m_\beta', m_\gamma'$ $(\{\alpha, \beta, \gamma\} \subset \{0, 1, 2, 3\})$ are obtained (consistency of shares is defined later). Reconstruct the secret with these shares.

In section 4, we handle an important family of graphs of which the above graph is a member. The general protocol Π_{sr} described in that section when applied to the graph above gives the protocol Π. Hence, the properties of Π_{sr} – perfect secrecy and reliability – are automatically inherited by Π. Therefore, a separate discussion of these properties for Π is not required.

1.3   Organization and Our Contribution

In Section 2, we discuss the model we are going to work with in detail, and provide rigorous definitions of the various terms we have been using informally so far. We first study the Las Vegas variant of $(0, \delta)$-SMT, and then the Monte Carlo variant. In Section 3, we characterize asynchronous digraphs for the possibility of Las Vegas $(0, \delta)$-SMT. We show that this characterization matches with the characterization of asynchronous digraphs for $(0, 0)$-SMT. Further, in Section 4, when the adversary is threshold Byzantine and PSMT tolerating this adversary is possible between every pair of nodes in a digraph, we provide a protocol – efficient in the size of the graph – for PSMT between any two nodes in the digraph. Finally, in Section 5, we find the minimum connectivity required in a digraph for a Monte Carlo $(0, \delta)$-SMT protocol to exist, and show that this requirement stays the same even when privacy is not required.

2   Model and Definitions

The network model: We model the underlying asynchronous network as a directed graph $\mathcal{N} = (V, \mathcal{E})$, where V denotes the set of nodes (or players) in the network and $\mathcal{E} \subseteq V \times V$ represents the channels available for communication between nodes. An important assumption we make in this paper is that all nodes know the topology of the network, i.e., all nodes know the digraph $\mathcal{N}$. Two special non-faulty nodes $\mathbf{S}, \mathbf{R} \in V$ denote the sender and receiver respectively. To distinguish between the message $\mathbf{S}$ intends to send to $\mathbf{R}$ through a protocol, and the messages exchanged between nodes during the execution of the protocol, we refer to the former as *secret*. In this work, a protocol is often composed of several sub-protocols; when this is the case, the secrets of sub-protocols would be referred to as *sub-secrets*.

The fault model: Fault in the network is modelled via an unbounded centralized fictitious entity called the adversary that can control a subset of nodes in the network, specified via an adversary structure (defined later), and make them behave in a Byzantine fashion [11]. We assume that the channels available between nodes, i.e. the set $\mathcal{E}$, cannot be corrupted by the adversary (similar to the *secure channels* setting). The adversary is *adaptive* and is allowed to dynamically corrupt nodes during protocol execution (and his choice may depend on the data seen so far). It knows the complete protocol specification as well as the topology of the network.

Additionally, since the network is asynchronous, computation proceeds in a sequence of steps controlled by the adversary. In a single step, the adversary activates a node by delivering some message to it, called an *event*, the node then performs local computation, changes its state and sends messages on its outgoing channels. A *schedule* is a finite or infinite sequence of events. See [5] for a detailed description of the asynchronous model.

A non-threshold adversary structure $\mathbb{A}$ is a set of subsets of the node set, i.e., $\mathbb{A} \subseteq \mathcal{P}(V \setminus \{S, R\})$, one of which may be corrupt during an execution ($\mathcal{P}$ denotes the power set). When an upper bound t is known on the number of faulty nodes in a network, the adversary structure contains all t-sized subsets of the node set and is referred to as t-threshold adversary. The adversary structures we consider have the property of *monotonicity*, i.e., whenever $B_1 \in \mathbb{A}$, then $\forall B_2$ such that $B_2 \subset B_1$, $B_2 \in \mathbb{A}$. We note that $\mathbb{A}$ can be uniquely represented by listing the elements in its maximal basis $\overline{\mathbb{A}}$ defined as follows.

Definition 1 (Maximal basis of $\mathbb{A}$). *For any monotone adversary structure $\mathbb{A}$, its maximal basis $\overline{\mathbb{A}}$ is defined as $\overline{\mathbb{A}} = \{B \mid B \in \mathbb{A} \text{ and } \nexists X \in \mathbb{A} \text{ s.t. } B \subset X\}$. Abusing the standard notation, we assume that $\mathbb{A}$ itself is a maximal basis.*

Let the message space be a large finite field $\langle \mathbb{F}, +, \cdot \rangle$ – all computations are done in this field. Also, in an execution of a protocol, let $\Gamma(m, r)$ denote the view of the adversary when **S** chooses to send the secret m and coin tosses of the adversary are r. We now formally define $(0, 0)$-SMT and the two variants of $(0, \delta)$-SMT along the lines of [3]. We express Monte Carlo $(0, \delta)$-SMT and Las Vegas $(0, \delta)$-SMT in short as $(0, \delta)$-SMT$_{MC}$ and $(0, \delta)$-SMT$_{LV}$ respectively.

Definition 2 ($(0, 0)$-SMT or PSMT). *In an asynchronous digraph $\mathcal{N} = (V, \mathcal{E})$ a protocol for transmitting any secret $m \in \mathbb{F}$ from **S** to **R** tolerating an adversary structure $\mathbb{A}$ is a $(0, 0)$-SMT protocol if for every Byzantine corruption $B \in \mathbb{A}$ and every schedule $\mathcal{D}$ the following two conditions are satisfied:*

1. ***Resiliency:*** **R** *always terminates with the secret m which **S** has chosen to send.*
2. ***Secrecy:*** $\forall r, \forall m_0, m_1$ *and every possible view c of the adversary it holds that $Pr[\Gamma(m_0, r) = c] = Pr[\Gamma(m_1, r) = c]$, where the probabilities are taken over the coin tosses of honest parties.*

Definition 3 ($(0, \delta)$-SMT). *Let $\delta < \frac{1}{2}$. In an asynchronous digraph $\mathcal{N} = (V, \mathcal{E})$ a protocol for transmitting any secret $m \in \mathbb{F}$ from **S** to **R** tolerating an adversary structure $\mathbb{A}$ is a Monte Carlo $(0, \delta)$-SMT or $(0, \delta)$-SMT$_{MC}$ protocol if for every Byzantine corruption $B \in \mathbb{A}$ and every schedule $\mathcal{D}$ the following two conditions are satisfied:*

1. ***Resiliency:*** $\forall m \; Pr[\textbf{R} \text{ outputs } m | \textbf{S} \text{ has sent } m] \geq (1 - \delta)$, *where the probability is taken over the coin tosses of all players.*
2. ***Secrecy:*** $\forall r, \forall m_0, m_1$ *and every possible view c of the adversary it holds that $Pr[\Gamma(m_0, r) = c] = Pr[\Gamma(m_1, r) = c]$, where the probabilities are taken over the coin tosses of honest parties.*

A Las Vegas $(0, \delta)$-SMT or $(0, \delta)$-SMT$_{LV}$ protocol satisfies the following stronger resiliency condition:

- **Resiliency**: $\forall m \; Pr[\boldsymbol{R} \text{ outputs } m | \boldsymbol{S} \text{ has sent } m] \geq (1 - \delta)$, where the probability is taken over the coin tosses of all players. Otherwise $\boldsymbol{R}$ outputs $\perp \notin \mathbb{F}$ or does not terminate.

With respect to a directed graph, we define the following terms which we will use throughout the paper. These definitions have been taken from [23].

Definition 4 (Strong path). A sequence of vertices $v_1, v_2, v_3, ..., v_k$ is said to be a strong path from v_1 to v_k in a graph $\mathcal{N} = (V, \mathcal{E})$ if for each $1 \leq i < k$, $(v_i, v_{i+1}) \in \mathcal{E}$. Furthermore, we assume that there vacuously exists a strong path from a node to itself.

Definition 5 (Weak path). A sequence of vertices $v_1, v_2, v_3, ..., v_k$ is said to be a weak path from v_1 to v_k in a graph $\mathcal{N} = (V, \mathcal{E})$ if for each $1 \leq i < k$, $(v_i, v_{i+1}) \in \mathcal{E}$ or $(v_{i+1}, v_i) \in \mathcal{E}$.

Definition 6 (Blocked node, Head node). A node u along a weak path p is called a blocked node if its out-degree along p is 0. A node y along a weak path p is called a head node if it is an intermediate node with out-degree 2 or a terminal node with out-degree 1.

The head nodes and blocked nodes along a weak path play a special role. A head node generates messages and forwards them to the two (or one) blocked nodes adjacent to it through the intermediate nodes. A blocked node receives messages from its two (or one) adjacent head nodes, performs operations on the messages received and forwards it to another node along a separate path. Hence, we look at a weak path p from $\boldsymbol{S}$ to $\boldsymbol{R}$ as an alternating sequence of blocked nodes u_i and head nodes y_i, starting with $\boldsymbol{S}$ as a head node denoted by y_0 and ending with $\boldsymbol{R}$ as a blocked denoted by u_{n+1}. In other words, the path p can be represented as $y_0, u_1, y_1, u_2, y_2, \ldots, u_n, y_n, u_{n+1}$ for some $n \geq 0$, such that y_0 has a strong path to u_1 along p, and y_i $(i > 0)$ has a strong path to u_i and u_{i+1} along p.

It may so happen that along the path p, $\boldsymbol{S}$ is not a head node. In this case, $\boldsymbol{S}$ can simulate two nodes s and u and a directed edge (s, u); the simulated node s becomes a virtual sender and a head node along p. Similarly, the case where $\boldsymbol{R}$ is not a blocked node can be handled. For a detailed discussion see [14].

Definition 7 (Authentication function). Let $\mathcal{K} = (K_1, K_2, K_3) \in \mathbb{F} \times \mathbb{F} \times \mathbb{F}$ and $m \in \mathbb{F}$. Authentication function χ is defined as $\chi(m; \mathcal{K}) = (\chi_1(m; \mathcal{K}), \chi_2(m; \mathcal{K}))$ where $\chi_1(m; \mathcal{K}) = m + K_1$ and $\chi_2(m; \mathcal{K}) = \chi_1(m; \mathcal{K}) \cdot K_2 + K_3$

Here K_1, K_2, K_3 are usually referred to as *keys*. Using χ_1 we blind the message and using χ_2 we authenticate the blinded message. Suppose a random triplet $\mathcal{K}$ unknown to the adversary is established between two nodes u and v in a network $\mathcal{N}$. The authentication function has the following important properties: (a) Even if u sends $\chi(m; \mathcal{K})$ along a faulty path to v, adversary will not know anything about m. (b) Node v will be able to detect any change in $\chi(m; \mathcal{K})$'s value except with an error probability of almost $\frac{1}{|\mathbb{F}|}$. (Proofs for the same appear in [20]).

Secret Sharing: We use the simple (k, n) *threshold scheme* $(n \geq k)$ from [22] to create n shares of a secret where knowledge of any set of atmost $k - 1$ shares reveals no information about the secret. The secret can be efficiently reconstructed using the Berlekamp-Welch (BW) algorithm [12] from any set of shares S of size m (where $k \leq m \leq n$) if it contains at most $\lfloor \frac{m-k}{2} \rfloor$ incorrect shares. Such a set S of shares is said to be consistent.

In the following, we only consider adversary structures of size greater than 1. If the adversary structure is of unit size, say $\mathbb{A} = \{\{B_1\}\}$, the adversary can always fail-stop every node in B_1. Hence, a strong path from **S** to **R** avoiding all nodes in B_1 is necessary to enable **S** to send messages to **R**; and therefore, it is necessary for any reliable protocol. It is easy to see that this is also sufficient.

3   Characterizing Asynchronous Networks for Las Vegas $(0, \delta)$-SMT

We first settle the straightforward case of adversary structure of size two.

Theorem 1. *In an asynchronous digraph $\mathcal{N}$, a $(0, \delta)$-SMT_{LV} protocol tolerating an adversary structure $\mathbb{A} = \{B_1, B_2\}$ exists if and only if there exists a strong path from **S** to **R** in the network avoiding all nodes in $B_1 \cup B_2$.*

Proof. Sufficiency is clear. For the necessity proof, please refer [14] where it is shown that a strong path is necessary even when privacy of the message is not required.   □

Let us now characterize asynchronous digraphs for adversary structures of size three. We would then show how the protocol presented here can be used as a building block to create protocols that tolerate adversary structures of higher size.

Theorem 2. *In an asynchronous digraph $\mathcal{N}$, a $(0, \delta)$-SMT_{LV} protocol from **S** to **R** tolerating $\mathbb{A} = \{B_1, B_2, B_3\}$ exists if and only if for each $\alpha \in \{1, 2, 3\}$, there exists a weak path q_α avoiding nodes in $B_1 \cup B_2 \cup B_3$ such that every node u along the path q_α has a strong path to **R** avoiding all nodes in $\bigcup_{\beta \in \{1,2,3\}-\{\alpha\}} B_\beta$. (Paths q_1, q_2, q_3 need not be distinct.)*

Sufficiency and necessity are proved in the following two sub-sections.

3.1   Sufficiency

Let $m \in \mathbb{F}$ be the secret **S** intends to send. In the special case when q_α is a strong path from **S** to **R**, for an $\alpha \in \{1, 2, 3\}$, **S** can trivially send m along q_α. Since q_α does not contain any corrupt node, **R** receives m privately and reliably.

For the rest of the cases, we construct a protocol $\Pi_{\{B_1, B_2, B_3\}}$, whose correctness is proved in the following Lemma. The protocol $\Pi_{\{B_1, B_2, B_3\}}$ is composed of three sub-protocols Π_1, Π_2, Π_3 which are run in parallel in the network, each one on m as the sub-secret. $\mathbf{R}$ first waits for any two of these three sub-protocols to terminate. If the same sub-secret is recovered from both these protocols, $\mathbf{R}$ outputs it. Otherwise, $\mathbf{R}$ waits for the third sub-protocol to terminate and outputs the majority of the outcome of the three sub-protocols.

We give a construction for Π_1, and the constructions of Π_2 and Π_3 follow by symmetry. The protocol Π_1 uses the honest weak path q_1. Recall that q_1 can be expressed as an alternating sequence of blocked nodes u_i and head nodes y_i – $y_0, u_1, y_1, u_2, y_2, \ldots, u_n, y_n, u_{n+1}$ for some $n > 0$, where y_0 denotes $\mathbf{S}$ and u_{n+1} denotes $\mathbf{R}$.

Π_1 proceeds as follows:

1. $\mathbf{S}$ sends m to u_1 along q_1. For $1 \leq j \leq n$, node y_j chooses a random key K_j and sends it to u_j and u_{j+1} along q_1.
2. Node u_1 sends $L_1 = m + K_1$ to $\mathbf{R}$ along a strong path avoiding $B_2 \cup B_3$ when it receives m from $\mathbf{S}$ and K_1 from y_1. For $1 < j \leq n$, u_j sends $L_j = K_{j-1} + K_j$ to $\mathbf{R}$ along a strong path avoiding $B_2 \cup B_3$ when it receives K_{j-1} from y_{j-1} and K_j from y_j.
3. $\mathbf{R}$ waits until it receives K'_n from y_n, and for $1 \leq j \leq n$, L'_j from u_j. It then does the following:
   for z in n to 2
   $\qquad K'_{z-1} = L'_z - K'_z.$
   Output $m'_1 = L'_1 - K'_1.$

This completes the description of Π_1, and hence of $\Pi_{\{B_1, B_2, B_3\}}$.

At this point, all we need to prove is that $\Pi_{\{B_1, B_2, B_3\}}$ is a $(0, \delta)$-SMT$_{LV}$ protocol. However, the protocol we have designed here not only achieves perfect secrecy, but also perfect reliability!

Lemma 1. *The protocol $\Pi_{\{B_1, B_2, B_3\}}$ is a $(0,0)$-SMT protocol tolerating $\mathbb{A} = \{B_1, B_2, B_3\}$*

Proof. W.l.o.g let us assume that the set B_1 is corrupt.

RESILIENCY: Since protocols Π_2 and Π_3 do not involve nodes in B_1, these protocols are bound to terminate (with the correct sub-secret) no matter how the adversary schedules messages in the asynchronous system. Hence, when $\mathbf{R}$ waits for at least two protocols to terminate, it will not wait indefinitely. If Π_2 and Π_3 indeed terminate before Π_1 does, $\mathbf{R}$'s output is correct.

However, there may exist a schedule in the network such that Π_1 terminates before both Π_2 and Π_3 terminate. Say Π_1 and Π_2 terminate before Π_3 does. If the same sub-secret is recovered from Π_1 and Π_2, $\mathbf{R}$'s output is correct as Π_2 terminates with the correct sub-secret. Otherwise, $\mathbf{R}$ waits for Π_3 to terminate. Again, $\mathbf{R}$ does not have to wait indefinitely. When Π_3 eventually terminates, we know that the majority will be the correct secret.

PRIVACY: Since protocols Π_2 and Π_3 do not involve nodes in B_1, none of the messages exchanged during these protocols is available to the adversary. Even in the case of protocol Π_1, adversary sees only $m + K_1, K_1 + K_2, \ldots, K_{n-1} + K_n$ where $K_1, K_2, \ldots, K_n$ are random numbers. This does not reveal any information about the secret m. □

3.2   Necessity

Consider an asynchronous digraph $\mathcal{N} = (V, \mathcal{E})$ with $\mathbf{S}, \mathbf{R} \in V$ as the sender and receiver respectively, and three subsets $B_1, B_2, B_3 \subseteq V \setminus \{S, R\}$ comprising the adversary structure $\mathcal{A}$. We show that if $\mathcal{N}$ does not satisfy the conditions of Theorem 2, $(0, \delta)$-SMT_{LV} tolerating $\mathcal{A}$ is impossible in $\mathcal{N}$. Without loss of generality, let us assume that the three sets comprising the adversary structure are disjoint. Let the path q_1 be not present between $\mathbf{S}$ and $\mathbf{R}$ in $\mathcal{N}$. (The case where path q_2 or q_3 is not present can be handled analogously.) Hence, every weak path between $\mathbf{S}$ and $\mathbf{R}$ avoiding $B_1 \cup B_2 \cup B_3$ has at least one node w such that every strong path from w to $\mathbf{R}$ passes through nodes in $B_2 \cup B_3$.

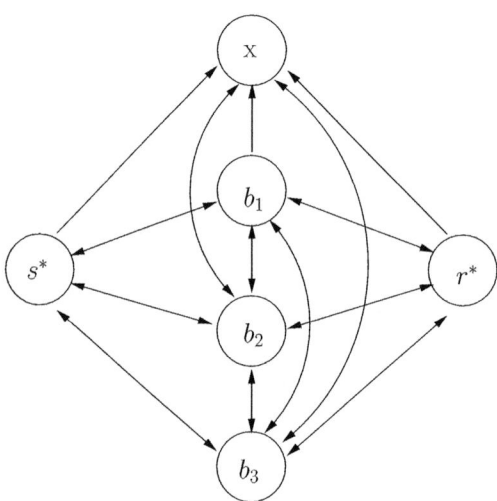

Fig. 2. The network $\mathcal{N}^*$

We first consider the simple asynchronous digraph $\mathcal{N}^* = (V^*, \mathcal{E}^*)$ shown in Figure 2 consisting of six nodes s^*, r^*, b_1, b_2, b_3 and x, where s^* is the sender and r^* is the receiver. Note that the edges (x, s^*), (x, r^*) and (x, b_1) are not present in the graph. Let the adversary structure be $A = \{\{b_1\}, \{b_2\}, \{b_3\}\}$. The only weak path avoiding b_1, b_2 and b_3 passes through x, and every strong path from x to r^* passes through either b_2 or b_3. Hence, since a weak path of type q_1 is not present, this network does not satisfy the connectivity requirements given in Theorem 2. In the following theorem, we show that no $(0, \delta)$-SMT_{LV} protocol from s^* to r^* tolerating A exists in $\mathcal{N}^*$. We then prove that the digraph $\mathcal{N}$ can be partitioned into disjoint sets whose connectivity properties are similar to the connectivity between nodes of the digraph $\mathcal{N}^*$. Now, if $(0, \delta)$-SMT_{LV} is possible in $\mathcal{N}$, it would also be possible in $\mathcal{N}^*$, which is a contradiction. This implies that the conditions mentioned in Theorem 2 are necessary.

Theorem 3. *In the asynchronous digraph* $\mathcal{N}^*$ *shown in Figure 2, no* $(0,\delta)$-*SMT$_{LV}$ protocol from* s^* *to* r^* *tolerating* $A = \{\{b_1\},\{b_2\},\{b_3\}\}$ *exists.*

Proof. W.r.t. to an execution E_i, we define the following: (a) The vector $C_i = (c_{s^*}^i, c_{r^*}^i, c_{b_1}^i, c_{b_2}^i, c_{b_3}^i, c_x^i)$ which denotes the coin tosses input to nodes, where c_z^i denotes the coin tosses of node z. (b) The time instant T_{E_i} at which r^* halts (T_{E_i} may not be finite). (c) The view of a node z, $view_z(E_i)$, which comprises of the internal coin tosses c_z^i of node z and the messages it receives during the execution E_i.

Till time T_{E_1} (defined later), adversary schedules events in the asynchronous network $\mathcal{N}^*$ in the following way: Any execution proceeds in a sequence of time periods. In any time period i, all nodes except b_3 are activated one by one; when a node z is active, all messages generated for node z in the previous time period $i-1$ are delivered in order.

Assume that a $(0,\delta)$-SMT$_{LV}$ protocol Π^* tolerating A exists in the network $\mathcal{N}^*$. Consider the following four executions of Π^*.

– Execution E_1: s^* chooses secret m_1 and the coin tosses of players are C_1. In this execution node b_3 fail-stops. Let r^* halt at time instant T_{E_1} outputting m_1.
– Execution E_2: s^* chooses secret m_1 and the coin tosses of players are $C_2 = C_1$. In this execution node b_1 is passively corrupt. As the view of r^* in this execution is same as in E_1 (b_3 is never active before T_{E_1}), it halts at time instant $T_{E_2} = T_{E_1}$ outputting m_1. Let the view at b_1 be v.

There must exist an execution E_3 such that the coin tosses of b_1 in E_3 are same as in E_2 (i.e., $c_{b_1}^3 = c_{b_1}^2$), and when s^* chooses to send a secret $m_2(\neq m_1)$ view at node b_1 is v. Else, view v would reveal information about m_1.

– Execution E_3: s^* chooses secret m_2 and the coin tosses of players are C_3 such that $c_{b_1}^3 = c_{b_1}^2$. In this execution node b_1 is passively corrupt. View at node b_1 is v.
– Execution E_4: s^* chooses secret m_2 and the coin tosses of players are C_4 such that $c_{s^*}^4 = c_{s^*}^3$, $c_{r^*}^4 = c_{r^*}^2$ and $c_{b_1}^4 = c_{b_1}^3 = c_{b_1}^2$. In this execution node b_2 is actively corrupt. Node b_2 ignores all messages that it receives, sends to s^* what it sent to s^* in E_3, sends to r^* what it sent to r^* in E_2, sends to b_1 what it sent in E_3 (or E_2), and does not send any message to any other node.

As proved in the following lemma, $view_{r^*}(E_4) = view_{r^*}(E_2)$ till time T_{E_1}, implying that r^* cannot distinguish between executions E_2 and E_4 till time T_{E_1}. Since it outputs m_1 at time instant T_{E_1} in execution E_2, it outputs m_1 at the same instant in E_4 too, where s^* chose to send m_2. But, Π^* is a Las Vegas protocol. We have a contradiction.

Lemma 2. *Till time T_{E_1}, the following equalities hold:* $view_{s^*}(E_4) = view_{s^*}(E_3)$, $view_{b_1}(E_4) = view_{b_1}(E_3) = view_{b_1}(E_2)$ *and* $view_{r^*}(E_4) = view_{r^*}(E_2)$.

Proof. Notice that the coin tosses of b_1 in all executions described above is $c_{b_1}^4$. Also, b_1 cannot distinguish between first three executions, it receives and sends same messages in all these executions. We already know that b_3 is never active before T_{E_1}. We give a proof of the lemma by induction on the number of *time-periods*:

Time period 1: The equalities obviously hold for this time-period.

Time period i: Assume that the equalities hold till time-period i–1. As $view_{b_1}(E_4)$ $= view_{b_1}(E_3) = view_{b_1}(E_2)$ till time-period $i - 1$, messages sent by b_1 to other nodes in time-period $i - 1$ are same in E_4, E_3 and E_2. Similar statements can be made for s^* and r^*.

In time-period i, s^* receives identical messages from b_1 in executions E_3 and E_4. Also, as described above, in execution E_4, b_2 sends to s^* what it sent to s^* in E_3. Till time T_{E_1}, since s^* does not receive input from any other node and its coin tosses are same in E_4 and E_3, $view_{s^*}(E_4) = view_{s^*}(E_3)$.

In time-period i, r^* receives identical messages from b_1 in executions E_2 and E_4. Also, in execution E_4, b_2 sends to r^* what it sent to r^* in E_2. Till time T_{E_1}, since r^* does not receive input from any other node and its coin tosses are same in E_4 and E_2, $view_{r^*}(E_4) = view_{r^*}(E_2)$.

In a manner similar to the above, it can be shown that the remaining equality also holds.   □

We now show how the graph $\mathcal{N}$ can be partitioned so that the connectivity among partitions of $\mathcal{N}$ *looks like* the connectivity between nodes of $\mathcal{N}^*$.

Theorem 4. *The set of nodes V in the graph $\mathcal{N}$ can be partitioned into 6 disjoint sets $S^*, R^*, B_1' \subseteq B_1, B_2, B_3$ and X' such that* $\mathbf{S} \in S^*$, $\mathbf{R} \in R^*$ *and* $\forall\, 1 \le i < j \le 6$ *an edge exists between a node of $F[i]$ and a node of $F[j]$ only if* $(f(i), f(j)) \in \mathcal{E}^*$, *where* $F = (S^*, R^*, B_1', B_2, B_3, X')$ *and* $f = (s^*, r^*, b_1, b_2, b_3, x)$ *are two vectors.*

Proof. In the network $\mathcal{N}$, every weak path between $\mathbf{S}$ and $\mathbf{R}$ avoiding $B_1 \cup B_2 \cup B_3$ has at least one node w such that every strong path from w to $\mathbf{R}$ passes through nodes in $B_2 \cup B_3$.

We partition the non-faulty nodes $H = V \setminus \{B_1 \cup B_2 \cup B_3\}$ into 3 disjoint sets. Let $R^* \subset H$ denote the set of all nodes that have a weak path to $\mathbf{R}$ (avoiding $B_1 \cup B_2 \cup B_3$) such that every node w in the weak path has a strong path to $\mathbf{R}$ avoiding nodes in $B_2 \cup B_3$. Divide the rest of non-faulty nodes in two disjoint sets S^* and X. Define $S^* = \{w \in H \setminus R^* \mid w$ has a strong path to $\mathbf{R}$ avoiding nodes in $B_2 \cup B_3\}$. Define $X = H \setminus \{S^* \cup R^*\}$. Clearly, $\mathbf{R} \in R^*$ and $\mathbf{S} \in S^*$ (otherwise even reliable message transmission would not be possible in $\mathcal{N}$). Moreover, if any node $w \in X$ has a strong path to $\mathbf{R}$, it passes through some node in $B_2 \cup B_3$. Otherwise w would belong to S^* itself.

Also, divide the set B_1 into two disjoint sets. Define $B_1^X = \{v \in B_1 | \exists\, u \in X$ such that there is a strong path from u to $v\}$. Let $B_1' = B_1 \setminus B_1^X$. Let us consider the two sets X and B_1^X together as a set X', i.e., $X' = X \cup B_1^X$.

The only edges missing from $\mathcal{N}^*$ are $(x, s^*), (x, r^*), (x, b_1)$ and $(s^*, r^*), (r^*, s^*)$. It easily follows from the definitions above that $\nexists\, (u, v) \in \mathcal{E}$ such that $u \in X'$

and $v \in S^* \cup R^* \cup B_1'$. Also, there cannot exist any directed edge between a node in S^* and a node in R^*. Hence proved. $\qquad\square$

Theorem 5. *In the asynchronous digraph $\mathcal{N} = (V, \mathcal{E})$, if a $(0, \delta)$-SMT$_{LV}$ protocol from $\mathbf{S}$ to $\mathbf{R}$ tolerating $\mathcal{A} = \{B_1, B_2, B_3\}$ exists, then a $(0, \delta)$-SMT$_{LV}$ protocol from s^* to r^* tolerating $A = \{\{b_1\}, \{b_2\}, \{b_3\}\}$ exists in the network $\mathcal{N}^*$.*

Proof. It is straightforward to prove the above theorem using standard player simulation technique. $\qquad\square$

However, from Theorem 3, we know that no $(0, \delta)$-SMT$_{LV}$ protocol tolerating A exists in the network $\mathcal{N}^*$. Hence, no $(0, \delta)$-SMT$_{LV}$ protocol tolerating $\mathcal{A}$ exists in the network $\mathcal{N}$.

3.3   Las Vegas $(0, \delta)$-SMT and PSMT

The sufficiency proof in Section 3.1 describes a protocol that provides both perfect privacy and perfect resiliency. We also know that if a PSMT protocol exists in a digraph, a $(0, \delta)$-SMT$_{LV}$ protocol obviously exists. Combining the two facts, we can say that:

Lemma 3. *In an asynchronous digraph, a $(0, \delta)$-SMT$_{LV}$ protocol tolerating a 3-sized adversary structure exists if and only if a PSMT protocol tolerating the same adversary structure exists too.*

We now show how to construct PSMT protocols that tolerate adversary structures of higher size.

Theorem 6. *In an asynchronous digraph $\mathcal{N} = (V, \mathcal{E})$, a PSMT protocol tolerating an adversary structure $\mathbb{A}$ ($|\mathbb{A}| > 3$) exists if and only if for every adversary structure $A \subseteq \mathbb{A}$ such that $|A| = 3$, a PSMT protocol tolerating A exists.*

Proof. Necessity is trivial. We give sufficiency proof here. We show how to construct a protocol tolerating an adversary structure $\mathcal{A}$ of size $n > 3$ from protocols tolerating adversary structures of smaller size. Using this technique, starting from protocols for adversary structures of size 3, we would be able to construct a protocol for an adversary structure of arbitrary size inductively.

Consider $\mathcal{A}_1$, $\mathcal{A}_2$, $\mathcal{A}_3$ and $\mathcal{A}_4$, four $\lceil \frac{3|\mathcal{A}|}{4} \rceil$-sized subsets of $\mathcal{A}$ such that each element of $\mathcal{A}$ occurs in at least three of the four sets. For $1 \leq i \leq 4$, let $\Pi_{\mathcal{A}_i}$ be the PSMT protocol tolerating $\mathcal{A}_i$. Let $m \in \mathbb{F}$ be the secret $\mathbf{S}$ intends to send. The PSMT protocol $\Pi_{\mathcal{A}}$ tolerating $\mathcal{A}$ proceeds as follows:

- $\mathbf{S}$ does a $(2, 4)$ secret sharing of m to obtain four shares m_1, m_2, m_3, m_4.
- The four protocols $\Pi_{\mathcal{A}_1}, \Pi_{\mathcal{A}_2}, \Pi_{\mathcal{A}_3}$ and $\Pi_{\mathcal{A}_4}$ are run in parallel; for $1 \leq i \leq 4$, $\Pi_{\mathcal{A}_i}$ is run on m_i as the sub-secret.
- $\mathbf{R}$ first waits for any three of the above four sub-protocols to terminate. If the sub-secrets received through these sub-protocols lead to the reconstruction of a unique secret, $\mathbf{R}$ outputs it. Otherwise, $\mathbf{R}$ further waits for another sub-protocol to terminate. It now applies the BW algorithm on the sub-secrets obtained through the four sub-protocols and outputs the outcome of the algorithm.

The correctness of this protocol is proved in the following lemma.   □

Lemma 4. *The protocol $\Pi_{\mathcal{A}}$ is a PSMT protocol tolerating $\mathcal{A}$.*

Proof. No matter which $B \in \mathcal{A}$ adversary chooses to corrupt, at least three out of the four sets $\mathcal{A}_1$, $\mathcal{A}_2$, $\mathcal{A}_3$ and $\mathcal{A}_4$ contain B. Hence, at least three out of the protocols $\Pi_{\mathcal{A}_1}, \Pi_{\mathcal{A}_2}, \Pi_{\mathcal{A}_3}$ and $\Pi_{\mathcal{A}_4}$ will be private and reliable. W.l.o.g assume that $\Pi_{\mathcal{A}_1}, \Pi_{\mathcal{A}_2}, \Pi_{\mathcal{A}_3}$ are those 3 protocols. Hence, for $1 \leq i \leq 3$, protocol $\Pi_{\mathcal{A}_i}$ terminates securely with $\mathbf{R}$ receiving the sub-secret m_i.

RESILIENCY: If the protocols $\Pi_{\mathcal{A}_1}$, $\Pi_{\mathcal{A}_2}$ and $\Pi_{\mathcal{A}_3}$ terminate before $\Pi_{\mathcal{A}_4}$ does, it is easy to see that $\mathbf{R}$ will output m. However, adversary may schedule events in the network such that $\Pi_{\mathcal{A}_4}$ terminates before all of $\Pi_{\mathcal{A}_1}, \Pi_{\mathcal{A}_2}, \Pi_{\mathcal{A}_3}$ do. When $\Pi_{\mathcal{A}_4}$ terminates, if $\mathbf{R}$ receives m_4 then we know that m will be reconstructed. However, since $\Pi_{\mathcal{A}_4}$ may not be tolerating the corrupt set B, $\mathbf{R}$ may receive m_4' ($\neq m_4$). But then $\mathbf{R}$ will wait for another protocol to terminate. Now, with only one incorrect share out of four, it is easy to see that BW algorithm will output m.

PRIVACY: Since adversary knows only m_4 which is a share of m obtained using $(2, 4)$ secret sharing, it does not reveal any information about m.   □

Let us see how the Theorem 6 proved above for PSMT can also be proved for $(0, \delta)$-SMT$_{LV}$. From Lemma 3, we know that if a $(0, \delta)$-SMT$_{LV}$ protocol tolerating a 3-sized adversary structure exists, a PSMT protocol tolerating the same adversary structure exists too. Hence, when we start with adversary structures of size three, instead of using $(0, \delta)$-SMT$_{LV}$ protocols, we will use corresponding PSMT protocols. Now, using the procedure described in the proof of Theorem 6 above, we will be able to construct a PSMT protocol for an adversary structure of arbitrary size, which will automatically be a $(0, \delta)$-SMT$_{LV}$ protocol.

Combining Theorem 2, Lemma 3, Theorem 6, and the discussion above, in a nutshell we have:

Lemma 5. *In an asynchronous digraph, a PSMT or a $(0, \delta)$-SMT$_{LV}$ protocol tolerating an adversary structure $\mathbb{A}$ ($|\mathbb{A}| \geq 3$) exists if and only if for every adversary structure $A \subseteq \mathbb{A}$ such that $|A| = 3$, Theorem 2 is satisfied.*

Notice that for $\Pi_{\{B_1, B_2, B_3\}}$, constructed in Section 3.1, both communication complexity of the protocol and computation complexity at every node is polynomial in the size of the network. If we denote the size of adversary structure $\mathbb{A}$ by N, from Theorem 6 we can see that $O(N^5)$[1] sub-protocols of the same complexity as of $\Pi_{\{B_1, B_2, B_3\}}$ need to be run to trasmit the message securely in the presence of $\mathbb{A}$. Starting with the output of these $O(N^5)$ sub-protocols[2], $\mathbf{R}$ can obtain the output of the protocol tolerating $\mathbb{A}$ in $O(N^5)$ computational steps. Hence, our PSMT protocol is efficient in the size of network and the size of adversary structure.

---

[1] To construct a protocol tolerating an adversary structure of size N, we use four sub-protocols which tolerate adversary structures of size roughly $3N/4$.

[2] Note that $\mathbf{R}$ need not wait for all sub-protocols to terminate.

4   All Pairs PSMT

Let $\mathcal{N} = (V, \mathcal{E})$ be an asynchronous digraph in which PSMT tolerating t-threshold adversary ($t \leq n - 2$, n is the number of nodes) is possible between every pair of nodes. Note that the adversary structure consists of all t-sized subsets of the node set. With the help of Theorem 5, we can draw two conclusions about $\mathcal{N}$:

- It is $3t + 1$ weakly connected. This is because if there are less than $3t + 1$ node disjoint weak paths between two nodes u and v, we can construct three disjoint sets, each of size t, such that there is no weak path between u and v that avoids all nodes in the union of the three sets. This would imply that PSMT from u to v is impossible in $\mathcal{N}$.
- It is $2t + 1$ strongly connected. If there are less than $2t + 1$ node disjoint strong paths between two nodes u and v, we can construct two disjoint sets B_1 and B_2, each of size t, such that every strong path from u to v passes through either B_1 or B_2. Let B_3 be any subset of nodes. Now, any weak path between u and v would obviously have the node u. But since every strong path from u passes through either B_1 or B_2, a weak path of type q_3, as required by Theorem 2, will not be present. This would imply that PSMT from u to v is impossible in $\mathcal{N}$. (A straightforward proof can also be obtained using Theorem 1.)

For such a graph, we give an efficient protocol Π_{sr} for PSMT tolerating t-threshold adversary from any node $s \in V$ to another node $r \in V$. This protocol is efficient in the size of the network.

Let m be the secret s intends to send to r. The node s makes $3t + 1$ shares of m, namely $m_1, m_2, \ldots, m_{3t+1}$, using $(t + 1, 3t + 1)$ secret sharing scheme. Now, sub-protocols $\Gamma_1, \Gamma_2, \ldots, \Gamma_{3t+1}$ (described later) are run in parallel in the network. Node r waits till at least $2t$ of these sub-protocols terminate. It sets a variable e to 0, then runs the following loop:

while (true)
  - Wait for another sub-protocol to terminate.
  - Let W denote the set of outcomes of the sub-protocols that have terminated so far. Run the Berlekamp-Welch (BW) algorithm on inputs t, e, W [12]. If it returns a polynomial, output its constant term, come out of the loop and halt.
  - $e := e + 1$

For $1 \leq i \leq 3t+1$, the sub-protocol Γ_i tries to securely communicate m_i assuming that weak path w_i does not contain any faulty node. If w_i is a strong path from s to r, the protocol Γ_i is simply: send m_i to r along w_i. Otherwise, w_i is a weak path expressed as $y_0, u_1, y_1, u_2, y_2, \ldots, u_n, y_n, u_{n+1}$ ($n \geq 0$), where u_i represent blocked nodes, y_i represent head nodes, y_0 represents s, and u_{n+1} represents r. The protocol Γ_i proceeds in the following steps:

1. s sends m_i to u_1 along w_i. For $1 \leq j \leq n$, node y_j chooses a random key K_j and sends it to u_j and u_{j+1} along w_i.

2. Node u_1 sends $L_1 = m_i + K_1'$ along $2t + 1$ node disjoint strong paths to r when it receives m_i from s and K_1' from y_1. For $1 < j \leq n$, u_j sends $L_j = K_{j-1}' + K_j'$ along $2t+1$ node disjoint strong paths to r when it receives K_{j-1}' from y_{j-1} and K_j' from y_j. (This ensures that for each j, $1 \leq j \leq n$, r receives L_j reliably).
3. r waits until it receives K_n'' from y_n, and for each j, $1 \leq j \leq n$, at least $t+1$ concurrent readings of L_j' from u_j. It then runs the following loop:

   for z in n to 2
   $$K_{z-1}'' = L_z' - K_z''.$$
   Output $m_i' = L_1' - K_1''$.

This completes the description of Γ_i, and hence of Π_{sr}.

Lemma 6. *The protocol Π_{sr} is an efficient PSMT protocol tolerating t-threshold adversary ($t \leq n - 2$).*

Proof. Consider a sub-protocol Γ_i where the weak path w_i does not contain any corrupted node. But the messages sent by blocked nodes in w_i may still be accessible to the adversary. In the worst case it may know all of $m_i + K_1, K_1 + K_2, \ldots, K_{n-1} + K_n$. Still this does not reveal any information about m_i. Moreover, every blocked node sends messages to r along $2t + 1$ vertex disjoint paths. Hence if r waits long enough it would receive all these messages reliably, and with the extra knowledge of K_n, recover m_i correctly.

Now, we know that at least $2t + 1$ weak paths are honest (since adversary can corrupt at most t nodes at a time). So at least $2t + 1$ Γ_i would eventually terminate with $m_i' = m_i$. Hence, in every iteration of the while loop the set W supplied to BW algorithm has at most t errors. So the algorithm produces an output in some iteration and we know that if it does produce an output, it is correct [12]. Therefore Π_{sr} is a reliable protocol. Moreover, at least $2t + 1$ shares of m are unknown to the adversary, making Π_{sr} private.

It is easy to see that the overall communication complexity of the protocol Π_{sr} and the computation complexity at each node is a polynomial in the size of network.   □

5   Characterizing Asynchronous Networks for Monte Carlo $(0, \delta)$-SMT

In this section we give a characterization of asynchronous digraphs over which $(0, \delta)$-SMT$_{MC}$ is possible. In this variant of $(0, \delta)$-SMT, **R** is allowed to output an incorrect message with small probability. We first present a theorem which shows how to reduce the problem of $(0, \delta)$-SMT$_{MC}$ tolerating an arbitrary sized adversary structure to tolerating all its two-sized subsets.

Theorem 7. *In an asynchronous digraph $\mathcal{N}$, a $(0, \delta)$-SMT$_{MC}$ protocol tolerating an adversary structure $\mathbb{A}$ exists if and only if for every adversary structure $A \subseteq \mathbb{A}$ such that $|A| = 2$, a $(0, \delta')$-SMT$_{MC}$ protocol exists.*

Proof. Necessity is trivial. We give sufficiency proof here. We show how to construct a protocol for an adversary structure $\mathcal{A}$ of size $n > 2$ from protocols for adversary structures of smaller size. Using this technique, starting from protocols for adversary structures of size 2, we would be able to construct a protocol for an adversary structure of arbitrary size inductively.

Consider $\mathcal{A}_1$, $\mathcal{A}_2$ and $\mathcal{A}_3$, three $\lceil \frac{2|\mathcal{A}|}{3} \rceil$-sized subsets of $\mathcal{A}$ such that each element of $\mathcal{A}$ occurs in at least two of the three sets. For $1 \le i \le 3$, let $\Pi_{\mathcal{A}_i}$ be the $(0, \delta')$-SMT$_{MC}$ protocol tolerating $\mathcal{A}_i$. Let m be the secret **S** intends to send. The $(0, \delta)$-SMT$_{MC}$ protocol tolerating $\mathcal{A}$ proceeds as follows:

1. For each $\beta \in \{1, 2, 3\}$, **S** chooses a three-tuple $\mathcal{K}_\beta = (K_{\beta,1}, K_{\beta,2}, K_{\beta,3})$ $\in_R \mathbb{F} \times \mathbb{F} \times \mathbb{F}$ and evaluates $\chi_1(m, \mathcal{K}_\beta) = m + K_{\beta,1}$ and $\chi_2(m, \mathcal{K}_\beta) = (m + K_{\beta,1}) \cdot K_{\beta,2} + K_{\beta,3}$.
2. Seven instances of the protocol $\Pi_{\mathcal{A}_1}$ are run in parallel in the network on the sub-secrets $K_{1,1}$, $K_{1,2}$, $K_{1,3}$, $\chi_1(m; \mathcal{K}_2)$, $\chi_2(m; \mathcal{K}_2)$, $\chi_1(m; \mathcal{K}_3)$ and $\chi_2(m; \mathcal{K}_3)$. Similarly, seven instances each of protocols $\Pi_{\mathcal{A}_2}$ and $\Pi_{\mathcal{A}_3}$ are run in parallel alongside the instances of protocol $\Pi_{\mathcal{A}_1}$. In essence, for each $\beta \in \{1, 2, 3\}$, the tuple of keys $\mathcal{K}_\beta$ is sent through three instances of protocol $\Pi_{\mathcal{A}_\beta}$ and the secret m authenticated with $\mathcal{K}_\beta$ is sent through both the other protocols.
3. **R** waits until for two *distinct* indices $x, y \in \{1, 2, 3\}$ it receives $K'_{x,1}$, $K'_{x,2}$, $K'_{x,3}$, $\chi_1(f, \mathcal{K}_y)'$, $\chi_2(f, \mathcal{K}_y)'$ through the protocol $\Pi_{\mathcal{A}_x}$ and $K'_{y,1}$, $K'_{y,2}$, $K'_{y,3}$, $\chi_1(m, \mathcal{K}_x)'$, $\chi_2(m, \mathcal{K}_x)'$ through the protocol $\Pi_{\mathcal{A}_y}$ such that the following conditions are satisfied:
   - $\chi_2(m, \mathcal{K}_x)' = \chi_1(m, \mathcal{K}_x)' \cdot K'_{x,2} + K'_{x,3}$
   - $\chi_2(m, \mathcal{K}_y)' = \chi_1(m, \mathcal{K}_y)' \cdot K'_{y,2} + K'_{y,3}$
   - $\chi_1(m, \mathcal{K}_x)' - K'_{x,1} = \chi_1(m, \mathcal{K}_y)' - K'_{y,1}$

   In simple words, **R** waits until two authenticated messages pass verification against corresponding keys and the same secret is recovered. **R** outputs $m' = \chi_1(f, \mathcal{K}_x)' - K'_{x,1}$.

The following Lemma proves the correctness of $\Pi_{\mathcal{A}}$. $\qquad\qquad\qquad\square$

Lemma 7. *The protocol $\Pi_{\mathcal{A}}$ is a $(0, \delta)$-SMT$_{MC}$ protocol tolerating $\mathcal{A}$, where* $\delta = 1 - (1 - \delta')^{10} \cdot (\frac{|\mathbb{F}| - 1}{|\mathbb{F}|})^2$.

Proof. No matter what $B \in \mathcal{A}$ adversary chooses to corrupt, at least two of the three sets $\mathcal{A}_1$, $\mathcal{A}_2$ and $\mathcal{A}_3$ contain B. Hence, the instances of at least two out of the three protocols $\Pi_{\mathcal{A}_1}$, $\Pi_{\mathcal{A}_2}$ and $\Pi_{\mathcal{A}_3}$ will be private and reliable. W.l.o.g assume that $\Pi_{\mathcal{A}_1}$ and $\Pi_{\mathcal{A}_2}$ are those two protocols. Therefore, the instances of protocols $\Pi_{\mathcal{A}_1}$ and $\Pi_{\mathcal{A}_2}$ do not reveal any information about the sub-secrets transmitted through them, and with high probability, they terminate with the correct output at **R**. We first prove that the protocol $\Pi_{\mathcal{A}}$ is perfectly private.

Privacy: Note that only the instances of protocol $\Pi_{\mathcal{A}_3}$ can reveal information about the sub-secrets transmitted through them. Hence, in the best case, adversary knows $K_{3,1}$, $K_{3,2}$, $K_{3,3}$, $\chi_1(m; \mathcal{K}_1)$, $\chi_2(m; \mathcal{K}_1)$, $\chi_1(m; \mathcal{K}_2)$ and $\chi_2(m; \mathcal{K}_2)$.

However, as it does not know $\mathcal{K}_1$ and $\mathcal{K}_2$, it does not know anything about the secret m.

Resiliency: Let $\psi_1 = (K_{1,1}, K_{1,2}, K_{1,3}, \chi_1(m, \mathcal{K}_2), \chi_2(m, \mathcal{K}_2))$ and $\psi_2 = (K_{2,1}, K_{2,2}, K_{2,3}, \chi_1(m, \mathcal{K}_1), \chi_2(m, \mathcal{K}_1))$ denote part of the sequence of sub-secrets sent through the instances of $\Pi_{\mathcal{A}_1}$ and $\Pi_{\mathcal{A}_2}$ respectively. We have ignored the sub-secrets obtained through the authentication of secret m using keys sent through $\Pi_{\mathcal{A}_3}$. Let ψ represent the combined sequence which has 10 elements. Also, let ζ represent a part of the sequence of sub-secrets sent through instances of $\Pi_{\mathcal{A}_3}$, $\zeta = (\chi_1(m, \mathcal{K}_1), \chi_2(m, \mathcal{K}_1), \chi_1(m, \mathcal{K}_2), \chi_2(m, \mathcal{K}_2))$. We do not consider the keys sent through $\Pi_{\mathcal{A}_3}$.

Suppose all sub-secrets in ψ are received reliably. To minimize the chances of reliable transmission of secret m, adversary would tamper with sub-secrets in ζ and schedule events in the network such that all sub-secrets in ζ are received before all sub-secrets in ψ are received. Now, consider the following event E: all sub-secrets in ψ are received reliably and any tampering in the sequence of sub-secrets ζ is detected. In such an event $\mathbf{R}$ recovers $m' = m$. The probability that the protocol $\Pi_{\mathcal{A}}$ produces correct output is at least the probability of the event E. Since adversary does not know $\mathcal{K}_1$ and $\mathcal{K}_2$, we know that $Pr(E) > (1 - \delta')^{10}(\frac{|\mathbb{F}|-1}{|\mathbb{F}|})^2$. Therefore, the probability of failure δ is atmost $1 - P(E)$. Note that since the error probability δ' is less than $1/2$, it can be brought to down to any small value we want. Once the value of δ' is low, $Pr(E)$ will be larger than $1/2$, and hence δ becomes less than $1/2$.   □

Having reduced the problem of characterizing $(0, \delta)$-SMT$_{MC}$ tolerating adversary structure $\mathbb{A}$ to all its 2-sized subsets, we now proceed to give a characterization of asynchronous digraphs tolerating a given 2-sized adversary structure.

Theorem 8. *In an asynchronous digraph $\mathcal{N}$, a $(0, \delta)$-SMT$_{MC}$ protocol from $\mathbf{S}$ to $\mathbf{R}$ tolerating $A = \{B_1, B_2\}$ exists if and only if for each $\alpha \in \{1, 2\}$, there exists a weak path q_α avoiding nodes in $B_1 \cup B_2$ such that every node u along the path q_α has a strong path to $\mathbf{R}$ avoiding all nodes in $B_{\bar{\alpha}}$[3] (Paths q_1, q_2 need not be distinct.)*

Proof. According to [14], same characterization holds for Monte Carlo URMT – where only reliability of the message is required – tolerating a 2-sized adversary structure. Hence, the characterization is obviously necessary for $(0, \delta)$-SMT$_{MC}$. Also, the protocol given for Monte Carlo URMT in [14] does not reveal any information about the secret being transmitted. Hence the same protocol can be used for $(0, \delta)$-SMT$_{MC}$.   □

References

1. Agarwal, S., Cramer, R., de Haan, R.: Asymptotically Optimal Two-Round Perfectly Secure Message Transmission. In: Dwork, C. (ed.) CRYPTO 2006. LNCS, vol. 4117, pp. 394–408. Springer, Heidelberg (2006)

---

[3] We denote $\bar{1} = 2$ and vice-versa.

2. Choudhary, A., Patra, A., Ashwinkumar, B.V., Srinathan, K., Rangan, C.P.: On Minimal Connectivity Requirement for Secure Message Transmission in Asynchronous Networks. In: Garg, V., Wattenhofer, R., Kothapalli, K. (eds.) ICDCN 2009. LNCS, vol. 5408, pp. 148–162. Springer, Heidelberg (2008)

3. Desmedt, Y.G., Wang, Y.: Perfectly Secure Message Transmission Revisited. In: Knudsen, L.R. (ed.) EUROCRYPT 2002. LNCS, vol. 2332, pp. 502–517. Springer, Heidelberg (2002)

4. Dolev, D., Dwork, C., Waarts, O., Yung, M.: Perfectly secure message transmission. In: Annual IEEE Symposium on Foundations of Computer Science, vol. 1, pp. 36–45 (1990)

5. Fischer, M.J., Lynch, N.A., Paterson, M.S.: Impossibility of distributed consensus with one faulty process. J. ACM 32(2), 374–382 (1985)

6. Franklin, M., Yung, M.: Secure Hypergraphs: Privacy from Partial Broadcast. In: Proceedings of 27th Symposium on Theory of Computing (STOC), pp. 36–44. ACM Press, New York (1995)

7. Franklin, M.K., Wright, R.N.: Secure communication in minimal connectivity models. J. Cryptology 13(1), 9–30 (2000)

8. Hirt, M., Maurer, U.: Player Simulation and General Adversary Structures in Perfect Multi-party Computation. Journal of Cryptology 13(1), 31–60 (2000)

9. Kumar, M., Goundan, P.R., Srinathan, K., Rangan, C.P.: On perfectly secure communication over arbitrary networks. In: Proceedings of the 21st Symposium on Principles of Distributed Computing (PODC), Monterey, California, USA, pp. 193–202. ACM Press (July 2002)

10. Kurosawa, K., Suzuki, K.: Truly efficient 2-round perfectly secure message transmission scheme. IEEE Trans. Inf. Theor. 55(11), 5223–5232 (2009)

11. Lamport, L., Shostak, R., Pease, M.: The Byzantine Generals Problem. ACM Transactions on Programming Languages and Systems 4(3), 382–401 (1982)

12. MacWilliams, F.J., Sloane, N.J.A.: The Theory of Error-Correcting Codes. North-Holland, Amsterdam (1977)

13. Mehta, A., Agrawal, S., Srinathan, K.: Brief Announcement: Synchronous las Vegas URMT iff Asynchronous Monte Carlo URMT. In: Lynch, N.A., Shvartsman, A.A. (eds.) DISC 2010. LNCS, vol. 6343, pp. 201–203. Springer, Heidelberg (2010)

14. Mehta, A., Agrawal, S., Srinathan, K.: Interplay between (im)perfectness, synchrony and connectivity: The case of probabilistic reliable communication. Cryptology ePrint Archive, Report 2010/392 (2010), http://eprint.iacr.org/

15. Nayak, M., Agrawal, S., Srinathan, K.: Minimal Connectivity for Unconditionally Secure Message Transmission in Synchronous Directed Networks. In: Fehr, S. (ed.) ICITS 2011. LNCS, vol. 6673, pp. 32–51. Springer, Heidelberg (2011)

16. Patra, A., Choudhary, A., Rangan, C.P.: Unconditionally Reliable and Secure Message Transmission in Directed Networks Revisited. In: Ostrovsky, R., De Prisco, R., Visconti, I. (eds.) SCN 2008. LNCS, vol. 5229, pp. 309–326. Springer, Heidelberg (2008)

17. Patra, A., Choudhary, A., Rangan, C.P.: On Communication Complexity of Secure Message Transmission in Directed Networks. In: Kant, K., Pemmaraju, S.V., Sivalingam, K.M., Wu, J. (eds.) ICDCN 2010. LNCS, vol. 5935, pp. 42–53. Springer, Heidelberg (2010)

18. Patra, A., Choudhary, A., Rangan, C.P.: Constant phase efficient protocols for secure message transmission in directed networks. In: Proceedings of the Twenty-Sixth Annual ACM Symposium on Principles of Distributed Computing, PODC 2007, pp. 322–323. ACM, New York (2007)

19. Patra, A., Choudhary, A., Srinathan, K., Rangan, C.P.: Perfectly Reliable and Secure Communication in Directed Networks Tolerating Mixed Adversary. In: Pelc, A. (ed.) DISC 2007. LNCS, vol. 4731, pp. 496–498. Springer, Heidelberg (2007)
20. Rabin, T., Ben-Or, M.: Verifiable secret sharing and multiparty protocols with honest majority. In: STOC 1989: Proceedings of the Twenty-First Annual ACM Symposium on Theory of Computing, pp. 73–85. ACM, New York (1989)
21. Sayeed, H.M., Abu-Amara, H.: Perfectly secure message transmission in asynchronous networks. In: SPDP 1995: Proceedings of the 7th IEEE Symposium on Parallel and Distributeed Processing, p. 100. IEEE Computer Society, Washington, DC, USA (1995)
22. Shamir, A.: How to Share a Secret. Communications of the ACM 22, 612–613 (1979)
23. Shankar, B., Gopal, P., Srinathan, K., Rangan, C.P.: Unconditionally reliable message transmission in directed networks. In: SODA 2008: Proceedings of the Nineteenth Annual ACM-SIAM Symposium on Discrete Algorithms, pp. 1048–1055. Society for Industrial and Applied Mathematics, Philadelphia (2008)
24. Srinathan, K., Narayanan, A., Rangan, C.P.: Optimal Perfectly Secure Message Transmission. In: Franklin, M. (ed.) CRYPTO 2004. LNCS, vol. 3152, pp. 545–561. Springer, Heidelberg (2004)
25. Srinathan, K., Raghavendra, P., Rangan, C.P.: On Proactive Perfectly Secure Message Transmission. In: Pieprzyk, J., Ghodosi, H., Dawson, E. (eds.) ACISP 2007. LNCS, vol. 4586, pp. 461–473. Springer, Heidelberg (2007)
26. Srinathan, K., Rangan, C.P.: Possibility and complexity of probabilistic reliable communications in directed networks. In: Proceedings of 25th ACM Symposium on Principles of Distributed Computing, PODC 2006 (2006)
27. Wang, Y.: Robust key establishment in sensor networks. SIGMOD Rec. 33(1), 14–19 (2004)
28. Yang, Q., Desmedt, Y.: Cryptanalysis of Secure Message Transmission Protocols with Feedback. In: Kurosawa, K. (ed.) Information Theoretic Security. LNCS, vol. 5973, pp. 159–176. Springer, Heidelberg (2010)

Towards a Provably Secure DoS-Resilient Key Exchange Protocol with Perfect Forward Secrecy

Lakshmi Kuppusamy, Jothi Rangasamy, Douglas Stebila, Colin Boyd, and Juan Gonzalez Nieto

Information Security Institute, Queensland University of Technology, GPO Box 2434, Brisbane, Queensland 4001, Australia
{l.kuppusamy,j.rangasamy,stebila,c.boyd,j.gonzaleznieto}@qut.edu.au

Abstract. Just Fast Keying (JFK) is a simple, efficient and secure key exchange protocol proposed by Aiello et al. (ACM TISSEC, 2004). JFK is well known for its novel design features, notably its resistance to denial-of-service (DoS) attacks. Using Meadows' cost-based framework, we identify a new DoS vulnerability in JFK. The JFK protocol is claimed secure in the Canetti-Krawczyk model under the Decisional Diffie-Hellman (DDH) assumption. We show that security of the JFK protocol, when re-using ephemeral Diffie-Hellman keys, appears to require the Gap Diffie-Hellman (GDH) assumption in the random oracle model. We propose a new variant of JFK that avoids the identified DoS vulnerability and provides perfect forward secrecy even under the DDH assumption, achieving the full security promised by the JFK protocol.

Keywords: Denial of service, Meadows' cost-based framework, Just Fast Keying, client puzzles, key agreement, perfect forward secrecy.

1 Introduction

Denial-of-service (DoS) attacks that are mounted to exhaust the processing, memory, or network resources of target systems have become a common occurrence. These attacks can easily disable servers, so it is important for the server to detect and filter bogus connection requests as early as possible. Cryptographic techniques such as authentication can assist a server in detecting bogus connections, but the computationally expensive operations involved in authentication may open a new DoS vulnerability. Hence care must be taken to design a protocol that implements defense strategies to efficiently tackle DoS attackers and to protect itself from exhausting the resources.

A number of DoS countermeasures such as client puzzles [3,11,20], stateless connections [2] and gradual authentication [14,19] are available for building DoS-resilient protocols. Resistance to DoS attacks is the main design goal for network protocols such as JFK [1], CA-RSA [8], IKE [10], IKEv2 [12], MIKE [13] and HIP [15]. Recently, Stebila et al. [23] described a model for assessing DoS-resistance in protocols by following the approach of Stebila and Ustaoglu [22]. They also gave a generic construction for DoS-resistant protocols.

D.J. Bernstein and S. Chatterjee (Eds.): INDOCRYPT 2011, LNCS 7107, pp. 379–398, 2011.
© Springer-Verlag Berlin Heidelberg 2011

Just Fast Keying (JFK) is a simple, efficient and secure key exchange protocol proposed by Aiello et al. [1]. JFK is well known for its novel design features, including its resistance to DoS attacks. The designers of JFK provided a careful analysis of its security properties but stopped short of providing a fully formal security proof. They pointed out that the basic structure of JFK, what they called the "cryptographic core", already has a proof of its key exchange security properties [6], and that the additional mechanisms of JFK beyond this core will not degrade key exchange security.

The JFK designers also discuss formal approaches to other protocol properties, namely privacy and forward secrecy. However, they do not provide any formal discussion of the DoS-resilience of JFK, one of its main properties. The main technique for achieving DoS-resilience in JFK is the reuse of ephemeral public keys, but this comes at the obvious expense of perfect forward secrecy.

Smith et al. [21] performed a detailed analysis of JFK's DoS resistance using Meadows' cost-based modelling framework [14] and identified two DoS attacks against the protocol that are possible when an attacker is willing to reveal the source IP address. The first attack was found with the direct application of Meadows' cost-based framework and the second attack was possible considering the presence of co-ordinated DoS attackers.

Contributions. The main goal of our work is to improve the DoS resistance of JFK, although our approach can be applied to any Diffie-Hellman (DH)-based protocols. Our contributions can be summarised as follows:

- We show that the security of JFK protocol with ephemeral DH reuse appears to require the GDH assumption in the random oracle model.
- We also analyse the JFK protocol in detail with the help of Meadows' cost-based framework and identify a new DoS vulnerability in JFK which is possible in the presence of co-ordinated DoS attackers.
- We propose a variant of JFK that efficiently generates fresh ephemeral keys from precomputed values using a technique of Boyko et al. [5] up to 3.4 times faster than in JFK. The proposed JFK variant not only avoids the identified DoS attack but also achieves both the security level promised by the original JFK protocol and perfect forward secrecy. We give a detailed yet simple security analysis of the resulting BPV-JFK protocol in the Canetti-Krawczyk model under the DDH assumption in the standard model.
- We implement the generic technique of Stebila et al. [23] in JFK and BPV-JFK to achieve strong DoS resilience in the Stebila et al. model.

A comparison of the security properties of JFK and the protocols proposed in this paper appears in Table 1.

2 Just Fast Keying Protocol and Its DoS Vulnerabilities

Aiello et al. [1] presented two variants of the JFK protocol namely JFKi and JFKr, both with same level of DoS resistance; we focus on the JFKi variant, which appears in Figure 1.

Table 1. Comparison of properties of JFK-based protocols

Protocol	Cost-based vulnerability	Security assumptions	Perfect Forward Secrecy	DoS-resilience
JFK [1]	Yes	GDH, ROM	Only with no reuse	No
DoS-JFK (§ 2.5)	No	GDH, ROM	Only with no reuse	Yes
BPV-JFK (§ 3)	No	DDH	Yes	No
DoS-BPV-JFK (§ 3.6)	No	DDH	Yes	Yes

---

1. $\mathbf{I} \rightarrow \mathbf{R} : N_{\mathbf{I}}', g^x, \mathsf{ID}_{\mathbf{R}}'$
2. $\mathbf{R} \rightarrow \mathbf{I} : N_{\mathbf{I}}', N_{\mathbf{R}}, g^y, \mathsf{grpinfo}_{\mathbf{R}}, \mathsf{ID}_{\mathbf{R}}, S_{k_{\mathbf{R}}}[g^y, \mathsf{grpinfo}_{\mathbf{R}}], \text{token}$
3. $\mathbf{I} \rightarrow \mathbf{R} : N_{\mathbf{I}}, N_{\mathbf{R}}, g^x, g^y, \{\mathsf{ID}_{\mathbf{I}}, \mathsf{sa}, S_{k_{\mathbf{I}}}[N_{\mathbf{I}}', N_{\mathbf{R}}, g^x, g^y, \mathsf{ID}_{\mathbf{R}}, \mathsf{sa}]\}_{K_a}^{K_e},$
   token
4. $\mathbf{R} \rightarrow \mathbf{I} : \{S_{k_{\mathbf{R}}}[N_{\mathbf{I}}', N_{\mathbf{R}}, g^x, g^y, \mathsf{ID}_{\mathbf{I}}, \mathsf{sa}, \mathsf{sa}'], \mathsf{sa}'\}_{K_a}^{K_e}$

$\text{token} = H_{\mathsf{HKR}}(g^y, N_{\mathbf{R}}, N_{\mathbf{I}}', IP_{\mathbf{I}}), \; K_e = H_{g^{xy}}(N_{\mathbf{I}}', N_{\mathbf{R}}, 1)$
$K_a = H_{g^{xy}}(N_{\mathbf{I}}', N_{\mathbf{R}}, 2), \; K_{xy} = H_{g^{xy}}(N_{\mathbf{I}}', N_{\mathbf{R}}, 0)$

$N_{\mathbf{I}}, N_{\mathbf{R}}$	Random nonces chosen by $\mathbf{I}$ and $\mathbf{R}$ respectively.
$H(\cdot)$	Unkeyed hash function.
$N_{\mathbf{I}}'$	Initiator's initial nonce, computed as $H(N_{\mathbf{I}})$.
g	Generator of a multiplicative group of order q.
g^x, g^y	Public key of the initiator and the responder.
HKR	Transient, hash key known only to the responder.
$H_K(\cdot)$	Keyed hash function (secure MAC) using key K.
$IP_{\mathbf{I}}$	Initiator's network address.
$\mathsf{ID}_{\mathbf{I}}, \mathsf{ID}_{\mathbf{R}}$	Information to identify $\mathbf{I}$ and $\mathbf{R}$ public keys.
$\mathsf{ID}_{\mathbf{R}}'$	Initiator indicates to the responder which authentication information (e.g. certificates) should be used.
$\mathsf{sa}, \mathsf{sa}'$	Information used to establish a security association.
$\mathsf{grpinfo}_{\mathbf{R}}$	Groups, algorithms and hash functions supported by the responder,.
$S_{k_{\mathbf{I}}}[\cdot]$	Digital signature computed by $\mathbf{I}$ using long-term secret key $k_{\mathbf{I}}$.
$S_{k_{\mathbf{R}}}[\cdot]$	Digital signature computed by $\mathbf{R}$ using long-term secret key $k_{\mathbf{R}}$.
$\{\cdot\}_{K_a}^{K_e}$	Encryption using key K_e and generating a message authentication code (MAC) over the resultant cipher text using key K_a.
k_{xy}	Session key computed using shared secret g^{xy}.

Fig. 1. JFKi protocol and its message components

The main technique for improving efficiency in the JFK protocols is for the participants to reuse their Diffie-Hellman (DH) exponentials (g^x and g^y) across multiple sessions with different parties. In the case of the responder, this technique drastically reduces the computational load by avoiding the exponentiation cost due to the generation of new g^y for each session and by allowing the signature $S_{k_{\mathbf{R}}}[g^y, \mathsf{grpinfo}_{\mathbf{R}}]$ to be kept constant as long as the DH exponential is reused. Unfortunately, this technique eliminates forward secrecy since y becomes part of the long-term key. Interestingly, JFK with ephemeral DH reuse achieves adaptive forward secrecy (AFS), the term defined by Aiello et al. [1] as below:

Definition 1 (AFS). *If an adversary knows the DH exponent during the time period t_{i+1} and if all the session keys generated during previous time periods $t_0, t_1, \ldots t_i$ remain secure, then the protocol is said to achieve AFS.*

2.1 Smith et al.'s Analysis of JFK in Meadows Framework

One of the design features of JFK is to delay the computational commitments at the responder until the initiator reveals the IP address. Smith et al. [21] argued that the protocols should resist DoS attacks given the ready availability of IP addresses and in the presence of the adversary willing to reveal the source IP address. A direct application of Meadows' cost-based framework to the JFK protocol captured the first attack. The second attack was identified when assessing DoS resistance of the protocol in the presence of coordinated attackers.

---

1. $\mathbf{I} \rightarrow \mathbf{R}$: compnonce1$(N_\mathbf{I})$, $N'_\mathbf{I} = \mathsf{hash1}(N_\mathbf{I})$, genexp1(g^x) ||
$\quad N'_\mathbf{I}, g^x$ || vergroup(g^x), accept1

2. $\mathbf{R} \rightarrow \mathbf{I}$: compnonce2$(N_\mathbf{R})$, token = genmac1$_\mathsf{HKR}(g^y, N_\mathbf{R}, N'_\mathbf{I}, IP_\mathbf{I})$ ||
$\quad N'_\mathbf{I}, N_\mathbf{R}, g^y$, grpinfo$_\mathbf{R}$, ID$_\mathbf{R}$, $sig1 = \mathsf{S}_{k_\mathbf{R}}[g^y, \mathsf{grpinfo_R}]$, token ||
$\quad$ versig1, accept2

3. $\mathbf{I} \rightarrow \mathbf{R}$: genDH1$(g^{xy})$, $K = \mathsf{compKeys1}(N'_\mathbf{I}, N_\mathbf{R}, g^{xy})$,
$\quad s = \mathsf{gensig2}(N'_\mathbf{I}, N_\mathbf{R}, g^x, g^y, \mathsf{ID_R}, \mathsf{sa})$, $C_1 = \mathsf{encrypt1}_K(\mathsf{ID_I}, s, \mathsf{sa})$,
$\quad C = \mathsf{genmac2}_K(C_1)$ || $N_\mathbf{I}$,$N_\mathbf{R}$, g^x, g^y, token, C_1, C ||
$\quad N'_\mathbf{I} = \mathsf{hash2}(N_\mathbf{I})$, ver1(token = genmac3$_\mathsf{HKR}(g^y, N_\mathbf{R}, N'_\mathbf{I}, IP_\mathbf{I})$),
$\quad$ genDH2(g^{xy}), $K = \mathsf{compKeys2}(N'_\mathbf{I}, N_\mathbf{R}, g^{xy})$,
$\quad$ ver2$(C = \mathsf{genmac4}_K(C_1))$, decrypt1$_K(C_1)$, versig2$(s)$, accept3

4. $\mathbf{R} \rightarrow \mathbf{I}$: $W = \mathsf{gensig3}(N'_\mathbf{I}, N_\mathbf{R}, g^x, g^y, \mathsf{ID_I}, \mathsf{sa}, \mathsf{sa'})$, $D_1 = \mathsf{encrypt2}_K(W, \mathsf{sa'})$,
$\quad D = \mathsf{genmac5}_K(D_1)$ || D_1, D || ver$(D = \mathsf{genmac6}_K(D_1))$,
$\quad$ decrypt2$_K(D_1)$, versig3(W), accept4

---

compnonce : selecting a random bit string. $\delta(\mathsf{compnonce}) = cheap$.
hash :   computing a hash value. $\delta(\mathsf{hash}) = cheap$.
genmac :   generating a MAC that is based on a keyed hash function.
$\quad\quad$ $\delta(\mathsf{generatemac}) = medium$.
gensig :   generating a digital signature. $\delta(\mathsf{gensig}) = expensive$.
genexp :   computing a DH ephemeral key. $\delta(\mathsf{genexp}) = expensive$.
genDH :   generating a shared secret. $\delta(\mathsf{genDH}) = expensive$.
versig :   verifying a digital signature. $\delta(\mathsf{versig}) = expensive$.
vergroup : verifying that a given DH value belongs to an acceptable group.
$\quad\quad$ $\delta(\mathsf{vergroup}) = medium$.
compKeys : computing the keys for encryption and authentication is based
$\quad\quad$ on a keyed hash function. $\delta(\mathsf{compKeys}) = medium$.
encrypt :   encrypt the message using symmetric key. $\delta(\mathsf{decrypt}) = medium$.
decrypt :   decrypt the message using symmetric key. $\delta(\mathsf{decrypt}) = medium$.

Fig. 2. Annotated Alice-and-Bob Specification of JFKi

Message Processing Costs in JFK. Meadows used the Alice-and-Bob specification style (for more details refer to Appendix A) to show how messages are processed during the protocol execution. In Figure 2, the annotated Alice-and-Bob specification of JFK protocol is provided [21]. The cost-based framework assigns cost to every action performed by the communicating principals. The cost for performing a particular action is defined to be an element in the set $S = \{cheap, medium, expensive\}$ [21]. The cost function δ (defined in [21]) associated with performing the following actions also appears in Figure 2.

The public-key validation (vergroup in Figure 2) should not be expensive to avoid DoS-attack. Since the adversary is capable of spoofing messages and signatures for which verifications result in failures we have $\delta(\text{spoof}) = cheap$.

2.2   A New DoS Vulnerability in JFK

The main motivation behind the reuse technique in JFK is to protect the responder from a certain type of DoS attack. However, the re-using technique introduces an attack we describe in this section. Smith et al. identified a second DoS attack on JFK under the assumption that both the initiator and the responder are reusing their ephemeral DH exponentials. However a similar type of DoS attack is possible in the case where the initiator does not reuse the ephemeral DH exponential while the responder reuses its ephemeral DH exponential.

To be precise, the initiator generates g^x for the first session and then computes $g^{2x}, g^{3x}, \ldots$ for further sessions using the previously computed DH values. In this way, the ephemeral DH values of the attackers looks different. On the other hand, the responder computes g^y once and reuses it for all sessions it participates during a time period. In this attack, the goal of the attackers is to trigger the responder to perform the expensive signature verification versig2.

The responder performs computationally expensive operations like genDH2 and versig2 only in step 3 of the protocol (Figure 2). To make the responder engage in signature verification versig2, an attacker must have performed medium to expensive operations such as genMac2, encrypt1, genexp1, genDH1 and gensig2.

Though a DoS attacker is capable of sending a spoof signature spoofgensig2 instead of gensig2, it must perform expensive computations such as genexp and genDH1 to convince the responder to perform up to versig2 computation. If an attacker can somehow decrease the cost of computing genexp and genDH1, then it can cause the responder to perform many expensive versig2 computations. Considering the presence of coordinated initiators, the cost for performing genexp and genDH1 computations can be amortised across all attackers resulting in many cheaply generated communications.

2.3   Cost Calculations

We follow Meadows' framework modified by Smith et al. [21] to demonstrate the impact of sharing Diffie-Hellman values by a group of coordinated initiators and analyse DoS susceptibility. For further details of the notations used in this section, refer to Appendix A.

Assume that there are n coordinated initiators/attackers who generate g^x and multiply g^x $n-1$ times to compute the remaining $n-1$ Diffie-Hellman values say $g^{2x}, g^{3x}, \ldots g^{nx}$. The coordinated initiators engage in protocol runs with the responder that is currently re-using its Diffie-Hellman value g^y for reducing its computational burden. The initiators compute shared secrets $g^{xy}, \ldots g^{nxy}$ in a similar way as above. As a result, the cost for computing genexp in message 1 and genDH in message 3 are amortised across all attackers. For larger values of n, the attackers cost function for sharing genexp denoted by $\phi(\text{shareexp})$ is:

$$\phi(\text{shareexp}) = \frac{1}{n} \times \phi(g^x) + (n-1) \times \phi(\text{modmul})$$
$$= \frac{1}{n} \times expensive + (n-1) \times cheap = cheap$$

as computing modular multiplication $\phi(\text{modmul})$ is *cheap*. The cost for sharing genDH is $\phi(\text{shareDH}) = cheap$.

Hence the attack cost function Θ for the attacker in constructing n valid-looking third messages to convince the responder to perform up to n decrypt1 operations and proceed with n versig2 operations is dominated by events of medium costs. Hence the attacker's total cost for performing a single decrypt1 operation in order to trigger the responder to proceed with versig2 is:

$$\Theta(\text{decrypt1}) = \phi(\text{shareDH}) + \delta(\text{compKeys1}_K) + \phi(\text{spoofsig})$$
$$+ \delta(\text{encrypt1}) + \delta(\text{genmac2}) = medium \quad (1)$$

On the other hand, to detect that the n signatures are spoofed, the responder does the following computation n times including hash2, genmac3, genDH2, decrypt1, and versig2. Hence a single message processing cost $\delta'(\text{versig2})$ for the responder is dominated by events of two expensive costs:

$$\delta'(\text{versig2}) = \delta(\text{hash2}) + \delta(\text{genmac3}) + \delta(\text{genmac4}) + \delta(\text{genDH2})$$
$$+ \delta(\text{compKeys2}) + \delta(\text{decrypt1}) + \delta(\text{versig2})$$
$$= expensive \quad (2)$$

From Equations (1) and (2), we see that $(\delta'(\text{versig2}), \Theta(\text{decrypt1}))$ which is equal to $(expensive, medium)$ does not belong to the tolerence relation $\mathcal{T}$ defined in [14] and in Appendix A; hence the protocol is vulnerable to a DoS attack.

2.4   Basic Security of the JFK Protocol

Aiello et al. [1] compared the basic JFK protocol with the ISO 9798-3 protocol. Since the ISO 9798-3 protocol is SK-secure in the UM model under the DDH assumption, they argued that the JFK protocol inherits the same security. The main design feature of the JFK protocol is that the responder can reuse its ephemeral DH key when it is under attack. If the security of the basic JFK (ISO 9798-3) protocol is analysed with the assumption that the responder reuses its ephemeral DH key, then its SK security may not be reduced to DDH assumption.

Moreover, Aiello et al. argued that JFK protocol with DH reuse achieves AFS security and identity protection under the combined H/DDH assumption (where H is a pseudo-random function) along with secure signature, encryption and MAC. Unfortunately, their assumptions are true only when they prove the security in a weakened model which considers only the presence of passive adversaries. In this case, JFK achieved AFS security only against passive adversaries. To prove JFK security against active adversaries, their assumptions are not enough as discussed above. We will see in section 3.4 that JFK can be shown secure using GDH assumption.

2.5   Resisting Other Type of DoS Attacks in JFK

Note that in JFK protocol, upon receiving third message, the responder first computes the shared secret g^{xy}, next verifies the MAC, and finally verifies the signature on the decrypted message. Therefore, as seen in the described DoS attack on JFK, if the responder is reusing its ephemeral DH key g^y, then attackers can compute valid shared secrets of the form g^{mxy} without performing exponentiations. With these efficiently generated keys they can trigger the responder perform two expensive operations, namely genDH2 and versig2 by sending fake signature with valid MACs. This attack could be thwarted if a responder can set a fresh ephemeral DH key in an efficient manner for each session which led the attacker to perform an exponentiation to compute the shared secret g^{xy} for each session in order to succeed in the attack.

Otherwise to avoid the exponentiation cost, the attacker can take another approach: do not compute the shared secrets and send bogus MACs in the third message. Here the attacker's goal is not to cause the responder perform up to versig2, but up to the expensive genDH2 operation. Unfortunately, JFK will be vulnerable to this DoS attack since the protocol requires the responder to compute the shared secret g^{xy} first, even to verify the MAC on the received message. Therefore in the JFK protocol the attacker, without computing the shared secret, can cause the responder to proceed up to the genDH2 operation. However it is easy to avoid such an attack by incorporating client puzzles in the protocol provided that the cost for computing the puzzle solution should be equal to or higher than the cost for performing genDH2 and versig2 operations. By verifying the puzzle solutions in an efficient manner the responder ensures that the cost for the initiator/attacker will be much higher than that of the responder at any stage of the protocol.

3   BPV-JFK

Although adding certain type of client puzzles with JFK (resulting in a protocol we call DoS-JFK) can thwart the above mentioned DoS attacks, we propose a new protocol, which we call BPV-JFK, for the following reasons. First, JFK with ephemeral key reuse can be proven under the GDH but not under the standard DDH assumption as claimed by the JFK designers; in this case, the security analysis will have to be in the random oracle model. Second, JFK with ephemeral key reuse does not achieve perfect forward secrecy, but gives AFS as per Definition 1. With BPV-JFK, we not only achieve the full security promised by the original JFK protocol but achieve perfect forward secrecy. Interestingly, by computing a fresh ephemeral DH value for each session in an efficient manner and by adding client puzzles to BPV-JFK, we avoid the above mentioned attacks.

3.1   BPV Generator

We use a technique due to Boyko et al. [5] that enables the responder to compute g^y with fewer modular multiplications (compared to a full exponentiation).

Definition 2 (BPV Generator). *Let p be a DSA modulus such that the prime q divides $p-1$. Select a random element g of order q in the multiplicative group $\mathbb{Z}_p^*$ For integer parameters $N \geq \ell \geq 1$, the BPV generator generates pairs of the form (i, g^i) as follows:*

BPV-Pre: Pre-processing. *Generate N random integers $x_1, x_2, \ldots x_N \in \mathbb{Z}_q$. Compute $X_i = g^{x_i} \mod n$ for each i and store the pair (x_i, X_i) in a table.*

BPV-Gen: Pair generation. *Whenever a pair (y, g^y) is needed, generate a random set $S \subseteq_R \{1, \ldots, N\}$ such that $|S| = \ell$. Compute $y = \sum_{j \in S} x_j \mod q$. If $y = 0$, stop and generate S again. Otherwise compute $g^y = \prod_{j \in S} g^{x_j} \mod n$ and return (y, g^y).*

In [16], Nguyen et al. presented the extended BPV generator (EBPV) which is exactly the BPV generator for certain parameter choices. Using the following theorem, they established the security of some discrete logarithm based signature schemes that use EBPV under adaptive chosen message attack . The theorem shows that for a fixed q, with overwhelming probability on the choice of x_i's, the distribution of the output y of the BPV generator is statistically close to the uniform distribution on $\mathbb{Z}_q$. Nguyen and Stern presented a similar result in [17]. In particular, a polynomial time adversary cannot distinguish the two distributions for appropriate choices of N and ℓ. Possible N and ℓ values are listed in Section 3.3.

Theorem 1. *Let q be a prime, and let $N \geq \ell \geq 1$. Then,*

$$\frac{1}{q^N} \sum_{\boldsymbol{x} \in \mathbb{Z}_q^N} \sum_{y \in \mathbb{Z}_q} \left| \Pr_{S \subseteq [1,N]:|S|=\ell} \left(\sum_{j \in S} x_j \equiv y \mod q \right) - \frac{1}{q} \right| \leq \sqrt{q / \binom{N}{\ell}} \qquad (3)$$

Moreover, there exists a $c > 0$, such that the following holds with probability at least $1 - 2^{-cN}$. If $x_1, \ldots, x_N$ are chosen independently and uniformly from $[0, q-1]$ and if $y = \sum_{j \in S} x_j \mod q$ is computed from a random set $S \subseteq [1, N]$ of ℓ elements, then the statistical distance between the distribution of y and the uniform distribution is bounded by $\sqrt{q / \binom{N}{\ell}}$.

Intuition on Theorem 1. Note that this result holds regardless of whether the pre-computed x_i's are known to a distinguisher or not. In other words, even if a distinguisher knew from the x_i's which elements of $\mathbb{Z}_q$ were more (or less) likely to be generated by the BPV generator, these elements are still generated only with negligibly more (or less) frequency compared to uniform. For example, the theorem implies that for appropriate choices of the N and ℓ values, the BPV generator outputs almost all the elements of $\mathbb{Z}_q$ and the proportion of elements not output by the BPV generator is very small.

To see this, fix random $x_1, \ldots, x_N$. Let z be the proportion of elements of $\mathbb{Z}_q$ which are output by the BPV generator for these x_i's. That is, for qz of the elements of $\mathbb{Z}_q$, there exists a subset of size ℓ of $\{x_1, \ldots, x_N\}$ that sum to that element, whereas for $q(1-z)$ of the elements no such subset exists. Suppose (for simplicity) that each one of the qz elements of $\mathbb{Z}_q$ that is output is output with equal frequency (namely, $1/qz$). Then consider inequality (3) and substitute:

$$\sqrt{q/\binom{N}{\ell}} \geq \frac{1}{q^N} \sum_{\boldsymbol{x} \in \mathbb{Z}_q^N} \sum_{y \in \mathbb{Z}_q} \left| \Pr\left(\sum_{j \in S : |S| = \ell} x_j \equiv y \mod q \right) - \frac{1}{q} \right|$$

$$= \frac{1}{q^N} \sum_{\boldsymbol{x} \in \mathbb{Z}_q^N} \left(qz \left| \frac{1}{qz} - \frac{1}{q} \right| + q(1-z) \left| 0 - \frac{1}{q} \right| \right) = 2(1-z)$$

$$\Rightarrow z \geq 1 - \frac{1}{2}\sqrt{q/\binom{N}{\ell}}$$

For appropriate choices of N and ℓ, $\sqrt{q/\binom{N}{\ell}}$ can be made negligibly small, so z can be made very close to 1. In other words, the proportion of elements of $\mathbb{Z}_q$ that are output by the BPV generator is very close to 1. Hence the theorem implies that the BPV generator outputs almost all the elements of $\mathbb{Z}_q$ with almost same probability.

3.2   The BPV-JFK Protocol and Its Security Analysis

We now replace the reuse technique in the JFK protocol with the BPV generator to generate new ephemeral DH value g^y for each session efficiently. We denote the resulting protocol, depicted in Figure 3, as BPV-JFK. The rest of the security features, such as inclusion of identity protection, addition of encryption to protocol and anti-replay cache, follow directly from the JFK design [1].

Aiello et al. [1] analysed JFKi protocol in two phases. In the first phase, Aiello et al. analysed the basic cryptographic core of the JFKi protocol and in the second phase, they analysed the additional security features implemented in the protocol. Since our BPV-JFK protocol is similar to the JFKi protocol, we follow the approach of Aiello et al. and separate the analysis of the basic cryptographic core of the BPV-JFK protocol from the analysis of the complete BPV-JFK protocol which has additional security features. As the security features are already well analysed by Aiello et al., we do not repeat their security analysis

Initiator I	**Responder R**
	Run BPV pre-processing step to generate the N pairs (x_i, X_i) Long term secrets: (x_i, X_i)
random $x, N_{\mathbf{I}}$ compute $N_{\mathbf{I}}', g^x$ $\xrightarrow{\quad g^x, N_{\mathbf{I}}', \mathsf{ID}_{\mathbf{R}}' \quad}$	random $N_{\mathbf{R}}$ Run BPV pair generation step to compute (y, g^y)
$\xleftarrow{\begin{array}{c} g^y, G, \\ N_{\mathbf{I}}', N_{\mathbf{R}}, \mathsf{ID}_{\mathbf{R}}, \mathsf{token} \end{array}}$	store y, compute token
compute $SH = H(g^{xy}, \mathbf{I}, \mathbf{R}), F,$ $\xrightarrow{\; N_{\mathbf{I}}, N_{\mathbf{R}}, g^x, g^y, F, \mathsf{token} \;}$ session key K_{xy}	compute $N_{\mathbf{I}}'$, verify token generate $SH = H(g^{xy}, \mathbf{I}, \mathbf{R})$, verify F
verify sig $\xleftarrow{\; N_{\mathbf{I}}, N_{\mathbf{R}}, g^x, g^y, \mathsf{sa}, sig \;}$	generate sig, session key K_{xy}

token $= \mathsf{MAC}_{\mathsf{HKR}}(g^y, N_{\mathbf{R}}, N_{\mathbf{I}}', IP_{\mathbf{I}})$, $K_e = H_{SH}(N_{\mathbf{I}}', N_{\mathbf{R}}, 1)$, $K_a = H_{SH}(N_{\mathbf{I}}', N_{\mathbf{R}}, 2)$,
$K_{xy} = H_{SH}(N_{\mathbf{I}}', N_{\mathbf{R}}, 0)$, $G = \mathsf{grpinfo}_{\mathbf{R}}$, $F = \{\mathsf{ID}_{\mathbf{I}}, \mathsf{sa}, \mathsf{S}_{k_{\mathbf{I}}}[N_{\mathbf{I}}', N_{\mathbf{R}}, g^x, g^y, ID_{\mathbf{R}}, \mathsf{sa}]\}_{K_a}^{K_e}$,
$sig = \{\mathsf{S}_{k_{\mathbf{R}}}[N_{\mathbf{I}}', N_{\mathbf{R}}, g^x, g^y, \mathsf{ID}_{\mathbf{I}}, \mathsf{sa}, \mathsf{sa}'], \mathsf{sa}'\}_{K_a}^{K_e}$

Fig. 3. BPV-JFK protocol

and analyse only the basic BPV-JFK protocol. In this section, we first analyse the security of the basic BPV-JFK protocol in Canetti-Krawzyck model [6] and show that the basic BPV-JFK is SK-secure in the unauthenticated link model (UM). Refer to [6] for a detailed analysis of the CK01 model.

To resist other type of DoS attacks described in Section 2.5, we follow the approach of Stebila et al. [23] to turn BPV-JFK into a DoS-resistant protocol. We call the resultant protocol DoS-BPV-JFK and show in Section 3.6 that DoS-BPV-JFK in Figure 5 satisfies the definition of Stebila et al..

Basic BPV-JFK protocol and its Security Analysis. The basic BPV-JFK protocol is depicted in Figure 4. The responder computes y and g^y using the ℓ random pairs (x_i, X_i) as in the BPV pair generation step. We prove in the following theorems that the basic BPV-JFK is SK-secure in the UM with PFS if the DDH assumption holds.

Definition 3 (Decisional Diffie-Hellman (DDH) assumption). *Let k be a security parameter. Let g be an element of order q in $\mathbb{Z}_p^*$ for primes p and q such that $q|p - 1$ Then the probability distributions of $\mathcal{D}_0 = \{\langle p, g, g^x, g^y, g^{xy}\rangle : x, y \leftarrow_R \mathbb{Z}_q\}$ and $\mathcal{D}_1 = \{\langle p, g, g^x, g^y, g^z\rangle : x, y, z \leftarrow_R \mathbb{Z}_q\}$ are computationally indistinguishable.*

Remark 1. Note that the responder requires only the value y to compute the shared secret g^{xy} in future. Hence it stores only y for each session but not the ℓ

---

1. $\mathbf{I} \to \mathbf{R} : \mathsf{N_I}, g^x, \mathsf{ID_I}$
2. $\mathbf{R} \to \mathbf{I} : \mathsf{N_I}, \mathsf{N_R}, g^y, S_{k_R}[\mathsf{N_I}, \mathsf{N_R}, g^x, g^y, \mathsf{ID_R}]$
3. $\mathbf{I} \to \mathbf{R} : \mathsf{N_I}, \mathsf{N_R}, S_{k_I}[\mathsf{N_I}, \mathsf{N_R}, g^x, g^y, \mathsf{ID_I}]$
shared session secret : $\sigma = g^{xy}$

Fig. 4. Basic BPV-JFK protocol

pairs (x_i, X_i) as they are already stored in the long-term table. Hence the session state reveal or session corruption query outputs the value y and session key (if it is computed) that are stored in the responder memory.

Theorem 2 (SK-security of BPV-JFK in AM). *If G is a group where the DDH assumption holds, then the basic BPV-JFK protocol (without signatures) is SK-secure in the authenticated links model (AM).*

Proof. During the protocol execution, if both the uncorrupted parties P_i and P_j complete the protocol, then they compute the same shared secret g^{xy}. Hence the first requirement of SK-security definition, namely correctness (refer to [6]) is satisfied.

Let $\mathcal{A}$ be an adversary against basic BPV-JFK protocol. Using $\mathcal{A}$, we can construct an adversary $\mathcal{B}$ that can distinguish between $\mathcal{D}_0$ and $\mathcal{D}_1$. The distinguishuer $\mathcal{B}$ is given a DDH challenge tuple (p, g, g^u, g^v, g^w) chosen from $\mathcal{D}_0$ or $\mathcal{D}_1$ with probability $1/2$ as input.

Assume that $\mathcal{B}$ creates an AKE experiment that involves n honest parties $P_i, i = 1, \ldots, n$ and the adversary $\mathcal{A}$. Each party can be activated to participate in at most s AKE sessions. Among all n parties, $\mathcal{B}$ randomly selects one party, call it P_j. Assume that P_j plays the role of the responder in all its sessions. For all parties activated as responders, $\mathcal{B}$ runs the BPV pre-processing step to compute the corresponding long-term secret pairs of $(x_i, X_i), i = 1 \ldots N$ as described in the protocol. Also $\mathcal{B}$ selects a random session denoted by s^* among the sessions where P_j plays the role of the responder. Let P_i be the initiator to the session s^* which could be either at P_i or P_j. Except for session s^*, $\mathcal{B}$ executes all session establishments by following the protocol specification.

When $\mathcal{A}$ corrupts a party other than P_i and P_j, $\mathcal{B}$ submits the secrets possessed by the party. If $\mathcal{A}$ issues a corrupt query to either of the parties P_i and P_j then $\mathcal{B}$ declares failure.

Assume that a session $s' \neq s^*$ is activated between two uncorrupted parties P_k (initiator) and P_j. In this case $\mathcal{B}$ can compute the shared session key (since the ephemeral DH public keys of P_k and P_j are set by $\mathcal{B}$) of s'. If s' is corrupted by $\mathcal{A}$, then the adversary $\mathcal{B}$ submits the session key to $\mathcal{A}$. Suppose that a session $s'' \neq s^*$ is activated between the adversary controlled initiator P_m (initiator) and the responder P_j. Now $\mathcal{B}$ can compute the session key g^{xy} even if it does not know x.

For the test session s^*, $\mathcal{B}$ assigns the DDH challenge values g^u and g^v in the ephemeral public keys of parties P_i and P_j respectively. $\mathcal{B}$ declares failure and stops the experiment if $\mathcal{A}$ tries to corrupt the party P_i or P_j and the session

s^*. For the test query issued by $\mathcal{A}$, $\mathcal{B}$ submits g^w to the adversary. $\mathcal{A}$ continues with the experiment even after the test query, but it is not allowed to expose the test session. At the end of the experiment $\mathcal{A}$ stops and returns a bit b' as its guess.

Analysis. Provided that parties the P_j or P_j or the session s^* are not corrupted, the AKE experiment simulated by $\mathcal{B}$ is perfect except with negligible probability. With probability $1/n$, $\mathcal{B}$ selects P_j as the responder of the session s^*. The event that $\mathcal{A}$ selects s^* as test session happens with probability at least $1/s$. For the test session, the ephemeral public keys of the parties are not generated using BPV pair generation step. By Theorem 1, the statistical distance between the computed y and ephemeral secret of the assigned challenge v is bounded by $\sqrt{q/\binom{N}{\ell}}$.

The adversary $\mathcal{A}$ was provided with g^w as answer to the test query. In this case, if the challenge tuple belongs to the distribution $\mathcal{D}_0$, then the actual session key was provided to $\mathcal{A}$. Otherwise, if the tuple belongs to the distribution $\mathcal{D}_1$, then the response was a random key. Whenever $\mathcal{A}$ wins the experiment with probability $1/2 + \epsilon$ for non-negligible ϵ, the adversary $\mathcal{B}$ also distinguishes between the two distributions $\mathcal{D}_0$ and $\mathcal{D}_1$ with probability $1/2 + \epsilon/ns + \sqrt{q/\binom{N}{\ell}}$ since the adversary chooses s^* as test session and P_j as the responder with probability $1/ns$.   □

Theorem 3 (SK-security of BPV-JFK in UM). *The basic BPV-JFK protocol is SK-secure in the unauthenticated links model (UM).*

Proof. We showed in Theorem 2 that the basic BPV-JFK protocol without signatures is SK-secure in the AM. Then the proof follows by Theorem 6 in [6] which states that any protocol which is SK-secure in the AM can be transformed in to a SK-secure protocol in the UM by including authenticators such as digital signatures.   □

Perfect Forward Secrecy (PFS). Except for the generation of g^y the basic BPV-JFK protocol in Figure 3 is the same as the ISO 9798-3 protocol (the basic JFK protocol without g^y reuse). In the basic BPV-JFK protocol, the BPV generator is used to compute g^y whereas in the ISO 9798-3 protocol a random y is used to compute g^y.

The ISO 9798-3 protocol is SK-secure [6] in the UM model with PFS if the signature scheme is secure, the pseudo-random function H is secure and the DDH assumption holds. We now provide a proof sketch using game hopping technique that the basic BPV-JFK is SK-secure in the UM model with PFS.

Let E_i be the event that the adversary wins in game G_i. Assume that the long-term secret N pairs (x_i, X_i) of the responder are revealed. Assume that G_0 is the game in which the BPV generator is used to compute g^y in the BPV-JFK protocol. Now we define game G_1 which is same as game G_0 except that a random y is used to compute g^y. An adversary that can distinguish G_0 from G_1

is effectively a distinguisher for the BPV generator in Theorem 1. Note that, whereas the proof of security in Theorem 2 did not have the x_i revealed to the adversary, they are revealed to the adversary in the PFS argument. Nonetheless, Theorem 1 still applies, and the games are statistically indistinguishable: $|\Pr(E_0) - \Pr(E_1)| \leq \sqrt{q/\binom{N}{\ell}}$, which is negligible for appropriate choices of N and ℓ.

Now, in G_1, the modified BPV-JFK protocol is same as the ISO 9798-3 protocol with the pre-computed N pairs which are independent of the different g^y values used to compute the session keys. If the adversary wins game G_1, then the challenger breaks the PFS security of the ISO protocol. The ISO protocol was proven to have PFS [6], so $\Pr(E_1)$ is negligible, and hence $\Pr(E_0) \leq$ $\mathrm{negl}(k) + \sqrt{q/\binom{N}{\ell}}$.

Remark 2. In the BPV-JFK protocol, the responder has to store the value y for each session. This small amount of storage for each session is not a resource constraint since a very small fraction (less than one one-millionth, say) of the responder's memory will be used to store such values when it executes thousands of key exchange sessions per second. Since the BPV-JFK protocol focus on minimizing the computational based expenditure to the responder, the amount of storage needed for each session is considered to be minimal.

3.3   Efficiency Comparison and Parameter Sizes

Table 2 compares the general efficiency of BPV-JFK with the JFK protocol, both when JFK reuses the ephemeral public keys and when it does not.

Table 2. Efficiency comparison of JFK and BPV-JFK

Protocol	Efficiency technique	Server operations in message 1	Perfect Forward Secrecy
JFK with no reuse	None	1 mod. exp.	Yes
JFK with reuse	Reuse of g^y	None	No
BPV-JFK	BPV generator	$\ell - 1$ mod. add., $\ell - 1$ mod. mul.	Yes

The specific efficiency of the BPV-JFK protocol depends on the number of elements ℓ in the random set S the responder choose to compute (y, g^y) in the BPV pair generation step. The responder may prefer to reduce the number of online modular multiplications required for generating g^y. Hence it might be appropriate for the responder to choose a bigger value of N (polynomial in $\log q$) to make ℓ smaller. Values for N and ℓ for a 160-bit q and their corresponding BPV-Pre and BPV-Gen running time are presented in Table 3. We ran the experiment on a single core of a 3.06 GHz Intel Core i3 with 4GB RAM, compiled using gcc -02 with architecture x86_64. The big integer arithmetic from OpenSSL 0.9.8r is used to implement the software.

Table 3. Parameter sizes, security bound for Theorem 1, and runtime for BPV generator with 1024-bit modulus and 160-bit exponent. For comparison, a single 160-bit modular exponentiation takes 0.461 ms.

N	ℓ	$\sqrt{q/\binom{N}{\ell}}$	Runtime	
			BPV-Pre (s)	BPV-Gen (ms)
$2^{11} = 2048$	48	2^{-82}	0.939	0.226
$2^{12} = 4096$	40	2^{-80}	1.892	0.196
$2^{13} = 8192$	35	2^{-81}	3.758	0.168
$2^{14} = 16384$	31	2^{-81}	7.527	0.156
$2^{16} = 65536$	26	2^{-83}	30.148	0.134

Our experimental results verify that computing g^y using the BPV-Pre and BPV-Gen algorithms is faster than performing a 160-bit modular exponentiation to compute g^y from a random y (which requires 0.461 ms). The advantage factor of BPV generation over modular exponentiation based on the parameter values listed in Table 3 is between 2 and 3.4. For example, if we choose $N = 2^{14}$ and $\ell = 31$, then g^y can be computed 3 times faster using the BPV generator.

3.4 On the Security of JFK

If the security of the basic JFK (ISO 9798-3) protocol is analysed with the assumption that the responder reuses its ephemeral DH key, then the security might not be reduced to DDH assumption for the following reason: consider the scenario as in session s'' of Theorem 2, where $\mathcal{A}$ controls the initiator P_m and activates the session s'' with responder P_j. Since the responder reuses the ephemeral DH key g^y for several sessions, the adversary $\mathcal{B}$ assigns one of the DDH challenge g^v in the place of g^y. If a session key reveal query is issued by $\mathcal{A}$, then $\mathcal{B}$ cannot compute the session key as it does not know both x and y.

The adversary $\mathcal{B}$ simulates the environment perfectly only if it can respond to this query consistently. Here the DDH assumption itself is not enough and therefore the security proof for the basic JFK with reuse technique may require the GDH assumption and the session key is set as $H(g^{xy}, \mathbf{I}, \mathbf{R})$. If H is a random oracle and the adversary has access to DDH oracle, then the session key reveal queries issued for session s'' could be answered correctly. Hence if H is a random oracle and $\mathbb{G}$ is a group where GDH assumption holds, then the basic JFK protocol (with g^y reuse) is SK-secure in the UM model.

3.5 Cost Calculations for BPV-JFK

Now, we calculate the cost functions for BPV-JFK protocol and show that the protocol is resistant to the DoS-attack described in Section 2.2. From Equations (1) and (2) it is clear that the attacker cost function for mounting a coordinated DoS-attack on JFK and the responder cost function to process the fake message does not satisfy the tolerance relation $\mathcal{T}$. In this section, we calculate the cost functions for both the attacker and the responder and show that our BPV-JFK protocol satisfies the tolerance relation and hence resistant to the attack.

Here the n coordinated initiators/attackers can compute the ephemeral DH values as $g^x, g^{2x}, g^{3x}, \ldots g^{nx}$ and can engage in protocol runs with the responder that uses the BPV generator to generate a new g^y for each session. Since g^y is unique for each session, the n shared secrets are independent of each other. That is for n different $g^{y_i}, i = 1, \ldots, n$, the attacker has to perform n exponentiations to compute the n shared secrets $g^{xy_1}, \ldots, g^{nxy_n}$. As a result, the cost for computing genexp in message 1 could be amortised across all attackers but not for the computation of genDH in message 3.

For large values of n, $\phi(\mathsf{shareexp})$, the attacker's cost for sharing genexp, is *cheap*. Hence the attack cost function Θ for the attacker in constructing n valid-looking third messages is dominated by events of expensive cost since the attacker must compute new shared secret for each session. Hence the attacker's total cost is $\Theta(\mathsf{decrypt1}) = expensive$. However, to process a single message $\delta'(\mathsf{versig2})$, the operations performed by the responder are dominated by events of two expensive costs. Hence, $\delta'(\mathsf{versig2}) = expensive$. Therefore, we see that $(\delta'(\mathsf{versig2}), \Theta(\mathsf{decrypt1})) = (expensive, expensive)$ which belongs to $\mathcal{T}$ and hence the protocol is resistant to the DoS attack.

3.6   Analysing the BPV-JFK Protocol in the Stebila et al. Model

Stebila et al. [23] gave a generic technique to transform any protocol into a DoS resistant protocol. Their technique uses strongly difficult interactive client puzzles as a DoS countermeasure and message authentication codes (MAC) for integrity of stateless connections. The server in the protocol must not perform any expensive operation until it verifies the MAC and the puzzle solution. Refer to [23] for a detailed description of the model.

In this section we follow the approach of Stebila et al. to turn BPV-JFK into a DoS-resistant protocol. We combine a strongly difficult interactive client puzzle with the BPV-JFK protocol and show that the resultant protocol, denoted DoS-BPV-JFK in Figure 5, satisfies the definition of Stebila et al.

DoS-BPV-JFK Protocol Specification. Let us assume Puz —consisting of Setup, GenPuz, FindSoln, and VerSoln algorithms — is a strongly difficult client puzzle [23]. Let DoS-BPV-JFK be the protocol consisting of the following algorithms:

- GlobalSetup(1^l): Set ρSpace $\leftarrow \{0,1\}^\ell$ and NonceSpace $\leftarrow \{0,1\}^\ell$.
- ServerSetup($\hat{S} \in$ Servers): Set HKR $\leftarrow_r \{0,1\}^\ell$ and $\rho_{\hat{S}} \leftarrow$ HKR
- CActionj$_{\mathsf{DoS-BPV-JFK}}(\cdots)$, SActionj$_{\mathsf{DoS-BPV-JFK}}(\cdots)$: As specified by the protocol in Figure 5.

Theorem 4. *Let* Puz *be an* $\epsilon_{\ell,Q,n}(t)$*-strongly difficult interactive puzzle and that MAC is a family of secure message authentication codes. Then DoS-BPV-JFK is an* $\epsilon'_{\ell,Q,n}(t)$*-denial-of-service-resistant protocol, for* $\epsilon'_{\ell,Q,n}(t) = \epsilon_{\ell,Q,n}(t +$

Client $\hat{C}$	Server $\hat{S}$
	long-term secrets: $\rho_{\hat{S}} = \mathsf{HKR}$,
	$((x_i, X_i))_{i=1}^N \leftarrow \mathsf{BPVPre}(g, n, N, q)$

Client $\hat{C}$		Server $\hat{S}$
CAction$1_{\mathsf{DoS-BPV-JFK}}$:		
1. $N_{\mathbf{I}} \leftarrow_r \mathsf{NonceSpace}$		
2. compute $N_{\mathbf{I}}'$		
3. CAction$1_{\mathsf{BPV-JFK}}$		
4. compute g^x	$\xrightarrow{\quad g^x, N_{\mathbf{I}}', \mathsf{ID}_{\mathbf{R}}' \quad}$	SAction$1_{\mathsf{DoS-BPV-JFK}}$:
5.		$N_{\mathbf{R}} \leftarrow_r \mathsf{NonceSpace}$
6.		$str \leftarrow (\hat{C}, \hat{S}, N_{\mathbf{I}}', N_{\mathbf{R}})$
7.		$puz \leftarrow GenPuz(Q, str)$
8.		SAction$1_{\mathsf{BPV-JFK}}$
9.		$(y, g^y) \leftarrow \mathsf{BPVGen}(g, n, \ell, q, (x_i, X_i))_{i=1}^N)$
10.	$\xleftarrow{\quad g^y, puz, \quad}$ $N_{\mathbf{I}}', N_{\mathbf{R}}, \mathsf{ID}_{\mathbf{R}}, G, \lambda$	$\lambda = \mathsf{MAC}_{\mathsf{HKR}}(g^y, N_{\mathbf{R}}, N_{\mathbf{I}}', IP_{\mathbf{I}}, str, puz)$
11. CAction$2_{\mathsf{DoS-BPV-JFK}}$:		store y
12. soln$\leftarrow$ FindSoln(str, puz)		
13. CAction$2_{\mathsf{BPV-JFK}}$:		
14. compute		
15. $SH = H(g^{xy}, \mathbf{I}, \mathbf{R}), F$	$\xrightarrow{\quad N_{\mathbf{I}}, N_{\mathbf{R}}, g^x, g^y, F, \quad}$ $puz, soln, str, \lambda$	
		SAction$2_{\mathsf{DoS-BPV-JFK}}$:
16.		compute $N_{\mathbf{I}}'$, reject if
17.		$\lambda \neq \mathsf{MAC}_{\mathsf{HKR}}(g^y, N_{\mathbf{R}}, N_{\mathbf{I}}', IP_{\mathbf{I}}, str, puz)$
18.		reject if $\neg$ VerSoln(str, puz, soln)
19.		$\tau \leftarrow (N_{\mathbf{I}}, N_{\mathbf{R}}, puz, soln)$
20.		verify no existing presession $[\hat{C}, \hat{S}, \tau]$
21.		accept and store presession $[\hat{C}, \hat{S}, \tau]$
22.		SAction$2_{\mathsf{BPV-JFK}}$
23.		generate $SH = H(g^{xy}, \mathbf{I}, \mathbf{R})$, verify F
24. verify sig	$\xleftarrow{\quad N_{\mathbf{I}}, N_{\mathbf{R}}, g^x, g^y, sa, sig \quad}$	generate sig

Fig. 5. DoS-BPV-JFK protocol

$cq) + negl(\ell)$, where c is a constant and q is the number of Send queries issued, assuming $t \in poly(\ell)$.

Proof. Since Puz is an $\epsilon_{\ell,Q,n}(t)$-strongly difficult interactive puzzle and MAC is a secure message authentication code, the proof for condition 1 of the DoS definition of Stebila et al. in [23] directly follows from Theorem 3 in [23]. It remains to show that SAction$1_{\mathsf{BPV-JFK}}$ does not involve any expensive operation. It is clear from Figure 5 that SAction$1_{\mathsf{BPV-JFK}}$ involves only a small number of modular additions and modular multiplications. $\qquad\square$

4   Conclusion and Future Work

Denial of service attacks are a growing concern to open networks. They may arise in a number of ways; in this work we focused on resource exhaustion DoS attacks (on network protocols) which cause a server to perform many expensive operations. We identified such an attack on the JFK protocol. Though there are many other security features offered by the JFK protocol, it achieves only adaptive forward secrecy. To achieve perfect forward secrecy and to resist the identified attack, we propose to use a technique introduced by Boyko et al. in place of ephemeral key reuse and show that the new protocol is SK-secure in CK01 model under the DDH assumption. The resultant protocol can easily be shown to be DoS-resistant after incorporating client puzzles and secure MACs. While there are many interesting features offered by the resultant protocol, it would be interesting to see if the proposed techniques can be embedded in any other DH based key exchange protocols.

Acknowledgements. The authors are grateful to Dr. Berkant Ustaoglu for his critical comments and to anonymous referees for helpful comments. This work is supported by Australia-India Strategic Research Fund project TA020002.

References

1. Aiello, W., Bellovin, S.M., Blaze, M., Canetti, R., Ioannidis, J., Keromytis, A.D., Reingold, O.: Just Fast Keying: Key agreement in a hostile Internet. ACM Transactions on Information and System Security 7(2), 1–30 (2004)
2. Aura, T., Nikander, P.: Stateless Connections. In: Han, Y., Quing, S. (eds.) ICICS 1997. LNCS, vol. 1334, pp. 87–97. Springer, Heidelberg (1997)
3. Aura, T., Nikander, P., Leiwo, J.: DOS-Resistant Authentication with Client Puzzles. In: Christianson, B., Crispo, B., Malcolm, J.A., Roe, M. (eds.) Security Protocols 2000. LNCS, vol. 2133, pp. 170–177. Springer, Heidelberg (2001)
4. Bellare, M., Rogaway, P.: Entity Authentication and Key Distribution. In: Stinson, D.R. (ed.) CRYPTO 1993. LNCS, vol. 773, pp. 232–249. Springer, Heidelberg (1994)
5. Boyko, V., Peinado, M., Venkatesan, R.: Speeding up Discrete Log and Factoring Based Schemes via Precomputations. In: Nyberg, K. (ed.) EUROCRYPT 1998. LNCS, vol. 1403, pp. 221–235. Springer, Heidelberg (1998)
6. Canetti, R., Krawczyk, H.: Analysis of Key-Exchange Protocols and Their Use for Building Secure Channels. In: Pfitzmann, B. (ed.) EUROCRYPT 2001. LNCS, vol. 2045, pp. 453–474. Springer, Heidelberg (2001)
7. Canetti, R., Krawczyk, H.: Security Analysis of IKE's Signature-Based Key-Exchange Protocol. In: Yung, M. (ed.) CRYPTO 2002. LNCS, vol. 2442, pp. 143–161. Springer, Heidelberg (2002)
8. Castelluccia, C., Mykletun, E., Tsudik, G.: Improving secure server performance by re-balancing SSL/TLS handshakes. In: Lin, F., Lee, D., Lin, B., Shieh, S., Jajodia, S. (eds.) ASIACCS 2006, pp. 26–34. ACM (2006)
9. Gong, L., Syverson, P.: Fail-Stop Protocols: An Approach to Designing Secure Protocols. In: Iyer, R.K., Morganti, M., Glogor, V., Fuchs, W.K. (eds.) Proc. Dependable Computing for Critical Applications, pp. 44–55. IEEE Computer Society (1998)
10. Harkins, D., Carrel, D., et al.: The Internet Key Exchange, IKE (1998), http://www.ietf.org/rfc/rfc2409

11. Juels, A., Brainard, J.: Client puzzles: A cryptographic countermeasure against connection depletion attacks. In: NDSS 1999, pp. 151–165. Internet Society (1999)
12. Kaufman, C.: Internet Key Exchange (IKEv2) protocol, RFC 4306 (December 2005)
13. Matsuura, K., Imai, H.: Modification of Internet Key Exchange resistant against Denial-of-Service. In: Pre-Proceeding of Internet Workshop (IWS) 2000, pp. 167–174 (February 2000)
14. Meadows, C.: A formal framework and evaluation method for network denial of service. In: CSFW 1999. IEEE (1999)
15. Moskowitz, R., Nikander, P., Jokela, P., Henderson, T.R.: Host Identity Protocol, RFC 5201 (April 2008)
16. Nguyen, P., Shparlinski, I., Stern, J.: Distribution of modular sums and the security of the server aided exponentiation. In: Proc. Workshop on Cryptography and Computational Number Theory (CCNT 1999), pp. 257–268. Birkhäuser (2001)
17. Nguyên, P.Q., Stern, J.: The Hardness of the Hidden Subset Sum Problem and Its Cryptographic Implications. In: Wiener, M. (ed.) CRYPTO 1999. LNCS, vol. 1666, pp. 31–46. Springer, Heidelberg (1999)
18. Okamoto, T., Tanaka, K., Uchiyama, S.: Quantum Public-Key Cryptosystems. In: Bellare, M. (ed.) CRYPTO 2000. LNCS, vol. 1880, pp. 147–165. Springer, Heidelberg (2000)
19. Rangasamy, J., Stebila, D., Boyd, C., González Nieto, J.: An integrated approach to cryptographic mitigation of denial-of-service attacks. In: Sandhu, R., Wong, D.S. (eds.) ASIACCS 2011, pp. 114–123. ACM (2011)
20. Rivest, R.L., Shamir, A., Wagner, D.A.: Time-lock puzzles and timed-release crypto. Technical Report TR-684, MIT Laboratory for Computer Science (March 1996)
21. Smith, J., González Nieto, J., Boyd, C.: Modelling denial of service attacks on JFK with Meadows's cost-based framework. In: Buyya, R., Ma, T., Safavi-Naini, R., Steketee, C., Susilo, W. (eds.) AISW-NetSec 2006. CRPIT, vol. 54, pp. 125–134. Australian Computer Society (2006)
22. Stebila, D., Ustaoglu, B.: Towards Denial-of-Service-Resilient Key Agreement Protocols. In: Boyd, C., González Nieto, J. (eds.) ACISP 2009. LNCS, vol. 5594, pp. 389–406. Springer, Heidelberg (2009)
23. Stebila, D., Kuppusamy, L., Rangasamy, J., Boyd, C., Gonzalez Nieto, J.: Stronger Difficulty Notions for Client Puzzles and Denial-of-Service-Resistant Protocols. In: Kiayias, A. (ed.) CT-RSA 2011. LNCS, vol. 6558, pp. 284–301. Springer, Heidelberg (2011)

A   Meadows' Cost-Based Framework [14]

Meadows [14] developed a cost-based framework for analysing vulnerabilities of DoS attacks in protocols. The model assigns cost to every action performed by the communicating principals (initiator and responder) to compare the amount of resources expended on both initiator and responder sides. In Meadows framework, a protocol is resistant to DoS attacks if the cost for an attacker to successfully interrupt the protocol at any stage exceeds the given threshold.

A.1   Notation

Meadows used the Alice-and-Bob specification style to specify the protocols for the framework to calculate the cost associated with every action that includes generating, verifying and accepting message. The following definitions appears in [14] and [21].

Definition 4. *An* Alice-and-Bob specification *is a sequence of statements of the form $A \rightarrow B : M$ where M represents the message is sent from A to B.*

The method of annotating an Alice-and-Bob specification to include message processing steps needed at both the communicating parties is described below.

Definition 5. *An* annotated Alice-and-Bob specification *is a sequence of statements of the form*
$A \rightarrow B : T_1, ..., T_k||M||O_1, ..., O_n$. *The operations performed by A to produce message M is represented by the ordered sequence T_i. Similarly, the ordered sequence O_j represents the operations performed by B to process and validate message M.*

Normally the events correspond to a sequence of generating, sending, receiving, validating and accepting a message. The different types of events are: (1) *normal,* (2) *verification,* and (3) *accept.*

A.2   Cost Sets and Cost Functions

The following definitions are used to compute the costs for participating in a protocol. Usually this is done by calculating the costs of individual events and summing them for each of the steps in a protocol.

Definition 6. *A cost set C is a monoid with the monoid operation $+$ and partial order $<$ such that $x \leq x + y$ and $y \leq x + y$ for all x, y in C*

Definition 7. *Assume that the events are defined in an Alice-and-Bob specification. A function δ that maps events to a cost set C and maps events to 0 on accept events is called an* event cost function.

Definition 8. *Let $\mathcal{P}$ be an annotated Alice-and-Bob specification protocol. The message processing cost function δ' associated with the event cost function δ on verification events following receipt of a message is as follows: If a line $A \rightarrow B : T_1, ..., T_k||M||V_1, ..., V_n$ appears in $\mathcal{P}$ an annotated Alice-and-Bob specification of a protocol, then for each verification event $V_j : \delta'(Vj) = \delta(V_1) + ... + \delta(V_j)$.*

We now define the costs to a responder for engaging in the protocol.

Definition 9. *Let $\mathcal{P}$ be an annotated Alice-and-Bob specification protocol. The protocol engagement cost function Δ associated with δ defined on accept events is as follows: If the line $A \rightarrow B : T_1, ..., T_k||M||V_1, ..., V_n$ appears in $\mathcal{P}$, and V_n is an accept event, then the protocol engagement cost $\Delta(V_n)$ is defined as the sum of the costs of all events occurring at B up to the accept event V_n plus cost of the event T_i.*

A DoS attacker has the capability to send bogus messages or to engage in bogus protocol runs by spoofing the IP address and hence the attacker is not restricted to the same events as a legitimate protocol participant. Therefore, it is necessary to define an independent set of cost functions to define the cost for actions performed by the attacker.

Definition 10. *Let us denote the attacker cost set by $\mathcal{G}$ and the set of actions performed by the attacker by $\mathcal{I}$. Let ϕ be the function that maps the attacker's actions to their cost set $\mathcal{G}$. Then the cost function denoted by Φ is defined on a sequence of actions performed by the attacker as, $\Phi(i_1, ..., i_n) = \phi(i_1) + ... + \phi(i_n)$ for $i_k \in \mathcal{I}$*

Definition 11. *Let Θ be the attacker cost function that maps events in $\mathcal{P}$ to a cost set $\mathcal{C}$. The protocol $\mathcal{P}$ is said to be* fail stop *if for any event $E \in \mathcal{P}$, if the attacker interferes with a message arriving before E, then the events arriving after E will not occur unless the cost expended by the attacker is $\Theta(E)$*

Definition 12. *Let us denote the responder and attacker cost sets by $\mathcal{C}$ and $\mathcal{G}$ respectively. A tolerance relation $\mathcal{T}$ is the subset of $\mathcal{C} \times \mathcal{G}$ consisting of all pairs (c, g) such that if the cost expended by the attacker is greater than g, then the protocol designer will tolerate the cost expended by the responder up to c. A tuple (c_0, g_0) is said to be* within $\mathcal{T}$ *if there exists $(c, g) \in \mathcal{T}$ with $c_0 \leq c$ and $c_0 \leq c$.*

A.3   Evaluating DoS Resistance

The procedure for assessing DoS resistance in a protocol using this framework is as follows [14]:

- determine the attacker's capabilities and the attack cost function Θ for each step of the protocol execution,
- decide on the tolerance relation $\mathcal{T}$, and
- verify that $(\delta'(E_2), \Theta(E_1)) \in \mathcal{T}$ for a verification event E_2 immediately followed by the event E_1, and that $(\Delta(E), \Theta(E)) \in \mathcal{T}$ if E is an accept event.

To analyse JFK using Meadows' cost based framework, we follow the definition of cost sets and tolerance relation proposed by Smith et al. [21]: the cost sets for both the responder and the attacker, denoted respectively by $\mathcal{C}$ and $\mathcal{G}$, consist of elements *cheap, medium* and *expensive* such that $0 < cheap < medium < expensive$. Adding two elements in the cost set results in assigning the maximum of two: $x + y = max(x, y)$. The tolerance relation $\mathcal{T}$ is defined as ([14], [21]):

$$\mathcal{T} = \begin{pmatrix} (cheap, cheap); & (cheap, medium); \\ (medium, cheap); & (cheap, expensive) \\ (medium, medium); & (medium, expensive); \\ (expensive, expensive); & \end{pmatrix}$$

A protocol may become vulnerable to a cheaply mounted DoS attack if the pair *(expensive, cheap)* and *(expensive, medium)* ever occurs in the analysis.

Software Optimizations for Cryptographic Primitives on General Purpose x86_64 Platforms

Shay Gueron[1,2]

[1] Department of Mathematics, University of Haifa, Israel
[2] Intel Architecture Group, Israel Development Center, Intel Corporation, Haifa, Israel

Abstract. The need for end-to-end security in the internet, constantly increases the world-wide number (and percentage) of SSL/TLS connections. As a result, the cryptographic algorithms that support such secure communications become a critical computational load for servers, and therefore an important target for optimization. We discuss here techniques for speeding up the software performance of several important cryptographic primitives on the ubiquitous x86_64 architectures that are used in most server platforms, and report new and improved results. A few examples are the following performance numbers, measured on the 2nd Generation Intel® Core™ processor: RSA1024/2048 implementation which is ~1.6x faster than the current OpenSSL version (1.0.0e), and SHA-1, SHA-256 and SHA-512 performing at, respectively, 5.75, 14, 9.71 cycles per byte.

Keywords: software optimization, x86_64 architectures, RSA, SHA-256, SHA512.

1 The Tutorial—Introduction

This tutorial reports on a comprehensive study on software optimization of several cryptographic primitives, which are part of SSL/TLS protocols, therefore being an important optimization goal for server platforms. We target superior software performance on the general purpose x86_64 platforms, and report performance results on both the previous generation 2010 Intel® Core™ processor and the 2nd Generation Intel® Core™ processor. Our study includes RSA1024, RSA2048, SHA-1, SHA-256, SHA-512, and AES in various modes of operation.

We describe various considerations involved with optimizing software implementations of these primitives. Some of the methods described here are improvements on the algorithmic side, combined with special code optimizations. In addition, we demonstrate some methods to take advantage of specific microarchitectural features of the discussed processors, to achieve improved performance.

Nowadays, software implementations of cryptographic algorithms are also required to be resistant to threats stemming from different types of software side channel analyses. We explain the concept of what we call an "inherently protected

D.J. Bernstein and S. Chatterjee (Eds.): INDOCRYPT 2011, LNCS 7107, pp. 399–400, 2011.

implementation", and show how to satisfy these requirements while at the same time, trying to minimize the associated performance cost of the mitigations.

The results reported here include software implementation of modular exponentiation, which we call "RSAZ" (short for RSA ZARIZ—Hebrew for "quick"). RSAZ is compatible with the OpenSSL interface, and can be integrated into its main path. Compared to OpenSSL 1.0.0e [1], RSAZ1024 and RSAZ2048 show, respectively, a speedup factor of 1.72 and 1.61 on the previous generation 2010 Intel® Core™ processor, and a speedup factor of 1.62 and 1.64 on the 2nd Generation Intel® Core™ processor. These results are a demonstration of the achievable performance, but can be (and already are) useful in real implementations. RSAZ details, and some performance results, were published in [1]. Per a subsequent request from the OpenSSL Team, RSAZ has been released as an OpenSSL patch, and is available in [3]. Some of the techniques that were described in [1] are already used in a development (unofficial) version of OpenSSL. RSAZ uses the technique described in [5], for big number squaring.

For secure hash algorithms, we report on implementations of SHA-1, SHA-256 and SHA-512 performing at, respectively, 5.75, 14, 9.71 cycles per byte [2,6]. Two of the methods described in [2] are already being used in a development (unofficial) version of OpenSSL. We note that the performance of SHA-256 is a comparison baseline for the SHA-3 competition. Recent standardization of SHA-512 truncation to a 256-bit digest, makes SHA-512 another comparison bar. Our new SHA-256 and SHA-512 performance numbers reflect on some of the SHA-3 finalists.

We also report on the performance of AES on the 2nd Generation Intel® Core™ processor, for various modes of operation. These implementations use the new Intel AES instructions, as well as the AVX architecture. The results include AES-128 in counter mode at 0.82 cycles per byte, and AES-GCM (128-bit keys) at 3.06 cycles per byte. We explain how these results were obtained.

Acknowledgements. This software optimization project is joint work with Vlad Krasnov.

References

1. OpenSSL: The Open Source toolkit for SSL/TLS, http://www.openssl.org/
2. Gueron, S.: Efficient Software Implementations of Modular Exponentiation (2011), http://eprint.iacr.org/2011/239
3. Gueron, S.: Speeding up SHA-1, SHA-256, SHA-512 on the 2nd Generation Intel Core™ Processors (manuscript 2011)
4. Gueron, S., Krasnov, V.: Efficient and side channel analysis resistant 512-bit and 1024-bit modular exponentiation for optimizing RSA1024 and RSA2048 on x86_64 platforms, OpenSSL #2582 patch, http://rt.openssl.org/Ticket/Display.html?id=2582&user=guest&pass=guest (posted August 2011)
5. Gueron, S., Krasnov, V.: Speeding up Big-Number Squaring (manuscript 2011)
6. Gueron, S., Krasnov, V.: Parallelizing message schedules to accelerate hash computations (manuscript 2011)

Author Index